READ AT A GLANCE

All Learning is Self-Teaching

Do you sometimes wish your calculus textbooks were more user-friendly?

1. (a) Do you find it difficult to construct $\varepsilon - \delta$ proofs and mathematical proofs?
  (b) If yes, see page 42-76; page 35-39

2. ((a) Do you find the topic **Simple *u*-Substitution** Techniques of Integration difficult to understand?
  (b) If yes, see page 210

3. . (a) Do you find the topic **Integration by Parts:** difficult to understand?
  (b) If yes, see page 244

4. . ((a) Do you find the topic **Trigonometric Substitution** Techniques difficult to understand?
  (b) If yes, see page 285

5. . ((a) Do you find it difficult to apply Integration to find **Volumes of Revolution?**
  (b) If yes, see page 354

6. ((a) Do you find the topic **Volumes of Solids** of known Cross-Sections difficult to understand?
  (b) If yes, see page 370

7. (a) Do you find the topic **L'Hopitals rule** difficult to understand?
  (b) If yes, see page 389

8. (a) Do you find the topic **Improper Integrals** difficult to understand?
  (b) If yes, see page 399

9. (a) Do you find the topic **Hyperbolic functions** difficult to understand?
  (b) If yes, see page 405

10. ((a) Do you want devices for **memorizing** the basic **differentiation** and **integration** formulas for **trigonometric** functions?
  (b) If yes, see pages 177-178; 236-237

11. ((a) Do you want devices for **memorizing** the basic **differentiation** and **integration** formulas for inverse **trigonometric** functions?
  (b) If yes, see pages 185, 188; 277

12. ((a) Do you want **guidelines** for overall approach to **integration** to help you save trial-and error time on examinations?
  (b) If yes, see page 320

13. (a) Do you find it difficult to find the **limits of functions?**
  (b) If yes, see page 7-39

14. ((a) Do you want a **memory reminder** for change of base formulas for logarithms and exponents.
  (b) If yes, see page 120

15. (a) Do you find the topic **Related Rate Problems** difficult to understand?
  (b) If yes, see page 124

16. ((a) Do you find the topic **Maximum and Minimum Values** difficult to understand?
  (b) If yes, see page 158

17. ((a) Do you find the topic **Relative Maxima and Minima** difficult to understand?
  (b) If yes, see page 151

18. ((a) Do you find the topic **Applied Maxima and Minima** Problems difficult to understand?
  (b) If yes, see page 163--176

19. (a) Do you find the topic **Integration of Radical Functions** difficult to understand?
  (b) If yes, see pages 225-231; 304--318

20. (a) Do you find the topic **Integration of Rational Functions** difficult to understand?
  (b) If yes, see pages 216-222; 280-298

If you can remember a needed information, you make decisions faster, you learn faster, you work faster, and you are more productive.

Sequence: A function whose domain consists of a set of consecutive positive integers.

Limit of a sequence: Given the sequence $\{a_n\}$, we say that $\{a_n\}$ approaches (converges to) the limit L as $n \to \infty$ (n goes to infinity or as n gets large) if for a chosen positive number ε, there exists a positive integer N such that if $n > N$, then $|a_n - L| < \varepsilon$ and we write $\lim_{n \to \infty} a_n = L$ ($n > N$ means after the Nth term)

Limit of a function: The value that the dependent variable approaches as the independent variable approaches a fixed value. Symbolically, we write $\lim_{x \to a} f(x) = L$ (read: the limit of f of x as x approaches a equals L). Lim $f(x)$ describes the behavior of f when x is near a, but different from a (as x squeezes in on a). Note that the value of $f(x)$ at a does not matter. However $f(x)$ may be equal to L, but it does not have to be equal to L.

Also: Given a function $y = f(x)$, we write $\lim_{x \to a} f(x) = L$ if for a chosen positive number ε, there exists a positive number δ such that if $0 < |x - a| < \delta$, then $|f(x) - L| < \varepsilon$.

Infinite Limit Calculations: You may use the following (1-7):

1. $\infty + a = a + \infty = \infty + \infty = \infty$;

2. $-\infty + a = a - \infty = -\infty - \infty = -\infty$

3. $a(\infty) = (-a)(-\infty) = +\infty$ if $a > 0$;

4. $a(-\infty) = (-a)(+\infty) = -\infty$ if $a > 0$

5. $(\infty)\infty = (-\infty)(-\infty) = +\infty$;

6. $(+\infty)(-\infty) = (-\infty)(+\infty) = -\infty$

7. $\dfrac{a}{+\infty} = \dfrac{a}{-\infty} = 0 \quad a \neq 0$

Do not use the following, since no meanings are assigned to them:

1. $\dfrac{0}{0}$; **2.** $0 \bullet \infty$; **3.** $\dfrac{\infty}{\infty}$; **4.** $(+\infty) + (-\infty)$. **5.** $\infty - \infty$; **6.** 0^0; **7.** ∞^0; **8.** 1^∞

Do not use them; Use **other techniques** such as L'Hôpital's Rule.

Infinity symbols $+\infty$ (plus infinity) and $-\infty$ (minus infinity): When we write $\lim_{x \to a} f(x) = +\infty$, we mean as x approaches a, $f(x)$ increases without bound, In this case, the limit does **not** exist and the "$+\infty$" is not a real number, but it is used to describe the behavior that as x approaches a, $f(x)$ increases without bound. However, when we write $\lim_{x \to a} f(x) = L$, where L is a real number, the limit **exists**.

Similarly, when we write $\lim_{x \to a} f(x) = -\infty$, we mean as x approaches a, $f(x)$ decreases without bound. In this case also, the limit does **not** exist and the "$-\infty$" is not a number, but shows behavior that as x approaches a $f(x)$ decreases without bound.

Continuous Function: A function is continuous at the point x_0 if $\lim_{x \to x_0} f(x) = f(x_0)$

The **derivative** of $f(x)$ at x_0, is given by $f'(x_0) = \lim_{h \to 0} \dfrac{f(x_0 + h) - f(x_0)}{h}$ if it exists.

Power Rule If $y = x^n$, n is a constant, then $\dfrac{dy}{dx} = nx^{n-1}$ ($y' = nx^{n-1}$ or $f'(x) = nx^{n-1}$)

Derivative: of b^x: $\frac{d}{dx}\left[b^x\right] = b^x \ln b$ **Derivative: of** e^x: $\frac{d}{dx}\left[e^x\right] = e^x$

Derivative of $\ln x$ $\frac{d}{dx}\left[\ln x\right] = \frac{1}{x}$;   **Derivative of** $\log_b x$ $\frac{d}{dx}\left[\log_b x\right] = \frac{1}{x}\log_b e = \frac{1}{x\ln b}$

Absolute maximum: Let x_0 be a point on an interval on which f is defined
Then if $f(x_0) \geq f(x)$ for all x on this interval, f has an **absolute maximum** at x_0,
and $f(x_0)$ is the absolute maximum of f on this interval; but if $f(x_0) \leq f(x)$ for all
x on this interval, then f has an **absolute minimum** at x_0, and $f(x_0)$ is the absolute
minimum of f on this interval.

Extreme Value Theorem: If f is continuous on the closed interval $[a, b]$ then f attains a
maximum at some point in $[a, b]$, and also a minimum at some point in $[a, b]$.

Critical x-value of $f(x)$ is any value of x in the domain of f at which $f'(x) = 0$

1. $\frac{d}{dx}\sin x = \cos x$;   **2.** $\frac{d}{dx}\cos x = -\sin x$;   **3.** $\frac{d}{dx}\tan x = \sec^2 x$;   **4.** $\frac{d}{dx}\cot x = -\csc^2 x$

5. $\frac{d}{dx}\sec x = \sec x \tan x$;   **6.** $\frac{d}{dx}\csc x = -\csc x \cot x$

1. $\frac{d}{dx}\operatorname{Sin}^{-1}x = \frac{1}{\sqrt{1-x^2}}$;   2. $\frac{d}{dx}\operatorname{Cos}^{-1}x = -\frac{1}{\sqrt{1-x^2}}$;   3. $\frac{d}{dx}\operatorname{Tan}^{-1}x = \frac{1}{1+x^2}$;

4. $\frac{d}{dx}\operatorname{Sec}^{-1}x = \frac{1}{x\sqrt{x^2-1}}$;   5. $\frac{d}{dx}\operatorname{Csc}^{-1}x = -\frac{1}{x\sqrt{x^2-1}}$;   6. $\frac{d}{dx}\operatorname{Cot}^{-1}x = -\frac{1}{1+x^2}$

Antiderivative: A function $g(x)$ is an antiderivative of a function $f(x)$ if
the derivative of $g(x)$ is $f(x)$. We symbolize the **indefinite integral or**

the antiderivative by: $\int f(x)\, dx = g(x) + C$ (where $\frac{d}{dx}\left[g(x)\right] = f(x)$)

Power Rule (Integration) $\int x^n dx = \frac{x^{n+1}}{n+1} + C$. ($C$ is the integration constant, and $n \neq -1$.)

$\int \frac{1}{x} dx = \ln x + C$ (Integral of $\frac{1}{x}$ is $\ln x + C$)

Integration by parts $\int u\, dv = uv - \int v\, du$ or $\int [uv] dx = u\int v\, dx - \int \left(\frac{du}{dx} \cdot \int v\, dx\right) dx$

$\int e^x dx = e^x + C$; $\int b^x dx = \frac{b^x}{\ln b} + C$ (b^x divided by the natural log of the base

1. $\int \frac{1}{\sqrt{1-x^2}} dx = \operatorname{Sin}^{-1}x + C$;   **2.** $\int -\frac{1}{\sqrt{1-x^2}} dx = \operatorname{Cos}^{-1}x + C$; $\int \operatorname{Sin}^{-1}x\, dx$.

3. $\int \frac{1}{1+x^2} dx = \operatorname{Tan}^{-1}x + C$;   **4.** $\int \frac{1}{x\sqrt{x^2-1}} dx = \operatorname{Sec}^{-1}x + C$; $= x\operatorname{Sin}^{-1}x + \sqrt{1-x^2} + C$

5. $\int -\frac{1}{x\sqrt{x^2-1}} dx = \operatorname{Csc}^{-1}x + C$;   **6.** $\int -\frac{1}{1+x^2} dx = \operatorname{Cot}^{-1}x + C$ $\int \operatorname{Cos}^{-1}x\, dx$.

     $= x\operatorname{Cos}^{-1}x - \sqrt{1-x^2} + C$.

Definite Integral If the function f is continuous on the closed interval $[a, b]$, then the definite integral from a to b, symbolized $\int_a^b f(x)\,dx$ *(a number)* is given by

$$\int_a^b f(x)\,dx = g(b) - g(a), \quad \text{where } \frac{d}{dx}\big[g(x)\big] = f(x)$$

L'Hôpital's Rule If $f(x)$ and $g(x)$ are two functions such that, **Case 1**: $\lim f(x) = 0$, $\lim g(x) = 0$ and $g'(x) \neq 0$ or **Case 2**: $\lim f(x) = \infty$; $\lim g(x) = \infty$; and $g'(x) \neq 0$, then $\lim \dfrac{f(x)}{g(x)} = \lim \dfrac{f'(x)}{g'(x)}$, if $\lim \dfrac{f'(x)}{g'(x)}$ exists or is infinite.(i.e., a real number, or $+\infty$ or $-\infty$)

Improper Integrals: Integrals with infinite limits or discontinuous integrand.

Trigonometric Functions and Derivatives

$y = f(x)$	Derivative: $\dfrac{dy}{dx}$	Use pattern recognition and the exceptions to help memorization
1. $y = \sin x$ 2. $y = \cos x$ 3. $y = \tan x$ 4. $y = \cot x$ 5. $y = \sec x$ 6. $y = \csc x$	1. $\cos x$ 2. $-\sin x$ 3. $\sec^2 x$ 4. $-\csc^2 x$ 5. $\sec x \tan x$ 6. $-\csc x \cot x$	The derivative of a co-named function (cos, cot, cosec) is minus something; and the derivative of a non co-named function (sine, secant and tangent) is plus something. With the exception of the sine and cosine, each derivative is either the product of two like trigonometric functions or two unlike trigonometric functions. **See more on p.177**

Trigonometric Functions and Antiderivatives

Function: $f(x)$	Antiderivative $\int f(x)dx$	
1. $y = \sin x$ 2. $y = \cos x$ 3. $y = \tan x$ 4. $y = \cot x$ 5. $y = \sec x$ 6. $y = \csc x$ 7. $y = \sec^2 x$ 8. $y = \csc^2 x$ 9. $y = \sec x \tan x$ 10. $y = \csc x \cot x$	1. $-\cos x$ 2. $\sin x$ 3. $\ln|\sec| \text{ or } -\ln|\cos x|$ 4. $-\ln|\csc x| \text{ or } \ln|\sin x|$ 5. $\ln|\sec x + \tan x|$ 6. $\ln|\csc x - \cot x|$ 7. $\tan x$ (from the derivative) 8. $-\cot x$ (from the derivative) 9. $\sec x$ (from the derivative) 10. $-\csc x$ (from the derivative)	Interchange of derivatives For those not obtained from this interchange, each begins with the natural log of something, and the signs of the derivatives are carried over to the antiderivatives **See more on p. 236**

Hyperbolic Functions

1. $\sinh x = \dfrac{e^x - e^{-x}}{2}$	2. $y = \cosh x = \dfrac{e^x + e^{-x}}{2}$	7. $\cosh x + \sinh x = e^x$
3. $\tanh x = \dfrac{\sinh x}{\cosh x} = \dfrac{e^x - e^{-x}}{e^x + e^{-x}}$	4. $\coth x = \dfrac{1}{\tanh x} = \dfrac{\cosh x}{\sinh x} = \dfrac{e^x + e^{-x}}{e^x - e^{-x}}$	8. $\cosh x - \sinh x = e^{-x}$
5. $\operatorname{sech} x = \dfrac{1}{\cosh x} = \dfrac{2}{e^x + e^{-x}}$	6. $\operatorname{csch} x = \dfrac{1}{\sinh x} = \dfrac{2}{e^x - e^{-x}}$	**See more on p.405**

Hyperbolic Functions and Derivatives

$y = f(x)$	Derivative: $\frac{dy}{dx}$	
1. $y = \sinh x$	1. $\cosh x$	Plus signs for basic functions
2. $y = \cosh x$	2. $\sinh x$	
3. $y = \tanh x$	3. $\operatorname{sech}^2 x$	
4. $y = \coth x$	4. $-\operatorname{csch}^2 x$	Minus signs for the reciprocal functions
5. $y = \operatorname{sech} x$	5. $-\operatorname{sech} x \tanh x$	
6. $y = \operatorname{csch} x$	6. $-\operatorname{csch} x \coth x$	**See more on p.408**

Inverse Hyperbolic Functions in Terms of Logarithms

1. $\operatorname{Sinh}^{-1} x = \ln(x + \sqrt{1+x^2})$; 2. $\operatorname{Cosh}^{-1} x = \ln(x + \sqrt{x^2-1})$ | 5. $\operatorname{Sech}^{-1} x = \ln(\frac{1+\sqrt{1-x^2}}{x})$;

3. $\operatorname{Tanh}^{-1} x = \frac{1}{2}\ln\left(\frac{1+x}{1-x}\right)$;   4. $\operatorname{Coth}^{-1} x = \frac{1}{2}\ln\left(\frac{x+1}{x-1}\right)$ | **6.** $\operatorname{Csch}^{-1} x = \ln(\frac{1}{x} + \frac{\sqrt{1+x^2}}{|x|})$

Hyperbolic Functions and Corresponding Antiderivatives

Function: $f(x)$	Antiderivative $\int f(x)dx$	
1. $y = \sinh x$	1. $\cosh x$	Similar to Trig. Functions except for $\operatorname{sech} x$ and $\operatorname{csch} x$
2. $y = \cosh x$	2. $\sinh x$	
3. $y = \tanh x$	3. $-\ln|\operatorname{sech} x|$ or $\ln|\cosh x|$	
4. $y = \coth x$	4. $-\ln|\operatorname{csch} x|$ or $\ln|\sinh x|$	
5. $y = \operatorname{sech} x$	5. $\tan^{-1}(\sinh x)$ or $\cot^{-1}(\operatorname{csch} x)$ *	
6. $y = \operatorname{csch} x$	6. $\ln\sqrt{\frac{\cosh x - 1}{\cosh x + 1}} = \ln|\tanh\frac{x}{2}|$ *	
7. $y = \operatorname{sech}^2 x$	7. $\tanh x$ (from the derivative)	
8. $y = \operatorname{csch}^2 x$	8. $-\coth x$ (from the derivative)	
9. $y = \operatorname{sech} x \tanh x$	9. $-\operatorname{sech} x$ (from the derivative)	
10. $y = \operatorname{csch} x \coth x$	10. $-\operatorname{csch} x$ (from the derivative)	**See more on p.423**

Inverse Trig Function Derivative: See more on p.185 & p.408	Inverse Hyperbolic Function Derivative:	Change signs as indicated (More on p.418)
1. $y = \operatorname{Sin}^{-1} x$; $\frac{dy}{dx} = \frac{1}{\sqrt{1-x^2}}$	1. $y = \operatorname{Sinh}^{-1} x$; $\frac{dy}{dx} = \frac{1}{\sqrt{1+x^2}}$	x^2–term only
2. $y = \operatorname{Cos}^{-1} x$; $\frac{dy}{dx} = -\frac{1}{\sqrt{1-x^2}}$	2. $y = \operatorname{Cosh}^{-1} x$; $\frac{dy}{dx} = \frac{1}{\sqrt{x^2-1}}$	Both terms
3. $y = \operatorname{Tan}^{-1} x$; $\frac{dy}{dx} = \frac{1}{1+x^2}$	3. $y = \operatorname{Tanh}^{-1} x$; $\frac{dy}{dx} = \frac{1}{1-x^2}, |x| > 1$	x^2–term only
4. $y = \operatorname{Cot}^{-1} x$; $\frac{dy}{dx} = -\frac{1}{1+x^2}$	4. $y = \operatorname{Coth}^{-1} x$; $\frac{dy}{dx} = \frac{1}{1-x^2}$	x^2–term only
5. $y = \operatorname{Sec}^{-1} x$; $\frac{dy}{dx} = \frac{1}{x\sqrt{x^2-1}}$	5. $y = \operatorname{Sech}^{-1} x$; $\frac{dy}{dx} = -\frac{1}{x\sqrt{1-x^2}}$	Both terms
6. $y = \operatorname{Csc}^{-1} x$; $\frac{dy}{dx} = -\frac{1}{x\sqrt{x^2-1}}$	6. $y = \operatorname{Csch}^{-1} x$; $\frac{dy}{dx} = -\frac{1}{|x|\sqrt{1+x^2}}$	Sign of "1" only

Inverse **Trig** Function Integral:	Inverse **Hyperbolic** Function Integral	Sign of
(Note; Integration by parts with $u = f(x)$)		
1. $\int \sin^{-1}x\,dx = x\sin^{-1}x + \sqrt{1-x^2}$	1. $\int \sinh^{-1}x\,dx = x\sinh^{-1}x - \sqrt{1+x^2}$	x^2
2. $\int \cos^{-1}x\,dx = x\cos^{-1}x - \sqrt{1-x^2}$	2. $\int \cosh^{-1}x\,dx = x\cosh^{-1}x - \sqrt{x^2-1}$	Both
3. $\int \tan^{-1}x\,dx = x\tan^{-1}x - \frac{1}{2}\ln(1+x^2)$	3. $\int \tanh^{-1}x\,dx = x\tanh^{-1}x + \frac{1}{2}\ln(1-x^2)$	x^2
4. $\int \cot^{-1}x\,dx == x\cot^{-1}x + \frac{1}{2}\ln(1+x^2)$	4. $\int \coth^{-1}x\,dx = x\coth^{-1}x + \frac{1}{2}\ln\lvert 1-x^2 \rvert$	x^2
5. $\int \sec^{-1}x\,dx = x\sec^{-1}x - \ln\lvert x+\sqrt{x^2-1} \rvert$	5. $\int \operatorname{sech}^{-1}x\,dx = x\operatorname{sech}^{-1}x \pm \sin^{-1}x$ *	
6. $\int \csc^{-1}x\,dx = x\csc^{-1}x + \ln\lvert x+\sqrt{x^2-1} \rvert$	6. $\int \operatorname{csch}^{-1}x\,dx = x\operatorname{csch}^{-1}x + \ln\lvert x+\sqrt{x^2+1} \rvert$	"1"
See more on p.274 & p.432	* If $\operatorname{sech}^{-1}x > 0$ use $+\sin^{-1}x$ but if $\operatorname{sech}^{-1}x < 0$, use $-\sin^{-1}x$	

Study the trig functions first and note the suggested changes to study and
memorize the analogous hyperbolic functions. Repeat the above in the
coming days, and thereafter, from time to time, test yourself by writing
the contents of the tables on paper, being guided by the mnemonic devices
suggested on various pages, and f you do not recall anything, review the
tables and the devices. Note that at least 60% of school and college work
is memory work, especially since almost every exam is a closed book exam.
The student who can recall accurately what he/she has learned will work
faster than a student who has to look up information from a book. Try to
find a way to remember what you have learned. Key: Critical observation
Become a math detective

CALCULUS
1 & 2

Second Edition

A Guided Approach

Excellent for Exam Preparation

Includes Sample Problems With

Step-by-Step Solutions
plus
Practice Problems With Answers

A.A. FREMPONG

Calculus 1 & 2 (A Guided Approach)

ISBN 978-1-946485-32-8

Printed in the United States of America

Faculty/Student Suggestions for Future Editions

Send Suggestions to the Publisher

Complete the following. This survey is to help improve future editions of this book.

1. Which Topics would you like to be added to future Editions? You may include sample problems (and solutions).

2. Which Topics in this book do you think need more or better coverage?.

3. Which coverage in this book impressed you most?

.

4. How useful was this book in preparing for class and taking the class and final exam?

   Check one: A (Excellent) ; B (Good) ; C (Average) ; D (Fair)

In Memory of the Giants

Archimedes (250 BC), Rene Descartes (1596-1650), Bonaventure Cavalieri (1598-1547),
 Pierre de Fermat (1601-1665); Evangelista Torricelli (1608-1647), John Wallis (1616-1703),
Robert Boyle (1627-1691), Christian Huygens (1629-1695), Isaac Barrow (1630-1677),
Robert Hooke (1635-1703), **Isaac Newton** (1642-1727), **Gottfried Leibniz** (1646-1716),
 Michel Rolle (1652-1719), Jacob Bernoulli (1654-1705), Marquis L'Hopital (1661-1704),
Johann Bernoulli (1667-1748), Brook Taylor (1685-1731), Colin Maclaurin (1698-1746),
Joseph Lagrange (1736-1813), Adrien Legendre (1752-1853), Carl F Gauss (1777-1855),
 Augustin Cauchy (1789-1857), August Mobius (1790-1868), Gustav Jacobi (1804-1851),
William Hamilton (1805-1865), Karl Weierstrass (1815-1897), Evariste Galois (1811-1832),
 Eduard Heine (1821-1881), Bernhard Riemann (1826-1866), and David Hilbert (1862-1943).

With the above and others, it all started; and with John Tate, Michael Atiyah,
 Singer Share, Timothy Gowers, Jean-Pierre Serre, and others, it continues..

In Memory of My Parents

Mom:
She was a devoted mother, sharing, kind, kinder to strangers and generous
to a fault. She never cursed, she never hated; she never cheated, and she never
envied. She never lied, and she never got angry. Once, she nursed an almost
dying stranger renting a room in her house back to good health to the extent
that the relatives of this renter later travelled one hundred miles just to thank
mom. She was always peaceloving and forever forgiving.
An angel once lived on this earth to serve others.

Dad:
A great dad, kind, generous and forgiving. He emphasized and was an example
of both formal education and self-education. A veterinarian, a bacteriologist,
an Associate of the Institute of Medical Laboratory Technology (UK), a Fellow
of the Royal Society of Health (UK); an incorruptible civil servant; his book on
ticks has always inspired me to write whenever the need arises.

NOTE TO THE STUDENT

This book was written with you in mind at all times. You may use this book as the course textbook or as a review book, since the book gets to the point quickly on all relevant topics, and yet covers these topics in detail.

Begin to master the definitions and the solutions of the sample problems thoroughly. (You have mastered a sample problem if you can solve the sample problem and similar problems without reference to this book or any other source.). For some problems, two or more methods are presented. Read the various methods and decide which methods you would like to remember; but always be aware of the existence of the other methods, in case the need arises. After having mastered the sample problems, try the exercise problems.
The answers to these problems are presented immediately after the problems. You may cover the answers with paper before you attempt these problems, if the answers are too obvious. You may refer back and forth to the solved problems when you do not remember how to proceed.

For good practice, always have paper and pen or pencil on your study table, since you must write when studying mathematics.

You may also attempt some of the sample problems first, if you have been exposed to the topics previously, before reading the solution methods, and in this approach, the sample problems become more practice problems for you.

As a reminder, in any book, do not dwell on the few inadvertent errors you may find, but rather concentrate on what is useful to you.

For this book to be useful both as the course textbook, as well as review for exams, it is **important** to **Understand, Remember, Apply**, and
Remember the material covered.

Wishing you Good Luck on all the exams
A.A.Frempong

Books in the series by the author: Integrated Arithmetic; Elementary Algebra; Intermediate Algebra, Elementary Mathematics; Intermediate Mathematics; Elementary & Intermediate Mathematics (combined); College Algebra; College Trigonometry; College Algebra & Trigonometry and **Calculus 1 & 2.**

PREFACE TO THE SECOND EDITION

This second edition retains the spirit of the first edition and continues to emphasize the importance of this level of mathematics for various branches of mathematics.

Some topics have been rewritten. However, the pagination of the first edition has been retained, for easy comparison with the first edition.

The topics rewritten include integration-by-parts (page 245), with the introduction of the "LIATE" order in choosing the u-part in applying

$$\int u\, dv = uv - \int v\, du \text{ or } \int [uv]dx = u\int v\, dx - \int \left(\frac{du}{dx} \cdot \int v\, dx \right) dx$$

Other topics include the inverse cofunction identities and their applications in the differentiation and the integration of inverse cofunctions (p.188; p 274, 277).

For example, knowing $\int \mathrm{Sin}^{-1}x\, dx$, you can readily find $\int \mathrm{Cos}^{-1}x\, dx$ by applying the inverse cofunction identity $\mathrm{Sin}^{-1}x + \mathrm{Cos}^{-1}x = \frac{\pi}{2}$.

We can also easily deduce that $\frac{d}{dx}(\mathrm{Cos}^{-1}x) = -\frac{d}{dx}(\mathrm{Sin}^{-1}x)$ by applying

$\mathrm{Sin}^{-1}x + \mathrm{Cos}^{-1}x = \frac{\pi}{2}$ ..

A. A. Frempong
New York,
July 2011

PREFACE TO THE FIRST EDITION

The idea of producing a calculus textbook had been on mind for some time, and so in June, 2002, I began this book. After about 3 months, I realized that there is so much room for improving on the user-friendly aspects of the current calculus textbooks.

This book is an attempt to help students who in spite of all the hard work and determination find their textbooks difficult to understand This book could be used as a course textbook as well as for self-study, especially by the working student or the continuing education student with so much demand on his or her time. It could also be used as a reference book.
This book was written with the student in mind at all times. Any time, two or more approaches can be used to a cover a topic, the author has done so.
For every topic, the author always looked for patterns to help understanding and recall.

The following topics have been covered in an analytical, methodical, and a guided way that, the author hopes, students and teachers will find them a great delight.

$\varepsilon - \delta$ proofs:

The formerly difficult to understand and construct $\varepsilon - \delta$ proofs have been presented in a step-by-step fashion that high school and first year college students can understand and apply. There is a review of operations on inequalities before these proofs. Two approaches for constructing these proofs have been covered. Examples include linear, quadratic, rational, radical, logarithmic and trigonometric functions.
Simple *u*-Substitution Technique of Integration: A generalized and a guided approach is used to quickly determine when this technique is possible. The author recommends that as early as possible, students should **master** simple *u*-substitution technique.

Integration by Parts: Two versions of the integration by parts formula have been presented. The processes and the functions involved have been classified and covered accordingly.

Trigonometric Substitution Techniques: The flexibility of various substitutions is covered. Discussed also is why we use particular identities.

Application of Integration to finding Volumes of Revolution
A visualized and a guided approach is used so that only two formulas have to be memorized; one formula for the disk method, and another formula for shell method.
Volumes of Solids of known Cross-Sections
Attention has been given to volumes of known cross sections because of the engineering applications.

Integration Techniques: After covering simple *u*-substitution, partial fraction decomposition, trigonometric substitution, and integration by parts, guidelines for overall approach to integration are presented to help the student save trial-and error time on examinations. See p.320
L'Hopital's Rule, Improper Integrals
A guided step-by-step approach has been used to cover applications of L'Hopital's Rule as well as various types and cases of improper integrals.

Memory Devices : Mnemonic devices have been presented to help the student memorize the basic differentiation and integration formulas, as well some trigonometric identities.

After rereading this book many times, I realized that I could always improve on the topics; and this book would never be published. The compromise: there are too many good things in this book to delay its publication; further improvements would be left for the next edition.

I gratefully acknowledge the help and encouragement of students, colleagues, and friends. My sincere appreciation goes to those who wrote before me.

It was a great joy writing this book.
Next: "Advanced Calculus made Elementary"

A. A. Frempong
New York,
August, 2006

CONTENTS

CHAPTER 16 405
Hyperbolic Functions I

CHAPTER 17 423
Hyperbolic Functions II

Appendix A 436
Infinite Series and Series Representations

Appendix B 437
Mathematical Induction

Appendix C 443

INDEX 451

CHAPTER 1
LIMITS

Lesson 1
Introduction to Sequences

Specification of a Sequence; Finite and Infinite Sequence

Definition 1: A **sequence** is an ordered set of numbers.

Examples: *(a)* $\{1, \ 4, \ 9, \ 16,...\ \}$

(b) $\left\{1, \ \frac{1}{2}, \ \frac{1}{3}, \ \frac{1}{4},...\right\}$

(c) $\{8, \ 11, \ 14,...\ \}$

Each number of a sequence is called a **term** of the sequence.

Definition 2: A **sequence** is also defined as a function whose domain consists of a set of consecutive positive integers.

The **range** of a sequence then consists of the terms of the sequence, and the domain consists of the term numbers. A sequence has a first term, a second term, a third term and so forth. In the sequence $1, 4, 9, 16,...$ the first term is 1, the second term is 4, and the 3rd term is 9.

Specification of sequences

A sequence may be specified in a number of ways:
(1) By listing its terms.
(2) By giving a formula for the nth term (general term).
(3) By stating the first term and a recursive formula for calculating the other terms of the sequence.

The nth term (also called the general term) is denoted by a_n. In this case, we can denote the sequence by the set notation as $\{a_n\}$. We can also write a_n as $a(n)$, but the subscript notation is preferred.

Given a general term of a sequence

Example 1 If the general term $a_n = n^2$ (where n is the term number.). Then

the 1st term (i.e., for $n = 1$) is $a_1 = (1)^2 = 1$;

the 2nd term (i.e. for $n = 2$) is $a_2 = (2)^2 = 4$;

the 3rd term (i.e. for $n = 3$) is $a_3 = (3)^2 = 9$;

the 4th term (i.e. for $n = 4$) is $a_4 = (4)^2 = 16$.

Thus, the first four terms of $\{a_n\}$ are $1, \ 4, \ 9,$ and 16.

Given a recursive formula of a sequence

Example 2   Using a recursive formula

$$\begin{cases} a_1 = 4 \\ a_{k+1} = 5a_k - 1 \qquad k \geq 1 \end{cases}$$

then $a_{1+1} = 5a_1 - 1$ when $k = 1$

$$a_2 = 5a_1 - 1$$

$$\qquad = 5(4) - 1 \qquad (a_1 = 4)$$

$$a_2 = 19$$

$$a_3 = 5a_2 - 1 \qquad (k = 2)$$

$$a_3 = 5(19) - 1 \qquad (a_2 = 19)$$

$$a_3 = 95 - 1$$

$$a_3 = 94$$

The first three terms of the given sequence are 4, 19, and 94.

Finite and Infinite Sequence

A sequence which has a last term or contains a finite number of terms is called a **finite sequence**.

Examples: (a)   1, 4, 9, 16.   (The last term is 16.)

               (b)   8, 11, 14, 17, ..., 35.   (The last term is 35.)

               (c) $1, \frac{1}{2}, \frac{1}{3}, \frac{1}{4}$.   (The last term is $\frac{1}{4}$.)

A sequence which does **not** have a last term (has an infinite number of terms) is called an **infinite sequence**. We usually indicate that a sequence is infinite by placing three dots at the end of the list.

Examples: (a)   8, 11, 14, 17, ...

               (b) $1, \frac{1}{2}, \frac{1}{3}, \frac{1}{4}$, ...

More Examples:

Finite sequence

If the *n*th term is given by $a_n = 2n + 1$ and $n = 1, 2,..., 8$ then the sequence $\{a_n\}$ has 8 terms.

$$a_1 = 2(1) + 1 = 3$$

$$a_2 = 2(2) + 1 = 5$$

$$a_3 = 2(3) + 1 = 7$$

$$.........................$$

$$a_8 = 2(8) + 1 = 17$$

We could write the above sequence as $3, 5, 7,..., 2n + 1, 17$. The above sequence is a finite sequence.

Infinite sequence

Example: If the nth term is $a_n = 2n + 1$ and $n = 1, 2, 3,...$ then we could specify the sequence as $3, 5, 7,..., 2n + 1,...$

Note that this is an infinite sequence, and in this case, the nth term is specified, followed by three dots.

Lesson 1 Exercises A

1. What is a sequence?

2. Describe the number of ways by which a sequence may be specified.

3. Determine the 4th term given the general tern $a_n = n^2$.

4. Given the recursive formula of a sequence determine the 3rd term.

$$\begin{cases} a_1 = 4 \\ a_{k+1} = 5a_k - 1 \qquad k \geq 1 \end{cases}$$

5. What is a finite sequence? Give an example.

6. What is an infinite sequence.? Give an example.

Answers: **3.** 16; **4.** 94;

Lesson 1 Exercises B

Find the first four terms of each sequence whose general term is given.

1. $a_n = 3n + 2$;

2. $a_n = 3^n$;

3. $a_n = (-1)(n + 5)$;

4. $a_n = \dfrac{n(n-1)}{2}$;

5. $a_n = (-1)^{n+2}$;

6. $a_n = \dfrac{(-1)^n (n-3)}{n}$.

What conditions must be stated in giving a recursive definition of a sequence.

7-11 Find the first four terms of each sequence whose recursive formula is given.

7. $a_1 = 3$

   $a_{n+1} = a_n - 2 \qquad (n \geq 1)$

8. $8a_1 = 3$

   $a_{n+1} + 1 = a_n + 4$

9. $a_1 = -3$

   $a_{n+1} = a_n + 7$

10. $a_1 = 2$

    $a_2 = 2$

    $a_n = a_{n-1} + a_{n-2} \ (n \geq 3)$

11. $a_1 = 1$

    $a_2 = 2$

    $a_{n+1} = a_n + a_{n-1} \ (n \geq 2)$

Answers: **1.** 5, 8, 11, 14; **2.** 3, 9, 27, 81; **3.** -6, -7, -8, -9; **4** .0, 1, 3, 6; **5.** -1, 1, -1, 1; **6.** 2, $-\dfrac{1}{2}$, 0, $\dfrac{1}{4}$; **7.** 3, 1, -1, -3; **8.** $\dfrac{3}{8}$, $\dfrac{27}{8}$, $\dfrac{51}{8}$, $\dfrac{75}{8}$; **9.** -3, 4, 11, 18; **10.** 2, 2, 4, 6; **11.** 1, 2, 3, 5.

Lesson 2
Limits of Sequences

Definition ($\varepsilon - N$ definition)

Given the sequence $\{a_n\}$, we say that $\{a_n\}$ approaches (converges to) the limit L as $n \to \infty$ (n goes to infinity or as n gets large) if for a chosen positive number ε, there exists a positive integer N such that if $n > N$, then $|a_n - L| < \varepsilon$ and we write $\lim\limits_{n \to \infty} a_n = L$ (note that $n > N$ means after the Nth term; that is $n = N + 1$, $n = N + 2,..., $)

A sequence that does not have a limit is said to diverge.

Example 1 Find $\lim\limits_{n \to \infty} \dfrac{n+1}{2n-1}$ (You may go to Lesson 4 and then return)

Solution

Step 1: Divide every term in the expression by the highest power of n in the **de**nominator. $$\lim_{n \to \infty} \frac{\frac{n}{n} + \frac{1}{n}}{\frac{2n}{n} - \frac{1}{n}} = \lim_{n \to \infty} \frac{1 + \frac{1}{n}}{2 - \frac{1}{n}}$$	**Step 2:** (For large values of n, $\frac{1}{n} = 0$) $$\lim_{n \to \infty} \frac{1 + \frac{1}{n}}{2 - \frac{1}{n}} = \frac{1+0}{2-0}$$ $$= \frac{1}{2}$$

Example 2 Prove that $\lim\limits_{n \to \infty} \dfrac{n+1}{2n-1} = \dfrac{1}{2}$

Definition: For a chosen or given number ε, if there exists a positive integer N such that if $n > N$ then $|a_n - L| < \varepsilon$

Step 1: If $n > N$,   (hypothesis)   (A)

then $\left| \dfrac{n+1}{2n-1} - \dfrac{1}{2} \right| < \varepsilon$ (conclusion)   (B)

Plan: The proof would be complete after showing that

If $n > N$, then $|a_n - L| < \varepsilon$. Here $a_n = \dfrac{n+1}{2n-1}$, $L = \dfrac{1}{2}$.

Step 2: We want a formula relating N and ε, and we begin with

$\left| \dfrac{n+1}{2n-1} - \dfrac{1}{2} \right| < \varepsilon$ (the conclusion)

and we solve for n.

$\left| \dfrac{2n+2-2n+1}{2(2n-1)} \right| < \varepsilon$

$\left| \dfrac{3}{4n-2} \right| < \varepsilon$

(Since this fraction is always positive, we ignore the negative part)

$\dfrac{3}{4n-2} < \varepsilon$

$\dfrac{4n-2}{3} > \dfrac{1}{\varepsilon}$ (inverting and reversing the sense)

$4n - 2 > \dfrac{3}{\varepsilon}$

$4n > \dfrac{3}{\varepsilon} + 2$

$$n > \frac{2\varepsilon + 3}{4\varepsilon} \qquad \text{(equivalent conclusion)} \qquad \text{(C)}$$

$$n > N \qquad \text{(hypothesis)} \qquad \text{(A)}$$

Step 3: Compare the right sides of (A) and (C), and take

$$N = \left[\frac{2\varepsilon + 3}{4\varepsilon}\right] \quad \text{(the greatest integer to the left of } \frac{2\varepsilon + 3}{4\varepsilon} \text{ on the number line)}$$

If $n > N$, then $\left|\frac{n+1}{2n-1} - \frac{1}{2}\right| < \varepsilon$, and $\lim\limits_{n \to \infty} \frac{n+1}{2n-1} = \frac{1}{2}$, and the proof is complete

(Note: $\left|\frac{n+1}{2n-1} - \frac{1}{2}\right| < \varepsilon$ is equivalent to (C) above)

After the $\left[\frac{2\varepsilon + 3}{4\varepsilon}\right]$th term, the desired closeness is achieved.

Example 3 In Example 1, above, given $\varepsilon = \frac{1}{1000}$, find N.

$$N = \left[\frac{2\varepsilon + 3}{4\varepsilon}\right] = \left[\frac{\frac{2}{1000} + 3}{\frac{4}{1000}}\right]$$
$$= [750.5]$$
$$N = 750.$$

After the 750th term, the desired closeness $\varepsilon = \frac{1}{1000}$, would be achieved.

(Same as beginning with the 751st term, the desired closeness would be achieved.) Let us check for a specific n, say $n = 751$.

Step 1: $\left|\frac{n+1}{2n-1} - \frac{1}{2}\right| \overset{?}{<} \frac{1}{1000}$

$\left|\frac{751+1}{2(751)-1} - \frac{1}{2}\right| \overset{?}{<} \frac{1}{1000}$

$\left|\frac{752}{1501} - \frac{1}{2}\right| \overset{?}{<} \frac{1}{1000}$

Step 2: $\left|\frac{1504 - 1501}{3002}\right| \overset{?}{<} \frac{1}{1000}$

$\frac{3}{3002} \overset{?}{<} \frac{1}{1000}$

$\frac{3}{3002} \overset{?}{<} \frac{1}{1000}$

$3000 \overset{?}{<} 3002$ Yes.

Lesson 2 Exercises A

1.. Find $\lim\limits_{n \to \infty} \frac{n+1}{2n-1}$;

2. Prove that $\lim\limits_{n \to \infty} \frac{n+1}{2n-1} = \frac{1}{2}$;

3. In problem 1, given $\varepsilon = \frac{1}{1000}$, find N

4.. Find $\lim\limits_{x \to 3} \frac{x-2}{(x-3)(x+4)}$;

5.. Find $\lim\limits_{x \to 1} \frac{x-2}{(x-3)(x+4)}$

Lesson 2 Exercises B

1.. Find $\lim\limits_{n \to \infty} \frac{n+2}{3n-1}$;

2. Prove that $\lim\limits_{n \to \infty} \frac{n+2}{3n-1} = \frac{1}{3}$;

3. In problem 1, given $\varepsilon = \frac{1}{1000}$, find N

4.. Find $\lim\limits_{x \to 2} \frac{x-3}{(x-1)(x+5)}$;

5.. Find $\lim\limits_{x \to 1} \frac{x-3}{(x-1)(x+5)}$

Answers: 1. $\frac{1}{2}$; **3** $N = 750$;

4. undefined.. **5.** $\frac{1}{10}$

Answers: 1. $\frac{1}{3}$; **4** $-\frac{1}{7}$; **5.** undefined..

About the previous lesson and the next two lessons

In Example 1 of the previous lesson, we considered large values of n in order to evaluate the limit. There, we agreed that for large values of n (as $n \to \infty$), $\frac{1}{n} \approx 0$. Similarly for large values of n, $\frac{1}{n^2} \approx 0, \frac{1}{n^3} \approx 0$. We will apply these approximations in Lesson 4.

In the next two lessons, Lessons 3 and 4, we cover more limit topics. In Lesson 3, we evaluate limits in which the variable approaches a real number.

1. For **polynomial functions** we can immediately substitute the value of the variable in the polynomial.

2. For **rational functions**, we can also immediately substitute the value of the variable, but if the resulting expression is undefined (denominator is zero), we will perform operations (such as factoring and dividing-out common factors in the numerator and denominator) on the expression before substituting, and hopefully, obtain defined results or otherwise. It is good practice to change any improper fraction to a polynomial plus a proper fraction.

3. For **radical functions**, we can immediately substitute the value of the variable, but if the resulting expression is undefined (denominator is zero) or indeterminate (both numerator and denominator are zero), we will perform operations (such as rationalizing the numerator or denominator; and simultaneous squaring and square root finding of a factor and dividing-out common factors in the numerator and denominator) on the expression before substituting.

4. In **Lesson 4** on infinite limit calculations, we will use an "Infinity Table" and perhaps a "Zero Table" to aid some of the calculations. In limits at infinity, if immediate substitution (of ∞ or $-\infty$) produces a permissible result in the infinity table, then we will do so; otherwise, we will perform operations (such as factoring) which will result in permissible operations in the infinity table. In addition, we will also use division by the highest power of the variable in the **denominator** (not in the numerator); since this avoids "zeros only" problems in the denominator and the subsequent reevaluation of a previous step.

Comparison: Finding Functional Values and Evaluating Limits

1. Given $f(x) = \frac{x}{5}$, when we write $f(10) = \frac{10}{5} = 2$, we mean when $x = 10$, $\frac{x}{5} = 2$. However, when we write $\lim\limits_{x \to 10} \frac{x}{5} = 2$, we mean when x approaches 10, $\frac{x}{5}$ approaches 2. Furthermore, $\lim\limits_{x \to 0} \frac{x}{5} = 0$ means when x approaches 0, $\frac{x}{5}$ approaches 0, either from the right or from the left. We must therefore note that if we evaluate a limit and we obtain a zero, we must note that we have not necessarily obtained a zero, but rather we have obtained a small positive or a small negative number. In problems in which say, $\lim\limits_{x \to 0} \frac{x}{5} = 0$ is a divisor for a subsequent step, we have to re-evaluate whether $f(x)$ is approaching zero through positive or negative values, and sometimes, with some modifications, use 0^+ (a small positive number) or 0^- (a small negative number).

Lesson 3

Limits of Functions (Calculations)

Case 1: Polynomial Functions; One-Sided Limits
Case 2: Rational Functions
Case 3: Radical Functions
Case 4: Logarithmic functions & Exponential Functions
Case 5: Trigonometric Functions
Case 6: Limits of Absolute-Value Functions
Case 7: Limits of the Greatest-Integer Function. $[x]$
Case 8: Limits of Trigonometric Functions

Definition

The limit of a function f is the value, L, that $f(x)$ approaches as x approaches the value a, and symbolically, we write

$$\lim_{x \to a} f(x) = L \quad \text{(read: the limit of } f \text{ of } x \text{ as } x \text{ approaches } a \text{ equals } L\text{ ")}$$

Lim $f(x)$ describes the behavior of f when x is near a, but different from a (as x squeezes in on a). Note that the value of $f(x)$ at a does not matter. However $f(x)$ may be equal to L, but it does not have to be equal to L.

Case 1: Polynomial Functions

For a polynomial $f(x)$, the limit of $f(x)$ as x approaches a is equal to the functional value $f(a)$ at a. To find the limit in this case, replace x by a in the equation defining the function.

Example 1 Find $\lim_{x \to 2} 3x$ **Solution** $\lim_{x \to 2} 3x = 3(2) = 6$	**Example 2** Find $\lim_{x \to 3} x^2 + 4x + 1$ **Solution** $\begin{aligned} \lim_{x \to 3} x^2 + 4x + 1 &= (3)^2 + 4(3) + 1 \\ &= 9 + 12 + 1 \\ &= 22 \end{aligned}$

Example 3 Given that $f(x) = x^2 + 4x$, find $\lim_{h \to 0} \dfrac{f(x+h) - f(x)}{h}$

Solution

$$\lim_{h \to 0} \frac{f(x+h) - f(x)}{h} = \lim_{h \to 0} \frac{[(x+h)^2 + 4(x+h)] - [x^2 + 4x]}{h}$$

$$= \lim_{h \to 0} \frac{x^2 + 2xh + h^2 + 4x + 4h - x^2 - 4x}{h}$$

$$= \lim_{h \to 0} \frac{2xh + h^2 + 4h}{h}$$

$$= \lim_{h \to 0} 2x + h^2 + 4$$

$$= 2x + 4 \qquad (h^2 = 0)$$

Therefore, given that $f(x) = x^2 + 4x$, $\lim_{h \to 0} \dfrac{f(x+h) - f(x)}{h} = 2x + 4$.

In Lesson 8, we will learn that the derivative of $x^2 + 4x$ is $2x + 4$.

One-sided Limits (Right and Left Limits)

When we write $\lim\limits_{x \to a} f(x) = L$ (the limit of f of x as x approaches a equals L), we imply that $\lim\limits_{x \to a+} f(x) = L$ and $\lim\limits_{x \to a-} f(x) = L$.

$\lim\limits_{x \to a+} f(x)$ is read "the limit of $f(x)$ as x approaches a over x-values greater than a " (from the right on the number line). Similarly, $\lim\limits_{x \to a-} f(x)$ is read "the limit of $f(x)$ as x approaches a over x-values less than a" (that is from the left on the number line). Lim $f(x)$ describes the behavior of f when x is near a (both from the right and from the left).

For a limit to exist at a point, the left limit must be equal to the right limit, except as below:

Exception: If a function f is defined on one side of a point, a, and if the corresponding one-sided limit exists, then $\lim\limits_{x \to a} f(x)$ is that of this one-sided limit. That is if $f(x)$ is defined on one side of a point, a, say, and $\lim\limits_{x \to a+} f(x)$ exists on this side, then $\lim\limits_{x \to a} f(x) = \lim\limits_{x \to a+} f(x)$, even though $\lim\limits_{x \to a-} f(x)$ does not exist.

An example of this exception is the square root function. If $f(x) = \sqrt{x}$ (with the domain $x \geq 0$), observe that this function is defined to the right of zero, but is not defined (not real) to the left of zero.. In this case, since

$$\lim\limits_{x \to 0+} \sqrt{x} = 0,$$

$$\lim\limits_{x \to 0} \sqrt{x} = \lim\limits_{x \to 0+} \sqrt{x} = 0.$$

More Examples

Example 1 $\lim\limits_{x \to 3} x^2 = 9$ implies $\lim\limits_{x \to 3^+} x^2 = 9$ (right limit) **and**

$$\lim\limits_{x \to 3^-} x^2 = 9 \text{ (left limit)}.$$

Example 2 $\lim\limits_{x \to 2} \sqrt{4 - x^2} = 0$ (That of the left limit $\lim\limits_{x \to 2-} \sqrt{4 - x^2}$)

Explanation: The domain of $f(x) = \sqrt{4 - x^2}$ is such that $4 - x^2 \geq 0$ or

$$-2 \leq x \leq 2$$

For the left limit: $\lim\limits_{x \to 2-} \sqrt{4 - x^2} = 0$, and the function is also defined from the left, and therefore $\lim\limits_{x \to 2} \sqrt{4 - x^2} = \lim\limits_{x \to 2-} \sqrt{4 - x^2} = 0$.even though

$\lim\limits_{x \to 2+} \sqrt{4 - x^2}$ does not exist, since for $x > 2$, $\sqrt{4 - x^2}$ is imaginary and not defined. The next example has no limit because the left limit is not equal to the right limit.

Example 3 Find $\lim\limits_{x \to 0} \dfrac{x}{|x|}$.

Solution: $\lim\limits_{x \to 0^+} f(x) = 1; \quad \lim\limits_{x \to 0^-} f(x) = -1$ and therefore the left and right limits are **not** equal, and the limit does not exist. (Figure below).

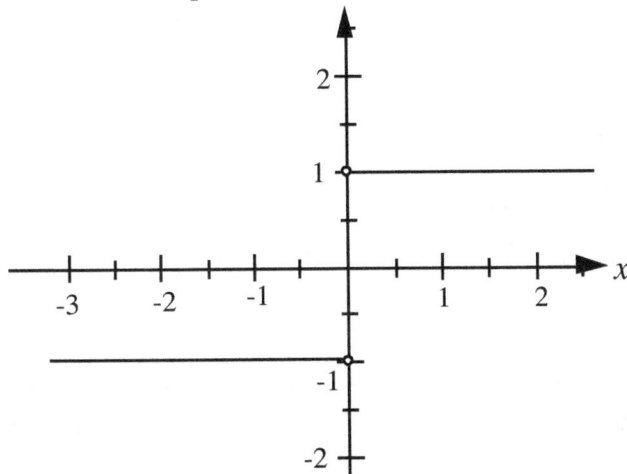

Figure: $f(x) = \dfrac{x}{|x|}$

For the left limit, choose say

$x = -0.001$

$\dfrac{x}{|x|} = \dfrac{-0.00001}{|-0.0001|}$

$= \dfrac{-0.00001}{0.0001}$

$= -1$

For the right limit, choose say

$x = 0.00001$

$\dfrac{x}{|x|} = \dfrac{0.00001}{|0.00001|}$

$= \dfrac{0.00001}{0.00001}$

$= 1$

Clarification of the application of one-sided limits

1. The left limit, $\lim\limits_{x \to a-} f(x) = \mathrm{L}$ (exists: real and defined), and the right limit $\lim\limits_{x \to a+} f(x) = \mathrm{L}$ (exists: real and defined), and therefore, the left limit equals the right limit, and hence the normal limit, $\lim\limits_{x \to a} f(x) = \mathrm{L}$. (exists)

2. The left limit, $\lim\limits_{x \to a-} f(x) = L_1$ exists (real and defined), and the right limit $\lim\limits_{x \to a+} f(x) = L_2$ (exists (real and defined), but $L_1 \neq L_2$ and hence the normal limit, $\lim\limits_{x \to a} f(x)$ does **not** exist.

3. Either $\lim\limits_{x \to a-} f(x)$ exists and f is defined from the left , but $\lim\limits_{x \to a+} f(x)$ does not exist, and in this case, the normal limit $\lim\limits_{x \to a} f(x) = \lim\limits_{x \to a-} f(x)$; or

$\lim\limits_{x \to a+} f(x)$ exists and f is defined from the right, but $\lim\limits_{x \to a-} f(x)$ does not exist, and in this case, the normal limit $\lim\limits_{x \to a} f(x) = \lim\limits_{x \to a+} f(x)$.

Note that for $\lim\limits_{x \to a} f(x)$ to exist, the function does **not** have to be defined at a.

Case 2: Rational Functions

Example 4 Find $\lim\limits_{x \to 2} \dfrac{3x^2 + 6}{x - 5}$ (see p.6, #2)

Solution $\lim\limits_{x \to 2} \dfrac{3x^2 + 6}{x - 5} = \dfrac{\lim\limits_{x \to 2} 3x^2 + 6}{\lim\limits_{x \to 2} x - 5} = \dfrac{3(2)^2 + 6}{2 - 5} = \dfrac{18}{-3} = -6$

(We substituted since the denominator is not zero when $x = 2$.)

Example 5 Find $\lim\limits_{x \to 3} \dfrac{x^2 - 9}{x - 3}$

Solution Observe that if we substitute for $x = 3$ in $\dfrac{x^2 - 9}{x - 3}$, this expression is indeterminate ($\frac{0}{0}$); and therefore, $x \neq 3$, We proceed as follows:

$$\lim\limits_{x \to 3} \dfrac{x^2 - 9}{x - 3} = \lim\limits_{x \to 3} \dfrac{(x + 3)(x - 3)}{(x - 3)} = \lim\limits_{x \to 3} x + 3 = 3 + 3 = 6$$

Note above that by factoring the numerator and dividing out the common factor $x - 3$, we obtain a new function $f(x) = x + 3$, whose limit is 6. The functions $f(x) = \dfrac{x^2 - 9}{x - 3}$ and $f(x) = x + 3$ are different functions. However, each has the limit 6. Furthermore, note that even though, $f(x) = \dfrac{x^2 - 9}{x - 3}$ is not defined when $x = 3$, the limit $\lim\limits_{x \to 3} \dfrac{x^2 - 9}{x - 3}$ exists and is 6.

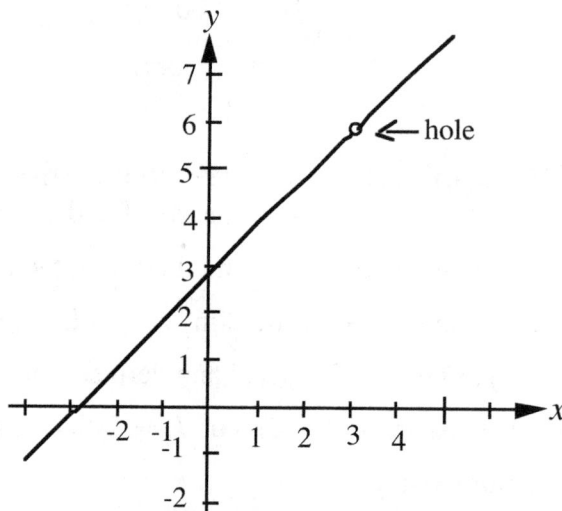

Figure

We can therefore say that the limit of a function at a point a is independent of whether or not the function is defined at the point a.

Example 5 Find $\lim\limits_{x \to 3} \dfrac{x-2}{(x-3)(x+4)}$

Solution: If we substitute 3 for x in the denominator, the expression obtained would be undefined, and therefore $x \neq 3$, and moreover, we cannot divide out any common factors. We proceed as follows:

Step 1: $\lim\limits_{x \to 3^+} \dfrac{x-2}{(x-3)(x+4)} = +\infty$ Try 3.0001 (right limit is plus infinity)

Step 2: $\lim\limits_{x \to 3^-} \dfrac{x-2}{(x-3)(x+4)} = -\infty$ Try 2.9999 (left) limit is minus infinity)

From Steps 1 and 2, $\lim\limits_{x \to 3} \dfrac{x-2}{(x-3)(x+4)}$ does not exist.

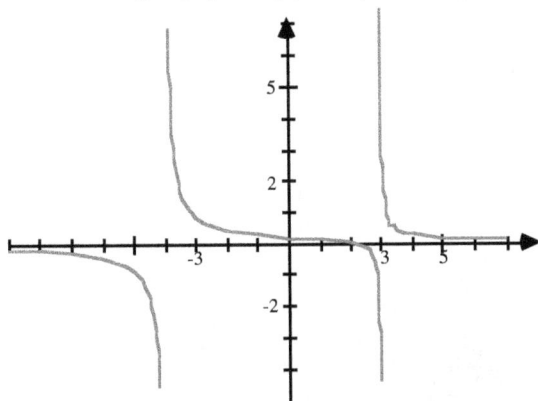

Figure: $f(x) = \dfrac{x-2}{(x-3)(x+4)}$

Case 3: Radical Functions   (see p.6, #3)

Example 6 Find $\lim\limits_{x \to 3} \sqrt{\dfrac{x^2+4x+3}{x+6}}$

Solution $\lim\limits_{x \to 3} \sqrt{\dfrac{x^2+4x+3}{x+6}} = \sqrt{\dfrac{(3)^2+3(4)+3}{3+6}}$ (substituting for x directly)

$$= \sqrt{\dfrac{9+12+3}{9}}$$

$$= \sqrt{\dfrac{24}{9}} = \dfrac{2\sqrt{6}}{3}$$

Example 7 Find $\lim\limits_{x \to 2} \dfrac{6-x^2}{2-\sqrt{x^2+1}}$

Solution $\lim\limits_{x \to 2} \dfrac{6-x^2}{2-\sqrt{x^2+1}} = \lim\limits_{x \to 2} \dfrac{6-(2)^2}{2-\sqrt{(2)^2+1}}$ (substituting for x directly)

$$= \dfrac{2}{2-\sqrt{5}}$$

$$= \dfrac{2}{(2-\sqrt{5})} \cdot \dfrac{(2+\sqrt{5})}{(2+\sqrt{5})}$$

$$= -2(2+\sqrt{5}) \text{ or } -4-2\sqrt{5}$$

Example 8 Find $\lim\limits_{x \to 3} \dfrac{9 - x^2}{4 - \sqrt{x^2 + 7}}$

Here, observe that if we substitute as is, we obtain zero in the denominator. We proceed as follows and begin by rationalizing the denominator:

$$\lim_{x \to 3} \frac{9 - x^2}{4 - \sqrt{x^2 + 7}}$$

$$= \lim_{x \to 3} \frac{(9 - x^2)}{(4 - \sqrt{x^2 + 7})} \cdot \frac{(4 + \sqrt{x^2 + 7})}{(4 + \sqrt{x^2 + 7})}$$

$$= \lim_{x \to 3} \frac{(9 - x^2)}{16 - x^2 - 7} \cdot \frac{(4 + \sqrt{x^2 + 7})}{1}$$

$$= \lim_{x \to 3} \frac{(9 - x^2)}{(9 - x^2)} \cdot \frac{(4 + \sqrt{x^2 + 7})}{1}$$

$$= \lim_{x \to 3} 4 + \sqrt{x^2 + 7}$$

$$= 4 + \sqrt{3^2 + 7}$$

$$= 4 + \sqrt{9 + 7}$$

$$= 8$$

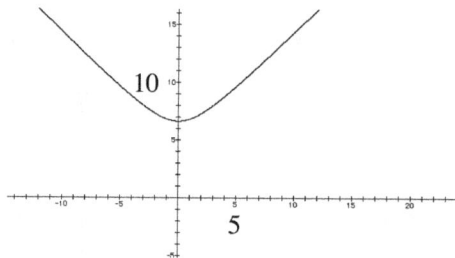

Example 9 Find $\lim\limits_{x \to 3} \dfrac{6 - x^2}{4 - \sqrt{x^2 + 7}}$

As in the previous example, we begin by rationalizing the denominator:

$$\lim_{x \to 3} \frac{6 - x^2}{4 - \sqrt{x^2 + 7}}$$

$$= \lim_{x \to 3} \frac{6 - x^2}{(4 - \sqrt{x^2 + 7})} \cdot \frac{(4 + \sqrt{x^2 + 7})}{(4 + \sqrt{x^2 + 7})}$$

$$= \lim_{x \to 3} \frac{(6 - x^2)}{16 - x^2 - 7} \cdot \frac{(4 + \sqrt{x^2 + 7})}{1}$$

$$= \lim_{x \to 3} \frac{(6 - x^2)}{9 - x^2} \cdot \frac{(4 + \sqrt{x^2 + 7})}{1}$$

$$= \frac{(6 - 3^2)}{9 - 3^2} \cdot \frac{(4 + \sqrt{3^2 + 7})}{1}$$

$$= \frac{-24}{0}, \text{ which is undefined}$$

Therefore $\lim\limits_{x \to 3} \dfrac{6 - x^2}{4 - \sqrt{x^2 + 7}}$ is undefined and there is **no** limit.

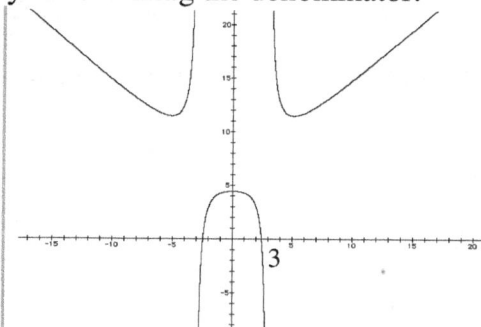

(A picture is worth thousand words)
Also, note below and see graph above:

$$\lim_{x \to 3^+} \frac{(6 - x^2)}{9 - x^2} \cdot \frac{(4 + \sqrt{x^2 + 7})}{1}$$

$$= \frac{-24}{0^-} = +\infty$$

$$\lim_{x \to 3^-} \frac{(6 - x^2)}{9 - x^2} \cdot \frac{(4 + \sqrt{x^2 + 7})}{1}$$

$$= \frac{-24}{0^+} = -\infty$$

Example 10 Find $\lim_{x \to 0} \dfrac{\sqrt{x+2} - \sqrt{2}}{x}$

Here, observe that if we substitute right away, we obtain zero in the denominator. We proceed as follows and begin by rationalizing the numerator.

$$\lim_{x \to 0} \frac{\sqrt{x+2} - \sqrt{2}}{x} = \lim_{x \to 0} \frac{\left(\sqrt{x+2} - \sqrt{2}\right)\left(\sqrt{x+2} + \sqrt{2}\right)}{x\left(\sqrt{x+2} + \sqrt{2}\right)}$$

$$= \lim_{x \to 0} \frac{x+2-2}{x\left(\sqrt{x+2} + \sqrt{2}\right)} \qquad \text{(Rationalizing the numerator)}$$

(Perhaps this is the first time you are rationalizing the numerator)

$$= \lim_{x \to 0} \frac{1}{\left(\sqrt{x+2} + \sqrt{2}\right)}$$

$$= \frac{1}{\left(\sqrt{2} + \sqrt{2}\right)}$$

$$= \frac{1}{2\sqrt{2}}$$

$$= \frac{\sqrt{2}}{4}.$$

Example 11 Find $\lim_{x \to 5} \sqrt{x-5}$

Solution The domain consists of all real values of x such that $x \geq 5$.
The function is not real when $x < 5$, because the square root would be that of a negative number, and the square root of a negative number is imaginary.
The right limit, $\lim_{x \to 5^+} \sqrt{x-5} = 0$, and the function is also defined from the

right and therefore, $\lim_{x \to 5} \sqrt{x-5} = \lim_{x \to 5^+} \sqrt{x-5} = 0$ even though the left limit

$\lim_{x \to 5^-} \sqrt{x-5}$ does not exist. See also one-sided limits in Case 1 of Lesson 3.

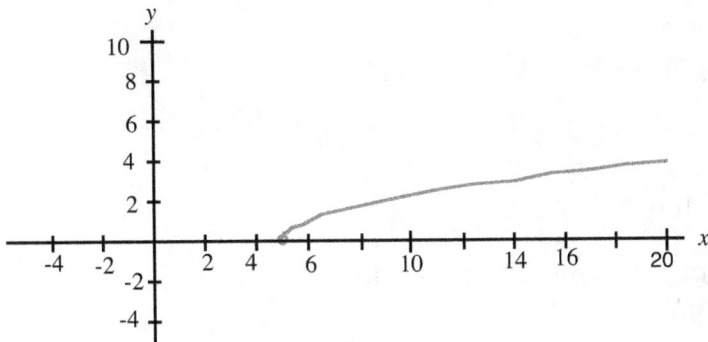

Figure: Graph of $f(x) = \sqrt{x-5}$

Case 4: Limits of logarithmic functions

Example 12   Find $\lim\limits_{x \to 3} \log_2\left(x^2 + 7\right)$

Solution: By substituting $x = 3$ in $\lim\limits_{x \to 3} \log_2\left(x^2 + 7\right)$, we obtain

$$\log_2\left(3^2 + 7\right) \text{ (by substitution)}$$

$$= \log_2(9 + 7)$$

$$= \log_2(16) \quad \text{<--We can evaluate this mentally, but let us exercise.}$$

Let $\log_2(16) = k$.

Then $2^k = 16$ (Applying the equivalent exponential definition)

$2^k = 2^4$ (Equating the exponents since the bases are equal)

$k = 4$

$\log_2(16) = 4$

Therefore $\lim\limits_{x \to 3} \log_2\left(x^2 + 7\right) = 4$

Example 13   Find $\lim\limits_{x \to 8} \log_2 x$

Solution: By substituting $x = 8$ in $\lim\limits_{x \to 8} \log_2 x$, we obtain $\log_2 8$.

(Question: What exponent must be placed on 2 to produce 8?)

Let $\log_2 8 = k$.

Then $2^k = 8$ (Applying the equivalent exponential definition)

$2^k = 2^3$ (Equating the exponents since the bases are equal)

$k = 3$, and

$\log_2 8 = 3$. Therefore, $\lim\limits_{x \to 8} \log_2 x = 3$

Case 5: Limits of Exponential Functions (see p.29 for the graphs)

Example 14   Find $\lim\limits_{x \to 0} e^x$

Solution: By substituting $x = 0$ in e^x, we obtain

$$e^0 = 1$$

Therefore, $\lim\limits_{x \to 0} e^x = 1$

Example 15   Find $\lim\limits_{x \to 3} 2^x$

Solution: By substituting $x = 3$ in 2^x, we obtain

$$2^3 = 8$$

Therefore, $\lim\limits_{x \to 3} 2^x = 8$

Case 6: Limits of Absolute-Value Functions

Example 16 $\lim\limits_{x \to 3} |x| = 3$ (by substitution)

Example 17 $\lim\limits_{x \to 0} |x| = 0$ (by substitution)

Case 7: Limits of the Greatest-Integer Function. $[x]$

The greatest integer function, denoted by $[x]$, is the largest integer that is less than or equal to x. Symbolically, $[x] = n$ if $n \leq x < n + 1$, where n is an integer. That is, if the given number is an integer, then the given number is the greatest integer; but if it is not an integer, then on the number line, the nearest integer on its left is the greatest integer.

Examples

$$[0] = 0; \ [2] = 2; \ [5] = 5; \ [1.5] = 1; \ [-1.5] = -2; \ [2.5] = 2 \ ; [-4.5] = -5$$

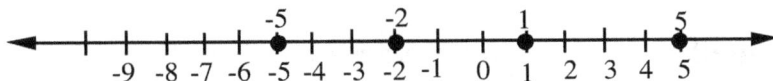

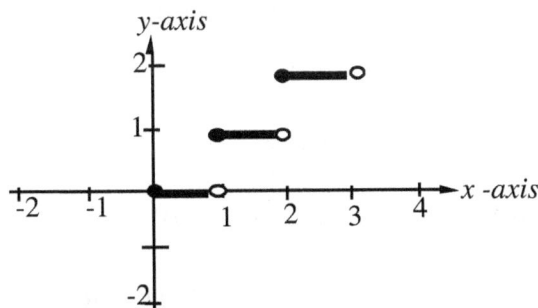

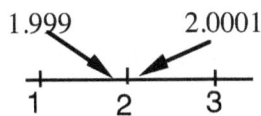

Fig. 2

Figure 1: Graph of $y = [x]$

Limits for $\lim\limits_{x \to n} [x]$ do not exist if n is an integer.

However $\lim\limits_{x \to n} [x]$ exists if n is a non-integral real number,

Examples: **1.** $\lim\limits_{x \to 1.2} [x] = 1$; **2.** $\lim\limits_{x \to 0.8} [x] = 0$; $\lim\limits_{x \to 2.8} [x] = 2$

However, $\lim\limits_{x \to 2} [x]$, does not exist since the right limit, 2, and the left limit, 1, are different. (Fig. 2, above).

Case 8: **Limits of Trigonometric and Inverse Trigonometric Functions** as $x \to 0$

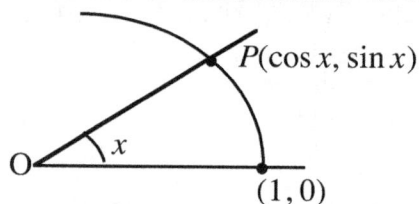

$P(\cos x, \sin x)$

x

O

$(1, 0)$

Figure

$\lim\limits_{x \to 0}$ trig function of x
(x in radians)

1. $\lim\limits_{x \to 0} \sin x = 0$ (by substitution)

2. $\lim\limits_{x \to 0} \cos x = 1$ (by substitution)

3. $\lim\limits_{x \to 0} \tan x = 0$ (by substitution)

4. $\lim\limits_{x \to 0} \csc x$ is undefined
((by substitution)

5. $\lim\limits_{x \to 0} \sec x = 1$

6 $\lim\limits_{x \to 0} \cot x$ is undefined

$\lim\limits_{x \to 0}$ inverse trig function of x
(x in radians)

1. $\lim\limits_{x \to 0} \arcsin x = 0$

2. $\lim\limits_{x \to 0} \arccos x = \frac{\pi}{2}$

3. $\lim\limits_{x \to 0} \arctan x = 0$

4. $\lim\limits_{x \to 0} arc \csc x$ is undefined

5. $\lim\limits_{x \to 0} arc \sec x$ is undefined

6 . $\lim\limits_{x \to 0} arc \cot x = \frac{\pi}{2}$

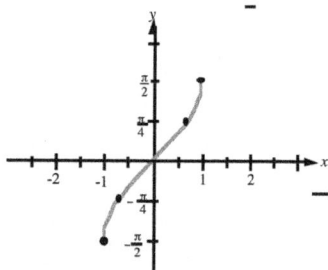

Graph $y = \text{Arc} \sin x$
or $\text{Sin}^{-1} x$

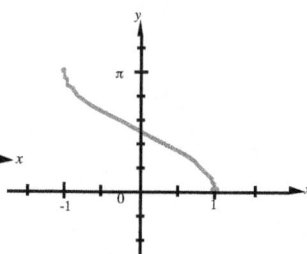

Graph of $y = \arccos x$

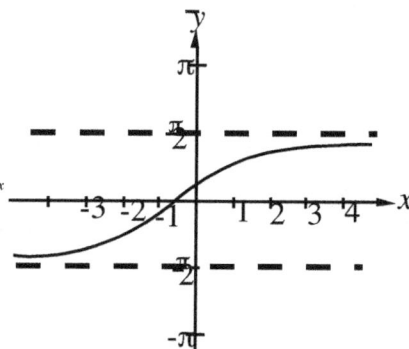

Graph of $y = \arctan x$

Practice sketching these graphs and include them among your "trigonometric vocabulary".

Theorem 3 $\lim\limits_{x \to 0} \dfrac{\sin x}{x} = 1$ (where x is in radians)

We prove Theorem 3 below as well as in Chapter 15 using L'Hôpital's Rule.

Theorem 4 $\lim\limits_{x \to 0} \dfrac{1 - \cos x}{x} = 0$:   Example $\lim\limits_{x \to 0} \dfrac{x}{\cos x} = \dfrac{0}{1} = 0$

Squeezing or Pinching Theorem for Limits

Sometimes, we can find the limit of a function by "squeezing" or pinching" the given function between other functions whose limits are known.

Squeezing Theorem

If $f(x)$, $g(x)$, and $h(x)$ are related by the inequality

$g(x) \leq f(x) \leq h(x)$ for all x on an open interval containing the point a (except possibly at a), and

if $\lim\limits_{x \to a} g(x) = L$, and $\lim\limits_{x \to a} h(x) = L$,

then $\lim\limits_{x \to a} f(x) = L$

In the above theorem, $f(x)$ is squeezed (or pinched) between $g(x)$ and $h(x)$.

Application of the Squeezing or Pinching Theorem

Example 1 Show that $\lim\limits_{x \to 0} \dfrac{\sin x}{x} = 1$ (where x is in radians) see also p. 391.

Step 1: Draw a unit circle with center.

at O , and radius = $OP = OA = 1$;

and extend $\overline{OP}$ to a point B such

that $\overline{BA} \perp \overline{OA}$. Also from P draw

PC perpendicular to $\overline{OA}$.

Step 2: Let x be the central angle at

O where $0 < x < \dfrac{\pi}{2}$

Length of $\overline{OP} = 1$

Length of $\overline{OA} = 1$

Length of $\overline{OC} = \cos x$

$(\cos x = \dfrac{OC}{OP} = \dfrac{OC}{1})$

Length of $\overline{PC} = \sin x$

$(\sin x = \dfrac{PC}{OP} = \dfrac{PC}{1})$

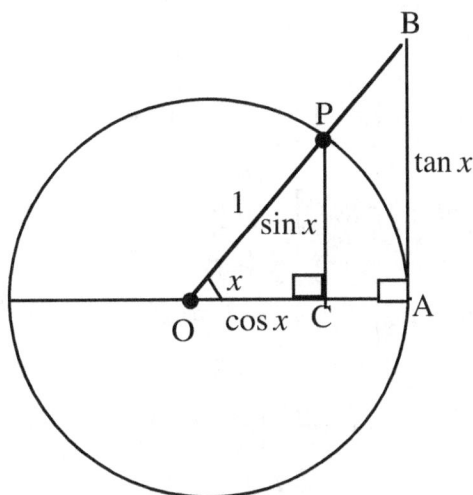

Step 3 : area of $\triangle POC$ < area of sector POA < area of $\triangle BOA$

$\dfrac{1}{2}\cos x \sin x < \dfrac{1}{2}r^2 x < \dfrac{1}{2}\tan x$ (area of triangle = $\dfrac{1}{2}bh$; area of a sector = $\dfrac{1}{2}r^2\theta$)

$\cos x \sin x < 1 \cdot x < \tan x$ (multiplying all sides by 2. $r = 1$)

$\cos x < \dfrac{x}{\sin x} < \dfrac{\tan x}{\sin x} = \dfrac{\sin x}{\cos x} \cdot \dfrac{1}{\sin x}$ (dividing by $\sin x$)

$\cos x < \dfrac{x}{\sin x} < \dfrac{1}{\cos x}$

$\dfrac{1}{\cos x} > \dfrac{\sin x}{x} > \cos x$ (inverting and reversing the direction of the inequality)

$\cos x < \dfrac{\sin x}{x} < \dfrac{1}{\cos x}$ (rewriting from right to left)

Step 4: $\lim\limits_{x \to 0} \cos x < \lim\limits_{x \to 0} \dfrac{\sin x}{x} < \lim\limits_{x \to 0} \dfrac{1}{\cos x}$ (taking limits as $x \to 0$)

$1 < \lim\limits_{x \to 0} \dfrac{\sin x}{x} < 1$ ($\lim\limits_{x \to 0} \cos x = 1$, $\lim\limits_{x \to 0} \dfrac{1}{\cos x} = 1$)

$\therefore \lim\limits_{x \to 0} \dfrac{\sin x}{x} = 1$ (by the Squeezing or Pinching Theorem)

(**Note** above: $g(x) = \cos x$, $f(x) = \dfrac{\sin x}{x}$; $h(x) = \dfrac{1}{\cos x}$)

Theorems on Limits (With Examples)

1. If $f(x) = L$, where L is a constant, then
$$\lim_{x \to a} f(x) = L$$
Example $\lim_{x \to 3} 5 = 5$

2. If $\lim_{x \to a} f(x) = L$, then
$$\lim_{x \to a} k \bullet f(x) = k \bullet \lim_{x \to a} f(x) = kL$$
Example $\lim_{x \to 2} 3x = 3 \lim_{x \to 2} x$

3-6 If $\lim_{x \to a} f(x) = L_1$ and $\lim_{x \to a} g(x) = L_2$ then

3. $\lim_{x \to a} \left[f(x) + g(x) \right] = \lim_{x \to a} f(x) + \lim_{x \to a} g(x)$
$$= L_1 + L_2$$
Example $\lim_{x \to 3} \left[x^2 + 5 \right] = \lim_{x \to 3} x^2 + \lim_{x \to 3} 5$
$$= 9 + 5 = 14$$
4. $\lim_{x \to a} \left[f(x) - g(x) \right] = \lim_{x \to a} f(x) - \lim_{x \to a} g(x)$
$$= L_1 - L_2$$
Example $\lim_{x \to 3} \left[x^2 - 5 \right] = \lim_{x \to 3} x^2 - \lim_{x \to 3} 5$
$$= 9 - 5 = 4$$
5. $\lim_{x \to a} \left[f(x) g(x) \right] = \left(\lim_{x \to a} f(x) \right) \left(\lim_{x \to a} g(x) \right)$
$$= L_1 L_2$$
6. $\lim_{x \to a} \dfrac{f(x)}{g(x)} = \dfrac{\lim_{x \to a} f(x)}{\lim_{x \to a} g(x)}$
$$= \dfrac{L_1}{L_2} \qquad (L_2 \ne 0)$$

7. $\lim_{x \to a} x^n = a^n$, where n is a positive integer.

Example $\lim_{x \to 3} x^2 = 3^2 = 9$

8. $\lim_{x \to a} \left[f(x) \right]^n = \left[\lim_{x \to a} f(x) \right]^n = L^n$

Example $\lim_{x \to 3} \left[4x \right]^2 = \left[\lim_{x \to 3} 4x \right]^2 = 12^2 = 144$ (since $\lim_{x \to 3} 4x = 12$)

9. $\lim_{x \to a} \sqrt[n]{f(x)} = \sqrt[n]{\lim_{x \to a} f(x)} = \sqrt[n]{L}$ (Where $L \ge 0$ if n is even)

Example $\lim_{x \to 4} \sqrt{x^2 + 9} = \sqrt{\lim_{x \to 4} x^2 + 9} = \sqrt{16 + 9} = \sqrt{25} = 5.$

10. $\lim\limits_{x \to 0} \dfrac{\sin ax}{x} = a$

(Proof: $\lim\limits_{x \to 0} \dfrac{\sin ax}{x} = \lim\limits_{x \to 0} \left(a \bullet \dfrac{\sin ax}{ax} \right) = a \left(\lim\limits_{x \to 0} \dfrac{\sin ax}{ax} \right) = a(1) = a$

↑ Multiply both numerator and denominator by a and apply $\lim\limits_{x \to 0} \dfrac{\sin ax}{ax} = \lim\limits_{u \to 0} \dfrac{\sin u}{u} = 1$

Example $\lim\limits_{x \to 0} \dfrac{\sin 3x}{2x}$

Step 1: We want to write $\lim\limits_{x \to 0} \dfrac{\sin 3x}{2x}$ in the form $\lim\limits_{x \to 0} \dfrac{\sin ax}{ax}$.

Multiply both the numerator and the denominator by $\dfrac{3}{2}$.

$$= \lim\limits_{x \to 0} \left(\dfrac{3}{2} \bullet \dfrac{\sin 3x}{2x \bullet \frac{3}{2}} \right)$$

Step 2: Factor out $\dfrac{3}{2}$

$$= \dfrac{3}{2} \left(\lim\limits_{x \to 0} \dfrac{\sin 3x}{3x} \right)$$

$$= \dfrac{3}{2}(1) \qquad\qquad (\lim\limits_{x \to 0} \dfrac{\sin 3x}{3x} = 1; \quad \lim\limits_{u \to 0} \dfrac{\sin u}{u} = 1)$$

Therefore, $\lim\limits_{x \to 0} \dfrac{\sin 3x}{2x} = \dfrac{3}{2}$.

Extra

Prove that $\lim\limits_{x \to 0} \dfrac{1 - \cos x}{x} = 0$

Tool box: $\boxed{\lim\limits_{x \to 0} \dfrac{\sin x}{x} = 1}$

Proof:

$\lim\limits_{x \to 0} \dfrac{1 - \cos x}{x}$

$= \lim\limits_{x \to 0} \dfrac{(1 - \cos x)(1 + \cos x)}{x(1 + \cos x)}$ (multiply numerator and denominator by $(1 + \cos x)$)

$= \lim\limits_{x \to 0} \dfrac{(1 - \cos^2 x)}{x(1 + \cos x)}$

$= \lim\limits_{x \to 0} \dfrac{\sin^2 x}{x(1 + \cos x)} \qquad (1 - \cos^2 x = \sin^2 x)$

$= \lim\limits_{x \to 0} \dfrac{\sin x}{x} \bullet \dfrac{\sin x}{1 + \cos x}$

$= \lim\limits_{x \to 0} \dfrac{\sin x}{x} \bullet \lim\limits_{x \to 0} \dfrac{\sin x}{1 + \cos x}$

$= 1 \bullet \dfrac{0}{1 + 1} \quad (1 \bullet \dfrac{0}{2} = 0) \quad (\lim\limits_{x \to 0} \dfrac{\sin x}{x} = 1)$

$= 0$

$\therefore \lim\limits_{x \to 0} \dfrac{1 - \cos x}{x} = 0$

Lesson 3 Exercises A

1. . Find $\lim\limits_{x \to 2} 3x$

2. Find $\lim\limits_{x \to 3} x^2 + 4x + 1$

3. Given that $f(x) = x^2 + 4x$, find

$$\lim\limits_{h \to 0} \frac{f(x+h) - f(x)}{h}$$

4. Find $\lim\limits_{x \to 2} \dfrac{3x^2 + 6}{x - 5}$

5. Find $\lim\limits_{x \to 3} \dfrac{x^2 - 9}{x - 3}$

6. Find $\lim\limits_{x \to 3} \dfrac{x - 2}{(x - 3)(x + 4)}$

7. Find $\lim\limits_{x \to 3} \sqrt{\dfrac{x^2 + 4x + 3}{x + 6}}$

8. Find $\lim\limits_{x \to 2} \dfrac{6 - x^2}{2 - \sqrt{x^2 + 1}}$

9. Find $\lim\limits_{x \to 3} \dfrac{9 - x^2}{4 - \sqrt{x^2 + 7}}$

10. Find $\lim\limits_{x \to 3} \dfrac{6 - x^2}{4 - \sqrt{x^2 + 7}}$

11. Find $\lim\limits_{x \to 0} \dfrac{\sqrt{x + 2} - \sqrt{2}}{x}$

12. Find $\lim\limits_{x \to 5} \sqrt{x - 5}$

13. Find $\lim\limits_{x \to 3} \log_2(x^2 + 7)$

14. Find $\lim\limits_{x \to 8} \log_2 x$

15. Find $\lim\limits_{x \to 0} e^x$

16. Find $\lim\limits_{x \to 3} 2^x$

17. . Find $\lim\limits_{x \to 3} |x|$

18. Find $\lim\limits_{x \to 0} |x|$

19. Find $[1.5]$

20. Find $[-1.5]$

21 Find $\lim\limits_{x \to n} [x]$ if n is an integer

22. Does the following limit exist if n is a non-integral real number,

$$\lim\limits_{x \to n} [x]$$

23. Find $\lim\limits_{x \to 1.2} [x]$

24. Find $\lim\limits_{x \to 2} [x]$

25. Find $\lim\limits_{x \to 0} \sin x$ (x in radians)

26. Find $\lim\limits_{x \to 0} \cos x$

27. Find $\lim\limits_{x \to 0} \dfrac{\sin x}{x}$ (x in radians)

28 Find $\lim\limits_{x \to 0} \dfrac{1 - \cos x}{x}$

29. Find $\lim\limits_{x \to 0} \dfrac{\sin 3x}{x}$

30. Find $\lim\limits_{x \to 0} \dfrac{x}{\cos x}$

31. Find $\lim\limits_{x \to 2} 3x$

32. Find $\lim\limits_{x \to a} x^n$, where n is a positive integer.

33. $\lim\limits_{x \to 3} [4x]^2$

34. $\lim\limits_{x \to 0} \dfrac{\sin ax}{x}$

35. Find the right and left limits:

$$\lim\limits_{x \to 2} \sqrt{4 - x^2}$$

36. . Find $\lim\limits_{x \to +\infty} \cos\left(\frac{1}{x}\right)$ (x in radians)

Answers: 1. 6; **2.** . 22; **3.** $2x + 4$; **4.** -6; **5.** 6; **6.** Limit does not exist..

7. $\dfrac{2\sqrt{6}}{3}$; **8.** $-4 - 2\sqrt{5}$; **9.** 8; **10.** There is **no** limit; **11.** $\dfrac{\sqrt{2}}{4}$;

12 0; **13.** 4; **14.** 3. **15.** 1; **16.** 8; **17.** 3; **18.** 0; **19.** 1; **20.** -2

21. Limit does not exist **22.** Yes; **23.** 1; **24.** Limit does not exist, **25.** 0

26. 1; **27.** 1; **28.** 0; **29.** 3; **30.** 0; **31.** 6; **32.** a^n **33.** 144; **34.** a; **35.** 0; **36.** 1.

Lesson 3 Exercises B

1. Find $\lim\limits_{x \to 3} 4x$

2. Find $\lim\limits_{x \to 2} x^2 - 6x + 1$

3. Given that $f(x) = x^2 + 3x$, find
$$\lim\limits_{h \to 0} \frac{f(x+h) - f(x)}{h}$$

4. Find $\lim\limits_{x \to 3} \dfrac{2x^2 + 3}{x - 2}$

5. Find $\lim\limits_{x \to 2} \dfrac{x^2 - 4}{x - 2}$

6. Find $\lim\limits_{x \to 2} \dfrac{x - 3}{(x - 2)(x + 4)}$

7. Find $\lim\limits_{x \to 3} \sqrt{\dfrac{x^2 + 5x + 3}{x + 2}}$

8. Find $\lim\limits_{x \to 3} \dfrac{8 - x^2}{3 - \sqrt{x^2 + 1}}$

9. Find $\lim\limits_{x \to 2} \dfrac{4 - x^2}{4 - \sqrt{x^2 + 7}}$

10. Find $\lim\limits_{x \to 3} \dfrac{6 - x^2}{4 - \sqrt{x^2 + 7}}$

11 Find $\lim\limits_{x \to 0} \dfrac{\sqrt{x + 3} - \sqrt{2}}{x}$

12 Find $\lim\limits_{x \to 2} \sqrt{x - 2}$

13. Find $\lim\limits_{x \to 4} \log_2(x^2 + 16)$

14. Find $\lim\limits_{x \to 2} \log_8 x$

15. Find $\lim\limits_{x \to 2} e^x$

16. Find $\lim\limits_{x \to 2} 3^x$

17. Find $\lim\limits_{x \to 2} |x|$

18. Find $\lim\limits_{x \to -2} |x|$

19. Find $[2.5]$

20. Find $[-2.5]$

21. Find $\lim\limits_{x \to n} [x]$ if n is an integer

22. Does the following limit exist if n is a non-integral real number,
$$\lim\limits_{x \to n} [x]$$

23. Find $\lim\limits_{x \to 1.4} [x]$

24. Find $\lim\limits_{x \to 3} [x]$

25. Find $\lim\limits_{x \to 0} \sin x$ (x in radians)

26. Find $\lim\limits_{x \to 0} \cos x$

27,. Find $\lim\limits_{x \to 0} \dfrac{\sin x}{x}$ (x in radians)

28, Find $\lim\limits_{x \to 0} \dfrac{1 - \cos x}{x}$

29. Find $\lim\limits_{x \to 0} \dfrac{\sin 2x}{x}$

30. Find $\lim\limits_{x \to 0} \dfrac{3x}{\cos x}$

31. Find $\lim\limits_{x \to 4} 5x$

32. Find $\lim\limits_{x \to b} x^n$, where n is a positive integer.

33. $\lim\limits_{x \to 3} [4x]^2$

34. $\lim\limits_{x \to 0} \dfrac{\sin ax}{x}$

35. Find $\lim\limits_{x \to 3} \sqrt{9 - x^2}$

Answers: 1. 12; **2.** -7; **3.** $2x + 3$; **4. 21** ; **5. 4**; **6.** Limit does not exist..(undefined)
7. $\frac{3}{5}\sqrt{15}$; **8.** $3 + \sqrt{10}$; **9.** 0; **10.** There is **no** limit (undefined); **11.** There is **no** limit
(undefined); ;**12** 0; **13.** 5 ; **14.** $\frac{1}{3}$. **15.** e^2 ; **16.** 9 ; **17.** 2 ; **18.** 2 ; **19.** 2; **20** -3.
21. Limit does not exist **22.** Yes; **23.** 1; **24.** Limit does not exist, **25.** 0
26. 1; **27. 1; 28.** 0; **29. 2** ; **30.** 0 ; **31.** 20 ; **32.** b^n **33.** 144; **34.** a ;. **35.** 0.

Lesson 4
Infinite Limits

Preliminaries

Infinity is an unbounded quantity, or a limitless quantity but infinity is **not** a number. Infinity is a quantity beyond any number.

In a quotient in which the numerator is not zero, as the denominator approaches zero, the quotient becomes larger and larger. As the denominator approaches but never reaches zero, the quotient approaches but never reaches infinity.

About the infinity symbols $+\infty$ (plus infinity) and $-\infty$ (minus infinity)

When we write $\lim\limits_{x \to a} f(x) = +\infty$, we mean as x approaches a, $f(x)$ increases without bound. In this case, the limit does **not** exist and the "$+\infty$" is not a real number, but it is used to describe the behavior that as x approaches a, $f(x)$ increases without bound. However, when we write $\lim\limits_{x \to a} f(x) = L$, where L is a real number, the limit exists.

Similarly, when we write $\lim\limits_{x \to a} f(x) = -\infty$, we mean as x approaches a, $f(x)$ decreases without bound, In this case also, the limit does **not** exist and the "$-\infty$" is not a number, but it is used to describe the behavior that as x approaches a, $f(x)$ decreases without bound.

Limits at Infinity

Example 1 In the graph below, as $x \to +\infty$ (x increases without bound), the graph of f approaches the line $y = 5$, and thus the value $f(x)$ approaches is 5. For this behavior, we symbolically write $\lim\limits_{x \to +\infty} f(x) = 5$.

Similarly, as $x \to -\infty$, the graph of f approaches the line $y = -3$, and thus the value $f(x)$ approaches is -3. For this behavior we symbolically write $\lim\limits_{x \to -\infty} f(x) = -3$.

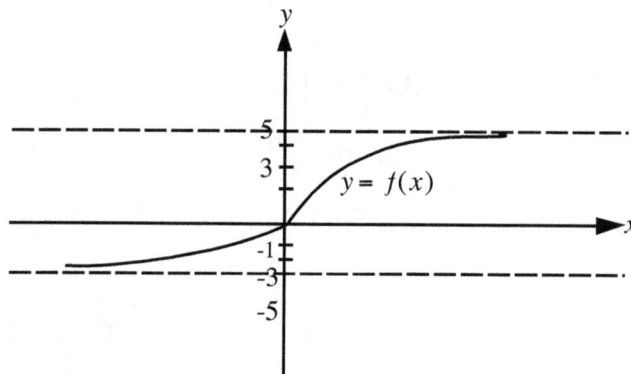

Figure 1

Example 2 In the graph below, as $x \to +\infty$ (x approaches $+\infty$), the graph of f increases without bound, and $\lim\limits_{x \to +\infty} f(x)$ does **not** exist . For this

behavior, we symbolically write $\lim\limits_{x \to +\infty} f(x) = +\infty$. However, as $x \to -\infty$, the

graph of f oscillates but approaches the line $y = 2$, and thus the value $f(x)$ approaches is 2. For this behavior, we symbolically write $\lim\limits_{x \to -\infty} f(x) = 2$.

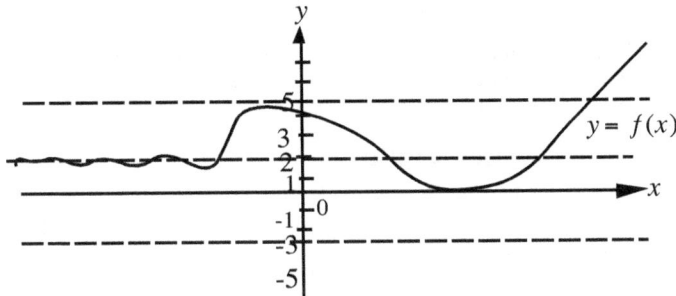

Figure 2

Similarly, below, we write $\lim\limits_{x \to -\infty} f(x) = -\infty$; but $\lim\limits_{x \to +\infty} f(x) = 0$,

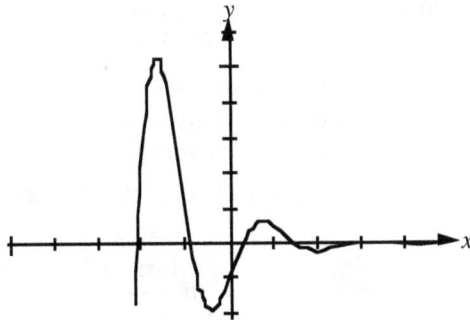

Figure 3

In the figure below, as x approaches $+\infty$, the graph of f oscillates between -1 and 1 and never approaches a single value, and consequently $\lim\limits_{x \to +\infty} f(x)$

does not exist, even though f does **not** increase without bound. For this oscillatory behavior, there is no special notation. The author suggests the following notation:

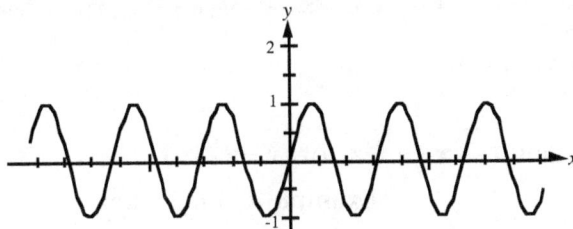

Figure 1

$\lim\limits_{x \to +\infty} f(x) = \text{osc}[-1, 1]$, which means the function oscillates between -1 and 1.

Infinite Limit Calculations ((see p.6, #4)   24

There are three main approaches for handling infinite limit calculations, namely
1. By direct substitution of $+\infty$ or $-\infty$, and applying the permissible operations as in the infinity table below.
2. By factoring and **3.** By dividing every term (for quotients) by the highest power of the variable in the **denominator** (not in the numerator); since this choice avoids "zeros only problems in the denominator and the subsequent reevaluation of a previous step. Operations involving infinity do not obey all the properties of the basic algebraic operations. Do not operate on infinity as you operate on real numbers. In calculations, you may apply the following in the Infinity Table below. Even though $+\infty$ and $-\infty$ are not real numbers, for calculations, it would be convenient to assume that $+\infty$ represents a very large positive number and $-\infty$ represents a very large negative number. (See also p.447 for analogy)

Infinity Table
For $+\infty$ and $-\infty$

Sum	Product	$9.(a)(+\infty) = +\infty$	Quotient
$1.+\infty + \infty = +\infty$	$5.\ (+\infty)(+\infty) = +\infty$	if $a > 0$	$12.\dfrac{a}{+\infty} = \dfrac{a}{-\infty} = 0$
$2.-\infty - \infty = -\infty$	$6.(-\infty)(-\infty) = +\infty$	$10.(a)(-\infty) = -\infty$	$a \neq 0$
$3.+\infty + a = \infty$	$7.(-\infty)(+\infty) = -\infty$	if $a > 0$	Example: $\dfrac{7}{+\infty} \approx 0$
$4.-\infty + a = -\infty$	$8.(+\infty)(-\infty) = -\infty$	$11.(-a)(-\infty) = +\infty$	

Do not use the following, since no meanings are assigned to them.
Try to convert the original expression to a form which on substitution will result in one of the above; or use other techniques such as L'Hôpital's Rule.

Indeterminate Table (Meaningless) Do not use.

Addition	Multiplication & Division	Powers
$13.\ (+\infty) + (-\infty)$ $14.\ \infty - \infty$	$15.\ 0 \bullet \infty$; $16.\ \dfrac{0}{0}$; $17.\ \dfrac{\infty}{\infty}$;	$18\ 0^0$; $19.\ \infty^0$; $20.\ 1^\infty$

Useful Generalizations: $\lim\limits_{n \to \pm\infty} \dfrac{1}{x^n} = 0$; $\lim\limits_{n \to +\infty} x^n = \infty$; $\lim\limits_{n \to -\infty} x^n = (-1)^n \infty$

where n is a positive integer. For large values of x, $\dfrac{1}{x} \approx 0$, $\dfrac{1}{x^2} \approx 0$, $\dfrac{1}{x^3} \approx 0$

Example 1 Find $\lim\limits_{x \to +\infty} (x^2 - 5x)$

Correct Method		Wrong Method
Method 1 $\lim\limits_{x \to +\infty} (x^2 - 5x)$ $= \lim\limits_{x \to +\infty} x(x - 5)$ $= (+\infty)(+\infty - 5)$ $= (+\infty)(+\infty)$ $= +\infty$ (#5 of Table)	**Method 2** $\lim\limits_{x \to +\infty} (x^2 - 5x)$ $\lim\limits_{x \to +\infty} x^2\left(1 - \dfrac{5}{x}\right)$ $= (+\infty)^2(1 - 0)$ $= +\infty$	$\lim\limits_{x \to +\infty} (x^2 - 5x)$ $(+\infty)^2 - 5(+\infty)$ $= \infty - \infty$ **<--**No meaning. #14 of Table

Case 1: Polynomial Functions (as $x \to +\infty$ or $x \to -\infty$)

Example 2 $\lim\limits_{x \to \infty} 5 = 5$

Example 3 $\lim\limits_{x \to +\infty} 3x = 3(+\infty) = +\infty$
(see #9 in above table)

Example 4 Find $\lim\limits_{x \to -\infty} 3x$

Solution $\lim\limits_{x \to -\infty} 3x = 3(-\infty)$
$\qquad\qquad\qquad = -\infty$
(see #10 in above table)

Example 5   Find $\lim\limits_{x \to +\infty} x^2 + 4x + 1$

Method 1 Factor-out the highest power of x	**Method 2**
$\lim\limits_{x \to +\infty} x^2\left(\dfrac{x^2}{x^2} + \dfrac{4x}{x^2} + \dfrac{1}{x^2}\right)$	$\lim\limits_{x \to +\infty} x^2 + 4x + 1$
$= \lim\limits_{x \to +\infty} x^2\left(1 + \dfrac{4}{x} + \dfrac{1}{x^2}\right)$	$= (+\infty)(+\infty) + 4(+\infty) + 1$
$= \lim\limits_{x \to +\infty} x^2 \bullet \lim\limits_{x \to +\infty} \left(1 + \dfrac{4}{x} + \dfrac{1}{x^2}\right)$	$= +\infty + \infty$
$= +\infty \bullet (1 + 0)$	$= +\infty$
$= +\infty \bullet 1 = +\infty$	

Case 2: Rational Functions (as $x \to +\infty$ or $x \to -\infty$)
   Divide every term in both the numerator and the denominator by the highest power of x in the **denominator**.

.**Example 6**   Find $\lim\limits_{x \to \infty} \dfrac{4x^2 + 3}{x^2 - 1}$

Method 1
Divide every term in both the numerator and the denominator by the highest power of x in the **denominator**.

Step 1: $\lim\limits_{x \to \infty} \dfrac{\dfrac{4x^2}{x^2} + \dfrac{3}{x^2}}{\dfrac{x^2}{x^2} - \dfrac{1}{x^2}}$

$= \lim\limits_{x \to \infty} \dfrac{4 + \dfrac{3}{x^2}}{1 - \dfrac{1}{x^2}}$

Step 2: As $n \to \infty$ (i.e., for large values of x), $\dfrac{3}{x^2} \approx 0$, $\dfrac{1}{x^2} \approx 0$, and

$\lim\limits_{x \to \infty} \dfrac{4 + \dfrac{3}{x^2}}{1 - \dfrac{1}{x^2}} = \dfrac{4 + 0}{1 + 0} = \dfrac{4}{1} = 4$

$\therefore \lim\limits_{x \to \infty} \dfrac{4x^2 + 3}{x^2 - 1} = 4$

Note that if we were given

$f(x) = \dfrac{4x^2 + 3}{x^2 - 1}$, the equation of the horizontal asymptote would be the line $y = 4$.

Method 2
Use long division to change the improper fraction to a polynomial plus a proper fraction: and apply Method 1 to the fractional part.

$\dfrac{4x^2 + 3}{x^2 - 1} = 4 + \dfrac{7}{x^2 - 1}$

$\lim\limits_{x \to \infty} \dfrac{4x^2 + 3}{x^2 - 1} = \lim\limits_{x \to \infty} 4 + \lim\limits_{x \to \infty} \dfrac{\dfrac{7}{x^2}}{\dfrac{x^2}{x^2} - \dfrac{1}{x^2}}$

$\lim\limits_{x \to \infty} \dfrac{4x^2 + 3}{x^2 - 1} = 4 + \dfrac{\dfrac{7}{x^2}}{1 - \dfrac{1}{x^2}}$

$\lim\limits_{x \to \infty} \dfrac{4x^2 + 3}{x^2 - 1} = 4 + \dfrac{0}{1}$

$\lim\limits_{x \to \infty} \dfrac{4x^2 + 3}{x^2 - 1} = 4$.

Example 7 Find $\lim\limits_{x \to +\infty} \dfrac{3x^2 + 6}{x - 5}$

Method 1	**Method 2**
Divide every term by the highest power of x in the denominator.	$\dfrac{3x^2 + 6}{x - 5} = 3x + 15 + \dfrac{81}{x - 5}$ (using long division)

Method 1

$$\lim_{x \to +\infty} \frac{\dfrac{3x^2}{x} + \dfrac{6}{x}}{\dfrac{x}{x} - \dfrac{5}{x}} \quad (A)$$

$$= \lim_{x \to +\infty} \frac{3x + \dfrac{6}{x}}{1 - \dfrac{5}{x}}$$

$$= \frac{3(\infty) + 0}{1 - 0}$$

$$= \frac{3(+\infty)}{1}$$

$$= +\infty$$

(see # 9 of Infinity Table. p.24)

Method 2

$$\frac{3x^2 + 6}{x - 5} = 3x + 15 + \frac{\dfrac{81}{x}}{\dfrac{x}{x} - \dfrac{5}{x}}$$

$$\frac{3x^2 + 6}{x - 5} = 3x + 15 + \frac{\dfrac{81}{x}}{1 - \dfrac{5}{x}}$$

$$\lim_{x \to +\infty} \frac{3x^2 + 6}{x - 5} = 3(+\infty) + 15 + \frac{0}{1 - 0}$$

$$= 3(+\infty) + 15 + \frac{0}{1}$$

$$= +\infty + 15 + 0$$

$$= +\infty$$

Example 8 Find $\lim\limits_{x \to \infty} \dfrac{x^2 - 9}{x - 3}$

Method 1 Factor and simplify first	**Method 2** Divide every term by the highest power of x in the **denominator**.

Method 1

$$\lim_{x \to +\infty} \frac{x^2 - 9}{x - 3}$$

$$= \lim_{x \to +\infty} \frac{(x + 3)(x - 3)}{(x - 3)}$$

$$= \lim_{x \to \infty} x + 3$$

$$= +\infty \quad \text{(see #3 of infinity table)}$$

Method 2

$$\lim_{x \to +\infty} \frac{\dfrac{x^2}{x} - \dfrac{9}{x}}{\dfrac{x}{x} - \dfrac{3}{x}} = \lim_{x \to +\infty} \frac{x - \dfrac{9}{x}}{1 - \dfrac{3}{x}}$$

$$= \frac{\infty - 0}{1 - 0} = \frac{+\infty}{1} = +\infty$$

Example 9 Find $\lim\limits_{x \to \infty} \dfrac{x - 2}{(x - 3)(x + 4)}$

Solution Multiply the factors in the denominator and divide the terms by highest power of x in the denominator

Step 1: $\quad \lim\limits_{x \to +\infty} \dfrac{x - 2}{x^2 + x - 12}$

Step 2: $\quad = \lim\limits_{x \to +\infty} \dfrac{\dfrac{1}{x} - \dfrac{2}{x^2}}{1 + \dfrac{1}{x} - \dfrac{12}{x^2}}$

$$= \frac{0 - 0}{1}$$

$$= \frac{0}{1}$$

$$\lim_{x \to \infty} \frac{x - 2}{(x - 3)(x + 4)} = 0 \, .$$

Case 3: Radical Functions, (as $x \to +\infty$ or $x \to -\infty$)

Example 10 Find $\displaystyle\lim_{x \to \infty} \frac{\sqrt{x+6}}{x-1}$

Solution Method 1 | Method 2

Method 1

Step 1: Square and find the square root of the denominator so that the square root sign covers both the numerator and the denominator. (see note below)

$$\lim_{x \to \infty} \frac{\sqrt{x+6}}{x-1} = \lim_{x \to \infty} \frac{\sqrt{x+6}}{\sqrt{(x-1)^2}}$$

$$= \lim_{x \to \infty} \sqrt{\frac{x+6}{x^2 - 2x + 1}}$$

$$= \lim_{x \to \infty} \sqrt{\frac{\dfrac{x}{x^2} + \dfrac{6}{x^2}}{\dfrac{x^2}{x^2} - \dfrac{2x}{x^2} + \dfrac{1}{x^2}}} \quad \text{(dividing)}$$

Step 2

$$= \lim_{x \to \infty} \sqrt{\frac{\dfrac{1}{x} + \dfrac{6}{x^2}}{1 - \dfrac{2}{x} + \dfrac{1}{x^2}}}$$

$$= \sqrt{\frac{0+0}{1-0+0}}$$

$$= \sqrt{\frac{0}{1}}$$

$$= 0.$$

Method 2

Step 1: Multiply by $\frac{1}{x}$ (which is equivalent to divide by x.)

$$\lim_{x \to \infty} \frac{\dfrac{1}{x}\sqrt{x+6}}{\dfrac{1}{x}(x-1)} = \lim_{x \to \infty} \frac{\sqrt{\dfrac{1}{x^2}(x+6)}}{\dfrac{1}{x}(x-1)}$$

(**Note:** We square $\frac{1}{x}$ before writing it as a factor of the radicand)

$$= \lim_{x \to \infty} \frac{\sqrt{\dfrac{x}{x^2} + \dfrac{6}{x^2}})}{\dfrac{x}{x} - \dfrac{1}{x}}$$

$$= \lim_{x \to \infty} \frac{\sqrt{\dfrac{1}{x} + \dfrac{6}{x^2}}}{1 - \dfrac{1}{x}}$$

$$= \frac{\sqrt{\lim_{x \to \infty}\left(\dfrac{1}{x} + \dfrac{6}{x^2}\right)}}{\lim_{x \to \infty}\left(1 - \dfrac{1}{x}\right)} \quad \left(\tfrac{1}{x} \approx 0, \ \tfrac{1}{x^2} \approx 0\right)$$

$$= \frac{\sqrt{0+0}}{1-0} = \frac{\sqrt{0}}{1} = 0$$

If $f(x) = \dfrac{\sqrt{x+6}}{x-1}$, the equation of the horizontal asymptote would be $y = 0$.

> **Note in Method 1**
>
> As $x \to \infty$ $x - 1 > 0$ and $x - 1 = \sqrt{(x-1)^2}$ .Example: $5 = \sqrt{(5)^2} = |5| = 5$; but as
>
> $x \to \infty$, $1 - x < 0$ and $1 - x = -\sqrt{(1-x)^2}$ .Example $-5 = -\sqrt{(-5)^2} = -|-5| = -5$
>
> In **Method 2**, If $n \to -\infty$, we would write a minus sign outside the sq, root. sign for $\frac{1}{x^2}$

Example 11 Find $\displaystyle\lim_{x \to \infty} \sqrt{\frac{x^2 + 4x + 3}{x + 6}}$

Method 1 Divide every term by the highest power of x in the **denominator**.

$$\sqrt{\lim_{x \to \infty} \frac{\dfrac{x^2}{x} + \dfrac{4x}{x} + \dfrac{3}{x}}{\dfrac{x}{x} + \dfrac{6}{x}}} = \sqrt{\lim_{x \to \infty} \frac{x + 4 + \dfrac{3}{x}}{1 + \dfrac{6}{x}}} = \sqrt{\frac{\infty + 4 + 0}{1 + 0}} = \sqrt{\frac{\infty}{1}} = +\infty$$

Method 2 First, by long division change radicand to a polynomial and a proper fraction

$$\sqrt{\lim_{x \to \infty} \frac{x^2 + 4x + 3}{x + 6}} = \sqrt{\lim_{x \to \infty} x + 6 + \frac{15}{x+6}} = \sqrt{\lim_{x \to \infty} x + 6 + \frac{\dfrac{15}{x}}{\dfrac{x}{x} + \dfrac{6}{x}}}$$

$$= \sqrt{+\infty + 6 + \frac{0}{1+0}} = \sqrt{+\infty + 6 + 0} = +\infty$$

Example 11b Find $\lim\limits_{x \to \infty} \sqrt{x^2 + x} - x$

$$\lim_{x \to \infty} \frac{(\sqrt{x^2+x}-x)}{1} \frac{(\sqrt{x^2+x}+x)}{\sqrt{x^2+x}+x} \quad \text{(Rationalize the numerator)}$$

$$= \lim_{x \to \infty} \frac{x^2+x-x^2}{\sqrt{x^2+x}+x)} = \lim_{x \to \infty} \frac{x}{\sqrt{x^2(1+\frac{1}{x})}+x)} = \lim_{x \to \infty} \frac{x}{x\sqrt{(1+\frac{1}{x})}+x}$$

$$= \lim_{x \to \infty} \frac{1}{\sqrt{1+\frac{1}{x}}+1} = \frac{1}{\sqrt{1+0}+1} = \frac{1}{\sqrt{1}+1} = \frac{1}{1+1} = \frac{1}{2}$$

Example 12 Find $\lim\limits_{x \to +\infty} \sqrt{x-5}$

Approach 1	**Approach 2**
$\sqrt{\lim\limits_{x \to +\infty} x(1-\frac{5}{x})}$ $= \sqrt{\infty(1-0)} \quad = \sqrt{+\infty} = +\infty$	$\lim\limits_{x \to +\infty} \sqrt{x-5} = \sqrt{\lim\limits_{x \to +\infty} x-5}$ $= \sqrt{+\infty-5} = \sqrt{+\infty} = +\infty.$

| **Example 13a** $x \to +\infty$

 Find $\lim\limits_{x \to \infty} \dfrac{\sqrt{x^2+3}}{x+5}$

 Approach 1

 $\lim\limits_{x \to \infty} \dfrac{\sqrt{x^2+3}}{x+5}$

 $= \lim\limits_{x \to \infty} \dfrac{\sqrt{x^2(1+\frac{3}{x^2})}}{x(1+\frac{5}{x})}$

 $= \lim\limits_{x \to \infty} \dfrac{|x|\sqrt{1+\frac{3}{x^2}}}{x(1+\frac{5}{x})}$

 $= \lim\limits_{x \to \infty} \dfrac{x\sqrt{1+\frac{3}{x^2}}}{x(1+\frac{5}{x})}$

 $= \lim\limits_{x \to \infty} \dfrac{\sqrt{1+\frac{3}{x^2}}}{1+\frac{5}{x}}$

 $= \dfrac{\sqrt{\lim\limits_{x \to \infty} 1+\frac{3}{x^2}}}{\lim\limits_{x \to \infty} 1+\frac{5}{x}}$

 $= \dfrac{\sqrt{1+0)}}{1+0} = 1$ | **Approach 2**
 Divide by x (equivalent to multiply by $\frac{1}{x}$)

 $\lim\limits_{x \to \infty} \dfrac{\sqrt{x^2+3}}{x+5}$

 $= \lim\limits_{x \to \infty} \dfrac{\frac{1}{x}\sqrt{x^2+3}}{\frac{1}{x}(x+5)}$

 $= \lim\limits_{x \to \infty} \dfrac{\sqrt{\frac{1}{x^2}(x^2+3)}}{\frac{1}{x}(x+5)}$

 $= \lim\limits_{x \to \infty} \dfrac{\sqrt{\frac{x^2}{x^2}+\frac{3}{x^2}}}{\frac{x}{x}+\frac{5}{x}}$

 $= \lim\limits_{x \to \infty} \dfrac{\sqrt{1+\frac{3}{x^2}}}{1+\frac{5}{x}}$

 $= \dfrac{\sqrt{\lim\limits_{x \to \infty} 1+\frac{3}{x^2}}}{\lim\limits_{x \to \infty} 1+\frac{5}{x}}$

 $= \dfrac{\sqrt{1+0}}{1+0} = \dfrac{\sqrt{1}}{1} = 1$

 Approach 3:
 Begin with the approach in Example 11b, above. | **Example 13b** $x \to -\infty$

 $\lim\limits_{x \to -\infty} \dfrac{\sqrt{x^2+3}}{x+5}$

 $= \lim\limits_{x \to -\infty} \dfrac{\sqrt{x^2(1+\frac{3}{x^2})}}{x(1+\frac{5}{x})}$

 $= \lim\limits_{x \to -\infty} \dfrac{|x|\sqrt{(1+\frac{3}{x^2})}}{x(1+\frac{5}{x})}$

 Step 3: $\lim\limits_{x \to -\infty} \dfrac{-x\sqrt{(1+\frac{3}{x^2})}}{x(1+\frac{5}{x})}$

 $= \lim\limits_{x \to -\infty} -\dfrac{\sqrt{1+\frac{3}{x^2}}}{1+\frac{5}{x}}$

 $= -\dfrac{\sqrt{\lim\limits_{x \to \infty} 1+\frac{3}{x^2}}}{\lim\limits_{x \to \infty} 1+\frac{5}{x}}$

 $= -\dfrac{\sqrt{1+0)}}{1+0} = -1$

 Note the minus sign in Step 3.

 $\sqrt{x^2} = |x| = \begin{cases} x \text{ if } x \geq 0 \\ -x \text{ if } x < 0 \end{cases}$

 and above, $x \to -\infty.$ |

Case 4: Limits of Exponential Functions

(as $x \to +\infty$ or $x \to -\infty$)

Here, we will be guided by the graphs of these functions.

"A picture is worth a thousand words" . Memorize these graphs.

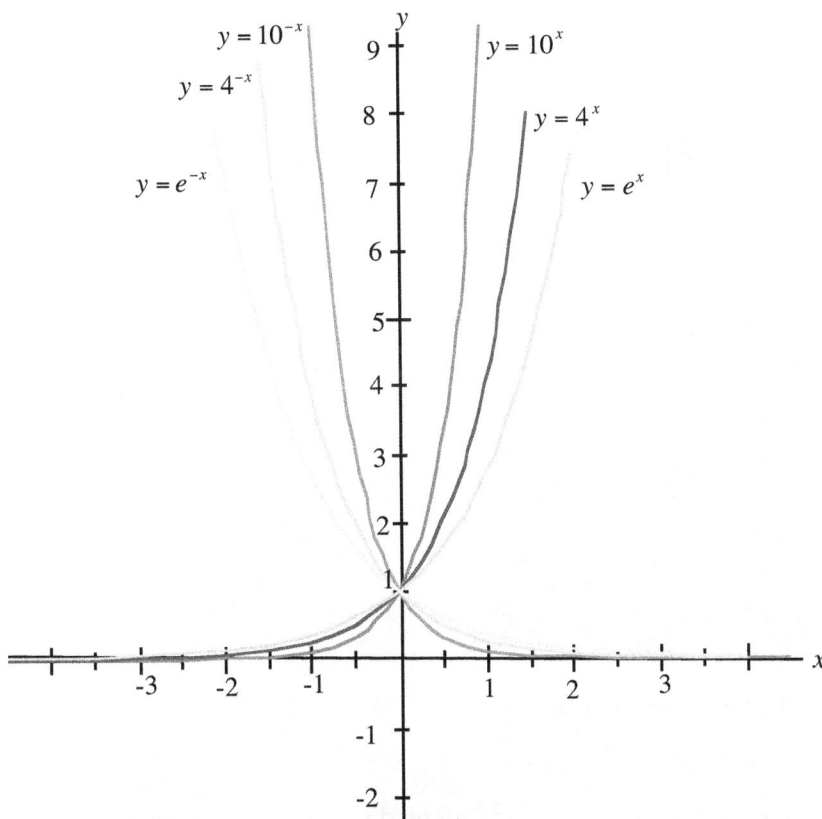

Example 13 $\displaystyle\lim_{x \to +\infty} e^x = +\infty$ (see graph of $y = e^x$)

Example 14 $\displaystyle\lim_{x \to -\infty} e^x = 0$ (see graph of $y = e^x$)

Example 15 $\displaystyle\lim_{x \to +\infty} e^{-x} = \lim_{x \to +\infty} \frac{1}{e^x} = 0$ (see graph of $y = e^{-x}$)

Example 16 $\displaystyle\lim_{x \to -\infty} e^{-x} = \lim_{x \to -\infty} \frac{1}{e^x} = +\infty$ (see graph of $y = e^{-x}$)

Example 17 $\displaystyle\lim_{x \to +\infty} 2^x = +\infty$

Example 18 $\displaystyle\lim_{x \to -\infty} 2^x = 0$

Case 5: Limits of logarithmic functions

(as $x \to +\infty$ or $x \to 0^+$)

For the natural logarithmic function, "one picture is worth a thousand words"

Example 19 $\lim\limits_{x \to \infty} \ln x = +\infty$ (From the graph: memorize this graph)

$$\lim\limits_{x \to 0^+} \ln x = -\infty$$

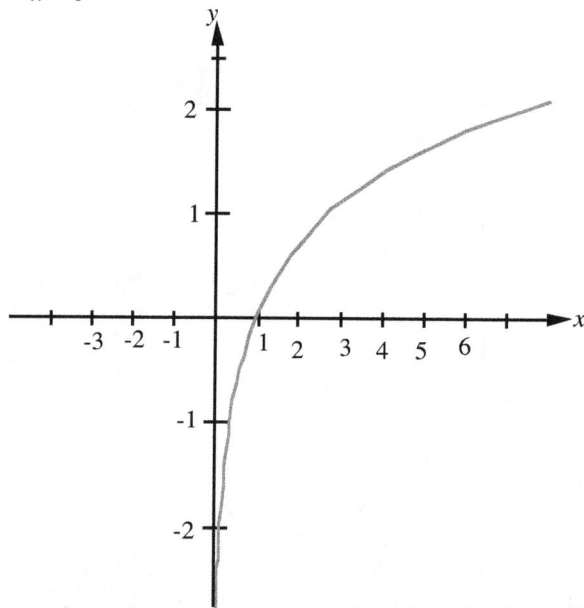

Figure: The graph of $y = \log_e x$ or $x = e^y$

Example 20 Explain why $\lim\limits_{x \to +\infty} \dfrac{e^x}{x^n} = +\infty$

Explanation: Even though as x increases, both e^x and x^n tend toward $+\infty$, because e^x in the numerator increases much more faster than x^n, in the denominator, the quotient tends towards $+\infty$;

Example 21 Explain why $\lim\limits_{x \to +\infty} \dfrac{x^n}{e^x} = 0$

Explanation: Even though as x increases, both e^x and x^n tend toward $+\infty$, because e^x in the denominator increases much more faster than x^n, in the numerator, the quotient tends towards 0.

Example 22 Explain why $\lim\limits_{x \to +\infty} \dfrac{\ln x}{x^n} = 0$

Note here that we **cannot** use $\dfrac{+\infty}{+\infty}$ since no meaning is given to this "quotient", and we proceed as follows (see also the infinity tables of Lesson 4):

Explanation: Even though as x increases, both $\ln x$ and x^n tend toward $+\infty$, because x^n in the denominator increases much more faster than $\ln x$, in the numerator, the quotient tends towards 0.

Example 23 Explain why $\lim\limits_{x \to +\infty} \dfrac{x^2}{\ln x} = +\infty$

Explanation: Even though as x increases, both x^2 and $\ln x$ tend toward $+\infty$, because x^2 in the numerator increases much more faster than $\ln x$, in the denominator, the quotient tends towards $+\infty$;

Example 24 Find $\lim\limits_{x \to +\infty} \log_2 x = +\infty$

Example 25 Find $\lim\limits_{x \to +\infty} \log_2\left(x^2 + 7\right) = +\infty$

Case 6: Limits of Trigonometric Functions
(as $x \to +\infty$ or $x \to -\infty$)
We list the interesting cases.

$\lim\limits_{x \to \infty}$ **trig function of** x	$\lim\limits_{x \to -\infty}$ **trig function of** x
1. $\lim\limits_{x \to \infty} \sin x$ does not exist (oscillation)	**1.** $\lim\limits_{x \to -\infty} \sin x$ does not exist (oscillation)
2. $\lim\limits_{x \to \infty} \cos x$ does not exist (oscillation)	**2.** $\lim\limits_{x \to -\infty} \cos x$ does not exist (oscillation)
3. $\lim\limits_{x \to \infty} \tan x$ is undefined, and does not exist	**3.** $\lim\limits_{x \to -\infty} \tan x$ undefined, does not exist.
4. $\lim\limits_{x \to \infty} \cot x$ is undefined	**4.** $\lim\limits_{x \to -\infty} \cot x$ is undefined

Also:

Theorem 1 $\lim\limits_{x \to \infty} \dfrac{\sin x}{x} = 0$ (where x is in radians)

Theorem 2 $\lim\limits_{x \to \infty} \dfrac{1 - \cos x}{x} = 0$

Case 7: Limits of Inverse Trigonometric Functions

as $x \to +\infty$ or $x \to -\infty$

We list the interesting cases.

$\lim\limits_{x \to +\infty}$ **inverse trig function of** x	$\lim\limits_{x \to -\infty}$ **inverse trig function of** x
1. $\lim\limits_{x \to +\infty} \arctan x = \frac{\pi}{2}$	**1.** $\lim\limits_{x \to -\infty} \arctan x = -\frac{\pi}{2}$
2. $\lim\limits_{x \to +\infty} arc\csc x = 0$	**2.** $\lim\limits_{x \to -\infty} arc\csc x = 0$
3. $\lim\limits_{x \to +\infty} arc\sec x = \frac{\pi}{2}$	**3.** $\lim\limits_{x \to -\infty} arc\sec x = \frac{\pi}{2}$
4. $\lim\limits_{x \to +\infty} arc\cot x = 0$	**4.** $\lim\limits_{x \to \infty} arc\cot x = \pi$

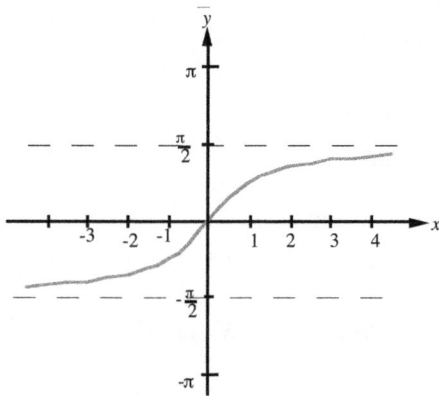

Fig. Graph of $\arctan x$

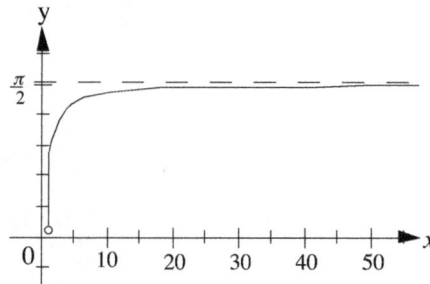

Fig. Graph of $arc\sec x$

Lesson 4 Exercises **A**

1. What do we mean when we write $\lim\limits_{x \to a} f(x) = +\infty$

2. Do the symbols $+\infty$ and $-\infty$ represent real numbers?

3. Which of the following is meaningless in calculations?

$$\frac{a}{+\infty} \text{ or } \frac{0}{0}$$

4 Find $\lim\limits_{x \to +\infty} (x^2 - 5x)$

5 Find $\lim\limits_{x \to \infty} 5$

6 Find $\lim\limits_{x \to +\infty} 3x$

7 Find $\lim\limits_{x \to -\infty} 3x$

8. Find $\lim\limits_{x \to \infty} x^2 + 4x + 1$

9. Find $\lim\limits_{x \to \infty} \frac{4x^2 + 3}{x^2 - 1}$

10 Find $\lim\limits_{x \to \infty} \frac{3x^2 + 6}{x - 5}$

11. Find $\lim\limits_{x \to \infty} \frac{x^2 - 9}{x - 3}$

12 Find $\lim\limits_{x \to \infty} \frac{x - 2}{(x - 3)(x + 4)}$

13 Find $\lim\limits_{x \to \infty} \frac{\sqrt{x + 6}}{x - 1}$

14 Find $\lim\limits_{x \to \infty} \sqrt{\frac{x^2 + 4x + 3}{x + 6}}$

15. Find $\lim\limits_{x \to \infty} \sqrt{x - 5}$

16. Find $\lim\limits_{x \to +\infty} e^x = +\infty$

17 Find $\lim\limits_{x \to -\infty} e^x$

18 Find $\lim\limits_{x \to +\infty} e^{-x}$

19 Find $\lim\limits_{x \to -\infty} e^{-x}$

20 Find $\lim\limits_{x \to +\infty} 2^x$

21. Find $\lim\limits_{x \to -\infty} 2^x$

22 $\lim\limits_{x \to \infty} \ln x$

23. $\lim\limits_{x \to 0^+} \ln x$

24. Explain why $\lim\limits_{x \to +\infty} \frac{e^x}{x^n} = +\infty$

25. Explain why $\lim\limits_{x \to +\infty} \frac{x^n}{e^x} = 0$

26. Explain why $\lim\limits_{x \to +\infty} \frac{\ln x}{x^n} = 0$

27. Explain why $\lim\limits_{x \to +\infty} \frac{x^n}{\ln x} = +\infty$

28. Find $\lim\limits_{x \to \infty} \log_2 x$

29. $\lim\limits_{x \to \infty} \sin x$

30. Find $\lim\limits_{x \to \infty} \cos x$

31. Find $\lim\limits_{x \to \infty} \frac{\sin x}{x}$ (x is in radians)

32. Find $\lim\limits_{x \to \infty} \frac{1 - \cos x}{x}$

Answers: 3. $\frac{0}{0}$; **4.** $+\infty$; **5.** 5 **6.** $+\infty$; **7.** $-\infty$; **8.** $+\infty$; **9.** 4; **10.** $+\infty$; **11.** $+\infty$

12. 0; **13.** 0; **14.** $+\infty$; **15.** $+\infty$; **16.** $+\infty$; **17.** 0; **18.** 0; **19.** $+\infty$;

20. $+\infty$ **21.** 0; **22.** $+\infty$; **23.** $-\infty$; **28.** $+\infty$

29. No limit, oscillation between -1 and 1.

30. No limit, oscillation between -1 and 1.; **31.** 0 ; **32** 0.

Lesson 4 Exercises **B**

1. What do we mean when we write $\lim\limits_{x \to a} f(x) = +\infty$

2. Do the symbols $+\infty$ and $-\infty$ represent real numbers? Elaborate

3. Which of the following is meaningless in calculations?

$$\frac{a}{+\infty} \quad \text{or} \quad \frac{0}{0}$$

4 Find $\lim\limits_{x \to +\infty} (x^2 + 3x)$

5 Find $\lim\limits_{x \to \infty} 3$

6 Find $\lim\limits_{x \to +\infty} 2x$

7 Find $\lim\limits_{x \to -\infty} x$

8. Find $\lim\limits_{x \to \infty} x^2 - 4x + 14$

9. Find $\lim\limits_{x \to \infty} \dfrac{2x^2 + 5}{x^2 - 1}$

10 Find $\lim\limits_{x \to \infty} \dfrac{2x^2 + 4}{x - 3}$

11. Find $\lim\limits_{x \to \infty} \dfrac{x^2 - 4}{x - 2}$

12 Find $\lim\limits_{x \to \infty} \dfrac{x - 1}{(x - 4)(x + 3)}$

13 Find $\lim\limits_{x \to \infty} \dfrac{\sqrt{x + 6}}{x - 2}$

14 Find $\lim\limits_{x \to \infty} \sqrt{\dfrac{x^2 + 4x - 3}{x + 2}}$

15. Find $\lim\limits_{x \to \infty} \sqrt{x - 3}$

16. Find $\lim\limits_{x \to +\infty} e^{-x}$

17 Find $\lim\limits_{x \to -\infty} e^{x}$

18 Find $\lim\limits_{x \to +\infty} e^{-x^2}$

19 Find $\lim\limits_{x \to -\infty} e^{-x}$

20 Find $\lim\limits_{x \to +\infty} 3^{x}$

21. Find $\lim\limits_{x \to -\infty} 3^{x}$

22 $\lim\limits_{x \to \infty} \ln x$

23. $\lim\limits_{x \to 0^+} \ln x$

24. Explain why $\lim\limits_{x \to +\infty} \dfrac{e^x}{x^n} = +\infty$

25. Explain why $\lim\limits_{x \to +\infty} \dfrac{x^n}{e^x} = 0$

26. Explain why $\lim\limits_{x \to +\infty} \dfrac{\ln x}{x^n} = 0$

27. Explain why $\lim\limits_{x \to +\infty} \dfrac{x^n}{\ln x} = +\infty$

28. Find $\lim\limits_{x \to \infty} \log_4 x$

29. $\lim\limits_{x \to \infty} \sin x$

30. Find $\lim\limits_{x \to \infty} \cos x$

31. Find $\lim\limits_{x \to \infty} \dfrac{\sin x}{x}$ (x is in radians)

32. Find $\lim\limits_{x \to \infty} \dfrac{1 - \cos x}{x}$

33. Find $\lim\limits_{x \to \infty} \dfrac{\cos x}{x}$

Answers: 3. $\frac{0}{0}$; **4.** $+\infty$; **5.** 3 **6.** $+\infty$; **7.** $-\infty$; **8.** ∞ ; **9.** 2; **10.** ∞ ; **11.** ∞;

12. 0 ; **13.** 0 ; **14.** ∞ ; **15.** ∞; **16.** 0 ; **17.** 0 ; **18.** 0 **19.** ∞ ;

20. ∞; **21.** 0 ; **22.** ∞; **23.** $-\infty$; **24.** $+\infty$; **25.** 0; **27.** $+\infty$; **28.** ∞;

29. No limit, oscillation between -1 and 1.;

30. No limit, oscillation between -1 and 1.: **31.** 0; **32** 0; **33.** 0.

Lesson 5A
Constructing Direct Mathematical Proofs
I want to use basic things to do big things.
The best things in life are free.

In constructing a mathematical proof, we show that a given mathematical statement is true. If the statement is in the "If-then" form, the **hypothesis** is the clause that follows the word "if" and the **conclusion** is the clause that follows the word "then". The hypothesis is what is given, and the conclusion is what is to be proved. Usually, we begin with the hypothesis and proceed to the conclusion by logically combining axioms, definitions, and already proved statements.

There are a number of strategies that can be used to construct a proof.
Strategies
As shown in the diagram below, we can view the construction of a proof as drawing a line between two points A and B, where we can start from A and proceed to B (This is the usual approach.) or start from B and proceed to A; or start from A to a point C between A and B, and then complete the line by starting from B and proceeding to C.

Let **A** represent the **hypothesis** and let **B** represent the **conclusion**.

Approach 1. Begin from A and proceed continuously and logically to B.
Approach 2: Begin from B and proceed continuously and logically to A.
Approach 3: Begin from A and proceed continuously and logically to any point C between A and B, followed by beginning from B and proceeding to C.
Approach 4: Begin from B and proceed continuously and logically to any point C between A and B, followed by beginning from A and proceeding to C. There are other strategies such as start from C.

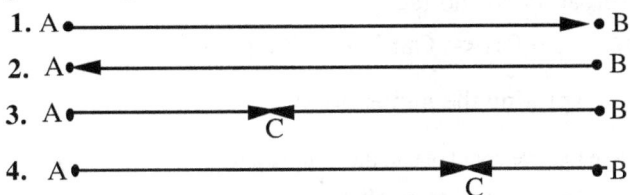

1. A •——————————————► • B

2. A •◄—————————————— • B

3. A •————————►►———————— • B
              C

4. A •—————————————►►◄———— • B
                        C

Finally, check to make sure all the statements flow logically from **A**, the hypothesis to **B**, the conclusion.
Note: From any step to the next step, first determine what you want for the next step, and perform a mathematical operation to get to the next step.

If you have experience in constructing proofs (e.g., proving theorems) in geometry, then you are ready, with some minor adjustments, to construct proofs in calculus and in mathematics in general.
In constructing proofs, we usually operate on expressions, on equations and on inequalities.

Case 1: Performing Operations on Expressions
On an expression, we perform two operations which are inverses of each other so that the resulting expression is equivalent to the original expression. Always, think of the properties of the basic operations in elementary mathematics such as **the multiplicative property of 1**, and the **additive property of zero.** These properties are applied to expressions.

Multiplicative property of 1: For any real number n, $n \times 1 = 1 \times n = n$

Example **1.** Forming equivalent fractions: Given the fraction $\frac{1}{2}$, we can do the

following: $\frac{1 \times 5}{2 \times 5} = \frac{5}{10}$ (We multiplied by $\frac{5}{5}$ which equals 1.

Example **2.** Rationalizing the denominator of a radical expression.

Given the radical $\frac{1}{\sqrt{2}}$, we can do the following

$$\frac{1}{\sqrt{2}} \cdot \frac{\sqrt{2}}{\sqrt{2}} = \frac{\sqrt{2}}{2} \quad (\frac{\sqrt{2}}{\sqrt{2}} = 1)$$

Example **3.** Dividing a complex number. Divide: $\frac{4}{3-2i}$

$$\frac{4}{3-2i} = \frac{4}{(3-2i)} \cdot \frac{(3+2i)}{(3+2i)} \qquad \left(\text{Note}: \frac{3+2i}{3+2i} = 1 \right)$$

$$= \frac{12}{13} + \frac{8}{13}i$$

Additive property of zero: for any real number n, $n + 0 = 0 + n = n$

Example 1 Complete the square: $x^2 - 12x + 8 = 0$

Solution: On the left side of the equation, **add** and **subtract** the square of $\frac{b}{2}$.

$$\boxed{x^2 - 12x + (-6)^2 - (-6)^2 + 8 = 0} \quad (b = -12, \frac{b}{2} = -6, \left(\frac{b}{2}\right)^2 = (-6)^2)$$

$$(x - 6)^2 - (-6)^2 + 8 = 0$$

Adding $(-6)^2$ and subtracting $(-6)^2$ on the left side of the equation is equivalent to adding **zero t**o the left side of the equation.

Example 2: Divide $\frac{x}{x + 2}$

Solution In the numerator, add and subtract 2.

$$\frac{x + 2 - 2}{x + 2} \quad ((2 - 2 = 0) \text{ <-- Our interest is in this step}$$

$$= \frac{x + 2}{x + 2} - \frac{2}{x + 2} \quad \text{(splitting the numerators)}$$

$$= 1 - \frac{2}{x + 2} \quad \text{(Note that we could obtain the same result by long division)}.$$

Case 2: Operating on equations and inequalities

On equations and inequalities, we perform the same operation on both sides of an equation or an inequality.

1. Completing the square: $x^2 - 12x = -8$

Step 1: Add the square of half the coefficient of the x-term to both sides of the

equation. (i.e., add the square of $\frac{b}{2}$ to both sides of the equation)

$$x^2 - 12x + \left(\frac{-12}{2}\right)^2 = -8 + \left(\frac{-12}{2}\right)^2 \quad (b = -12, \left(\frac{b}{2}\right)^2 = (-6)^2$$

$$x^2 - 12x + (-6)^2 = -8 + (-6)^2$$

Step 2: Complete the square on the left-hand side of the equation

$$(x - 6)^2 = -8 + 36$$
$$(x - 6)^2 = 28$$

Our interest here is in **Step 1**, where we added $(-6)^2$ to both sides of the equation.

The following may be useful in proofs:

Equivalents in Trigonometry

Quantities = 1	Quantities = −1	Quantities = 0
$\sin\frac{\pi}{2} = 1$	$\sin\frac{3\pi}{2} = -1$	$\sin 0 = 0$
$\cos 0 = 1$	$\cos\pi = -1$	$\sin\pi = 0$
$\csc\frac{\pi}{2} = 1$	$\sec\pi = -1$	$\cos\frac{\pi}{2} = \cos\frac{3\pi}{2} = 0$
$\tan\frac{\pi}{4} = 1$	$\csc\frac{3\pi}{2} = -1$	$\cot\frac{\pi}{2} = \cot\frac{3\pi}{2} = 0$
$\cot\frac{\pi}{4} = 1$	-----------------------	$\tan\pi = 0$
$\sin^2\theta + \cos^2\theta = 1$	Quantities = $\frac{1}{2}$	$\tan 0 = 0$
$\sec^2\theta - \tan^2\theta = 1$	$\sin\frac{\pi}{6} = \frac{1}{2}$; $\cos\frac{\pi}{3} = \frac{1}{2}$	-----------------------
$\csc^2\theta - \cot^2\theta = 1$	-----------------------	Quantities = $\sqrt{3}$
$\cos 2\theta + 2\sin^2\theta = 1$	Quantities = 2	$\tan\frac{\pi}{3} = \sqrt{3}$
$2\cos^2\theta - \cos 2\theta = 1$	$\csc\frac{\pi}{6} = 2$	**Note:** also that:
	$\sec\frac{\pi}{3} = 2$	If $A = B$ then $A - B = 0$

Example 1 Using $\tan\frac{\pi}{4} = 1$ and $\tan(A + B) = \frac{\tan A + \tan B}{1 - \tan A \tan B}$, show that

$$\int \sec x\, dx = \ln\left|\frac{1+\tan\frac{x}{2}}{1-\tan\frac{x}{2}}\right| + C = \ln\left|\tan\left(\frac{\pi}{4} + \frac{x}{2}\right)\right| + C$$

Solution By applying $1 = \tan\frac{\pi}{4}$ to

$$\ln\left|\frac{1+\tan\frac{x}{2}}{1-\tan\frac{x}{2}}\right|,$$

$$\ln\left|\frac{1+\tan\frac{x}{2}}{1-\tan\frac{x}{2}}\right| = \ln\left|\frac{\tan\frac{\pi}{4}+\tan\frac{x}{2}}{1-\tan\frac{x}{2}\tan\frac{\pi}{4}}\right| \quad (A)$$

Similarly,, by applying

$$\tan(A + B) = \frac{\tan A + \tan B}{1 - \tan A \tan B} \quad \text{to}$$

$$\ln\left|\tan\left(\frac{\pi}{4} + \frac{x}{2}\right)\right|$$

$$\ln\left|\tan\left(\frac{\pi}{4} + \frac{x}{2}\right)\right| = \ln\left|\frac{\tan\frac{\pi}{4}+\tan\frac{x}{2}}{1-\tan\frac{\pi}{4}\tan\frac{x}{2}}\right| \quad (B)$$

Since the right sides of (A) and (B) are identical, the left sides are equal, and

$$\ln\left|\frac{1+\tan\frac{x}{2}}{1-\tan\frac{x}{2}}\right| = \ln\left|\tan\left(\frac{\pi}{4} + \frac{x}{2}\right)\right|$$

$$\therefore \int \sec x\, dx = \ln\left|\frac{1+\tan\frac{x}{2}}{1-\tan\frac{x}{2}}\right| + C = \ln\left|\tan\left(\frac{\pi}{4} + \frac{x}{2}\right)\right| + C$$

Example 2
To exaggerate the substitution axiom:

If the cost of 2 books = \$100,
then, the cost of $\sin\frac{\pi}{2}$ book = **\$ 50**
(same as the cost of one book = \$50;
since $\sin\frac{\pi}{2} = 1$)
We could also say that
the cost of $\tan\frac{\pi}{4}$ book = \$50, since

$\tan\frac{\pi}{4} = 1$.

Similarly, the cost of x^0 book = \$50.

Also $x^0 = \sin\frac{\pi}{2} = \tan\frac{\pi}{4} = 7 - 6 = \frac{5}{5} = 1$.

$(x \neq 0)$

Extra: Similarly,
$(\sin^2\theta + \cos^2\theta)$ book costs \$50.
since $\sin^2\theta + \cos^2\theta = 1$.

Some Axioms for Operating on Equations

Axioms are general mathematical statements that we accept as true, without any proof, in order to deduce other less obvious statements. The following are very useful in constructing proofs and solving equations.

1. A quantity is equal to itself. (reflexive property of equality, also identity principle)

$$a = a$$

2. An equality may be reversed. (symmetric property of equality)

If $a = b$, then
$b = a$

3. Quantities equal to the same quantity are equal to each other.(transitive property of equality)

If $a = b$, and $b = c$, then
$a = c$.

4. A quantity may be substituted for its equal in any expression or equation. (substitution axiom)

5. A whole equals the sum of all its parts. (partition axiom)

6. If equal quantities are added to equal quantities, the sums are equal (addition axiom)

If $a = b$,
then $a + c = b + c$ **ALSO**   (If $a = b$, and $c = d$, then $a + c = b + d$)

7. If equal quantities are subtracted from equal quantities, the differences are equal.(subtraction axiom)

If $a = b$,
then $a - c = b - c$ **ALSO**   (If $a = b$, and $c = d$, then $a - c = b - d$)

8. If equal quantities are multiplied by equal quantities, the products are equal. (multiplication axiom)

If $a = b$,
then $ac = bc$ **ALSO**   (If $a = b$, and $c = d$, then $ac = bd$.)

9. If equal quantities are divided by equal quantities (not zero), the quotients are equal. (division axiom)

If $a = b$,
then $\dfrac{a}{c} = \dfrac{b}{c}$ **ALSO**   (If $a = b$, and $c = d$, then $\dfrac{a}{c} = \dfrac{b}{d}$)

10. Like powers of equals are equal . (powers axiom)

Example: If $a = 3$, then
$a^2 = 3^2$ or $a^2 = 9$.

11. Like roots of equals are equal. (roots axiom)

Example: if $a^3 = 8$, then

$$\sqrt[3]{a^3} = \sqrt[3]{8}$$
$$a = 2.$$

As elementary as the properties we have covered so far may sound, these properties are very useful in constructing proofs at the advanced level. The advice is that when constructing proofs at the advanced level, keep in mind all the operations you learned at the elementary level. You would be amazed in the future, how a basic elementary operation would make a seemingly difficult situation easy to handle.

Example 1

Prove that if $f(x)$ has a derivative at x_0, then $f(x)$ is continuous at x_0.

Solution

Given: $f(x)$ has a derivative at x_0 (hypothesis)

Required: To prove that $f(x)$ is continuous at x_0 (conclusion)

Plan: If $f(x)$ is continuous at x_0, then $\lim\limits_{x \to x_0} f(x) = f(x_0)$ (Continuity definition)

The proof would be complete after showing that $\lim\limits_{x \to x_0} f(x) = f(x_0)$.

Proof:

Statement	Reason
1. $f(x)$ has a derivative at x_0	1. Given
2.. $\lim\limits_{h \to 0} \dfrac{f(x_0 + h) - f(x_0)}{h}$ exists	2. Definition of derivative at x_0
3. $f(x_0 + h) = f(x_0 + h)$	3. A quantity is equal to itself
4. $f(x_0 + h) = f(x_0 + h) - f(x_0) + f(x_0)$	5. Adding zero to right-hand side
5. $\dfrac{f(x_0 + h)}{h} = \dfrac{f(x_0 + h) - f(x_0)}{h} + \dfrac{f(x_0)}{h}$	(dividing by h)
7. $h \cdot \dfrac{f(x_0 + h)}{h} = h \cdot \dfrac{f(x_0 + h) - f(x_0)}{h} + h \cdot \dfrac{f(x_0)}{h}$	(multiplying by h)
8. $f(x_0 + h) = h \cdot \dfrac{f(x_0 + h) - f(x_0)}{h} + f(x_0)$	(canceling some h 's)
9. $\lim\limits_{h \to 0} f(x_0 + h) = \lim\limits_{h \to 0} h \cdot \lim\limits_{h \to 0} \dfrac{f(x_0 + h) - f(x_0)}{h} + \lim\limits_{h \to 0} f(x_0)$	(taking limits)
10. $\lim\limits_{h \to 0} f(x_0 + h) = 0 \cdot f^{'}(x_0) + f(x_0)$	($\lim\limits_{h \to 0} \dfrac{f(x_0 + h) - f(x_0)}{h} = f'(x_0)$)
11. $\lim\limits_{h \to 0} f(x_0 + h) = f(x_0)$	

12. Let $h = x - x_0$ or let $x = x_0 + h$

Then $\lim\limits_{h \to 0} f(x_0 + h) = f(x_0)$ becomes

$\lim\limits_{x - x_0 \to 0} f(x_0 + x - x_0) = f(x_0)$ and

$\qquad \lim\limits_{x \to x_0} f(x) = f(x_0)$ (definition of continuity)

The proof is complete.

Lesson 5B
Review of Inequality Operations

The operations on inequalities are similar to the operations on equations (equalities), except for: **1.** When both sides of an inequality are multiplied by or divided by a **negative** number, or if both sides (having the same sign) of an inequality are inverted, the sense of the inequality must be reversed.

2. To operate on **two inequalities** simultaneously, the sense (direction or order) of both inequalities must be the same.

The properties below are useful for the epsilon-delta proofs in the next lesson.

Transitivity	Addition	Multiplication
1. If $a > b$ and $b > c$. then $a > c$ Example: If $7 > 5$ and $5 > 2$ then $7 > 2$ (true)	**2.** If $a > b$ and $c > d$. then $a + c > b + d$ Example: If $8 > 4$ and $3 > 2$ then $8 + 3 > 4 + 2$ or $11 > 6$ (True)	**2.** If $a > b$ and $c > 0$. then $ac > bc$ but If $a > b$ and $c < 0$. then $ac < bc$ (sense reversed)

Subtraction	Multiplication	Division
4. If $a > b$ and $c > d$. We **cannot** subtract as we do with **equations**. Step 1: $-c < -d$ (reverse sense) Step 2: $-d > -c$ (rewriting) If $a > b$ and $-d > -c$ then (same sense) $a - d > b - c$ (adding) -------------------------- **Inversion** of both sides of the same sign. If diffent signs, **no** inversion Invert both sides and reverse the sense of the inequality. If $a > b > 0$ e.g.$(10 > 2)$ Then $\frac{1}{a} < \frac{1}{b}$ $(\frac{1}{10} < \frac{1}{2})$ **No** inversion: $10 > -2$; $\frac{1}{10} > -\frac{1}{2}$	**5.** If $a > b > 0$ and $c > d > 0$. then $ac > bd$ **Note:** $a, b, c, d,$ are all positive **Example 1:** If $8 > 4$ and $3 > 2$ then $8 \times 3 > 4 \times 2$ or $24 > 8$ (True) **Example 2:** If $|x - 3| < \frac{\varepsilon}{7}$ and $|x + 3| < 7$, then $|x - 3||x + 3| < \frac{\varepsilon}{7}(7)$ $|(x + 3)(x - 3)| < \varepsilon$	**5.** If $a > b > 0$ and $c > d > 0$. We **cannot** divide as we do with **equations**. Step 1: If $\frac{1}{c} < \frac{1}{d}$ Step 2: $\frac{1}{d} > \frac{1}{c}$.(rewriting) If $a > b$ and $\frac{1}{d} > \frac{1}{c}$ $\frac{a}{d} > \frac{b}{c}$ (Multiplying) **Note**: In Step 1, we inverted both sides and reversed the sense of the inequality.

Powers:	Roots:
Powers: Given $a > b > 0$ If m and n are positive integers then **1.** Then $a^m > b^m$ Like positive powers of unequal quantities are unequal in the same sense. **Example** We can square both sides of an inequality. If $4 > 3$, then $4^2 > 3^2$ or $16 > 9$ Note above that a aand b re positive.	**Roots:** If $a > b > 0$, then $\sqrt[n]{a} > \sqrt[n]{b}$ or $a^{\frac{1}{n}} > b^{\frac{1}{n}}$ Like positive roots of unequal quantities are unequal in the same sense **Example** We can take the square root both sides of an inequality **Example 1.** If $16 > 9$, then $\sqrt{16} > \sqrt{9}$ and $4 > 3$ (true).

More Examples

Roots:

Example

If $\left|(x-3)^2\right| < \varepsilon$ then

$|(x-3)| < \sqrt{\varepsilon}$.

Inversion of three sides of a continued inequality (condensed method)

Given $a < b < c$

We can invert all three sides and simultaneously reverse the sense
(or direction or order) of the inequality.

Then $\frac{1}{a} > \frac{1}{b} > \frac{1}{c}$ and we can rewrite as

$$\frac{1}{c} < \frac{1}{b} < \frac{1}{a}.$$

> **Example** If $2 < 4 < 7$
>
> then $\frac{1}{2} > \frac{1}{4} > \frac{1}{7}$
>
> $\frac{1}{7} < \frac{1}{4} < \frac{1}{2}$ (rewriting)

Extra: Solve for x: $|3 - 7x| < 17$ (We cover two methods)

Method 1: (Rewriting after solving)

$|3 - 7x| < 17$

$-17 < 3 - 7x < 17$

$\underline{ -3 \quad -3 \qquad -3}$

$-20 < -7x < 14$

$\frac{-20}{-7} > \frac{-7}{-7}x > \frac{14}{-7}$

(Reversing the direction of the inequality)

$\frac{20}{7} > x > -2$

($x < \frac{20}{7}$ and $x > -2$)

Rewriting the lower limit first,

$-2 < x < \frac{20}{7}$

$\left\{ x \mid -2 < x < \frac{20}{7} \right\}.$

Method 2: Relatively, a popular method (Rewriting the absolute value term before solving) Note for example that $|7 - 2| = |2 - 7|$, and generally, $|b - a| = |a - b|$.

Replacing $|3 - 7x|$ by $|7x - 3|$, the original inequality becomes

$|7x - 3| < 17$

Now, we solve the equivalent inequality, $|7x - 3| < 17$

$-17 < 7x - 3 < 17$

$\underline{+3 \qquad\quad +3 \quad +3}$

$-14 < 7x < 20$

$\frac{-14}{7} < \frac{7}{7}x < \frac{20}{7}$

$-2 < x < \frac{20}{7}$

$\left\{ x \mid -2 < x < \frac{20}{7} \right\}$

Lesson 5B Exercises

1. If $9 < k$, then $\sqrt{9} < ?$ and

2. If $\left|(x-3)^2\right| < \varepsilon$ then $|(x-3)| < ?$

3. Solve for x: $|3 - 7x| < 17$

4. Solve for x: $|4 - 2x| < 10$

Answers: **1.** $\sqrt{k}$; **2.** $\sqrt{\varepsilon}$. **3.** $\left\{ x \mid -2 < x < \frac{20}{7} \right\}$, **4.** $\left\{ x \mid -3 < x < 7 \right\}$

Lesson 5C
Limits of Functions
(Epsilon-Delta Proofs)

About Epsilon-Delta Proofs

A number of textbooks have given the impression that epsilon-delta proofs are difficult to construct. After very careful analysis of these proofs, the author has come to the conclusion that the difficulty has always been due to how the proofs are presented in textbooks and consequently how they may be taught in the classroom. In this lesson, a step-by-step approach is used to the extent that even high school students with a good background in solving inequalities, will be able to handle the proofs involving linear, quadratic, rational, radical, logarithmic, and trigonometric functions. The prerequisite here is an elementary-to-intermediate background in solving compound inequalities involving a single variable.
 Two approaches for constructing these proofs are covered.

Definition

The limit of a function f is the value, L, that $f(x)$ approaches as x approaches the value a. Symbolically, we write

$$\lim_{x \to a} f(x) = L \text{ (read: the limit of "} f \text{ of } x \text{ " as } x \text{ approaches } a \text{ equals } L \text{"}$$

Lim $f(x)$ describes the behavior of f when x is near a, but different from a (as x squeezes in on a). Note that the value of $f(x)$ at a does not matter. However, $f(x)$ may be equal to L, but it does not have to be equal to L. We must also note that the limit may not exist.

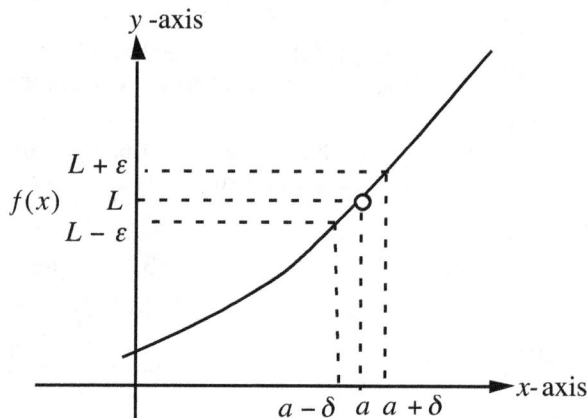

More rigorous definition

If L is the limit of $f(x)$ as x approaches the value a, we write

$$\lim_{x \to a} f(x) = L \text{ if for any positive number } \varepsilon, \text{ there exists a positive}$$

number δ such that if $0 < |x - a| < \delta$, then $|f(x) - L| < \varepsilon$.

Geometrically, this definition means that $f(x)$ will lie between $L - \varepsilon$ and $L + \varepsilon$ whenever x lies between $a - \delta$ and $a + \delta$. Thus by restricting x between $a - \delta$ and $a + \delta$, we can restrict $f(x)$ between $L - \varepsilon$ and $L + \varepsilon$. Note that we chose ε arbitrarily as small as we please, and that δ depends on ε: the smaller the value of ε, the smaller the corresponding δ – value.

Guidelines for Epsilon-Delta Proofs

The author's philosophy is that some helpful guidelines are better than no guidelines at all.

In the examples covered in this lesson, guidelines for two main methods for constructing epsilon-delta proofs are covered, namely,

1. The **delta assumption** method in which we assume $\delta \le 1$, $\delta \le \frac{1}{2}$ and so on. This method has been the traditional approach for some functions.
2. The **continued inequality** or condensed method in which no assumption of the value of δ is made for all functions covered.
Note that some of the steps in both methods may overlap.

Note also that for each method, we will also practice some of the strategies (such as from the hypothesis to the conclusion or from the conclusion to the hypothesis) that were preached in Lesson 5A. Students are encouraged to experiment with various strategies in mathematical proofs, and not to be satisfied with only those in print; since by so doing, perhaps, better alternatives may be produced.

Guidelines for the delta assumption method

Main strategy: The main strategy here is to relate the hypothesis and the conclusion and then reproduce the conclusion from this relationship. All the manipulations are done using the hypothesis and the conclusion just like how we "juggle" a system of two linear equations. However, we must always be aware that we are dealing with **inequalities** and not equalities.

Step 1: Begin with the conclusion inequality and if the left side is factorable, do so, otherwise leave this side as is and let the result of this step be (C).

Step 2: Compare the factorization (if any) from Step 1 with the hypothesis and note the factors of (C) which are different from the left side of the hypothesis. If there is no factorization, go to Step 3.

Step 3: Begin with the hypothesis, assuming $\delta \le 1$, and change the left side to that of any factor in the conclusion (C) different from that of the hypothesis. Let the resulting inequality be (D), This step will produce an upper bound for this factor. We may repeat this process, beginning with the hypothesis, for other factors.

Step 4: Check to see if by multiplying the hypothesis by inequality (D), we produce the left side of the factored conclusion (C), and if this is the case, multiply them and let this inequality be (F).

Step 5: The left sides of (C) and (F) must be identical now; and equate their right sides and let this equation be (G).

Step 6: Solve equation (G) for δ in terms of ε, and replace the δ of the hypothesis (A) by the right side of (G), and let the new inequality be (H).

Step 7: **Proof:** Multiply inequalities (D) and (G) and solve for ε to obtain the conclusion. Therefore if $0 < |x - a| < \delta$, then $|f(x) - L| < \varepsilon$, and the proof is complete.

Guidelines for the Continued Inequality Method

Main strategy: Begin with the conclusion, and change the middle term of the conclusion to middle term of the hypothesis, and thereby relating the hypothesis and the conclusion; followed by the reproduction of the conclusion from this relationship. We may also begin with the hypothesis and change its middle term to that of the conclusion.

Step 1: Write the conclusion as a continued inequality.

Step 2: Solve for x.

Step 3: Make the middle term identical with the middle term of the hypothesis (This is usually accomplished by adding or subtracting a constant) Let the resulting inequality be inequality (C).

Step 4: Write the hypothesis as a continued inequality, say, inequality (D).

Step 5: Compare the left sides of (C) and (D) and equate these sides. and let this equation be (E) Similarly, equate the right sides of (C) and (D) and let this equation be (F).

Step 6: Solve equations (E) and (F) for δ. Let the solutions be δ_1 and δ_2, respectively. Note that δ_1 and δ_2 may or may not be equal.

Step 7: **Proof**: Break up inequality (C) into two simple inequalities and solve each inequality for $-\varepsilon$ or ε.

Step 8: Since we have shown that if $-\delta < x - a$ and if $x - a < \delta$, $|f(x) - L| < \varepsilon$, we have proved that if $0 < |x - a| < \delta$, then $|f(x) - L| < \varepsilon$.

For a chosen ε, $\delta = \min(\delta_1, \delta_2)$ or any number less than $\min(\delta_1, \delta_2)$.

Examples on Epsilon-delta Proofs

Case 1: Linear Functions

Example 1 Prove that $\lim\limits_{x \to 2} 4x + 1 = 9$

Method 1 (delta assumption method)

Definition: If $0 < |x - a| < \delta$ then $|f(x) - L| < \varepsilon$

Step 1: If $0 < |x - 2| < \delta$,   (hypothesis)   (A)

then $|4x + 1 - 9| < \varepsilon$. (conclusion)   (B)

Plan: The proof would be complete after showing that if $0 < |x - 2| < \delta$, then $|4x + 1 - 9| < \varepsilon$

Step 2: We want a formula relating δ and ε, and we begin with

$|4x + 1 - 9| < \varepsilon$ (the conclusion)

$|4x - 8| < \varepsilon$

$4|x - 2| < \varepsilon$

$|x - 2| < \dfrac{\varepsilon}{4}$ (C)

> **Alternatively,** we could change the left side of the hypothesis from $|x - 2| < \delta$ to $|4x - 8| < 4\delta$, compare with the right side of $|4x - 8| < \varepsilon$ to obtain $\delta = \dfrac{\varepsilon}{4}$

We have made the left side the same as the left side of the hypothesis)

Step 3: Now, compare the equivalent conclusion $|x - 2| < \dfrac{\varepsilon}{4}$ (C) with the

hypothesis $0 < |x - 2| < \delta$ and

take $\delta = \dfrac{\varepsilon}{4}$ or less (which means you can take $\delta = \dfrac{\varepsilon}{5}, \dfrac{\varepsilon}{6}, \dfrac{\varepsilon}{8}$ etc.)

Proof: Step 4: We replace the δ on the right-hand side of the hypothesis

$0 < |x - 2| < \delta$ by $\frac{\varepsilon}{4}$ to obtain $0 < |x - 2| < \frac{\varepsilon}{4}$ or $|x - 2| < \frac{\varepsilon}{4}$ and solve for ε.

Multiply both sides of the above inequality by 4.

Then, $4|x - 2| < 4\left(\frac{\varepsilon}{4}\right)$

$4|x - 2| < \varepsilon$, which is equivalent to the conclusion in (B)

$$(|4x + 1 - 9| = |4x - 8| = 4|x - 2|).$$

We have therefore shown that if $0 < |x - 2| < \delta$, **then** $|4x + 1 - 9| < \varepsilon$,

Therefore, $\lim\limits_{x \to 2} 4x + 1 = 9$ and the proof is complete.

Method 2 (Continued inequality approach)

Example 1 Prove that $\lim\limits_{x \to 2} 4x + 1 = 9.$

Definition: If $0 < |x - a| < \delta$ then $|f(x) - L| < \varepsilon$

Step 1: If $0 < |x - 2| < \delta$,   (hypothesis)   (A)

then $|4x + 1 - 9| < \varepsilon$.   (conclusion)   (B)

Plan: The proof would be complete after showing that

if $0 < |x - 2| < \delta$, then $|4x + 1 - 9| < \varepsilon$

Step 2: The hypothesis $|x - 2| < \delta$ is equivalent to

$$\boxed{-\delta < x - 2 < \delta}\quad\text{(C)}$$

We want a formula relating δ and ε, and we begin with

$|4x + 1 - 9| < \varepsilon$ (conclusion)

$|4x - 8| < \varepsilon$ is equivalent to

$-\varepsilon < 4x - 8 < \varepsilon$

Step 3: Make the middle term the same as the middle term of the hypothesis (C).	**Alternative Step 3:**
Solve for x $8 - \varepsilon < 4x < \varepsilon + 8$	$\lvert 4x + 1 - 9\rvert < \varepsilon$ (conclusion)
$2 - \frac{\varepsilon}{4} < x < \frac{\varepsilon}{4} + 2$ (solving for x)	$\lvert 4x - 8\rvert < \varepsilon$
$2 - \frac{\varepsilon}{4} - 2 < x - 2 < \frac{\varepsilon}{4} + 2 - 2$ or	$-\varepsilon < 4x - 8 < \varepsilon$
$\boxed{-\frac{\varepsilon}{4} < x - 2 < \frac{\varepsilon}{4}}$ (D)	$\frac{-\varepsilon}{4} < x - 2 < \frac{\varepsilon}{4}$
(Making middle term the same as the middle term of the hypothesis (C).	Dividing all three sides by 4)
	$\boxed{\frac{-\varepsilon}{4} < x - 2 < \frac{\varepsilon}{4}}$ **(D1)**
	(Note the equality of the middle terms in (C) and (D1)

We have now connected the hypothesis and the conclusion)
Note the equality of the middle terms in (C) and (D))

Step 4: Compare the left sides of (C) and (D) and equate these sides. Similarly, compare the right sides of (C) and (D) and equate these sides.

For the left sides	For the right sides
Take $-\delta = -\frac{\varepsilon}{4}$ and solving, $\delta = \frac{\varepsilon}{4}$	$\delta = \frac{\varepsilon}{4}$

In either case, $\delta = \frac{\varepsilon}{4}$ or less (which means you can take $\delta = \frac{\varepsilon}{5}, \frac{\varepsilon}{6}, \frac{\varepsilon}{8}$ etc.)

Step 7: **Proof**: Replace the left and right sides of (C) by $-\frac{\varepsilon}{4}$

and $\frac{\varepsilon}{4}$, respectively to obtain $-\frac{\varepsilon}{4} < x - 2 < \frac{\varepsilon}{4}$ (E)

Step 8: Multiply inequality (E) by 4. (Note that (E) is now the same as (D))

Then $-\frac{\varepsilon}{4} < x - 2 < \frac{\varepsilon}{4}$ becomes

$-\varepsilon < 4x - 8 < \varepsilon$ which is equivalent to

$|4x - 8| < \varepsilon$ (the equivalent conclusion)

We have therefore shown that if $0 < |x - 2| < \delta$, then $|4x + 1 - 9| < \varepsilon$,

Therefore, $\lim_{x \to 2} 4x + 1 = 9$, and the proof is complete.

Extra

Note above that in Steps 2 & 3, we could begin with the hypothesis as follows

$|x - 2| < \delta$ is equivalent to

Step 9: 1b: $-\delta < x - 2 < \delta$

Multiply this inequality by 4.

$\boxed{-4\delta < 4x - 8 < 4\delta}$ (C2) (changing the middle term to that of the conclusion

$|4x - 8| < \varepsilon$ is equivalent to

$\boxed{-\varepsilon < 4x - 8 < \varepsilon}$ (D2)

(Note the equality of the middle terms in (C2) D2))

Now, equating the left sides and equating the right sides

For the left sides	For the right sides
$-4\delta = -\varepsilon$	$4\delta = \varepsilon$
$\delta = \frac{\varepsilon}{4}$	$\delta = \frac{\varepsilon}{4}$

Example 2 Prove that $\lim_{x \to 3} 3x + 2 = 11$

Method 1 (delta assumption method)

Definition: If $0 < |x - a| < \delta$ then $|f(x) - L| < \varepsilon$

Step 1: If $0 < |x - 3| < \delta$,   (hypothesis)   (A)

then $|3x + 2 - 11| < \varepsilon$.   (conclusion)   (B)

Plan: The proof would be complete after showing that

If $0 < |x - 3| < \delta$, then $|3x + 2 - 11| < \varepsilon$

Step 2: We want a formula relating δ and ε, and we begin with

$|3x + 2 - 11| < \varepsilon$ (the conclusion)

$|3x - 9| < \varepsilon$

$3|x - 3| < \varepsilon$

$|x - 3| < \frac{\varepsilon}{3}$ (C)

Step 3: Now, compare the equivalent conclusion $|x - 3| < \frac{\varepsilon}{3}$ with the hypothesis

$0 < |x - 3| < \delta$, and

take $\delta = \frac{\varepsilon}{3}$ or less (which means you can take $\delta = \frac{\varepsilon}{4}, \frac{\varepsilon}{5}, \frac{\varepsilon}{6}, \frac{\varepsilon}{8}$ etc.

(Note the equality of the left sides in (A) and (C))

Step 4: **Proof**: We replace δ on the right-hand side of the hypothesis

$$0 < |x - 3| < \delta \text{ with } \frac{\varepsilon}{3} \text{ to obtain } |x - 3| < \frac{\varepsilon}{3} \text{ and solve for } \varepsilon.$$

Then $3|x - 3| < 3\left(\frac{\varepsilon}{3}\right)$ Multiply both sides of the above inequality by 3.

$3|x - 3| < \varepsilon$, which is equivalent to the conclusion in (B)

We have shown that if $0 < |x - 3| < \delta$, then $|3x + 2 - 11| < \varepsilon$ ($|3x - 9| = 3|x - 3|$)

Therefore, $\lim_{x \to 3} 3x + 2 = 11$, and the proof is complete.

Method 2 (continued inequality approach)

Example 2 Prove that $\lim_{x \to 3} 3x + 2 = 11$

Definition: If $0 < |x - a| < \delta$ then $|f(x) - L| < \varepsilon$

Step 1: If $0 < |x - 3| < \delta$,   (hypothesis)   (A)

then $|3x + 2 - 11| < \varepsilon$.   (conclusion)   (B)

Plan: The proof would be complete after showing that

If $0 < |x - 3| < \delta$, then $|3x + 2 - 11| < \varepsilon$

Step 2: $|x - 3| < \delta$ (hypothesis) is equivalent $\boxed{-\delta < x - 3 < \delta}$ (C)

We want a formula relating δ and ε, and we begin with

$|3x + 2 - 11| < \varepsilon$ (conclusion)

$|3x - 9| < \varepsilon$, which is equivalent to

$-\varepsilon < 3x - 9 < \varepsilon$ (continued inequality)

Step 3: Make the middle term the same as the middle term of the hypothesis (C).	**Alternative Step 3**
Solve for x $9 - \varepsilon < 3x < \varepsilon + 9$	$|3x + 2 - 11| < \varepsilon$ (conclusion)
$3 - \frac{\varepsilon}{3} < x < \frac{\varepsilon}{3} + 3$ (solving for x)	$|3x - 9| < \varepsilon$
	$-\varepsilon < 3x - 9 < \varepsilon$
$3 - \frac{\varepsilon}{3} - 3 < x - 3 < \frac{\varepsilon}{3} + 3 - 3$ or	$\frac{-\varepsilon}{3} < x - 3 < \frac{\varepsilon}{3}$
	Dividing all three sides by 3)
$\boxed{-\frac{\varepsilon}{3} < x - 3 < \frac{\varepsilon}{3}}$ (D)	$\boxed{\frac{-\varepsilon}{3} < x - 3 < \frac{\varepsilon}{3}}$ (D1)
(Making middle term the same as the middle term of the hypothesis (C).	(Note the equality of the middle terms in (C) and (D1)

(We have now connected the hypothesis and the conclusion))

Step 4: Compare the left sides of (C) and (D) and equate these sides. Similarly, compare the right sides of (C) and (D) and equate these sides.

For the left sides	For the right sides
Take $-\delta = -\frac{\varepsilon}{3}$ and $\delta = \frac{\varepsilon}{3}$ (solving)	$\delta = \frac{\varepsilon}{3}$

Step 5: **Proof**: Replace the left and right sides of (C) by $-\frac{\varepsilon}{4}$ and $\frac{\varepsilon}{4}$,

respectively to obtain $-\frac{\varepsilon}{3} < x - 3 < \frac{\varepsilon}{3}$ (E). Multiply inequality (E) by 3

Then $-\frac{\varepsilon}{3} < x - 3 < \frac{\varepsilon}{3}$ becomes $-\varepsilon < 3x - 9 < \varepsilon$ which is

equivalent to $|3x - 9| < \varepsilon$ (the equivalent conclusion)

Step 6: We have shown that if $0 < |x - 3| < \delta$, then $|3x + 2 - 11| < \varepsilon$,

Therefore, $\lim_{x \to 3} 3x + 2 = 11$, and the proof is complete.

Generalization for the linear function $f(x) = mx + b$, follows.

Example 3

Prove that $\lim\limits_{x \to a} mx + b = ma + b$

Method 1 (delta assumption method)

Definition: If $0 < |x - a| < \delta$ then $|f(x) - L| < \varepsilon$

Step 1: If $0 < |x - a| < \delta$,   (hypothesis)   (A)

then $|mx + b - (ma + b)| < \varepsilon$. (conclusion)   (B)

Plan: The proof would be complete after showing that

If $0 < |x - a| < \delta$, then $|mx - ma| < \varepsilon$

Step 2: We want a formula relating δ and ε, and we begin with

$|m| |x - a| < \varepsilon$ (the conclusion)

$$|x - a| < \frac{\varepsilon}{|m|} \qquad (C)$$

(Note the equality of the left sides in (A) and (C))

Step 3: Now, compare the equivalent conclusion $|x - a| < \frac{\varepsilon}{|m|}$ with the

hypothesis $0 < |x - a| < \delta$, and

take $\delta = \frac{\varepsilon}{|m|}$ or less (which means you can take $\delta = \frac{\varepsilon}{|2m|}, \frac{\varepsilon}{|3m|}$, etc.)

Step 4: We replace δ on the right-hand side of the hypothesis $0 < |x - a| < \delta$ by

$\frac{\varepsilon}{|m|}$ to obtain $|x - a| < \frac{\varepsilon}{|m|}$.

Step 5: Multiply both sides of the above inequality $|x - a| < \frac{\varepsilon}{|m|}$ by $|m|$

Then $|m| |x - a| < \varepsilon$, which is equivalent to the conclusion inequality

$|mx + b - (ma + b)| < \varepsilon$

Therefore, if $0 < |x - a| < \delta$, them $|m| |x - a| < \varepsilon$, and the proof is complete.

From above, if the function is linear, $\delta = \frac{\varepsilon}{|m|}$, or less.

Application

Which of the following choices for δ can be used in an $\varepsilon - \delta$ proof of

$\lim\limits_{x \to 3} 4x + 1$?

(a) $\delta = \frac{\varepsilon}{3}$;   **(b)** $\delta = \frac{\varepsilon}{2}$;   **(c)** $\delta = \varepsilon$ (d) $\delta = \frac{\varepsilon}{5}$

Solution If $m = 4$, $\delta = \frac{\varepsilon}{4}$

Therefore the correct answer is **(d)** $\delta = \frac{\varepsilon}{5}$, since $\frac{\varepsilon}{5} < \frac{\varepsilon}{4}$.

Method 2 (continued inequality approach)
Generalization for the linear function $f(x) = mx + b$, follows.
Example 3 Prove that $\lim_{x \to a} mx + b = ma + b$

Step 1: If $0 < |x - a| < \delta$,   (hypothesis)   (A)

then $|mx + b - (ma + b)| < \varepsilon$. (conclusion)   (B)

Plan: The proof would be complete after showing that

If $0 < |x - a| < \delta$, then $|mx - ma| < \varepsilon$

Step 2: The hypothesis $|x - a| < \delta$ is equivalent to

$$\boxed{-\delta < x - a < \delta}\qquad\text{(C)}$$

We want a formula relating δ and ε, and we begin with

$|mx - ma| < \varepsilon$ (the conclusion)

$-\varepsilon < mx - ma < \varepsilon$

Step 3: Make the middle term the same as the middle term of the hypothesis (C).	**Alternative Step 3**
Solve for x $|m|a - \varepsilon < |m|x < \varepsilon + |m|a$	$|mx - ma| < \varepsilon$ (conclusion) $-\varepsilon < mx - ma < \varepsilon$
$a - \dfrac{\varepsilon}{|m|} < x < \dfrac{\varepsilon}{|m|} + a$ (solving for x)	$-\dfrac{\varepsilon}{|m|} < x - a < \dfrac{\varepsilon}{|m|}$
$a - \dfrac{\varepsilon}{|m|} - a < x - a < \dfrac{\varepsilon}{|m|} + a - a$ or	(Dividing all three sides by $|m|$) $-\varepsilon < mx - ma < \varepsilon$
$\boxed{-\dfrac{\varepsilon}{|m|} < x - a < \dfrac{\varepsilon}{|m|}}\qquad\text{(D)}$	$-\dfrac{\varepsilon}{|m|} < x - a < \dfrac{\varepsilon}{|m|}$ **(D1)**
We have now connected the hypothesis and the conclusion)	(Note the equality of the middle terms in (C) and (D1)

(Note the equality of the middle terms in (C) and (D))

Step 4: Compare the left sides of (C) and (D) and equate these sides. Similarly, compare the right sides of (C) and (D) and equate these sides.

For the left sides	For the right sides
Take $-\delta = -\dfrac{\varepsilon}{|m|}$ and $\delta = \dfrac{\varepsilon}{|m|}$ (solving)	$\delta = \dfrac{\varepsilon}{|m|}$

In either case, $\delta = \dfrac{\varepsilon}{|m|}$ or less (you can take $\delta = \dfrac{\varepsilon}{|2m|}, \dfrac{\varepsilon}{|3m|}$, etc.)

Step 5: **Proof**: Replace the left and right sides of (C) by $-\dfrac{\varepsilon}{|m|}$ and $\dfrac{\varepsilon}{|m|}$,

respectively to obtain $-\dfrac{\varepsilon}{|m|} < x - a < \dfrac{\varepsilon}{|m|}$ (E)

Multiply inequality (E) by $|m|$ to obtain

$-\varepsilon < mx - ma < \varepsilon$ which is equivalent to

$|mx - ma| < \varepsilon$ (the equivalent conclusion)

Step 6: We have shown that if $0 < |x - a| < \delta$, **then** $|mx + b - (ma + b)| < \varepsilon$,

Therefore, $\lim_{x \to a} mx + b = ma + b$, and the proof is complete.

Case 2: Quadratic Functions

Example 4   Prove that $\lim\limits_{x \to 3} x^2 = 9$

Method 1 (delta assumption method)

Definition: If $0 < |x - a| < \delta$ then $|f(x) - L| < \varepsilon$

Step 1: If $0 < |x - 3| < \delta$,   (hypothesis)   (A)

then $|x^2 - 9| < \varepsilon$.   (conclusion)   (B)

Plan: The proof would be complete after showing that

If $0 < |x - 3| < \delta$, then $|(x + 3)(x - 3)| < \varepsilon$ B)

:Step 2: We want a formula relating δ and ε, and we begin with

$|x^2 - 9| < \varepsilon$ (the conclusion)

$|(x + 3)(x - 3)| < \varepsilon$ (C)

By comparing the inequality $|(x + 3)(x - 3)| < \varepsilon$ (the conclusion) with (the

$0 < |x - 3| < \delta$ (the hypothesis) in (B), we observe that the factor $|x - 3|$ in the

conclusion is the same as the term $|x - 3|$ in the hypothesis ($0 < |x - 3| < \delta$); and

this situation is desirable.. However, for the factor $|x + 3|$, we find an upper

bound for it. We also note that when x is close to 3, $x + 3$ is close to 6. so an

upper bound close to 6 would be a good upper bound.

Step 3: If we assume that $\delta \le 1$, then the inequality $|x - 3| < \delta$ (hypothesis)

becomes $|x - 3| < 1$. We begin with this inequality and change the

$|x - 3|$ to $|x + 3|$ as follows: $|x - 3| < 1$ is equivalent to $-1 < x - 3 < 1$.

$-1 < x - 3 < 1$

$\underline{+6 \quad\quad + 6 \quad + 6}$ (We change the middle term to $|x + 3|$ by adding 6)

$5 < x + 3 < 7$

An upper bound for $|x + 3|$ is 7; and we can write

$|x + 3| < 7$ (D)

Step 4: $|x - 3| < \delta$ (hypothesis)   (E)

Multiply the left sides and multiply the right sides of (D) and (E)

$|(x + 3)(x - 3)| < 7\delta$ (F)

$|(x + 3)(x - 3)| < \varepsilon$ (C)

(Note the equality of the left sides in (F) and (C))

Step 5: Comparing the left-hand-sides and the right-hand sides of (F) and (C)

$7\delta = \varepsilon$, and from which $\delta = \frac{\varepsilon}{7}$

Step 6: Replace the δ on the right-hand side of (E), hypothesis by $\frac{\varepsilon}{7}$ to obtain

$|x - 3| < \frac{\varepsilon}{7}$ (G)

Proof:: $|x + 3| < 7$ (D)

Multiply the left sides and multiply the right sides of (G) and (D), above.

$|(x - 3)(x + 3)| < \frac{\varepsilon}{7}(7)$

$|(x + 3)(x - 3)| < \varepsilon$ which is equivalent to the conclusion $|x^2 - 9| < \varepsilon$.

We have shown that if $0 < |x - 3| < \delta$, then $|x^2 - 9| < \varepsilon$.

Therefore, $\lim_{x \to 3} x^2 = 9$, and the proof is complete

In the above proof, we could use $\delta = 1$ if $\frac{\varepsilon}{7} > 1$ Symbolically, we write

$\delta = \min(\frac{\varepsilon}{7}, 1)$ We can also use any number less than both $\frac{\varepsilon}{7}$ and 1.

Method 2 (continued inequality approach)

Example 4 Prove that $\lim_{x \to 3} x^2 = 9$

Definition: If $0 < |x - a| < \delta$ then $|f(x) - L| < \varepsilon$

Step 1: If $0 < |x - 3| < \delta$,   (hypothesis)   (A)

       then $|x^2 - 9| < \varepsilon$.   (conclusion)   (B)

       Plan: The proof would be complete after showing that

       If $0 < |x - 3| < \delta$, then $|(x + 3)(x - 3)| < \varepsilon$ B)

Step 2 The hypothesis $|x - 3| < \delta$ is equivalent to

    $\boxed{-\delta < x - 3 < \delta}$ (C)

: We want a formula relating δ and ε, and we begin with

       $|x^2 - 9| < \varepsilon$ (conclusion) and solve for x.

Solve for x: $-\varepsilon < x^2 - 9 < \varepsilon$

       $9 - \varepsilon < x^2 < \varepsilon + 9$

       $\sqrt{9 - \varepsilon} < x < \sqrt{\varepsilon + 9}$ (off course, $0 < \varepsilon < 9$)

Make the middle term the same as that of the hypothesis (C)

Step 3: $\boxed{\sqrt{9 - \varepsilon} - 3 < x - 3 < \sqrt{\varepsilon + 9} - 3}$ (D)

       We have now connected the hypothesis and the conclusion)

Step 4: Compare the left sides of (C) and (D) and equate these sides. Similarly,
       compare the right sides of (C) and (D) and equate these sides.

For the left sides	For the right sides
Take $-\delta = \sqrt{9 - \varepsilon} - 3$ and	$\delta = \sqrt{9 + \varepsilon} - 3 = \delta_2$. say.
$\delta = 3 - \sqrt{9 - \varepsilon} = \delta_1$, say. (solving)	

Step 5: **Proof:** Given $\varepsilon > 0$, choose $\delta = \min(\delta_1, \delta_2)$,

       $|x - 3| < \delta$ implies $-\delta_1 \le -\delta < x - 3 < \delta \le \delta_2$ (E)

Step 6: Replace the left and right sides of (E) by $\sqrt{9 - \varepsilon} - 3$ and $\sqrt{9 + \varepsilon} - 3$,

       respectively to obtain $\sqrt{9 - \varepsilon} - 3 < x - 3 < \sqrt{\varepsilon + 9} - 3$ (F)

       Break up (F) into two simple inequalities and solve each for $-\varepsilon$ or ε.

       $\sqrt{9 - \varepsilon} - 3 < x - 3 < \sqrt{\varepsilon + 9} - 3$ is equivalent to

$\sqrt{9 - \varepsilon} - 3 < x - 3$ (E)	$x - 3 < \sqrt{\varepsilon + 9} - 3$ (F)
$\sqrt{9 - \varepsilon} < x$	$x < \sqrt{9 + \varepsilon}$
$-\varepsilon < x^2 - 9$	$x^2 < 9 + \varepsilon$ or $x^2 - 9 < \varepsilon$

Step 7: $-\varepsilon < x^2 - 9$ and $x^2 - 9 < \varepsilon$ is equivalent to

       $|x^2 - 9| < \varepsilon$

Step 8: We have shown that if $0 < |x - 3| < \delta$, then $\left|x^2 - 9\right| < \varepsilon$.

Therefore, $\lim\limits_{x \to 3} x^2 = 9$, and the proof is complete. In the above proof, take δ to be the smaller of δ_1 and δ_2 or any number less than the smaller of δ_1 and δ_2.

Let us compare some values of δ for $\varepsilon = .0030$ using the first method and the condensed method

First method (delta assumption)	**Second method** (continued inequality)
$\delta = \min\left(\frac{\varepsilon}{7}, 1\right)$	$\delta_1 = 3 - \sqrt{9 - \varepsilon}$
$\delta = \frac{\varepsilon}{7}$	$ = 3 - \sqrt{9 - .003}$
$ = \frac{.003}{7}$	$ = 0.000500042 = 0.0005$
$ = 0.000428571 = 0.0004$	$\delta_2 = \sqrt{9 + \varepsilon} - 3$
$ = 0.0004$	$\delta_2 = \sqrt{9 + \varepsilon} - 3$
	$ = \sqrt{9 + .003} - 3$
	$ = 0.000499958 = 0.0005$

Now, let us also check in $\left|x^2 - 9\right| < \varepsilon$. Let us try $\delta = 0.0005$; $x = 3 \pm 0004$

$x = 3.0004$	Similarly, for
If $|x - 3| < 0.0005$.	$x = 3 - 0.0004 = 2.9996$
Then $\left|(3.0004)^2 - 9\right| \overset{?}{<} .0030$	If $|x - 3| < 0.0005$
	Then $\left|(2.9996)^2 - 9\right| \overset{?}{<} .0030$
$\left|9.0024 - 9\right| \overset{?}{<} .0030$	$\left|8.9976 - 9\right| \overset{?}{<} .0030$
$\left|0.0024\right| \overset{?}{<} .0030$ Yes.	$\left|0.0024\right| \overset{?}{<} .0030$ Yes.

Example 5 Prove that $\lim\limits_{x \to 2} x^2 + 3x = 10$

Method 1 (delta assumption method)

Definition: If $0 < |x - a| < \delta$ then $|f(x) - L| < \varepsilon$

Step 1: If $0 < |x - 2| < \delta$, $\qquad$ (hypothesis) $\qquad\qquad$ (A)

$\qquad$ then $\left|x^2 + 3x - 10\right| < \varepsilon$. $\qquad$ (conclusion) $\qquad\qquad$ (B)

$\qquad$ Plan: The proof would be complete after showing that

$\qquad$ If $0 < |x - 2| < \delta$, then $\left|(x + 5)(x - 2)\right| < \varepsilon$

Step 2: We want a formula relating δ and ε, and we begin with

$\qquad$ $\left|x^2 + 3x - 10\right| < \varepsilon$ (the conclusion)

$\qquad$ $\left|(x + 5)(x - 2)\right| < \varepsilon$ $\qquad\qquad\qquad\qquad\qquad\qquad\qquad\qquad$ (C)

$\qquad$ By comparing the inequality $\left|(x + 5)(x - 2)\right| < \varepsilon$ (the conclusion) with

$\qquad$ $0 < |x - 2| < \delta$ (the hypothesis)m we observe that the factor $|x - 2|$ in

$\qquad$ $\left|(x + 5)(x - 2)\right| < \varepsilon$ is the same as the term $|x - 2|$ in $0 < |x - 2| < \delta$, and

$\qquad$ this situation is desirable. However, for the factor $|x + 5|$, we find an

$\qquad$ upper bound for it. We note that when x is close to 2, $x + 5$ is close to 7,

$\qquad$ and therefore, an upper bound close to 7 would be a good upper bound.

Step 3: Assume that $\delta \leq 1$. Then the inequality $|x - 2| < \delta$ becomes

$|x - 2| < 1$. We begin this inequality and change $|x - 2|$ to $|x + 5|$ as

follows. $|x - 2| < 1$ is equivalent to $-1 < x - 2 < 1$.

$-1 < x - 2 < 1$

$\underline{+7 \quad +7 \quad +7}$ (We change the middle term to $x + 5$ by adding 7)

$6 < x + 5 < 8$

An upper bound for $|x + 5|$ is 8; and we can write

$|x + 5| < 8$ (D)

Step 4: $|x - 2| < \delta$ (hypothesis)   (E)

Multiply the left sides and multiply the right sides of (D) and (E)

$|(x + 5)(x - 2)| < 8\delta$ (F)

$|(x + 5)(x - 2)| < \varepsilon$ (C)

(Note the equality of the left sides in (F) and (C))

Step 5: Comparing the left-hand-sides and the right-hand sides of (F) and (C)

take $8\delta = \varepsilon$, and from which $\delta = \frac{\varepsilon}{8}$

Step 6: Replace the δ on the right-hand side of (E), hypothesis, by $\frac{\varepsilon}{8}$ to

obtain $|x - 2| < \frac{\varepsilon}{8}$.   (G)

Proof

Step 7: $|x + 5| < 8$ (D)

$|x - 2| < \frac{\varepsilon}{8}$ (G)

Multiply the left sides and multiply the right sides of (D) and (G)

Then $|(x + 5)(x - 2)| < \frac{\varepsilon}{8}(8)$

$|(x + 5)(x - 2)| < \varepsilon$. which is equivalent to the conclusion ,

$|x^2 + 3x - 10| < \varepsilon$

We have shown that if $0 < |x - 2| < \delta$, then $|x^2 + 3x - 10| < \varepsilon$.

Therefore, $\lim_{x \to 2} x^2 + 3x = 10$, and the proof is complete.

In the above proof, take δ to be the smaller of δ_1 and δ_2 or any number less than the smaller of δ_1 and δ_2 ..

Method 2 (continued inequality approach)

Example 5 Prove that $\lim\limits_{x \to 2} x^2 + 3x = 10$

Definition: If $0 < |x - a| < \delta$ then $|f(x) - L| < \varepsilon$

Step 1: If $0 < |x - 2| < \delta$, (hypothesis) (A)

   then $|x^2 + 3x - 10| < \varepsilon$. (conclusion) (B)

   Plan: The proof would be complete after showing that

   If $0 < |x - 2| < \delta$, then $|x^2 + 3x - 10| < \varepsilon$

Step 2: The hypothesis $|x - 2| < \delta$ is equivalent to

$$\boxed{-\delta < x - 2 < \delta} \quad (C)$$

We want a formula relating δ and ε, and we begin with

$\quad |x^2 + 3x - 10| < \varepsilon$ (conclusion), and solve for x. by completing the square;

$$-\varepsilon < x^2 + 3x - 10 < \varepsilon$$

$$-\varepsilon < x^2 + 3x + \left(\frac{3}{2}\right)^2 - \left(\frac{3}{2}\right)^2 - 10 < \varepsilon$$

$$-\varepsilon < \left(x + \frac{3}{2}\right)^2 - \frac{9}{4} - 10 < \varepsilon$$

$$-\varepsilon < \left(x + \frac{3}{2}\right)^2 - \frac{49}{4} < \varepsilon$$

$$\frac{49}{4} - \varepsilon < \left(x + \frac{3}{2}\right)^2 < \varepsilon + \frac{49}{4}$$

$$\sqrt{\frac{49}{4} - \varepsilon} < x + \frac{3}{2} < \sqrt{\varepsilon + \frac{49}{4}}$$

$$-\frac{3}{2} + \sqrt{\frac{49}{4} - \varepsilon} < x < \sqrt{\varepsilon + \frac{49}{4}} - \frac{3}{2}$$

Make the middle term the same as that of the hypothesis (C)

Step 3: $-\frac{3}{2} - 2 + \sqrt{\frac{49}{4} - \varepsilon} < x - 2 < \sqrt{\varepsilon + \frac{49}{4}} - \frac{3}{2} - 2$ or (D)

Making the middle term the same as the middle term of the hypothesis.
We have now connected the hypothesis and the conclusion)

$$\boxed{-\frac{7}{2} + \sqrt{\frac{49}{4} - \varepsilon} < x - 2 < \sqrt{\varepsilon + \frac{49}{4}} - \frac{7}{2}} \quad (D)$$

(Note the equality of the middle terms in (C) and (D))

Step 4: Compare the left sides of (C) and (D) and equate these sides. Similarly, compare the right sides of (C) and (D) and equate these sides.

For the left sides

$$-\delta = -\frac{7}{2} + \sqrt{\frac{49}{4} - \varepsilon}$$

$$\delta = \frac{7}{2} - \sqrt{\frac{49}{4} - \varepsilon}, \text{ say } \delta_1$$

For the right sides

$$\delta = \sqrt{\varepsilon + \frac{49}{4}} - \frac{7}{2}, \text{ say } \delta_2$$

Proof Step 5: Given $\varepsilon > 0$, choose $\delta = \min(\delta_1, \delta_2)$,

$$|x - 2| < \delta \text{ implies } -\delta_1 \le -\delta < x - 2 < \delta \le \delta_2 \quad (E)$$

Step 6: Replace the left and right sides of (E) by $-\frac{7}{2} + \sqrt{\frac{49}{4} - \varepsilon}$ and

$\sqrt{\varepsilon + \frac{49}{4}} - \frac{7}{2}$, respectively to obtain $-\frac{7}{2} + \sqrt{\frac{49}{4} - \varepsilon} < x - 2 < \sqrt{\varepsilon + \frac{49}{4}} - \frac{7}{2}$ (F)

Break up (F) into two simple inequalities and solve each for $-\varepsilon$ or ε.

For the left side	For the right side
$-\frac{7}{2} + \sqrt{\frac{49}{4} - \varepsilon} < x - 2$	$x - 2 < \sqrt{\varepsilon + \frac{49}{4}} - \frac{7}{2}$
$\sqrt{\frac{49}{4} - \varepsilon} < x - 2 + \frac{7}{2}$	$x - 2 + \frac{7}{2} < \sqrt{\varepsilon + \frac{49}{4}}$
$\sqrt{\frac{49}{4} - \varepsilon} < x + \frac{3}{2}$	$x + \frac{3}{2} < \sqrt{\varepsilon + \frac{49}{4}}$
$\left(\sqrt{\frac{49}{4} - \varepsilon}\right)^2 < \left(x + \frac{3}{2}\right)^2$	$\left(x + \frac{3}{2}\right)^2 < \left(\sqrt{\varepsilon + \frac{49}{4}}\right)^2$
$\frac{49}{4} - \varepsilon < x^2 + 3x + \frac{9}{4}$	$x^2 + 3x + \frac{9}{4} < \varepsilon + \frac{49}{4}$
$\frac{40}{4} - x^2 - 3x < \varepsilon$	$x^2 + 3x + \frac{9}{4} - \frac{49}{4} < \varepsilon$
$-\varepsilon < x^2 + 3x - 10$	$x^2 + 3x - 10 < \varepsilon$

Step 7: $-\varepsilon < x^2 + 3x - 10$ and $x^2 + 3x - 10 < \varepsilon$ is equivalent to

$$\left|x^2 + 3x - 10\right| < \varepsilon$$

Step 8: We have shown that if $0 < |x - 2| < \delta$, then $\left|x^2 + 3x - 10\right| < \varepsilon$.

Therefore, $\lim_{x \to 2} x^2 + 3x = 10$, and the proof is complete.

Let us find some numerical approximations for δ for $\varepsilon = 0.003$

Continued inequality approach	Method 1 (Delta assumption Method)
$\delta_1 = \frac{7}{2} - \sqrt{\frac{49}{4} - \varepsilon}$	$\delta_1 = \frac{\varepsilon}{8} = \frac{.003}{8} = 0.0004$
$\quad = \frac{7}{2} - \sqrt{\frac{49}{4} - .003} = 0.0004286$	
$\quad = 0.0004$ (to four decimal places)	
$\delta_2 = \sqrt{\varepsilon + \frac{49}{4}} - \frac{7}{2}$	
$\quad = \sqrt{.003 + \frac{49}{4}} - \frac{7}{2} = 0.0004285$	
$\quad = 0.0004$ (to four decimal places)	

If $\delta = 0.0004$. let us check for $x = 2 \pm 0.0003$ in $\left|x^2 + 3x - 10\right| < \varepsilon$.

If $|x - 2| < 0.0004$	Similarly for
$|2.0003 - 2| < 0.0004 \quad (x = 2.0003$	$x = 2 - 0.0003 = 1.9997,$
$\left|(2.0003)^2 + 3(2.0003) - 10\right| \overset{?}{<} 0.003$	$\left|(1.9997)^2 + 3(1.9997) - 10\right| \overset{?}{<} 0.003$
$\left|4.0012 + 6.0009 - 10\right| \overset{?}{<} 0.003$	$\quad\quad\quad |0.002| \overset{?}{<} 0.003$ Yes
$\quad\quad\quad |0.002| \overset{?}{<} 0.003$ Yes	

Case 3: Rational Functions

Example 6   Prove that $\lim\limits_{x \to 7} \dfrac{8}{x-3} = 2$

Method 1 (delta assumption method)

Definition: If $0 < |x - a| < \delta$ then $|f(x) - L| < \varepsilon$

Step 1: If $0 < |x - 7| < \delta$,   (hypothesis)   (A)

then $\left| \dfrac{8}{x-3} - 2 \right| < \varepsilon$ (conclusion)   (B)

Plan: The proof would be complete after showing that

If $0 < |x - 7| < \delta$, then $\left| \dfrac{8}{x-3} - 2 \right| < \varepsilon$.

Step 2: We want a formula relating δ and ε, and we begin with

$\left| \dfrac{8}{x-3} - 2 \right| < \varepsilon$ (the conclusion)

$\left| \dfrac{2x - 14}{x - 3} \right| < \varepsilon$

$\left| \dfrac{x - 7}{x - 3} \right| < \dfrac{\varepsilon}{2}$.   (C)

By comparing the inequality $2\left| \dfrac{x-7}{x-3} \right| < \varepsilon$ (an equivalent conclusion) with

$0 < |x - 7| < \delta$ (the hypothesis),we observe the factor $|x - 7|$ in $\left| \dfrac{x - 7}{x - 3} \right| < \dfrac{\varepsilon}{2}$ is the

same as the term $|x - 7|$ in $0 < |x - 7| < \delta$; and this situation is desirable..

However, for the factor $\left| \dfrac{1}{x-3} \right|$, we find an upper bound for it.

Step 3: Assume that $\delta \le 1$. Then, the inequality $0 < |x - 7| < \delta$ becomes

$|x - 7| < 1$. We begin with this inequality and change the $|x - 7|$ to $\left| \dfrac{1}{x - 3} \right|$

as follows. $|x - 7| < 1$ is equivalent to $-1 < x - 7 < 1$.

$-1 < x - 7 < 1$ (We change the middle term to $|x - 3|$ by adding 4)

$\dfrac{+4 \quad +4 \quad +4}{3 < x - 3 < 5}$

We do not have $\left| \dfrac{1}{x - 3} \right|$ yet. So we find the reciprocals of all the three sides of

the inequality and reverse the directions of the inequality symbols to obtain

$\dfrac{1}{3} > \dfrac{1}{x - 3} > \dfrac{1}{5}$, which is equivalent to (with the upper bound on the right)

$\dfrac{1}{5} < \dfrac{1}{x - 3} < \dfrac{1}{3}$. The upper bound for $\left| \dfrac{1}{x - 3} \right|$ is $\dfrac{1}{3}$, and we can write

$\left| \dfrac{1}{x - 3} \right| < \dfrac{1}{3}$ (D)

Step 4: Now, $|x - 7| < \delta$,   (E)

Multiply the left sides and multiply the right sides of (D) and (E)

$\left| \dfrac{x - 7}{x - 3} \right| < \dfrac{\delta}{3}$ (F)

$\left| \dfrac{x - 7}{x - 3} \right| < \dfrac{\varepsilon}{2}$ (C)

(Note the equality of the left sides in (F) and (C)

Step 5: Comparing the right-hand sides and the left-hand sides of (F) and (C)

take $\frac{\delta}{3} = \frac{\varepsilon}{2}$, and from which $\delta = \frac{3\varepsilon}{2}$. We take $\delta = \frac{3\varepsilon}{2}$ or less.

Step 6: We replace δ on the right-hand side of the hypothesis $|x - 7| < \delta$ by

$\frac{3\varepsilon}{2}$ to obtain $|x - 7| < \frac{3\varepsilon}{2}$ (G)

Proof: $|x - 7| < \frac{3\varepsilon}{2}$ (G)

$\left|\frac{1}{x - 3}\right| < \frac{1}{3}$ (D)

Multiply the left sides and multiply the right sides

Then $\left|\frac{x - 7}{x - 3}\right| < \frac{3\varepsilon}{2} \cdot \frac{1}{3}$

$\left|\frac{x - 7}{x - 3}\right| < \frac{\varepsilon}{2}$

Multiply both sides by 2 to obtain

$2\left|\frac{x - 7}{x - 3}\right| < \varepsilon$ which is equivalent to the conclusion $\left|\frac{8}{x - 3} - 2\right| < \varepsilon$ (see step 2)

We have shown that if $0 < |x - 7| < \delta$, then $\left|\frac{8}{x - 3} - 2\right| < \varepsilon$.

Therefore, $\lim\limits_{x \to 7} \frac{8}{x - 3} = 2$, and the proof is complete.

In the above proof, we could take $\delta = \frac{3\varepsilon}{2}$ or 1 (whichever is the smaller of the values)

We can also use any number less than both $\frac{3\varepsilon}{2}$ and 1.

      Symbolically $\delta = \min\left(\frac{3\varepsilon}{2}, 1\right)$

Method 2   (continued inequality approach)

Example 6   Prove that $\lim\limits_{x \to 7} \frac{8}{x - 3} = 2$

Definition: If $0 < |x - a| < \delta$ then $|f(x) - L| < \varepsilon$

Step 1: If $0 < |x - 7| < \delta$,   (hypothesis)   (A)

     then $\left|\frac{8}{x - 3} - 2\right| < \varepsilon$ (conclusion)   (B)

     Plan: The proof would be complete after showing that

     If $0 < |x - 7| < \delta$, then $\left|\frac{8}{x - 3} - 2\right| < \varepsilon$.

Step 2a: The hypothesis $|x - 7| < \delta$ is equivalent to

     $\boxed{-\delta < x - 7 < \delta}$ (C)

     We want a formula relating δ and ε, and we begin with

     $\left|\frac{8}{x - 3} - 2\right| < \varepsilon$ (conclusion) and solve for x

Step 2b

$\left|\frac{2x - 14}{x - 3}\right| < \varepsilon$

$\left|\frac{x - 7}{x - 3}\right| < \frac{\varepsilon}{2}$.

$-\frac{\varepsilon}{2} < \frac{x - 7}{x - 3} < \frac{\varepsilon}{2}$

Step 2C

$-\frac{\varepsilon}{2} < 1 - \frac{4}{x - 3} < \frac{\varepsilon}{2}$

$-1 - \frac{\varepsilon}{2} < -\frac{4}{x - 3} < \frac{\varepsilon}{2} - 1$

Step 2d. $\dfrac{-2-\varepsilon}{2} < -\dfrac{4}{x-3} < \dfrac{\varepsilon-2}{2}$

$\dfrac{2+\varepsilon}{2} > \dfrac{4}{x-3} > \dfrac{2-\varepsilon}{2}$

$\dfrac{2+\varepsilon}{8} > \dfrac{1}{x-3} > \dfrac{2-\varepsilon}{8}$

$\dfrac{8}{2+\varepsilon} < x-3 < \dfrac{8}{2-\varepsilon}$

$\dfrac{8}{2+\varepsilon} + 3 < x < \dfrac{8}{2-\varepsilon} + 3$

Step 3: $\dfrac{8}{2+\varepsilon} + 3 - 7 < x - 7 < \dfrac{8}{2-\varepsilon} + 3 - 7$

Making middle term the same as the middle term of the hypothesis.
We have now connected the hypothesis and the conclusion)

$\boxed{\dfrac{8}{2+\varepsilon} - 4 < x - 7 < \dfrac{8}{2-\varepsilon} - 4}$ (D)   (simplifying)

(Note the equality of the middle terms in (C) and (D))

Step 5: Compare the left sides of (C) and (D) and equate these sides. Similarly, compare the right sides of (C) and (D) and equate these sides.

For the left sides	For the right sides
$-\delta = \dfrac{8}{2+\varepsilon} - 4$ or $\delta = 4 - \dfrac{8}{2+\varepsilon}$, say δ_1	$\delta = \dfrac{8}{2-\varepsilon} - 4$. say δ_2

Proof:

Step 6 Given $\varepsilon > 0$, choose $\delta = \min(\delta_1, \delta_2)$,

$|x-7| < \delta$ implies $-\delta_1 \le -\delta < x - 7 < \delta \le \delta_2$ (E)

Step 7: Replace the left and right sides of (E) by $\dfrac{8}{2+\varepsilon} - 4$ and $\dfrac{8}{2-\varepsilon} - 4$

respectively to obtain $\boxed{\dfrac{8}{2+\varepsilon} - 4 < x - 7 < \dfrac{8}{2-\varepsilon} - 4}$ (F)

Break up (F)) into two simple inequalities an solve each for $-\varepsilon$ or ε.

For the left side	For the right side
$\dfrac{8}{2+\varepsilon} - 4 < x - 7$	$x - 7 < \dfrac{8}{2-\varepsilon} - 4$
$\dfrac{8}{2+\varepsilon} - 4 + 4 < x - 7 + 4$	$x - 3 < \dfrac{8}{2-\varepsilon}$
$\dfrac{8}{2+\varepsilon} < x - 3$	$\dfrac{x-3}{8} < \dfrac{1}{2-\varepsilon}$ (dividing by 8)
$\dfrac{2+\varepsilon}{8} > \dfrac{1}{x-3}$ (invert and reverse sense)	$\dfrac{8}{x-3} > 2 - \varepsilon$ (invert and reverse sense)
$2 + \varepsilon > \dfrac{8}{x-3}$ (multiplying by 8)	$-\dfrac{8}{x-3} < -2 + \varepsilon$ (Multiplying -1)
$\dfrac{8}{x-3} - 2 < \varepsilon$	$-\varepsilon < \dfrac{8}{x-3} - 2$

Step 7: $\dfrac{8}{x-3} - 2 < \varepsilon$ and $-\varepsilon < \dfrac{8}{x-3} - 2$ is equivalent to $\left|\dfrac{8}{x-3} - 2\right| < \varepsilon$.

Step 8: We have shown that if $0 < |x-7| < \delta$, then $\left|\dfrac{8}{x-3} - 2\right| < \varepsilon$.

Therefore, $\lim\limits_{x \to 7} \dfrac{8}{x-3} = 2$, and the proof is complete.

In the above proof, take δ to be the smaller of δ_1 and δ_2 or any number less than the smaller of δ_1 and δ_2.

In the above proof, $\delta = \min(\delta_1, \delta_2)$: $\delta_1 = 4 - \dfrac{8}{2 + \varepsilon}$ and $\delta_2 = \dfrac{8}{2 - \varepsilon} - 4$.

Let us find some numerical approximations for δ for $\varepsilon = 0.0030$

Continued inequality approach	Method 1 (delta assumption)
$\delta_1 = 4 - \dfrac{8}{2 + \varepsilon} = 4 - \dfrac{8}{2 + .003}$ $= 4 - 3.9940$ $= 0.006$ $\delta_2 = \dfrac{8}{2 - \varepsilon} - 4 = \dfrac{8}{2 - 0.003} - 4 = 0.006$	$\delta = \dfrac{3\varepsilon}{2}$ $\delta = \dfrac{3(.003)}{2}$ $= 0.005$

If $\delta = 0.006$. we check for $x = 7 \pm .005$ in

$\left| \dfrac{8}{x - 3} - 2 \right| < \varepsilon$ (If $|x - 7| < 0.006$, $|7.005 - 7| < 0.006$; $|6.995 - 7| < 0.006$

If $|7.005 - 7| < 0.006$	If $|6.995 - 7| < 0.006$
$\left\| \dfrac{8}{7.005 - 3} - 2 \right\| \overset{?}{<} .0030$ $(x = 7.005)$	$\left\| \dfrac{8}{6.995 - 3} - 2 \right\| \overset{?}{<} .0030$ $(x = 7 - .005)$
$\left\| \dfrac{8}{4.005} - 2 \right\| \overset{?}{<} .0030$	$\left\| \dfrac{8}{3.995} - 2 \right\| \overset{?}{<} .0030$
$\|1.997503 - 2\| \overset{?}{<} .0030$	$\|2.00250 - 2\| \overset{?}{<} .0030$
$\|0.0025\| \overset{?}{<} 0.0030$ Yes .	$\|0.0025\| \overset{?}{<} 0.0030$ Yes .

Case 4 Radical Functions

Example 7   Prove that $\lim\limits_{x \to 4} \sqrt{x + 5} = 3$

Method 1 (delta assumption method)

Definition: If $0 < |x - a| < \delta$ then $|f(x) - L| < \varepsilon$

Step 1: If $0 < |x - 4| < \delta$,   (hypothesis)   (A)

      then $\left| \sqrt{x + 5} - 3 \right| < \varepsilon$ (conclusion)   (B)

      Plan: The proof would be complete after showing that

      If $0 < |x - 4| < \delta$, then $\left| \sqrt{x + 5} - 3 \right| < \varepsilon$.

Step 2: We want a formula relating δ and ε, and we begin with

      $\left| \sqrt{x + 5} - 3 \right| < \varepsilon$ (the conclusion)

      $\left| \dfrac{\left(\sqrt{x + 5} - 3 \right)}{1} \cdot \dfrac{\left(\sqrt{x + 5} + 3 \right)}{\left(\sqrt{x + 5} + 3 \right)} \right| < \varepsilon$ (Rationalizing the numerator)

      $\left| \dfrac{x - 4}{\sqrt{x + 5} + 3} \right| < \varepsilon$ (C)

By comparing the inequality (C) (an equivalent conclusion) with $0 < |x - 4| < \delta$

(the hypothesis), we observe the factor $|x - 4|$ in $\left| \dfrac{x - 4}{\sqrt{x + 5} + 3} \right| < \varepsilon$

is the same as the term $|x - 4|$ in $0 < |x - 4| < \delta$; and this situation is desirable.

For the factor $\dfrac{1}{\sqrt{x+5}+3}$, we find an upper bound.

Step 4: Find an upper bound for $\dfrac{1}{\sqrt{x+5}+3}$

$\qquad$ Begin with $|x-4| < 1$

$\qquad -1 < x - 4 < 1$

$\qquad \underline{+9 \qquad +9 \ +9}$

$\qquad 8 < x + 5 < 10$

$\qquad \sqrt{8} < \sqrt{x+5} < \sqrt{10}$

$\qquad \sqrt{8} + 3 < \sqrt{x+5} + 3 < \sqrt{10} + 3$

$\qquad \dfrac{1}{\sqrt{8}+3} > \dfrac{1}{\sqrt{x+5}+3} > \dfrac{1}{\sqrt{10}+3}$

$\qquad \dfrac{1}{\sqrt{10}+3} < \dfrac{1}{\sqrt{x+5}+3} < \dfrac{1}{\sqrt{8}+3}$

$\qquad$ (An upper bound for $\dfrac{1}{\sqrt{x+5}+3}$ is $\dfrac{1}{\sqrt{8}+3}$ and

$\qquad \left| \dfrac{1}{\sqrt{x+5}+3} \right| < \dfrac{1}{\sqrt{8}+3}$

Step 5: $\left| \dfrac{1}{\sqrt{x+5}+3} \right| < \dfrac{1}{\sqrt{8}+3}$ $\hfill$ (D)

$|x-4| < \delta$ $\hfill$ (A)

Multiply the left sides and multiply the right sides of (A) and (D)

$\left| \dfrac{x-4}{\sqrt{x+5}+3} \right| < \dfrac{\delta}{\sqrt{8}+3}$ $\hfill$ (E)

$\qquad$ (Note the equality of the left sides in (C) and (E))

Step 6: Comparing the right-hand sides and the left-hand sides of (C) and (E)

$\qquad$ take $\varepsilon = \dfrac{\delta}{\sqrt{8}+3}$, and from which

$\qquad \delta = \left(\sqrt{8} + 3 \right) \varepsilon$

$\qquad \quad = \left(2\sqrt{2} + 3 \right) \varepsilon$

Step 7: We replace δ on the right-hand side of the hypothesis $|x-4| < \delta$ by

$\qquad \left(2\sqrt{2} + 3 \right) \varepsilon$ to obtain

$\qquad |x-4| < \left(2\sqrt{2} + 3 \right) \varepsilon$ $\hfill$ (F)

Proof: $\quad |x-4| < \left(2\sqrt{2} + 3 \right) \varepsilon$ $\hfill$ (F)

$\qquad \left| \dfrac{1}{\sqrt{x+5}+3} \right| < \dfrac{1}{2\sqrt{2}+3}$ $\hfill$ **(G)**

$\qquad$ Multiply the left sides and multiply the right sides of (F) and (G)

$\qquad |x-4| \bullet \left| \dfrac{1}{\sqrt{x+5}+3} \right| < \dfrac{1}{\left(2\sqrt{2}+3 \right)} \bullet \left(2\sqrt{2} + 3 \right) \varepsilon$,, which reduces to

$\qquad \left| \dfrac{x-4}{\sqrt{x+5}+3} \right| = \left| \sqrt{x+5} - 3 \right| < \varepsilon$ $\quad$ (Rationalizing the denominator)

We have shown that if $0 < |x - 4| < \delta$, then $\left|\sqrt{x + 5} - 3\right| < \varepsilon$

Therefore, $\lim\limits_{x \to 4} \sqrt{x + 5} = 3$, and the proof is complete.

In the above proof, we could take $\delta = \left(2\sqrt{2} + 3\right)\varepsilon$ or 1 (whichever is the smaller of the values) or any number less than both $\delta = \left(2\sqrt{2} + 3\right)\varepsilon$ and 1.

Method 2 (Using continued inequality)

Example 7   Prove that $\lim\limits_{x \to 4} \sqrt{x + 5} = 3$

Definition: If $0 < |x - a| < \delta$ then $|f(x) - L| < \varepsilon$

Step 1: If $0 < |x - 4| < \delta$,   (hypothesis)   (A)

then $\left|\sqrt{x + 5} - 3\right| < \varepsilon$ (conclusion)   (B)

Plan: The proof would be complete after showing that

If $0 < |x - 4| < \delta$, then $\left|\sqrt{x + 5} - 3\right| < \varepsilon$.

Step 2: The hypothesis $|x - 4| < \delta$ is equivalent to

$$\boxed{-\delta < x - 4 < \delta}$$ (C)

We want a formula relating δ and ε, and we begin with

$\left|\sqrt{x + 5} - 3\right| < \varepsilon$ (the conclusion) which is equivalent to

$$-\varepsilon < \sqrt{x + 5} - 3 < \varepsilon$$

Step 3: Solve for x:

$$3 - \varepsilon < \sqrt{x + 5} < \varepsilon + 3$$
$$(3 - \varepsilon)^2 < \left(\sqrt{x + 5}\right)^2 < (\varepsilon + 3)^2$$
$$(3 - \varepsilon)^2 < x + 5 < (\varepsilon + 3)^2$$
$$(3 - \varepsilon)^2 - 5 < x < (\varepsilon + 3)^2 - 5$$

Step 4: Make the middle term the same as that of the hypothesis (C)

$$(3 - \varepsilon)^2 - 5 - 4 < x - 4 < (\varepsilon + 3)^2 - 5 - 4$$

(Subtracting 4 and making middle term the same as the middle term of the hypothesis. We have now connected the hypothesis and the conclusion)

$$\boxed{(3 - \varepsilon)^2 - 9 < x - 4 < (\varepsilon + 3)^2 - 9}$$ (D)

(Note the equality of the middle terms in (C) and (D))

Step 5: Compare the left sides of (C) and (D) and equate these sides. Similarly, compare the right sides of (C) and (D) and equate these sides.

For the left sides	For the right sides
$-\delta = (3 - \varepsilon)^2 - 9$ or $\delta = 9 - (3 - \varepsilon)^2$ say δ_1	$\delta = (\varepsilon + 3)^2 - 9$. say δ_2

Proof

Step 6 Given $\varepsilon > 0$, choose $\delta = \min(\delta_1, \delta_2)$,

$$|x - 4| < \delta \text{ implies } -\delta_1 \leq -\delta < x - 4 < \delta \leq \delta_2 \quad \text{(E)}$$

Step 7: Replace the left and right sides of (E) by $9 - (3 - \varepsilon)^2$ and

$(\varepsilon + 3)^2 - 9$, respectively to obtain $(3 - \varepsilon)^2 - 9 < x - 4 < (\varepsilon + 3)^2 - 9$ (F)

Break up inequality (F) into two simple inequalities an solve each inequality for ε or $-\varepsilon$.

For the left side	For the right side
$(3 - \varepsilon)^2 - 9 < x - 4$	$x - 4 < (\varepsilon + 3)^2 - 9$
$(3 - \varepsilon)^2 < x + 5$	$x + 5 < (\varepsilon + 3)^2$
$3 - \varepsilon < \sqrt{x + 5}$	$\sqrt{x + 5} < \sqrt{(\varepsilon + 3)^2}$
$-\varepsilon < \sqrt{x + 5} - 3$ (transposing)	$\sqrt{x + 5} < \varepsilon + 3$
	$\sqrt{x + 5} - 3 < \varepsilon$

Step 8: $-\varepsilon < \sqrt{x + 5} - 3$ and $\sqrt{x + 5} - 3 < \varepsilon$ is equivalent to

$$\left| \sqrt{x + 5} - 3 \right| < \varepsilon$$

We have shown that if $|x - 4| < \delta$, then $\left| \sqrt{x + 5} - 3 \right| < \varepsilon$.

Therefore,, $\lim\limits_{x \to 4} \sqrt{x + 5} = 3$, the proof is complete.

In the above proof, for δ, we use the smaller of $\delta_1 = 9 - (3 - \varepsilon)^2$ and

$\delta_2 = (\varepsilon + 3)^2 - 9$ or any number less than the smaller of δ_1 and δ_2.

Let us find some numerical approximation for δ for $\varepsilon = 0.01$

Continued inequality approach	Method 1 (delta assumption method)
$\delta_1 = 9 - (3 - 0.01)^2$	$\delta = \left(2\sqrt{2} + 3 \right)\varepsilon$
$\delta_1 = 9 - 8.940$	$\delta = \left(2\sqrt{2} + 3 \right)(0.01)$
$\delta_1 = 0.06$ (to two decimal places)	$= (2.82843 + 3)(0.01)$
$\delta_2 = (0.01 + 3)^2 - 9$	$= (5.82843)(0.01)$
$\delta_2 = (3.01)^2 - 9$	$= 0.06$ (to two decimal places)
$= 9.0601 - 9$	
$= 0.06$ (to two decimal places)	

Example 8 Prove that $\lim\limits_{x \to 1} \sqrt{x^2 + 3} = 2$

Method 1 (delta assumption method)

Definition: If $0 < |x - a| < \delta$ then $|f(x) - L| < \varepsilon$

Step 1: If $0 < |x - 1| < \delta$, (hypothesis) (A)

then $\left| \sqrt{x^2 + 3} - 2 \right| < \varepsilon$ (conclusion) (B)

Plan: The proof would be complete after showing that

If $\boxed{0 < |x - 1| < \delta}$, then $\boxed{\left| \sqrt{x^2 + 3} - 2 \right| < \varepsilon}$.

Step 2: We want a formula relating δ and ε, and we begin with

$$\left| \sqrt{x^2 + 3} - 2 \right| < \varepsilon \text{ (the conclusion)}$$

$$\left| \frac{\left(\sqrt{x^2 + 3} - 2 \right)}{1} \cdot \frac{\left(\sqrt{x^2 + 3} + 2 \right)}{\left(\sqrt{x^2 + 3} + 2 \right)} \right| < \varepsilon$$

$$\left| \frac{x^2 - 1}{\sqrt{x^2 + 3} + 2} \right| < \varepsilon \qquad \text{(C)}$$

By comparing the inequality $\left|\dfrac{(x+1)(x-1)}{\sqrt{x^2+3}+2}\right| < \varepsilon$ (an equivalent conclusion) with

$0 < |x-1| < \delta$ (the hypothesis), we observe the factor $|x-1|$ in $\left|\dfrac{(x+1)(x-1)}{\sqrt{x^2+3}+2}\right| < \varepsilon$

is the same as the term $|x-1|$ in $0 < |x-1| < \delta$; and this situation is desirable..

For the factors $|x+1|$ and $\sqrt{x^2+3}+2$ we find an upper bound for each.

Step 3: Find am upper bound for $|x+1|$.

Assume that $\delta \le 1$. Then, the inequality $0 < |x-1| < \delta$ becomes
$|x-1| < 1$. We begin with this inequality and change the $|x-1|$ to $|x+1|$
as follows. $|x-1| < 1$ is equivalent to $-1 < x-1 < 1$.

$-1 < x-1 < 1$

$\underline{+2 \quad\ +2 \ +2}$ (We change the middle term to $|x+1|$ by adding 2)

$1 < x+1 < 3$

The upper bound for $|x+1|$ is 3, and we can write

$|x+1| < 3$ $\hspace{6cm}$ (D)

Step 4: Find an upper bound for $\dfrac{1}{\sqrt{x^2+3}+2}$

We note that when x is close to 1, $\dfrac{1}{\sqrt{x^2+3}+2}$ is close to $\dfrac{1}{4}$

Begin with $|x-1| < 1$ (the hypothesis with $\delta = 1$)

$\quad -1 < x-1 < 1$

$\quad\ 0 < x < 2$ (solving)

$\quad\ 0 < x^2 < 4$.   (squaring)

$\ 3 < x^2+3 < 7$ (adding 3 to all three sides)

$\sqrt{3} < \sqrt{x^2+3} < \sqrt{7}$

$\sqrt{3}+2 < \sqrt{x^2+3}+2+2 < \sqrt{7}+2$

$\dfrac{1}{\sqrt{3}+2} > \dfrac{1}{\sqrt{x^2+3}+2} > \dfrac{1}{\sqrt{7}+2}$

$\dfrac{1}{\sqrt{7}+2} < \dfrac{1}{\sqrt{x^2+3}+2} < \dfrac{1}{\sqrt{3}+2}$

(An upper bound for $\dfrac{1}{\sqrt{x^2+3}+2}$ is $\dfrac{1}{\sqrt{3}+2}$

$\dfrac{1}{\sqrt{3}+2} = 2 - \sqrt{3}$

$\quad\ \approx 0.27$ (which is close to the ≈ 0.25 or $\dfrac{1}{4}$

Step 5: $\left|\dfrac{1}{\sqrt{x^2+3}+2}\right| < 2-\sqrt{3}$ (E)

$|x+1| < 3$ (D)

$|x-1| < \delta$ (A)

Multiply the left sides and multiply the right sides of (A), (D) and (E)

$\left|\dfrac{(x+1)(x-1)}{\sqrt{x^2+3}+2}\right| < 3\delta(2-\sqrt{3})$ (F)

$\left|\dfrac{(x+1)(x-1)}{\sqrt{x^2+3}+2}\right| < \varepsilon$ (C)

(Note the equality of the left sides in (F) and (C))

Step 6: Comparing the right-hand sides and the left-hand sides of (F) and (C) take $\varepsilon = 3\delta(2-\sqrt{3})$, and from which

$\delta = \dfrac{\varepsilon}{3(2-\sqrt{3})}$, We take $\delta = \dfrac{\varepsilon}{3(2-\sqrt{3})}$ or less.

Step 7: We replace δ on the right-hand side of the hypothesis $|x-1| < \delta$ by

$\dfrac{\varepsilon}{3(2-\sqrt{3})}$ to obtain

$|x-1| < \dfrac{\varepsilon}{3(2-\sqrt{3})}$ (G)

Proof: $|x-1| < \dfrac{\varepsilon}{3(2-\sqrt{3})}$ (G)

$|x+1| < 3$ (D)

$\left|\dfrac{1}{\sqrt{x^2+3}+2}\right| < 2-\sqrt{3}$ (E)

Multiply the left sides and multiply the right sides of (G), (D) and (E)

$\left|\dfrac{x-1)(x+1)}{\sqrt{x^2+3}+2}\right| < \dfrac{\varepsilon}{3(2-\sqrt{3})} \bullet 3(2-\sqrt{3})$, which reduces to

$\left|\dfrac{(x+1)(x-1)}{\sqrt{x^2+3}+2}\right| = \left|\sqrt{x^2+3}-2\right| < \varepsilon$ (Rationalizing the denominator)

We have shown that if $0 < |x-1| < \delta$, then $\left|\sqrt{x^2+3}-2\right| < \varepsilon$.

Therefore, $\lim\limits_{x\to 1}\sqrt{x^2+3} = 2$, and the proof is complete.

In the above proof, we could take $\delta = \dfrac{\varepsilon}{3(2-\sqrt{3})} = \dfrac{(2+\sqrt{3})\varepsilon}{3}$ or 1 (whichever is the smaller of the values) or any number less than both $\dfrac{(2+\sqrt{3})\varepsilon}{3}$ and 1.

Method 2 (Using continued inequality)

Example 8 Prove that $\lim\limits_{x \to 1} \sqrt{x^2 + 3} = 2$

Definition: If $0 < |x - a| < \delta$ then $|f(x) - L| < \varepsilon$

Step 1: If $0 < |x - 1| < \delta$,   (hypothesis)   (A)

   then $\left| \sqrt{x^2 + 3} - 2 \right| < \varepsilon$ (conclusion)   (B)

   Plan: The proof would be complete after showing that

   If $\boxed{0 < |x - 1| < \delta}$, then $\boxed{\left| \sqrt{x^2 + 3} - 2 \right| < \varepsilon}$.

Step 2: We want a formula relating δ and ε, and we begin with

   $\left| \sqrt{x^2 + 3} - 2 \right| < \varepsilon$ (the conclusion) which is equivalent to

   $-\varepsilon < \sqrt{x^2 + 3} - 2 < \varepsilon$

   $2 - \varepsilon < \sqrt{x^2 + 3} < \varepsilon + 2$

   $(2 - \varepsilon)^2 < x^2 + 3 < (\varepsilon + 2)^2$

   $(2 - \varepsilon)^2 - 3 < x^2 < (\varepsilon + 2)^2 - 3$

   $\sqrt{(2 - \varepsilon)^2 - 3} < x < \sqrt{(\varepsilon + 2)^2 - 3}$

Step 3: $\sqrt{(2 - \varepsilon)^2 - 3} - 1 < x - 1 < \sqrt{(\varepsilon + 2)^2 - 3} - 1$

   (Subtracting 4 and making middle term the same as the middle term of
   the hypothesis. We have now connected the hypothesis and the conclusion)

   $\sqrt{(2 - \varepsilon)^2 - 3} - 1 < x - 1 < \sqrt{(\varepsilon + 2)^2 - 3} - 1$ (C)

Step 4 $|x - 1| < \delta$ (hypothesis) is equivalent to

   $\boxed{-\delta < x - 1 < \delta}$ (D)

   (Note the equality of the middle terms in (C) and (D))

Step 5: Compare the left sides of (C) and (D) and equate these sides. Similarly,
   compare the right sides of (C) and (D) and equate these sides.

For the left sides	For the right sides
$-\delta = \sqrt{(2 - \varepsilon)^2 - 3} - 1$	$\delta = \sqrt{(\varepsilon + 2)^2 - 3} - 1.$ say δ_2
$\delta = 1 - \sqrt{(2 - \varepsilon)^2 - 3}$, say δ_1	

Proof

Step 6 Given $\varepsilon > 0$, choose $\delta = \min(\delta_1, \delta_2)$,

   $|x - 1| < \delta$ implies $-\delta_1 \le -\delta < x - 1 < \delta \le \delta_2$

Step 7: Break up inequality (C) into two simple inequalities and solve
   each inequality for $-\varepsilon$ or ε ..

   $\sqrt{(2 - \varepsilon)^2 - 3} - 1 < x - 1 < \sqrt{(\varepsilon + 2)^2 - 3} - 1$ is equivalent to.

For the left side	For the right side
$\sqrt{(2-\varepsilon)^2 - 3} - 1 < x - 1$	$x - 1 < \sqrt{(\varepsilon + 2)^2 - 3} - 1$
$\sqrt{(2-\varepsilon)^2 - 3} < x$	$x < \sqrt{(\varepsilon + 2)^2 - 3}$
$(2-\varepsilon)^2 - 3 < x^2$	$x^2 < (\varepsilon + 2)^2 - 3$
$(2-\varepsilon)^2 < x^2 + 3$	$x^2 + 3 < (\varepsilon + 2)^2$
$2 - \varepsilon < \sqrt{x^2 + 3}$	$\sqrt{x^2 + 3} < \varepsilon + 2$
$-\varepsilon < \sqrt{x^2 + 3} - 2$ (transposing)	$\sqrt{x^2 + 3} - 2 < \varepsilon$

Step 8: $-\varepsilon < \sqrt{x^2 + 3} - 2$ and $\sqrt{x^2 + 3} - 2 < \varepsilon$ is equivalent to

$$\left| \sqrt{x^2 + 3} - 2 \right| < \varepsilon$$

We have shown that if $|x - 1| < \delta$, then $\left| \sqrt{x^2 + 3} - 2 \right| < \varepsilon$

Therefore, $\lim_{x \to 1} \sqrt{x^2 + 3} = 2$, and the proof is complete.

In the above proof, for δ, we use the smaller of $\delta_1 = 1 - \sqrt{(2-\varepsilon)^2 - 3}$ and $\delta_2 = \sqrt{(\varepsilon + 2)^2 - 3} - 1$ or any number less than the smaller of δ_1 and δ_2.

Let us find some numerical approximation for δ for $\varepsilon = 0.01$

Continued inequality approach	Method 1 (delta assumption)
$\delta_1 = 1 - \sqrt{(2 - 0.01)^2 - 3}$ $\quad = 1 - \sqrt{3.9601 - 3}$ $\quad = 1 - \sqrt{0.9601}$ $\quad = 1 - 0.9798$ $\quad = 0.020$ $\delta_2 = \sqrt{(\varepsilon + 2)^2 - 3} - 1$ $\quad = 0.020$	$\delta = \dfrac{(2 + \sqrt{3})\varepsilon}{3}$ $\quad = \dfrac{(2 + \sqrt{3})(0.01)}{3}$ $\quad = \dfrac{(3.732)(0.01)}{3}$ $\quad = 0.012$

If $\delta = 0.020$. and we check for this value by substituting $x = 1 \pm 0.010$ in $\left| \sqrt{x^2 + 3} - 2 \right| < \varepsilon$

$\qquad x = 1 + .010 = 1.010$

If $|x - 1| < 0.020$

Then $\left| \sqrt{(1.01)^2 + 3} - 2 \right| \overset{?}{<} 0.010$

$\left| \sqrt{(1.01)^2 + 3} - 2 \right| \overset{?}{<} 0.010$

$|0.005| \overset{?}{<} 0.010$. Yes .

For $x = 1 - 0.010 = 0.990$

$\left| \sqrt{(0.990)^2 + 3} - 2 \right| \overset{?}{<} 0.010$

$\left| \sqrt{0.9801 + 3} - 2 \right| \overset{?}{<} 0.010$

$|0.005| \overset{?}{<} 0.010$. Yes

Example 9   Prove that $\lim\limits_{x \to 24} \sqrt{x+1} = 5$

Method 1 (delta assumption method)

Definition: If $0 < |x - a| < \delta$ then $|f(x) - L| < \varepsilon$

Step 1: If $0 < |x - 24| < \delta$,   (hypothesis)   **(A)**

   then $\left|\sqrt{x+1} - 5\right| < \varepsilon$ (conclusion)   **(B)**

   Plan: The proof would be complete after showing that

   If $0 < |x - 24| < \delta$, then $\left|\sqrt{x+1} - 5\right| < \varepsilon$.

Step 2: We want a formula relating δ and ε, and we begin with

   $\left|\sqrt{x+1} - 5\right| < \varepsilon$ (the conclusion)

   $\left|\dfrac{\left(\sqrt{x+1} - 5\right)}{1} \bullet \dfrac{\left(\sqrt{x+1} + 5\right)}{\left(\sqrt{x+1} + 5\right)}\right| < \varepsilon$ (Rationalizing the numerator)

   $\left|\dfrac{x - 24}{\sqrt{x+1} + 5}\right| < \varepsilon$ **(C)**

By comparing the inequality $\left|\dfrac{x - 24}{\sqrt{x+1} + 5}\right| < \varepsilon$ (an equivalent conclusion) with

$0 < |x - 24| < \delta$ (the hypothesis), we observe the factor $|x - 24|$ in

$\left|\dfrac{x - 24}{\sqrt{x+1} + 5}\right| < \varepsilon$ is the same as the term $|x - 24|$ in $0 < |x - 24| < \delta$; and this

situation is desirable. For the factor $\dfrac{1}{\sqrt{x+1} + 5}$ we find an upper bound for it.

Step 3: Find an upper bound for $\dfrac{1}{\sqrt{x+1} + 5}$

   Begin with $|x - 24| < 1$ (assume $\delta = 1$. Later, we will try $\delta = \frac{1}{2}$ and $\delta = \frac{1}{4}$)

   $-1 < x - 24 < 1$

   $\underline{+25 \quad + 25 \ + 25}$

   $24 < x + 1 < 26$

   $\sqrt{24} < \sqrt{x+1} < \sqrt{26}$

   $\sqrt{24} + 5 < \sqrt{x+1} + 5 < \sqrt{26} + 5$

   $\dfrac{1}{\sqrt{24} + 5} > \dfrac{1}{\sqrt{x+1} + 5} > \dfrac{1}{\sqrt{26} + 5}$ (inverting and reversing the sense)

   $\dfrac{1}{\sqrt{26} + 5} < \dfrac{1}{\sqrt{x+1} + 5} < \dfrac{1}{\sqrt{24} + 5}$ (rewriting)

   (An upper bound for $\dfrac{1}{\sqrt{x+1} + 5}$ is $\dfrac{1}{\sqrt{24} + 5}$ and

   $\left|\dfrac{1}{\sqrt{x+1} + 5}\right| < \dfrac{1}{\sqrt{24} + 5}$

Step 4: $\left|\dfrac{1}{\sqrt{x+1}+5}\right| < \dfrac{1}{\sqrt{24}+5}$ (D)

$|x-24| < \delta$ (A)

Multiply the left sides and multiply the right sides of (A) and (D)

$\left|\dfrac{x-24}{\sqrt{x+1}+5}\right| < \dfrac{\delta}{\sqrt{25}+5}$ (E)

Step 5: Comparing the right-hand sides and the left-hand sides of (C) and (E)

take $\varepsilon = \dfrac{\delta}{\sqrt{24}+5}$, and from which

$\delta = \left(\sqrt{24}+5\right)\varepsilon$.

$= \left(2\sqrt{6}+5\right)\varepsilon$

Step 6: We replace δ on the right-hand side of the hypothesis $|x-24| < \delta$ by

$\left(2\sqrt{6}+5\right)\varepsilon$ to obtain

$|x-24| < \left(2\sqrt{6}+5\right)\varepsilon$ (F)

Proof: $|x-24| < \left(2\sqrt{6}+5\right)\varepsilon$ (F)

$\left|\dfrac{1}{\sqrt{x+1}+5}\right| < \dfrac{1}{2\sqrt{6}+5}$ **(G)**

Multiply the left sides and multiply the right sides of (F) and (G)

$|x-24| \bullet \left|\dfrac{1}{\sqrt{x+1}+5}\right| < \dfrac{1}{\left(2\sqrt{6}+5\right)} \bullet \left(2\sqrt{6}+5\right)\varepsilon$,, which reduces to

$\left|\dfrac{x-24}{\sqrt{x+1}+5}\right| = \left|\sqrt{x+1}-5\right| < \varepsilon$ (Rationalizing the denominator)

We have shown that if $0 < |x-24| < \delta$, then $\left|\sqrt{x+1}-5\right| < \varepsilon$.

Therefore, $\displaystyle\lim_{x\to24} \sqrt{x+1} = 5$, and the proof is complete.

In the above proof, we could take $\delta = \left(2\sqrt{6}+5\right)\varepsilon$ or 1 (whichever is the

smaller of the values) or any number less than both $\delta = \left(2\sqrt{6}+5\right)\varepsilon$ and 1

In particular, if $\varepsilon = 0.1$. $\delta = \left(2\sqrt{6}+5\right)\varepsilon$

$= (2(2.4495)+5)(0.1)$ ($\sqrt{6} \approx 2.4495$)

$\delta = 0.99$

Note: in Step 4 above, if we let $\delta = \frac{1}{2}$, and begin with $|x-24| < \frac{1}{2}$, we obtain

in Step 6, $\delta = \left(\sqrt{24.5}+5\right)\varepsilon$, and if $\varepsilon = 0.1$, we find

$\delta = \left(\sqrt{24.5}+5\right)(0.1) = 0.99$ (to two decimal places,

Similarly, if we let $\delta = \frac{1}{4}$ in Step 4, in Step 6, $\delta = \left(\sqrt{24.75}+5\right)\varepsilon$, and if

$\varepsilon = 0.1$, $\delta = \left(\sqrt{24.75}+5\right)(0.1) = 0.99$ (to two decimal laces)

We compare some assumptions of δ in Step 4 and δ in Step 6, with $\varepsilon = 0.1$.
If we assume $\delta = 1$ in Step 4, we obtain in Step 6, $\delta = 0.9898997$.

If we assume $\delta = \frac{1}{2}$ in Step 4, we obtain in Step 6, $\delta = 0.9949747$.

If we assume $\delta = \frac{1}{4}$ in Step 4, we obtain in Step 6, $\delta = 0.9974937$.

If we assume $\delta = \frac{1}{8}$ in Step 4, we obtain in Step 6, $\delta = 0.9987484$.

In the assumptions of $\delta = 1$, and $\delta = \frac{1}{2}$ in Step 4, if we round-off δ in Step 6 two decimal places with $\varepsilon = 0.1$, we obtain $\delta = 0.99$.

Method 2 (Using continued inequality)

Example 9   Prove that $\lim\limits_{x \to 24} \sqrt{x + 1} = 5$

Definition: If $0 < |x - a| < \delta$ then $|f(x) - L| < \varepsilon$

Step 1: If $0 < |x - 24| < \delta$,   (hypothesis)   (A)

then $|\sqrt{x + 1} - 5| < \varepsilon$ (conclusion)   (B)

Plan: The proof would be complete after showing that

If $0 < |x - 24| < \delta$, then $|\sqrt{x + 1} - 5| < \varepsilon$.

Step 2: We want a formula relating δ and ε, and we begin with

$|\sqrt{x + 1} - 5| < \varepsilon$ (the conclusion) which is equivalent to

$-\varepsilon < \sqrt{x + 1} - 5 < \varepsilon$

$5 - \varepsilon < \sqrt{x + 1} < \varepsilon + 5$

$(5 - \varepsilon)^2 < \left(\sqrt{x + 1}\right)^2 < (\varepsilon + 5)^2$

$(5 - \varepsilon)^2 < x + 1 < (\varepsilon + 5)^2$

$(5 - \varepsilon)^2 - 1 < x < (\varepsilon + 5)^2 - 1$

Step 3: $(5 - \varepsilon)^2 - 1 - 24 < x - 24 < (\varepsilon + 5)^2 - 1 - 24$

(Subtracting 24 and making middle term the same as the middle term of the hypothesis. We have now connected the hypothesis and the conclusion)

$(5 - \varepsilon)^2 - 25 < x - 24 < (\varepsilon + 5)^2 - 25$ (C)

Step 4 $|x - 24| < \delta$ (hypothesis) is equivalent to

$-\delta < x - 24 < \delta$ (D)

(Note the equality of the middle terms in (C) and (D))

Step 5: Compare the left sides of (C) and (D) and equate these sides. Similarly, compare the right sides of (C) and (D) and equate these sides.

For the left sides	For the right sides
$-\delta = (5 - \varepsilon)^2 - 25$	$\delta = (\varepsilon + 5)^2 - 25$. say δ_2
$\delta = 25 - (5 - \varepsilon)^2$, say δ_1	

Proof: Given $\varepsilon > 0$, choose $\delta = \min(\delta_1, \delta_2)$,

$|x - 24| < \delta$ implies $-\delta_1 \le -\delta < x - 24 < \delta \le \delta_2$

Step 6: Break up inequality (C) into two simple inequalities an solve

each inequality for $-\varepsilon$ or ε. $(5 - \varepsilon)^2 - 25 < x - 24 < (\varepsilon + 5)^2 - 25$ is equiv. to

For the left side	For the right side
$(5 - \varepsilon)^2 - 25 < x - 24$	$x - 24 < (\varepsilon + 5)^2 - 25$
$(5 - \varepsilon)^2 < x + 1$	$x + 1 < (\varepsilon + 5)^2$
$\sqrt{(5 - \varepsilon)^2} < \sqrt{x + 1}$	$\sqrt{x + 1} < \sqrt{(\varepsilon + 5)^2}$
$5 - \varepsilon < \sqrt{x + 1}$	$\sqrt{x + 1} < \varepsilon + 5$
$-\varepsilon < \sqrt{x + 1} - 5$ (transposing)	$\sqrt{x + 1} - 5 < \varepsilon$

Step 8: $-\varepsilon < \sqrt{x + 1} - 5$ and $\sqrt{x + 1} - 5 < \varepsilon$ is equivalent to

$$\left| \sqrt{x + 1} - 5 \right| < \varepsilon$$

We have shown that if $|x - 24| < \delta$, then $\left| \sqrt{x + 1} - 5 \right| < \varepsilon$.

Therefore, $\lim_{x \to 24} \sqrt{x + 1} = 5$, and the proof is complete.

.For δ, we use the smaller of $\delta_1 = 25 - (5 - \varepsilon)^2$; $\delta_2 = (\varepsilon + 5)^2 - 25$.

Let us find some numerical approximation for δ for $\varepsilon = 0.100$

Continued inequality approach	**Method 1** (delta assumption)
$\delta_1 = 25 - (5 - 0.1)^2$	$\delta = (2\sqrt{6} + 5)\varepsilon$
$\quad = 25 - (4.9)^2$	$\quad = (2(2.4495) + 5)(0.1)$
$\quad = 25 - 24.01 = 0.99$	$\delta = 0.99$ (see two pages back)
$\delta_2 = (0.1 + 5)^2 - 25 = 1.01$	
If $\delta = 0.99$, we check for $x = 24.98$ in $\left\| \sqrt{x + 1} - 5 \right\| < \varepsilon$	We also check for $x = 23.02$ in $\left\| \sqrt{x + 1} - 5 \right\| < \varepsilon$
If $|x - 24| < 0.99$, $x = 24.98$)	If $|x - 24| < 0.99$, $(x = 23.02)$
$\left\| \sqrt{24.98 + 1} - 5 \right\| \overset{?}{<} 0.100$	$\left\| \sqrt{23.02 + 1} - 5 \right\| \overset{?}{<} 0.100$
$\left\| \sqrt{25.98} - 5 \right\| \overset{?}{<} 0.100$	$\left\| \sqrt{24.02} - 5 \right\| \overset{?}{<} 0.100$
$\left\| 5.097 - 5 \right\| \overset{?}{<} 0.100$	$\left\| 4.901 - 5 \right\| \overset{?}{<} 0.100$
$\left\| 0.097 \right\| \overset{?}{<} 0.100$ Yes	$\left\| 0.099 \right\| \overset{?}{<} 0.100$ Yes
	Therefore $\delta = 0.99$ is accepted.
If $\delta = 1.01$. and we check for $x = 24 \pm 1$ in $\left\| \sqrt{x + 1} - 5 \right\| < \varepsilon$	
If $|x - 24| < 1.01$	If $|x - 24| < 1.01$
$\left\| \sqrt{25 + 1} - 5 \right\| \overset{?}{<} 0.100 \quad x = 25$	$\left\| \sqrt{23 + 1} - 5 \right\| \overset{?}{<} 0.100 \quad x = 23$
$\left\| 0.099 \right\| \overset{?}{<} 0.100$ Yes	$0.101 \overset{?}{<} 0.100$ No
	Therefore, $\delta = 1.01$ is rejected.

Case 5: Logarithmic Functions

Example 10   Prove that $\lim\limits_{x \to 3} \log_2\left(x^2 + 7\right) = 4$

Method 1 (delta assumption method)

Definition: If $0 < |x - a| < \delta$ then $|f(x) - L| < \varepsilon$

Step 1: Recall that when asked to graph the logarithmic function, we usually graph the equivalent exponential form.. Here also, we similarly prove the limit of the equivalent exponential function, since this form is more convenient to deal with. Therefore, equivalently, we will prove

$$\lim\limits_{x \to 3} x^2 + 7 = 2^4$$

If $0 < |x - 3| < \delta$,	(hypothesis)	(A)
then $\left|x^2 + 7 - 2^4\right| < \varepsilon$	(conclusion)	(C)

$$\left|x^2 - 9\right| < \varepsilon$$

Plan: The proof would be complete after showing that

If $0 < |x - 3| < \delta$, then $\left|x^2 - 9\right| < \varepsilon$.

Step 2: We want a formula relating δ and ε, and we begin wit
(the equivalent exponential conclusion)

$$\left|x^2 - 9\right) < \varepsilon$$

$$|(x - 3)(x + 3)| < \varepsilon \qquad\qquad\qquad\qquad (D)$$

By comparing the inequality $|(x + 3)(x - 3)| < \varepsilon$ with $0 < |x - 3| < \delta$ (the hypothesis) in (A), we observe that the factor $|x - 3|$ in the conclusion (D) is the same as the term $|x - 3|$ in the hypothesis ($0 < |x - 3| < \delta$); and this situation is desirable.. However, for the factor $|x + 3|$, we find an upper bound for it. We also note that when x is close to 3, $x + 3$ is close to 6. so an upper bound close to 6 would be a good upper bound.

Step 3: If we assume that $\delta \le 1$, then the inequality $|x - 3| < \delta$ (hypothesis) becomes $|x - 3| < 1$. We begin with this inequality and change the $|x - 3|$ to $|x + 3|$ as follows.

$|x - 3| < 1$ is equivalent to $-1 < x - 3 < 1$.

$-1 < x - 3 < 1$
    (We change the middle term to $|x + 3|$ by adding 6)
$5 < x + 3 < 7$

An upper bound for $|x + 3|$ is 7; and we can write
$$|x + 3| < 7 \qquad\qquad\qquad (E)$$

Step 4: $|x - 3| < \delta$ (hypothesis)   (A)
Multiply the left sides and multiply the right sides of (E) and (A)

$$|(x + 3)(x - 3)| < 7\delta \qquad\qquad\qquad\qquad (F)$$

$$|(x + 3)(x - 3)| < \varepsilon \qquad\qquad\qquad\qquad (D)$$

(Note the equality of the left sides in (F) and (D))

Step 5: Comparing the left-hand-sides and the right-hand sides of (F) and (D)

$7\delta = \varepsilon$, and from which $\delta = \frac{\varepsilon}{7}$.

Step 6: Replace the δ on the right-hand side of (A), hypothesis by $\frac{\varepsilon}{7}$ to obtain

$$|x - 3| < \frac{\varepsilon}{7} \qquad\qquad\qquad\qquad (G)$$

Proof:

Step 7: $|x + 3| < 7$ $\qquad\qquad\qquad\qquad\qquad\qquad$ (E)

     Multiply the left sides and multiply the right sides of (G) and (E), above.

$$\left|(x - 3)(x + 3)\right| < \frac{\varepsilon}{7}(7)$$

$\left|(x + 3)(x - 3)\right| < \varepsilon$ which is equivalent to $\left|x^2 - 9\right| < \varepsilon$., and also to the

equivalent exponential form of $\left|x^2 + 7 - 2^4\right| < \varepsilon$

We have shown that if $0 < |x - 3| < \delta$, then $\left|x^2 + 7 - 2^4\right| < \varepsilon$.

Therefore, $\lim\limits_{x \to 3} \log_2\left(x^2 + 7\right) = 4$, and the proof is complete.

In the above proof , we could use $\delta = 1$ if $\frac{\varepsilon}{7} > 1$ Symbolically , we write

$\delta = \min\left(\frac{\varepsilon}{7}, 1\right)$. We can also use any number less than both $\frac{\varepsilon}{7}$ and 1.

Note: above that the steps in Example 10 are the same as those in Example 4.

Example 11 Prove that $\lim\limits_{x \to 8} \log_2 x = 3$

Method (delta assumption method)

Definition: If $0 < |x - a| < \delta$ then $|f(x) - L| < \varepsilon$

Step 1: Recall that when asked to graph the logarithmic function, we usually graph the
     equivalent exponential form.. Here also, we similarly prove the limit of the
     equivalent exponential function, since this form is more convenient to deal with.
     Therefore, equivalently, we will prove

$$\lim\limits_{x \to 8} x = 2^3$$

    If $0 < |x - 8| < \delta$, $\qquad\qquad$ (hypothesis) $\qquad$ (A)

    then $\left|x - 2^3\right| < \varepsilon$ $\qquad\qquad\qquad$ (conclusion) $\qquad$ (C)

Plan: The proof would be complete after showing that

    If $0 < |x - 8| < \delta$, then $|x - 8| < \varepsilon$.

Step 2: We want a formula relating δ and ε , and we begin wit

     (the equivalent exponential conclusion)

    $\left|x - 2^3\right| < \varepsilon$ $\qquad\qquad\qquad\qquad\qquad$ (C)

    $|x - 8| < \varepsilon$ $\qquad\qquad\qquad\qquad\qquad\qquad$ (D)

Step 3: $|x - 8| < \delta$ $\qquad\qquad$ (hypothesis) $\qquad\qquad$ (A)

    $|x - 8| < \varepsilon$ $\qquad\qquad\qquad\qquad\qquad\qquad$ (D)

    (Note the equality of the left sides in (A) and (D))

Step 4: Comparing the left-hand-sides and the right-hand sides of (A) and (D)

    take $\delta = \varepsilon$

Step 5: Replace the δ on the right-hand side of (A), hypothesis by ε to obtain

    $|x - 8| < \varepsilon$ $\qquad\qquad\qquad\qquad\qquad\qquad$ (E)

If we assume that $\delta \le 1$, then the inequality $|x - 8| < \delta$ (hypothesis) becomes

$\qquad |x - 8| < 1$. $\qquad\qquad\qquad\qquad$ (F)

Proof:

Step 6: $\quad |x - 8| < 1$ $\qquad\qquad\qquad\qquad\qquad\qquad\qquad$ (F)

$\qquad\quad |x - 8| < \varepsilon$ $\qquad\qquad\qquad\qquad\qquad\qquad\qquad$ (D)

We have shown that if $0 < |x - 8| < \delta$, then $|x - 8| < \varepsilon$.

Therefore, $\lim\limits_{x \to 8} \log_2 x = 3$, and the proof is complete

In the above proof , we could use $\delta = 1$. If $\delta > 1$ Symbolically , we write

$\delta = \min\left(\frac{\varepsilon}{7}, 1\right)$. We can also use any number less than both $\frac{\varepsilon}{7}$ and 1.

Case 6: Trigonometric Functions

Example 12   Prove that $\lim\limits_{x \to \frac{\pi}{4}} \sin x = \frac{\sqrt{2}}{2}$

Solution   **(continued inequality approach)**

Definition: If $0 < \left|x - \frac{\pi}{4}\right| < \delta$ then $\left|\sin x - \frac{\sqrt{2}}{2}\right| < \varepsilon$

Step 1: If $0 < \left|x - \frac{\pi}{4}\right| < \delta$, $\qquad\qquad$ (hypothesis) $\qquad\qquad$ (A)

$\qquad\quad$ then $\left|\sin x - \frac{\sqrt{2}}{2}\right| < \varepsilon$ $\qquad\qquad$ (conclusion) $\qquad\qquad$ (B)

$\qquad\quad$ Plan: The proof would be complete after showing that

$\qquad\quad$ If $0 < \left|x - \frac{\pi}{4}\right| < \delta$, then $\left|\sin x - \frac{\sqrt{2}}{2}\right| < \varepsilon$.

Step 2: We want a formula relating δ and ε, and we begin with

$\qquad\qquad \left|\sin x - \frac{\sqrt{2}}{2}\right| < \varepsilon$, which is equivalent to

$\qquad\qquad -\varepsilon < \sin x - \frac{\sqrt{2}}{2} < \varepsilon$

$\qquad\qquad \frac{\sqrt{2}}{2} - \varepsilon < \sin x < \varepsilon + \frac{\sqrt{2}}{2}$

$\qquad\qquad \text{Sin}^{-1}\left(\frac{\sqrt{2}}{2} - \varepsilon\right) < x < \text{Sin}^{-1}\left(\varepsilon + \frac{\sqrt{2}}{2}\right) \quad$ (solving for x)

$\qquad\qquad$ (Note that $\sin x$ is increasing $0 < x < \frac{\pi}{2}$)

$\text{Sin}^{-1}\left(\frac{\sqrt{2}}{2} - \varepsilon\right) - \frac{\pi}{4} < x - \frac{\pi}{4} < \text{Sin}^{-1}\left(\varepsilon + \frac{\sqrt{2}}{2}\right) - \frac{\pi}{4}$ $\quad$ (C)

(Making middle term the same as the middle term of the hypothesis.
We have now connected the hypothesis and the conclusion)

Step 3: From (A) $\left|x - \frac{\pi}{4}\right| < \delta$ is equivalent to

$\qquad\qquad \boxed{-\delta < x - \frac{\pi}{4} < \delta}$ $\qquad\qquad\qquad$ (D)

$\qquad\quad$ (Note the equality of the middle terms in (C) and (D))

Step 4: Comparing the left and right sides of (C) and (D)

For the left sides	For the right sides
Take $-\delta = \operatorname{Sin}^{-1}\left(\frac{\sqrt{2}}{2} - \varepsilon\right) - \frac{\pi}{4}$ and	Take $\delta = \operatorname{Sin}^{-1}\left(\varepsilon + \frac{\sqrt{2}}{2}\right) - \frac{\pi}{4} = \delta_2$ .say.
$\delta = -\operatorname{Sin}^{-1}\left(\frac{\sqrt{2}}{2} - \varepsilon\right) + \frac{\pi}{4} = \delta_1$, say.	

Proof: Given $\varepsilon > 0$, choose $\delta = \min(\delta_1, \delta_2)$,

$$|x - 1| < \delta \quad \text{implies} \quad -\delta_1 \le -\delta < x - 1 < \delta \le \delta_2$$

Break-up inequality (C) into two simple inequalities and solve for $-\varepsilon$ and ε .

Step 5: For the left part	For the right part
$\operatorname{Sin}^{-1}\left(\frac{\sqrt{2}}{2} - \varepsilon\right) - \frac{\pi}{4} < x - \frac{\pi}{4}$	$x - \frac{\pi}{4} < \operatorname{Sin}^{-1}\left(\varepsilon + \frac{\sqrt{2}}{2}\right) - \frac{\pi}{4}$
$\operatorname{Sin}^{-1}\left(\frac{\sqrt{2}}{2} - \varepsilon\right) < x$	$x < \operatorname{Sin}^{-1}\left(\varepsilon + \frac{\sqrt{2}}{2}\right)$
$\frac{\sqrt{2}}{2} - \varepsilon < \sin x$	$\sin x < \varepsilon + \frac{\sqrt{2}}{2}$
$-\varepsilon < \sin x - \frac{\sqrt{2}}{2}$	$\sin x - \frac{\sqrt{2}}{2} < \varepsilon$

Step 6: $-\varepsilon < \sin x - \frac{\sqrt{2}}{2}$ and $\sin x - \frac{\sqrt{2}}{2} < \varepsilon$ is equivalent to $\left|\sin x - \frac{\sqrt{2}}{2}\right| < \varepsilon$

We have shown that if $\left|x - \frac{\pi}{4}\right| < \delta$, **then** $\left|\sin x - \frac{\sqrt{2}}{2}\right| < \varepsilon$.

Therefore, $\lim\limits_{x \to \frac{\pi}{4}} \sin x = \frac{\sqrt{2}}{2}$, and the proof is complete.

In the above proof, for δ, we use the smaller of $\delta_1 = \frac{\pi}{4} - \operatorname{Sin}^{-1}\left(\frac{\sqrt{2}}{2} - \varepsilon\right)$ and

$\delta_2 = \operatorname{Sin}^{-1}\left(\varepsilon + \frac{\sqrt{2}}{2}\right) - \frac{\pi}{4}$ or any number less than the smaller of δ_1 and δ_2.

Let us find some numerical values for δ_1 and δ_2 for $\varepsilon = 0.002$

$\delta_1 = \frac{\pi}{4} - \operatorname{Sin}^{-1}\left(\frac{\sqrt{2}}{2} - \varepsilon\right)$

$= .76539816 - \operatorname{Sin}^{-1}(.707107 - .002)$

$= .76539816 - \operatorname{Sin}^{-1}(.7051068)$

$= .76539816 - .78257372$

$= 0.002824$

$\delta_2 = \operatorname{Sin}^{-1}\left(\varepsilon + \frac{\sqrt{2}}{2}\right) - \frac{\pi}{4}$

$= \operatorname{Sin}^{-1}(.002 + .707107) - .78539816$

$= \operatorname{Sin}^{-1}(.709107) - .78539816$

$= .788230916 - .7853981$

$= 0.0028323$

(Assume that $\varepsilon + \frac{\sqrt{2}}{2} \le 1$)

Example 13 Prove that $\lim\limits_{x \to \frac{\pi}{3}} \cos x = \frac{1}{2}$

Solution (**continued inequality approach**)

Definition: If $0 < \left| x - \frac{\pi}{3} \right| < \delta$ then $\left| \cos x - \frac{1}{2} \right| < \varepsilon$

Step 1: If $0 < \left| x - \frac{\pi}{3} \right| < \delta$,   (hypothesis)   (A)

then $\left| \cos x - \frac{1}{2} \right| < \varepsilon$ (conclusion)   (B)

Plan: The proof would be complete after showing that

If $0 < \left| x - \frac{\pi}{3} \right| < \delta$, then $\left| \cos x - \frac{1}{2} \right| < \varepsilon$.

Step 2: We want a formula relating δ and ε, and we begin with

$\left| \cos x - \frac{1}{2} \right| < \varepsilon$, which is equivalent to

$-\varepsilon < \cos x - \frac{1}{2} < \varepsilon$

$\frac{1}{2} - \varepsilon < \cos x < \varepsilon + \frac{1}{2}$

$\mathrm{Cos}^{-1}\left(\frac{1}{2} - \varepsilon \right) > x > \mathrm{Cos}^{-1}\left(\varepsilon + \frac{1}{2} \right)$ (solving for x)

(Reversing the sense since $\cos x$ is decreasing for $0 < x < \frac{\pi}{2}$)

$\mathrm{Cos}^{-1}\left(\varepsilon + \frac{1}{2} \right) < x < \mathrm{Cos}^{-1}\left(\frac{1}{2} - \varepsilon \right)$ (Rewriting)

$\mathrm{Cos}^{-1}\left(\varepsilon + \frac{1}{2} \right) - \frac{\pi}{3} < x - \frac{\pi}{3} < \mathrm{Cos}^{-1}\left(\frac{1}{2} - \varepsilon \right) - \frac{\pi}{3}$ (C)

(Making middle term the same as the middle term of the hypothesis.)

Step 3: From (A) $\left| x - \frac{\pi}{3} \right| < \delta$ is equivalent to

$-\delta < x - \frac{\pi}{3} < \delta$ (D)

Step 4: Comparing the left and right sides of (C) and (D)

Equating the left sides	Equating the right sides
Take $-\delta = \mathrm{Cos}^{-1}\left(\varepsilon + \frac{1}{2} \right) - \frac{\pi}{3}$	Take $\mathrm{Cos}^{-1}\left(\frac{1}{2} - \varepsilon \right) - \frac{\pi}{3} = \delta$,say δ_2
$\delta = \frac{\pi}{3} - \mathrm{Cos}^{-1}\left(\varepsilon + \frac{1}{2} \right) = \delta_1$, say.	

Proof: Given $\varepsilon > 0$, choose $\delta = \min(\delta_1, \delta_2)$,

$\left| x - \frac{\pi}{3} \right| < \delta$ implies $-\delta_1 \le -\delta < x - \frac{\pi}{3} < \delta \le \delta_2$

Break-up inequality (C) into two simple inequalities and solve for $-\varepsilon$ and ε

For the left side	For the right side
$\mathrm{Cos}^{-1}\left(\varepsilon + \frac{1}{2} \right) - \frac{\pi}{3} < x - \frac{\pi}{3}$,	$x - \frac{\pi}{3} < \mathrm{Cos}^{-1}\left(\frac{1}{2} - \varepsilon \right) - \frac{\pi}{3}$ ..

$$\text{Cos}^{-1}\left(\varepsilon + \tfrac{1}{2}\right) < x \qquad\qquad x < \text{Cos}^{-1}\left(\tfrac{1}{2} - \varepsilon\right)$$

$$\varepsilon + \tfrac{1}{2} > \cos x \quad \text{(also reversing the sense)} \qquad \cos x > \tfrac{1}{2} - \varepsilon \quad \text{(also reversing the sense)}$$

$$-\varepsilon < \tfrac{1}{2} - \cos x \qquad\qquad \tfrac{1}{2} - \cos x < \varepsilon \quad \text{(transposing)}$$

Step 6: $-\varepsilon < \tfrac{1}{2} - \cos x$ and $\tfrac{1}{2} - \cos x < \varepsilon$ is equivalent to

$$\left|\cos x - \tfrac{1}{2}\right| < \varepsilon \qquad\qquad \textbf{Note: } |a - b| = |b - a|$$

We have shown that if $0 < \left|x - \tfrac{\pi}{3}\right| < \delta$, then $\left|\cos x - \tfrac{1}{2}\right| < \varepsilon$

Therefore, $\lim\limits_{x \to \frac{\pi}{3}} \cos x = \tfrac{1}{2}$, and the proof is complete.

In the above proof, for δ, we use the smaller of $\delta_1 = \tfrac{\pi}{3} - \text{Cos}^{-1}\left(\varepsilon + \tfrac{1}{2}\right)$ and

$\delta_2 = \text{Cos}^{-1}\left(\tfrac{1}{2} - \varepsilon\right) - \tfrac{\pi}{3}$ or any number less than the smaller of δ_1 and δ_2.

Let us find some numerical values for δ_1 and δ_2 for $\varepsilon = 0.002$

$\delta_1 = \tfrac{\pi}{3} - \text{Cos}^{-1}\left(\varepsilon + \tfrac{1}{2}\right)$	$\delta_2 = \text{Cos}^{-1}\left(\tfrac{1}{2} - \varepsilon\right) - \tfrac{\pi}{3}$
$= 1.047198 - \text{Cos}^{-1}(0.002 + .5)$	$= \text{Cos}^{-1}(0.5 - 0.002) - 1.047198$
$= 1.047198 - \text{Cos}^{-1}(0.502)$	$= \text{Cos}^{-1}(0.498) - 1.047198$
$= 1.047198 - 1.04488$	$= 1.0495054 - 1.047198$
$= 0.00232$	$= 0.002301$

Lesson 5C Exercises A

State the $\varepsilon-\delta$ definition of the limit of a function.

Prove the following

1. $\lim\limits_{x \to 2} 4x + 1 = 9$

2. $\lim\limits_{x \to 3} 3x + 2 = 11$

3. $\lim\limits_{x \to a} mx + b = ma + b$

4. Which of the following choices for δ can be used in an $\varepsilon - \delta$ proof of

$\lim\limits_{x \to 3} 4x + 1$? (a) $\delta = \frac{\varepsilon}{3}$; **(b)** $\delta = \frac{\varepsilon}{2}$;

(c) $\delta = \varepsilon$ (d) $\delta = \frac{\varepsilon}{5}$

5. Prove that $\lim\limits_{x \to 3} x^2 = 9$

6. Prove that $\lim\limits_{x \to 2} x^2 + 3x = 10$

7. Prove that $\lim\limits_{x \to 7} \dfrac{8}{x - 3} = 2$

8. Prove that $\lim\limits_{x \to 4} \sqrt{x + 5} = 3$

9. Prove that $\lim\limits_{x \to 1} \sqrt{x^2 + 3} = 2$

10. Prove that $\lim\limits_{x \to 24} \sqrt{x + 1} = 5$

11. Prove that $\lim\limits_{x \to 3} \log_2\left(x^2 + 7\right) = 4$

12. Prove that $\lim\limits_{x \to 8} \log_2 x = 3$

13. Prove that $\lim\limits_{x \to \frac{\pi}{4}} \sin x = \dfrac{\sqrt{2}}{2}$

Answer **4.** (d)

Lesson 5C Exercises B

State the $\varepsilon-\delta$ definition of the limit of a function

Prove the following

1. $\lim\limits_{x \to 1} 5x - 2 = 3$

2. $\lim\limits_{x \to 3} 4x + 3 = 15$

3. $\lim\limits_{x \to c} mx + b = mc + b$

4. Which of the following choices for δ can be used in an $\varepsilon - \delta$ proof of

$\lim\limits_{x \to 3} 4x + 1$? {a} $\delta = \frac{\varepsilon}{3}$; (b) $\delta = \frac{\varepsilon}{2}$;

(c) $\delta = \varepsilon$ (d) $\delta = \frac{\varepsilon}{6}$

5. Prove that $\lim\limits_{x \to 2} x^2 = 4$

6. Prove that $\lim\limits_{x \to 3} x^2 - 1 = 8$

7. Prove that $\lim\limits_{x \to 3} \dfrac{10}{x - 1} = 5$

8. Prove that $\lim\limits_{x \to 3} \sqrt{x + 1} = 2$

9. Prove that $\lim\limits_{x \to 3} \sqrt{x^2 + 7} = 4$

10. Prove that $\lim\limits_{x \to 15} \sqrt{x + 1} = 4$

11. Prove that $\lim\limits_{x \to 6} \log_2\left(x^2 - 4\right) = 5$

12. Prove that $\lim\limits_{x \to 8} \log_2 x = 3$

13. Prove that $\lim\limits_{x \to \frac{\pi}{3}} \cos x = \dfrac{1}{2}$

Answer **4.** (d)

CHAPTER 2
Lesson 6
Continuity and Continuous Functions
Introduction: **Continuous and Discontinuous Functions**
Case 1: **Continuity of Polynomial Functions**
Case 2: **Continuity of Rational Functions**
Case 3: **Continuity of Radical Functions**
Case 4: **Continuity of Exponential Functions**
Case 5: **Continuity of Logarithmic Functions**
Case 6: **Continuity of Trigonometric Functions**

Introduction: Continuous and Discontinuous Functions

Precalculus Treatment
We use the concepts of continuous and discontinuous functions in sketching the graphs of functions. The graphs of polynomial, sine, and cosine functions are continuous functions. The graphs of rational functions, the tangent ,the cosecant and cotangent functions as well as the graphs of the hyperbolas are generally discontinuous functions.

Continuous Functions

Graphically, a function is continuous at a point or on an interval if the graph (curve or line) has no breaks, "jumps" or "holes" in it at that point or on that interval. We can also view a continuous function as one whose line or curve can be drawn without lifting the pencil from the paper.

Calculus Definition
A function is **continuous** at the point x_0 if the following three conditions are satisfied simultaneously.

1. $f(x_0)$ is defined;   2. $\lim_{x \to x_0} f(x)$ exists; and **3.** $\lim_{x \to x_0} f(x) = f(x_0)$

Sometimes, we write only $\lim_{x \to x_0} f(x) = f(x_0)$ (A)

for 1, 2, and 3

Another form of $\lim_{x \to x_0} f(x) = f(x_0)$ (A)

is $\lim_{h \to 0} f(x_0 + h) = f(x_0)$ (B)

(which can be obtained by letting $x - x_0 = h$ and $x = x_0 + h$ in (A))

Note that in the definition of a limit of a function $f(x)$ as x approaches the point x_0, the limit of the function is independent of whether or not the function is defined at the point x_0. However, in the definition of continuity at a point x_0, the function must be defined at the point x_0.

Definition: **Continuity from the left and from the right**
A function f **is** continuous from the left at the point c if $\lim_{x \to c^-} f(x) = f(c)$.

A function f is continuous from the right at the point c if $\lim_{x \to c^+} f(x) = f(c)$.

Note that each of the above definitions implies three conditions.

Continuity on a Closed Interval

A function f is continuous on the closed interval $[a, b]$ if f is continuous on (a, b) and continuous from the right at a, and continuous from the left at b.

Discontinuous Functions

Graphically, a function may be discontinuous at a point or on an interval if it has any of the "defective" properties of breaks, jumps, or holes in its curve or line. The breaks represent the excluded values in the domain or range of the function. There are two main types of discontinuities, namely finite discontinuity and infinite discontinuity.

Finite Discontinuity, Holes

A **finite discontinuity** is a discontinuity which can be removed by redefining the function. Sometimes, we call a finite discontinuity removable or temporary discontinuity. (By redefining the function at the point of discontinuity, we bypass or go around the hole")

Hole: A hole is a circle representing a break in a line or curve at a finite discontinuity (Figure).

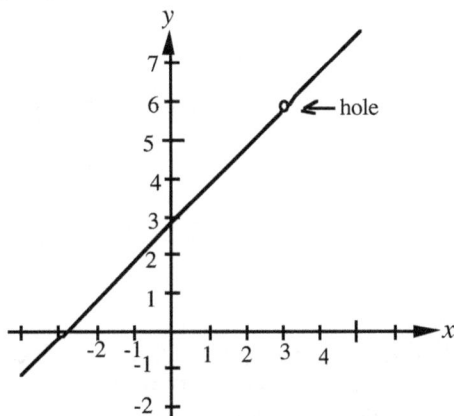

Figure

In a finite discontinuity, if c is a point of removable discontinuity, then $\lim_{x \to c} f(x)$ exists.

Example Observe that in $f(x) = \dfrac{x^2 - 9}{x - 3}$, if $x = 3$, the right side of this equation is indeterminate ($\frac{0}{0}$); and therefore $f(3)$ is not defined. However, if we redefine the value of this function at $x = 3$ as 6, an extended function (a new function), say, $g(x)$ would be continuous at $x = 3$.

Note above that by factoring the numerator and dividing out the common factor $x - 3$, we obtain the function $g(x) = x + 3$, whose limit is 6. The functions

$f(x) = \dfrac{x^2 - 9}{x - 3}$ and $g(x) = x + 3$ are different functions. However, each has the

limit 6. Even though, $f(x) = \dfrac{x^2 - 9}{x - 3}$ is not defined when $x = 3$, $\lim_{x \to 3} \dfrac{x^2 - 9}{x - 3}$

exists and is 6. Therefore the limit of a function at a point a is independent of whether or not the function is defined at the point a.

Jump discontinuity

At a jump discontinuity, both $\lim\limits_{x \to c^-} f(x)$ (left limit) and $\lim\limits_{x \to c^+} f(x)$ (right limit) exist but are not equal to each other, and therefore $\lim\limits_{x \to c} f(x)$ does not exist

.**Example** Given $f(x) = \frac{x}{|x|}$, $\lim\limits_{x \to 0^+} f(x) = 1$, $\lim\limits_{x \to 0^-} f(x) = -1$ and therefore the left and right limits are not equal (Figure below).

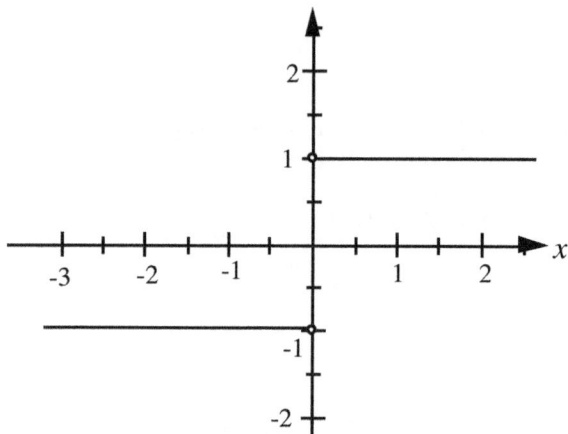

Note:

$x > 0, \ f(x) = \frac{x}{x} = 1$

$x < 0, \ f(x) = \frac{x}{-x} = -1$

Figure: $f(x) = \frac{x}{|x|}$

Infinite (Essential) Discontinuity

An **infinite discontinuity** is a discontinuity (a break in a curve) which cannot be removed by redefining the function. For an infinite discontinuity at a point c, either $\lim\limits_{x \to c^-} f(x)$ or $\lim\limits_{x \to c^+} f(x)$ does not exist.

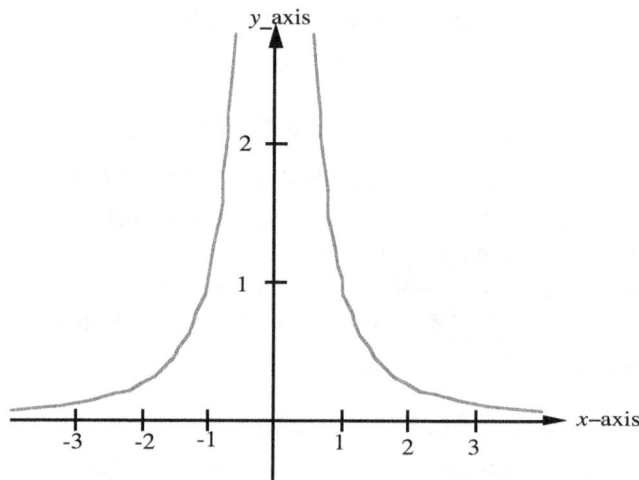

Figure: Function has infinite discontinuity at $x = 0$

Case 1: Continuity of Polynomial Functions

Example 1 $f(x) = x^2 - 3x + 1$

Solution: All polynomial functions are continuous for all real values of the
independent variable.
We may note that the right-hand side of the equation does not involve
denominators or square roots (or even roots) involving the
independent variable.
For a polynomial. $\lim\limits_{x \to x_0} f(x) = f(x_0)$.

Case 2: Continuity of Rational Functions

In calculations, we sometimes meet ratios in which certain values when
substituted for a variable in the denominator make the denominator zero, and
consequently the value of each such ratio (fraction) is undefined.

Examples are **1.**. $f(x) = \dfrac{3x}{2x + 1}$ which is undefined at $x = -\frac{1}{2}$.

2. $f(x) = \dfrac{3x^2 - 7x + 2}{x - 1}$ which is undefined at $x = 1$.

We call such ratios of polynomial functions **rational functions**. A rational
function is therefore the ratio of two polynomial functions.

A **rational function** is **continuous** at all real values of the independent
variable, except at where the values of the independent variable make the
denominator zero.

Example 2 At what values of x is the following function continuous?

Solution $\qquad\qquad f(x) = \dfrac{3x - 2}{x - 1}$

Step 1: Setting the denominator to zero, and solving, $x = 1$

The function is not defined when $x = 1$, and therefore it is continuous at all real
values of x, except at $x = 1$.

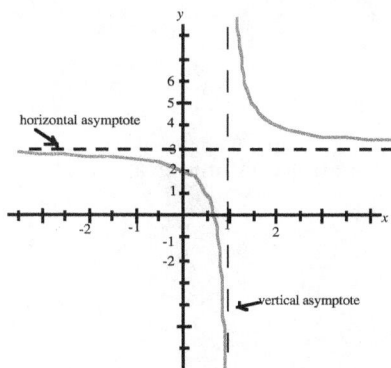

Graph of $f(x) = \dfrac{3x - 2}{x - 1}$; $\qquad$ Graph of $f(x) = \dfrac{2(x - 1)}{(x - 2)(x + 4)}$.

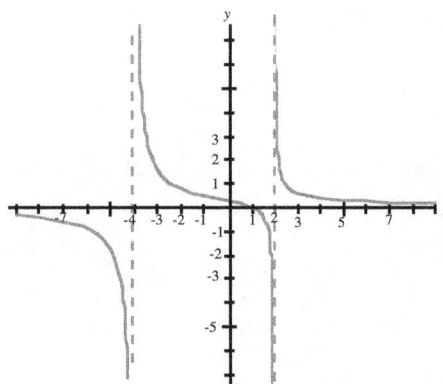

Example 3 For what values of x is the given function continuous?

$$f(x) = \frac{2(x-1)}{(x-2)(x+4)}$$

Solution Step 1: Setting the denominator to zero, $(x-2)(x+4) = 0$

Step 2: Solving for x, $x = 2$, or $x = -4$.

The function is not defined when $x = 2$ and -4 (Figure: previous page) and therefore it is continuous at all real values of x, except at 2 and -4.

Example 4 (a) For what values of x is the given function continuous?

$$f(x) = \frac{1}{x}$$

Solution Setting the denominator to zero, $x = 0$.

(a) The function is not defined when $x = 0$, and therefore it is continuous at all real values of x, except at 0.

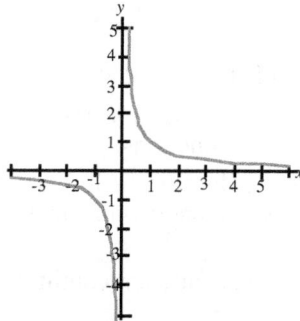

Figure: Graph of $f(x) = \frac{1}{x}$

Example 6 For what values of x is the given function continuous?

$$f(x) = \frac{8}{x^2 - 4}$$

Solution Step 1: Setting the denominator to zero.

$$x^2 - 4 = 0$$
$$(x+2)(x-2) = 0$$

Step 2: Solving for x, $x = 2$, or $x = -2$.

The function is not defined when $x = 2$ and -2 and therefore it is continuous at all real values of x, except at 2 and -2.

Figure: Graph of $f(x) = \frac{8}{x^2 - 4}$:

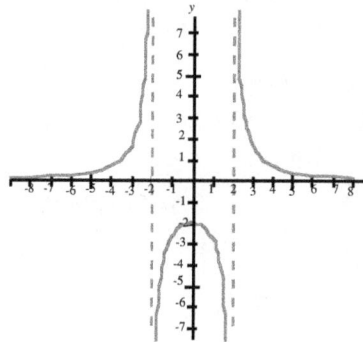

Example 7 For what values of x is the given function continuous?

$$f(x) = \frac{8}{x^2 + 4}$$

Solution   Setting the denominator to zero, and solving, we obtain non-real values. Since we are dealing with real-valued functions, there are **no** excluded values, and $x^2 + 4$ is positive for all real values of x and never zero (the square of any nonzero real number is always positive).

The function is therefore continuous at all real values of x.

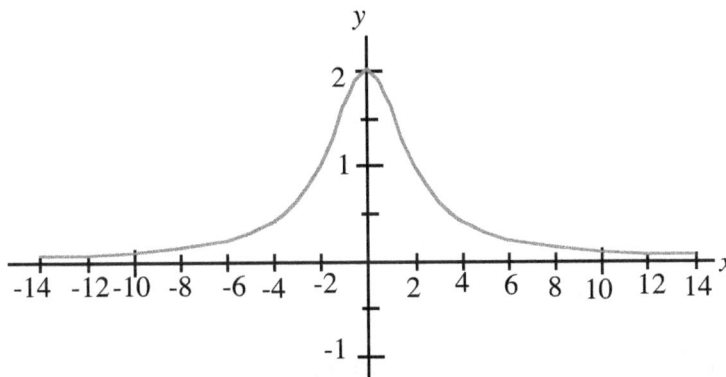

Figure: Graph of $f(x) = \dfrac{8}{x^2 + 4}$

Case 3: Continuity of Radical Functions

Example 10 Determine the continuity of $f(x) = \sqrt{x - 5}$

Solution The domain consists of all real values of x such that $x \geq 5$. The function is not real when $x < 5$.

However, since the right limit, $\lim\limits_{x \to 5^+} \sqrt{x - 5} = 0$, and $f(5) = 0$,

$\lim\limits_{x \to 5^+} \sqrt{x - 5} = f(5)$, $f(x)$ is continuous from the right.

The given function is continuous on $[5, +\infty)$.

or the function is continuous for all real values of x such that $x \geq 5$.

See also p 8-9; and also p.78-79.

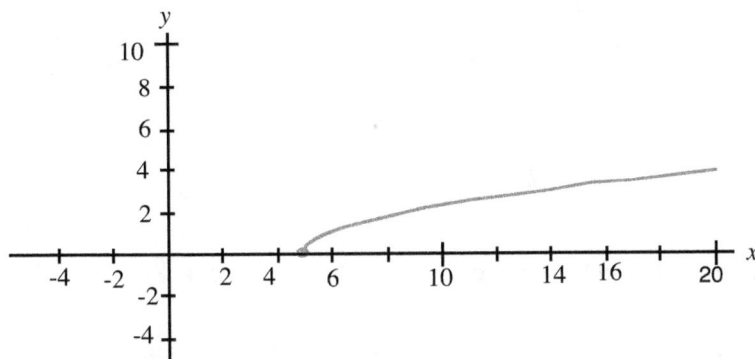

Figure: Graph of $f(x) = \sqrt{x - 5}$

Case 4: Continuity of Exponential Functions

The exponential function $f(x) = b^x$ ($b > 0,\ b \neq 1$.) is continuous.
(that is, for all real x).

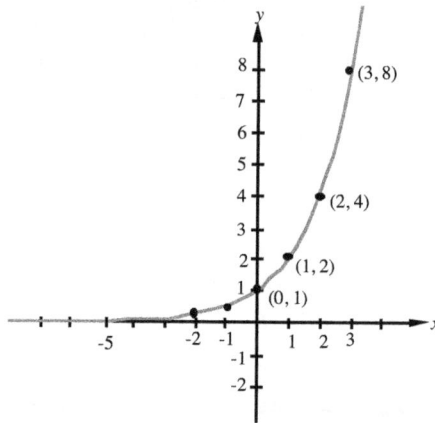

Figure: $f(x) = b^x$

Case 5:: Continuity of Logarithmic Functions

The Logarithmic Function $f(x) = \log_b x$ is continuos for $x > 0$.

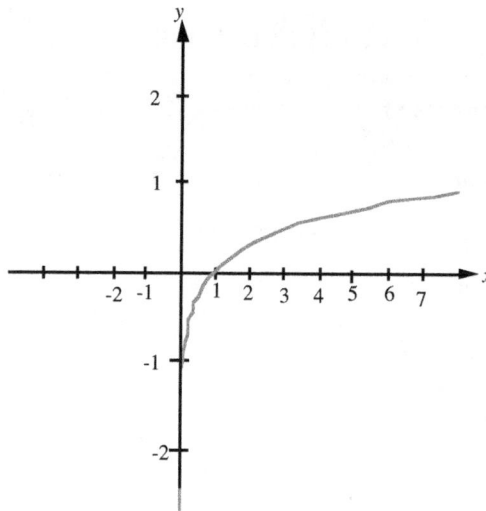

Figure: $f(x) = \log_2 x$

Memorize the above graphs and those in Case 6, because a picture is worth a thousand words.

Case 6: Continuity of Trigonometric Functions

Continuous Trigonometric Functions

The sine, and cosine functions are continuous (that is, for all real x).

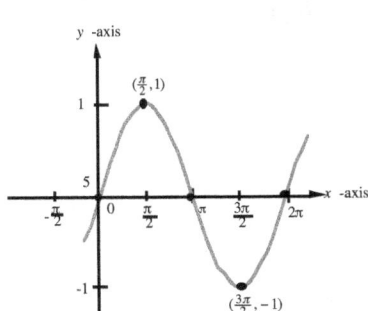

Figure: The graph of $y = \sin x$

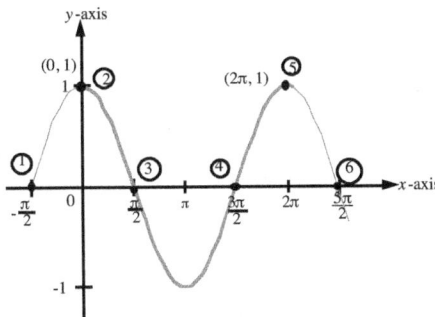

Figure The graph of $y = \cos x$

Generally, Discontinuous Trigonometric Functions

The functions, tangent , secant, cosecant and cotangent are generally discontinuous.

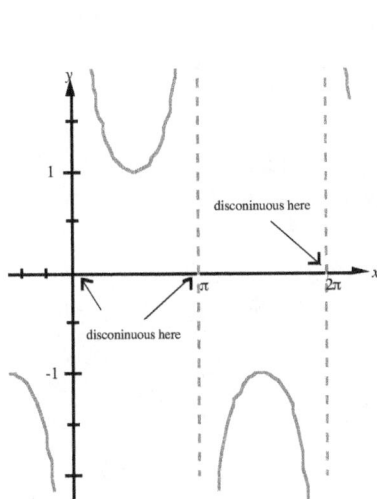

Figure : The graph of $\csc x$

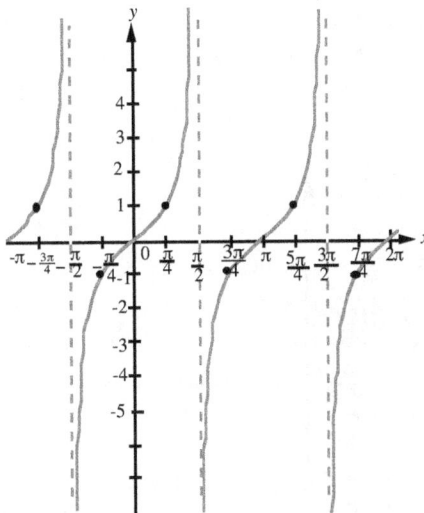

Figure : The graph of $\tan x$

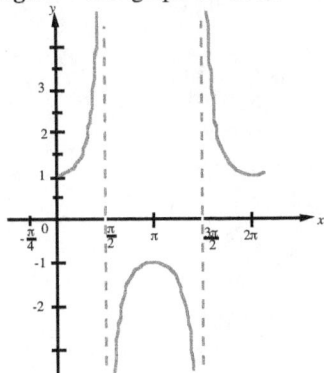

Figure : The graph of $y = \sec x$

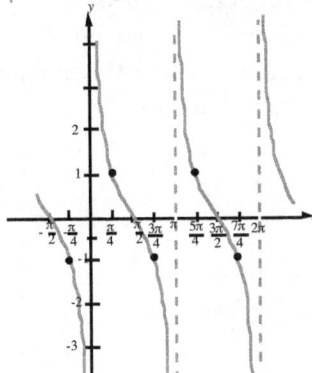

Figure : The graph of $\cot x$

Theorems on Continuity

Intermediate Value Theorems

If f is continuous on the closed interval $a \leq x \leq b$ and k is any number between $f(a)$ and $f(b)$, inclusive, then there exists at least one number c in the interval $a \leq x \leq b$ such that $f(c) = k$. Geometrically, the line $y = k$ must intersect the curve given by $y = f(x)$.

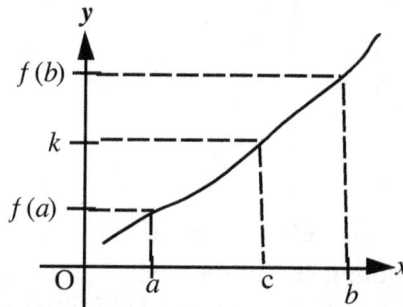

A Consequence Theorem of the Intermediate Value Theorem

If f is continuous on the closed interval $a \leq x \leq b$ and if $f(a)$ and $f(b)$ have opposite signs, then there is at least one solution of the equation $f(x) = 0$ in the open interval $a < x < b$, and the graph of f crosses the x-axis at least once between a and b.

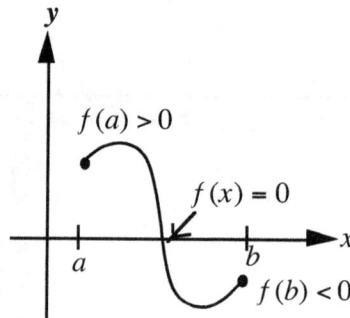

Extreme Value Theorem

If f is continuous on the closed interval $[a, b]$ then f has a maximum at some point in $[a, b]$, and also a minimum at some point in $[a, b]$.

That is, a continuous function defined on a closed interval $[a, b]$ has a maximum and a minimum on this interval. **Note:** $[a, b] = a \leq x \leq b$.

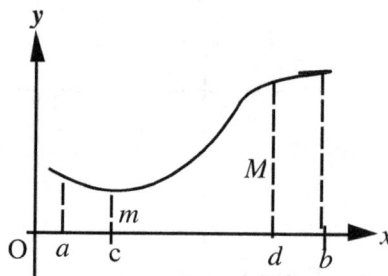

Lesson 6 Exercises A

1. What is meant by a function is continuous at x_0

2, Distinguish between a finite discontinuity essential discontinuity., with examples.

3. Where is $f(x) = x^2 - 3x + 1$ continuous?

4. At what values of x is the following function continuous? $f(x) = \dfrac{3x - 2}{x - 1}$,

5. For what values of x is the given function continuous?

$$f(x) = \frac{2(x - 1)}{(x - 2)(x + 4)}$$

6. For what values of x is the given function continuous? $f(x) = \dfrac{1}{x}$

7. For what values of x is the given function continuous? $f(x) = \dfrac{8}{x^2 - 4}$

8 For what values of x is the given function continuous? $f(x) = \dfrac{8}{x^2 + 4}$

9. Determine the continuity of $f(x) = \sqrt{x - 5}$

10. Determine the continuity of $f(x) = b^x$ $(b > 0, \, b \neq 1.$

11. What is the continuity of the logarithmic function $f(x) = \log_2 x$?

12. Discuss the continuity of the following

(a) $y = \sin x$; **(b)** $y = \cos x$; **(c)** $y = \sec x$ **(d)** $y = \tan x$; **(e)** $y = \csc x$

Answers: 3. All real values of x, **4.** All real values of x, except at $x = 1$
5. All real values of x, except at 2 and -4; **6.** All real values of x, except at 0.
7. Continuous at all real values of x, except at 2 and -2.; **8.** continuous at all real values of x.; **9.** Continuous at all real values of x such that $x \geq 5$. **10.** All real x..

or

3. $\{x \mid x \in R\}$ or $(-\infty, \infty)$. **4.** $\{x \in R \mid x \neq 1\}$ or $(-\infty, 1) \cup (1, \infty)$
5. $\{x \in R \mid x \neq -4, x \neq 2\}$; **6.** $\{x \in R \mid x \neq 0\}$; **7.** $\{x \in R \mid x \neq -2, x \neq 2\}$
8. $\{x \mid x \in R\}$ or $(-\infty, \infty)$; **9.** $\{x \in R \mid x \geq 5\}$;
10. $\{x \mid x \in R\}$ or $(-\infty, \infty)$; **11.** $\{x \in R \mid x > 0\}$.

Lesson 6 Exercises **B**

1. What is meant by a function is continuous at x_0

2, Distinguish between a finite discontinuity essential discontinuity., with examples.

3. Where is $f(x) = x^2 + 2x + 1$ continuous?

4. At what values of x is the following function continuous? $f(x) = \dfrac{x-2}{x+1}$,

5. For what values of x is the given function continuous?

$$f(x) = \frac{3(x-2)}{(x-3)(x+2)}$$

6. For what values of x is the given function continuous? $f(x) = \dfrac{1}{2x}$

7. For what values of x is the given function continuous? $f(x) = \dfrac{2}{x^2 - 1}$

8 For what values of x is the given function continuous? $f(x) = \dfrac{2}{x^2 + 1}$

9. Determine the continuity of $f(x) = \sqrt{x-3}$

10. Determine the continuity of $f(x) = a^x$ ($a > 0$, $a \neq 1$.

11. What is the continuity of the logarithmic function $f(x) = \log_2 x$?

12. Discuss the continuity of the following

(a) $y = \sin x$; **(b)** $y = \cos x$; **(c)** $y = \sec x$ **(d)** $y = \tan x$; **(e)** $y = \csc x$

13. Determine the continuity of $f(x) = \sqrt{3-x}$

Answers: 3. All real values of x, **4.** All real values of x, except at $x = -1$
5. All real values of x, except at -2 and 3; **6.** All real values of x, except at 0.
7. Continuous at all real values of x, except at 1 and -1.; **8.** Continuous at all real values of x.; **9.** Continuous at all real values of x such that $x \geq 3$. **10.** All real x..
13. Continuous at all real values of x such that $x \leq 3$.

or

3. $\{x \mid x \in \mathrm{R}\}$ or $(-\infty, \infty)$. **4.** $\{x \in R \mid x \neq -1\}$ or $(-\infty, -1) \cup (-1, \infty)$
5. $\{x \in R \mid x \neq -2, x \neq 3\}$; **6.** $\{x \in R \mid x \neq 0\}$; **7.** $\{x \in R \mid x \neq -1, x \neq 1\}$
8. $\{x \mid x \in \mathrm{R}\}$ or $(-\infty, \infty)$; **9.** $\{x \in R \mid x \geq 3\}$ or $[3, \infty)$;
10. $\{x \mid x \in \mathrm{R}\}$ or $(-\infty, \infty)$; **11.** $\{x \in R \mid x > 0\}$.
13. $\{x \in R \mid x \leq 3\}$ or $(-\infty, 3]$

CHAPTER 3
Derivatives

Lesson 7: **Derivatives: Definition of the Derivative**
Lesson 8: **Derivatives: Using the Basic Definition**

Lesson 7
Definition of the Derivative

We will derive the basic formulas in this chapter. In the next chapter, we will learn faster methods of performing the differentiation of functions.

Algebraic and Geometric Interpretation of the Derivative

As shown in the figure below, let $P(x, f(x))$ and $Q(x + h, f(x + h))$ be two points on the curve given by $y = f(x)$, where h is a small change in x from P to Q. Also let the secant so formed be $\overline{PQ}$.

Then the **average rate of change or the slope** of the secant line is the ratio

$$\frac{\text{change in } y}{\text{change in } x} = \frac{\Delta y}{\Delta x} = \frac{f(x + h) - f(x)}{h} \qquad (\Delta x = x + h - x = h)$$

Assuming that the point P is fixed, let the point Q move (slide) along the curve so as to approach P; and as Q approaches P, h, the change in x will continuously decrease (shrink) and approach zero. The line $\overleftrightarrow{PQ}$ will reach a limiting position to form the tangent line $\overleftrightarrow{PT}$ at the point $P(x, f(x))$ on the curve

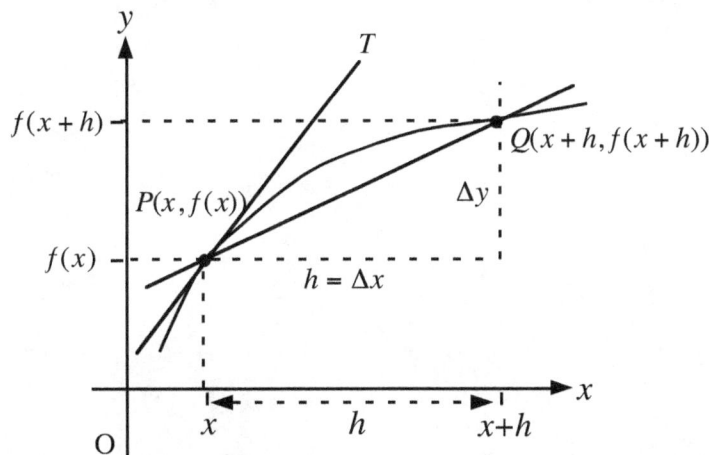

Instantaneous rate of change or the slope of the tangent line is given by

$$\lim_{h \to 0} \frac{f(x + h) - f(x)}{h}, \text{ if this limit exists.}$$

This limit is also known as the **derivative** of $f(x)$ with respect to x at x. We may symbolize the derivative as $\frac{dy}{dx}$, $f'(x)$, y' or $\frac{d}{dx} f(x)$. Each notation has its relative merits. We will use these notations interchangeably.

Using the symbol $f'(x)$ (read " f prime of x"), the **derivative** of $f(x)$ at x in the domain of f is given by

$$f'(x) = \lim_{h \to 0} \frac{f(x+h) - f(x)}{h}, \text{ if it exists.} \qquad (A)$$

Now, if we want the derivative of $f(x)$ at x_0, we will replace x by x_0 in equation (A) to obtain

$$f'(x_0) = \lim_{h \to 0} \frac{f(x_0 + h) - f(x_0)}{h} \quad \text{if it exists.}$$

If the derivative of $f(x)$ exists at x_0, then we say that f is differentiable at x_0.

Lesson 7 Exercises

1. Distinguish between the average rate of change and the instantaneous rate of change.
2. Give both the algebraic and geometric interpretations of the derivative.

Lesson 8
Derivative Using the Basic Definition

Example 1: Find $f'(x)$ given $f(x) = x^2 - 3x + 1$

Solution: Step 1: $\quad f'(x) = \lim\limits_{h \to 0} \dfrac{f(x+h) - f(x)}{h} \quad$ (definition)

Step 2: $\quad = \lim\limits_{h \to 0} \dfrac{\left[(x+h)^2 - 3(x+h) + 1\right] - \left[x^2 - 3x + 1\right]}{h}$

$\quad = \lim\limits_{h \to 0} \dfrac{x^2 + 2xh + h^2 - 3x - 3h + 1 - x^2 + 3x - 1}{h}$

$\quad = \lim\limits_{h \to 0} \dfrac{2xh + h^2 - 3h}{h}$

Step 3: $\quad = \lim\limits_{h \to 0} 2x + h - 3$

Step 4: $\quad f'(x) = 2x - 3$

Example 2: Find $f'(x_0)$ if $f(x) = x^2 - 3x + 1$ (That is find $f'(x)$, if $x = x_0$)

Solution: We could replace x by x_0 in all the steps in Example 1. However, it is more convenient to substitute x_0 in the last step (Step 4) of Example 1.
Then we obtain $f'(x_0) = 2x_0 - 3$.

Example 3: Find $f'(5)$ if $f(x) = x^2 - 3x + 1$ (That is, find $f'(x)$, if $x = 5$)

Solution: We could replace x by 5 in all the steps in Example 1. However, it is more convenient to substitute 5 in the last step (Step 4) of Example 1.
Then we obtain $\quad f'(5) = 2(5) - 3 \qquad (f'(x) = 2x - 3)$

$\qquad\qquad\qquad = 10 - 3$

$\qquad\quad f'(5) = 7$

Example 4: Find $f'(-2)$ if $f(x) = 2x^2 + 5x - 3$.

In Method 1, we substitute -2 in last step. In **Method 2**, we substitute -2 in the first Step 1 (in the basic definition).

Method 1 $\qquad f'(x) = \lim\limits_{h \to 0} \dfrac{f(x+h) - f(x)}{h}$

$\qquad\qquad = \lim\limits_{h \to 0} \dfrac{\left[2(x+h)^2 + 5(x+h) - 3\right] - \left[2x^2 + 5x - 3\right]}{h}$

$\qquad\qquad = \lim\limits_{h \to 0} \dfrac{2x^2 + 4hx + 2h^2 + 5x + 5h - 3 - 2x^2 - 5x + 3}{h}$

$\qquad\qquad = \lim\limits_{h \to 0} \dfrac{4hx + 2h^2 + 5h}{h}$

$\qquad\qquad = \lim\limits_{h \to 0} 4x + 2h + 5$

$\qquad f'(x) = 4x + 5$

$\qquad f'(-2) = 4(-2) + 5$

$\qquad\qquad = -8 + 5$

$\qquad\qquad = -3$.

Method 2 Substituting -2 for x from the beginning $(f(x) = 2x^2 + 5x - 3)$

Solution: $f'(x) = \lim_{h \to 0} \dfrac{f(x + h) - f(x)}{h}$

$$f'(-2) = \lim_{h \to 0} \dfrac{\left[2(-2 + h)^2 + 5(-2 + h) - 3 \right] - \left[2(-2)^2 + 5(-2) - 3 \right]}{h}$$

$$= \lim_{h \to 0} \dfrac{8 - 8h + 2h^2 - 10 + 5h - 3 - (8 - 13)}{h}$$

$$= \lim_{h \to 0} \dfrac{-3h + 2h^2}{h}$$

$$= \lim_{h \to 0} -3 + 2h$$

$$f'(-2) = -3.$$

Other Forms of the Derivative Definition

Form 1: $f'(x_1) = \lim_{\Delta x \to 0} \dfrac{f(x_1 + \Delta x) - f(x_1)}{\Delta x}$

$f'(x) = \lim_{\Delta x \to 0} \dfrac{f(x + \Delta x) - f(x)}{\Delta x}$

Form 2: $f'(x) = \lim_{h \to 0} \dfrac{f(x + h) - f(x)}{h}$

$f'(x_1) = \lim_{h \to 0} \dfrac{f(x_1 + h) - f(x_1)}{h}$

$f'(a) = \lim_{h \to 0} \dfrac{f(a + h) - f(a)}{h}$

Form 3: Using the two points $P_1(x_1, f(x_1))$ and $P_2(x_2, f(x_2))$
(where from elementary math, $y_1 = f(x_1)$ $y_2 = f(x_2)$)

$$f'(x_1) = \lim_{x_2 \to x_1} \dfrac{f(x_2) - f(x_1)}{x_2 - x_1} = f'(x_1) = \lim_{x_2 - x_1 \to 0} \dfrac{f(x_2) - f(x_1)}{x_2 - x_1}$$

Form 4: $f'(x_1) = \lim_{x \to x_1} \dfrac{f(x) - f(x_1)}{x - x_1}$

Comparison of Definitions:
Limit, Continuity, and Derivative

Limit	Continuity	Derivative
$\lim_{x \to a} f(x) = L$	$\lim_{x \to a} f(x) = f(a)$	$f'(a) = \lim_{h \to 0} \dfrac{f(a + h) - f(a)}{h}$

More On Proofs
Direct Proofs

In constructing proofs, we logically combine axioms, definitions, and already proved theorems.

Example 1 (This is a repetition from Lesson 5A)

Prove that if $f(x)$ has a derivative at x_0, then $f(x)$ is continuous at x_0.

Given: $f(x)$ has a derivative at x_0 (hypothesis)

Required: To prove that $f(x)$ is continuous at x_0 (conclusion)

Plan: If $\lim\limits_{x \to x_0} f(x) = f(x_0)$, then $f(x)$ is continuous at x_0 (Continuity definition)

The proof would be complete after showing that $\lim\limits_{x \to x_0} f(x) = f(x_0)$.

Proof:

Statement	Reason
1. $f(x)$ has a derivative at x_0	1. Given
2. $\lim\limits_{h \to 0} \dfrac{f(x_0 + h) - f(x_0)}{h}$ exists	2. Definition of derivative at x_0
3. $f(x_0 + h) = f(x_0 + h)$	3. A quantity is equal to itself
4. $f(x_0 + h) = f(x_0 + h) - f(x_0) + f(x_0)$	5. Adding zero to right-hand side
5. $\dfrac{f(x_0 + h)}{h} = \dfrac{f(x_0 + h) - f(x_0)}{h} + \dfrac{f(x_0)}{h}$	(dividing by h)
7. $h \cdot \dfrac{f(x_0 + h)}{h} = h \cdot \dfrac{f(x_0 + h) - f(x_0)}{h} + h \cdot \dfrac{f(x_0)}{h}$	(multiplying by h)
8. $f(x_0 + h) = h \cdot \dfrac{f(x_0 + h) - f(x_0)}{h} + f(x_0)$	(canceling some h's)
9. $\lim\limits_{h \to 0} f(x_0 + h) = \lim\limits_{h \to 0} h \cdot \lim\limits_{h \to 0} \dfrac{f(x_0 + h) - f(x_0)}{h} + \lim\limits_{h \to 0} f(x_0)$	(taking limits)
10. $\lim\limits_{h \to 0} f(x_0 + h) = 0 \cdot f'(x_0) + f(x_0)$	($\lim\limits_{h \to 0} \dfrac{f(x_0 + h) - f(x_0)}{h} = f'(x_0)$)
11. $\lim\limits_{h \to 0} f(x_0 + h) = f(x_0)$	

12. Let $h = x - x_0$ or let $x = x_0 + h$

Then $\lim\limits_{h \to 0} f(x_0 + h) = f(x_0)$ becomes

$\lim\limits_{x - x_0 \to 0} f(x_0 + x - x_0) = f(x_0)$ and

$\lim\limits_{x \to x_0} f(x) = f(x_0)$ (definition of continuity)

The proof is complete.

Some Theorems on Derivatives

Mean Value Theorem (Law of the Mean)

If f is differentiable on the open interval $a < x < b$, and continuous on the closed interval $a \leq x \leq b$,. then there exists a point c in the open interval $a < x < b$ such that $f'(c) = \dfrac{f(b) - f(a)}{b - a}$.

Geometrically, the Mean Value Theorem means that if P_1 and P_2 are two points on the graph of a continuous curve which has a tangent at every point between these two points, then there is at least one point on the curve between P_1 and P_2 at which the slope of the curve equals the slope of the line connecting the points P_1 and P_2. The tangent to the curve is parallel to the line connecting P_1 and P_2.

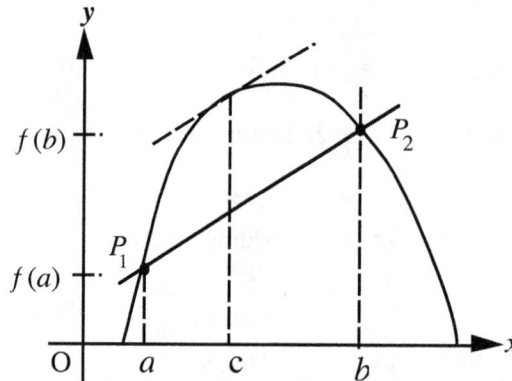

When $f(a) = f(b) = 0$ in the Mean Value Theorem, we obtain Rolle's Theorem.

Rolle's Theorem

If f is differentiable on the open interval $a < x < b$, and continuous on closed interval $a \leq x \leq b$. and if $f(a) = f(b) = 0$, then there exists a point c in the open interval such that $f'(c) = 0$.

Geometrically, Rolle's Theorem means that if the graph of a continuous function intersects the x-axis at $x = a$ and $x = b$, and has tangent at every point between a and b, then there is at least one point on the graph where the tangent line is horizontal. (that is the tangent is parallel to the x-axis).

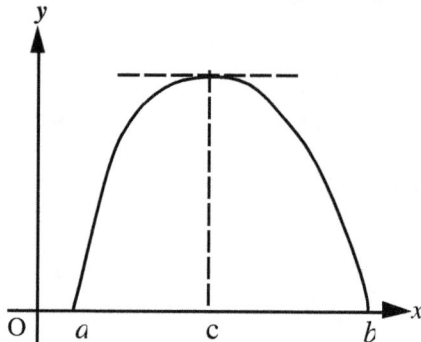

Note above the conditions in the Mean Value Theorem and in Rolle's Theorem:

Continuity on a **closed** interval but **differentiability** on an **open** interval.
Note: Differentiability here pertains to having derivatives.

Generalized Mean Value Theorem (Extended Law of the Mean)

If $f(x)$ and $g(x)$ are differentiable on the open interval $a < x < b$, and continuous on closed interval $a \le x \le b$, then there exists a point c in the open interval $a < x < b$ such that $\dfrac{f(b)-f(a)}{g(b)-g(a)} = \dfrac{f'(c)}{g'(c)}$, assuming $g(b) \ne g(a)$ **and** $f'(x)$ and $g'(x)$ are not simultaneously zero. Note that when $g(x) = x$, we obtain the Mean Value Theorem

Theorem See page 39 or 93 for the proof of the following theorem

If $f(x)$ has a derivative at x_0, then $f(x)$ is continuous at x_0.

Lesson 8 Exercises **A**

1. Using the basic definition, find $f'(x)$ given $f(x) = x^2 - 3x + 1$

2. Find $f'(x_0)$ if $f(x) = x^2 - 3x + 1$.

3, Find $f'(-2)$ if $f(x) = 2x^2 + 5x - 3$.

4. Give three forms of the derivative definition.

5. Compare and contrast the definitions of the limit,, continuity, and the derivative.

Answers: **2.** $f'(x_0) = 2x_0 - 3$; **3.** -3;.

Lesson 8 Exercises **B**

1. Using the basic definition find $f'(x)$ given $f(x) = x^3 - 4x + 1$

2. Find $f'(x_0)$ if $f(x) = x^3 - 4x + 1$

3, Find $f'(-3)$ if $f(x) = x^2 + 4x - 3$

4. Give three forms of the derivative definition

5. Compare and contrast the definitions of the limit, continuity, and the derivative.

Answers: 1. $3x^2 - 4$; **2.** $f'(x_0) = 3x_0^2 - 4$; **3.** -2.

CHAPTER 4

Differentiation Techniques I

Lesson 9: **Differentiation of Polynomial Functions**
Lesson 10: **Implicit Differentiation**
Lesson 11: **Differentiation of Rational Functions**
Lesson 12: **Differentiation of Radical Functions**

Types of Functions

In calculus we perform differentiation and integration on functions. We also cover the applications of differentiation and integration in various fields. The basic functions we will differentiate or integrate include the following:

A. Polynomial Functions

Examples **1.** $f(x) = x^2$; **2.** $y = x^5$; **3.** $y = x^4$; **4.** $y = x^6$

4. $f(x) = x^2 - 3x + 1$; **5.** $f(x) = 2x + 3$; **6.** $f(x) = 3x^2 + 8x - 1$;;

7. $f(x) = x^4 + x^2 + 5$;; **8.** $f(x) = -x^5 + 2$; **9.** $f(x) = (x+1)^2$; **10.** $y = (x+3)^2$;

11. $f(x) = (x-1)^2(x+3)$

B. Rational functions

Examples   **1.** $f(x) = \frac{1}{x}$ **2.** $f(x) = \frac{3x-2}{x-1}$; 3. $f(x) = \frac{x^3 + x^2 + 2}{x^2 - 16}$;

4. $f(x) = \frac{8}{x^2 - 4}$.

C.. Radical Functions

Examples   **1.** $f(x) = \sqrt{x-5}$;   **2.** $f(x) = \frac{\sqrt{x+6}}{x-1}$.

D. Exponential and Logarithmic Functions (See Chapter 5)

Examples: **1.** $f(x) = 2^x$; **2.** $f(x) = 2^{-x}$; **3.** $y = \ln x$; **4.** $y = \log_2 x$; **5.** $f(x) = 4^x$;

6. $f(x) = 4^{-x}$; **7.** $y = e^x$; **8.** $f(x) = e^{-x}$; **9.** $y = 10^x$; $y = 10^{-x}$;

E. Trigonometric Functions

Examples:

1. $y = \sin x$; **2.** $y = \cos x$; **3.** $y = \tan x$; **4.** $y = \csc x$; **5.** $y = \sec x$;

6. $y = \cot x$; **7.** $y = \sin 3x$; **8.** $y = 4\sin(2x + \pi)$; **9.** $f(x) = 3\tan\left(x - \frac{\pi}{4}\right)$

F. Inverse Trigonometric Functions

Examples   **1.** $y = \text{Arcsin}\, x$ or $\text{Sin}^{-1}x$;   **2.** $y = \text{Arccos}\, x$ or $\text{Cos}^{-1}x$;

3. $y = \text{Arctan}\, x$ or $\text{Tan}^{-1}x$; **4.** $y = \text{Arccot}\, x$ or $\text{Cot}^{-1}x$;

5. $y = \text{Arcsec}\, x$ or $\text{Sec}^{-1}x$;   **6.** $y = \text{Arccsc}\, x$ or $\text{Csc}^{-1}x$

We can always apply the basic definition $f'(x) = \lim_{h \to 0} \frac{f(x+h) - f(x)}{h}$ to find the derivatives of functions. However, there are short-cut rules for finding derivatives of functions. These rules can be justified using the above basic definition. For the rest of this chapter, we state and apply these rules.

Lesson 9
Differentiation of Polynomial Functions

For a function $y = f(x)$, we symbolize the derivative of y with respect to x as either y' or $f'(x)$; or as $\dfrac{dy}{dx}$, $\dfrac{d}{dx} f(x)$. Each notation has its relative merits. We will use these notations interchangeably.

For polynomial, radical and some rational functions we will use the Power Rule for differentiation.

Power Rule: If $y = x^n$, where n is a constant, then

$$\frac{dy}{dx} = nx^{n-1} \ \ (\text{same as } y' = nx^{n-1} \ \text{ or } \ f'(x) = nx^{n-1})$$

Case 1: Derivative of a monomial

Example 1 If $f(x) = x^2$, find $f'(x)$

Solution We apply $f'(x) = nx^{n-1}$.

$f'(x) = 2x^{2-1}$ (Exponent multiplies the base and the exponent decreases by 1)

$\qquad = 2x$

Example 2. If $y = x^4$, find $\dfrac{dy}{dx}$

Solution We apply $\dfrac{dy}{dx} = nx^{n-1}$.

$\dfrac{dy}{dx} = 4x^{4-1}$ (Exponent multiplies the base and the exponent decreases by 1)

$\qquad = 4x^3$

Example 3. If $y = x^5$, find $\dfrac{dy}{dx}$

Solution $\dfrac{dy}{dx} = 5x^{5-1}$(Exponent multiplies the base and the exponent decreases by 1)

$\qquad = 5x^4$

Example 4 If $y = x^6$, find $\dfrac{dy}{dx}$

Solution We apply $\dfrac{dy}{dx} = nx^{n-1}$.

$\dfrac{dy}{dx} = 6x^{6-1}$ (Exponent multiplies the base and the exponent decreases by 1)

$\qquad = 6x^5$

For a **monomial constant** c (such as 4, or 9):
If $y = f(x) = c$ for all x, and where c is a constant, then

$$\frac{dy}{dx} = 0 \quad \text{(the derivative of a constant is zero)}$$

(We could consider for example the monomial constant 4 as $4x^0$. Then if $f(x) = 4x^0$, then $f'(x) = 0(4)x^{0-1} = 0x^{-1} = 0$

Example 5 If $f(x) = -6$, find $f'(x)$.

Solution $f'(x) = 0$.

Note the steps in the above differentiation process:

Step 1: The exponent (in x^n) **multiplies** the base.

Step 2: The exponent (in x^n) **decreases** by 1.

In the antidifferentiation process (integration, chapter 10), we will reverse the steps in differentiation.

Case 2: Derivative of the product of a constant and a function

Using functional notation: $(cf)' = cf'$ or

Using $\frac{d}{dx}$-notation: $\qquad \frac{d}{dx}(cf) = c\frac{df}{dx}$ (c is any constant)

Example 6 If $f(x) = 8x^2$, find $f'(x)$
Solution

$$f'(x) = 8\frac{d}{dx}(x^2)$$
$$= 8(2x)$$
$$= 16x$$

Case 3: Derivative of a Polynomial function

In this case, apply the differentiation process to each term (monomial)

Sum Rule: If $f(x)$ and $g(x)$ are differentiable on an interval,

$$\frac{d}{dx}(f+g) = \frac{d}{dx}(f) + \frac{d}{dx}(g)$$

Example 7 If $f(x) = x^2 - 3x + 1$; find $f'(x)$
Solution

$$f'(x) = \frac{d}{dx}(x^2) - \frac{d}{dx}(3x) + \frac{d}{dx}(1)$$

$$f'(x) = 2x - 3 + 0$$
$$= 2x - 3 \quad \text{(see also Example 1 of Lesson 8)}$$

Example 8 If $f(x) = x^4 + x^2 + 5$, find $f'(x)$
Solution

$$f'(x) = 4x^3 + 2x + 0$$
$$= 4x^3 + 2x$$

Example 9 If $f(x) = x^4 + x^2 + 12$, find $f'(x)$
Solution $\qquad f'(x) = 4x^3 + 2x + 0$
$$= 4x^3 + 2x$$

(Note: Examples 8 & 9 have the same derivative. In Lesson 28, we will see more of this.)

Product Rule: $\frac{d}{dx}(fg) = f\frac{d}{dx}(g) + g\frac{d}{dx}(f)$ or

$$(fg)' = f \bullet g' + g \bullet f'$$

In words, the **derivative of a product of two functions** is equal to the product of the first function and the derivative of the second function **plus** the product of the second function and the derivative of the first function.

Example 10   If $f(x) = (x - 1)(x^2 + 3)$, find $f'(x)$
Solution
Method 1

$$f'(x) = (x - 1)(x^2 + 3)$$

$$= (x - 1)\frac{d}{dx}(x^2 + 3) + (x^2 + 3)\frac{d}{dx}(x - 1)$$

$$= (x - 1)(2x) + (x^2 + 3)(1)$$

$$= 2x^2 - 2x + x^2 + 3$$

$$= 3x^2 - 2x + 3$$

Note above that we could also have multiplied to simplify the expression and then differentiate, as in Method 2 below.
Method 2

$$f(x) = (x - 1)(x^2 + 3)$$

$$= x^3 + 3x - x^2 - 3$$

$$= x^3 - x^2 + 3x - 3$$

Now, we differentiate; then

$f'(x) = 3x^2 - 2x + 3$, which is the same derivative as before.

Case 4: Derivative of a Composite Function

(Function "inside" another function)

Composite functions are functions of the form $f[g(x)]$, where $g(x)$ is the inside function.

Examples of composite functions are   **1.** $f(x) = (x + 1)^2$ and **2.** $f(x) = \sqrt{x + 3}$
(Of course, Example **2** is a **not** a polynomial function, but a radical function)

We will apply the chain rule to differentiate composite functions.

Version 1 of the chain rule (Leibniz's notation)
If $y = f(u)$ and $u = g(x)$, then

$$\frac{dy}{dx} = \frac{dy}{du} \cdot \frac{du}{dx}$$

Note: If u is an expression in terms of x. and $y = u^n$, then

$$\frac{dy}{dx} = \frac{dy}{du} \cdot \frac{du}{dx}$$

$$= nu^{n-1}\frac{du}{dx}$$

Version 2 of the chain rule (Lagrange's notation) Also, **power chain rule**

$$\frac{d}{dx}[f(g(x))] = f'[g(x)] \cdot g'(x)$$

$\frac{d}{dx}[f(g(x))] =$ The derivative of the outside function multiplied by the derivative

of the inside function.<----This notion speeds-up the writing.

Example 12 If $y = (x^2 + 1)^3$, find $\frac{dy}{dx}$.

Solution

Approach I: Using version 1 of the chain rule:

Let $u = x^2 + 1$, Then

$$y = u^3$$

$$\frac{du}{dx} = 2x$$

$$\frac{dy}{du} = 3u^2 = 3(x^2 + 1)^2$$

Substituting for $\frac{dy}{du}$ and $\frac{du}{dx}$ in

$$\frac{dy}{dx} = \frac{dy}{du} \cdot \frac{du}{dx}, \text{ we obtain}$$

$$\frac{dy}{dx} = 3(x^2 + 1)^2 \cdot (2x)$$

$$= 6x(x^2 + 1)^2$$

Approach 2: Using version 2 of the chain rule: $\frac{d}{dx}[f(g(x))] = f'[g(x)] \cdot g'(x)$

$\frac{d}{dx}[f(g(x))] = f'[g(x)] \cdot g'(x)$	<-- (Lagrange's notation)
$\frac{d}{dx}(x^2 + 1)^3 = 3(x^2 + 1)^2(2x)$	<--(Derivative of the outside function multiplied by the derivative of the inside function)
$\qquad = 6x(x^2 + 1)^2$	

Chain rule for multiple substitutions

If $y = f(v)$, $v = g(u)$, and $u = h(x)$, then $\frac{dy}{dx} = \frac{dy}{dv} \cdot \frac{dv}{du} \cdot \frac{du}{dx}$

Example 13 If $y = \sin^3(x^4)$, find $\frac{dy}{dx}$, given that the derivative of $\sin u$ is $\cos u$.

Step 1: Let $u = x^4$ (1)	**Step 2:**
Then $y = \sin^3 u$, and	From (2) $\frac{dv}{du} = \cos u$
Let $v = \sin u$ (2)	$\qquad = \cos(x^4)$ (in x)
Then $y = v^3$	From (1). $\frac{du}{dx} = 4x^3$
$\frac{dy}{dv} = 3v^2$ (in terms of v)	$\frac{dy}{dx} = \frac{dy}{dv}\frac{dv}{du}\frac{du}{dx}$
$= 3(\sin u)^2$ (in terms of u)	$= 3\sin^2(x^4) \bullet \cos(x^4) \bullet 4x^3$
$= 3\sin^2 u$	$= 12x^3 \sin^2(x^4)\cos(x^4)$.
$= 3\sin^2(x^4)$ (in terms of x)	

Some more formulas

1. $\boxed{\dfrac{dy}{dx} = \dfrac{1}{\frac{dx}{dy}}}$; 2. $\boxed{\dfrac{dy}{dx} = \dfrac{dy}{du} \Big/ \dfrac{dx}{du}}$ or $\dfrac{dy}{dx} = \dfrac{\frac{dy}{du}}{\frac{dx}{du}}$

Lesson 9 Exercises A

1. If $y = x^n$ then $\dfrac{dy}{dx} = ?$

2. If $f(x) = x^2$, find $f'(x)$

3. If $y = x^4$, find $\dfrac{dy}{dx}$

4. If $y = x^5$, find $\dfrac{dy}{dx}$

5. If $y = x^6$, find $\dfrac{dy}{dx}$

6.. If $y = f(x) = c$ for all x, and where c is a constant, find $\dfrac{dy}{dx}$

7. If $f(x) = -6$, find $f'(x)$

8. If $f(x) = 8x^2$, find $f'(x)$

9. If $f(x) = x^2 - 3x + 1$; find $f'(x)$

10. If $f(x) = x^4 + x^2 + 5$, find $f'(x)$

11.. If $f(x) = x^4 + x^2 + 12$, find $f'(x)$

12.. If $f(x) = (x-1)(x^2 + 3)$, find $f'(x)$

13. If $y = (x^2 + 1)^3$, find $\dfrac{dy}{dx}$

Answers: 1. nx^{n-1}; **2.** $2x$; **3.** $4x^3$; **4.** $5x^4$; **5.** $6x^5$; **6.** 0; **7.** 0; **8** $16x$; **9.** $2x - 3$; **10.** $4x^3 + 2x$; **11.** $4x^3 + 2x$; ; **12.** $3x^2 - 2x + 3$; **13.** $6x(x^2 + 1)^2$

Lesson 9 Exercises B

Find $\dfrac{dy}{dx}$

1. If $y = x^4 - 5x^2 + 6x - 3$

2 If $y = 5x^3 - 2x^2 + 7x - 2$

3. If $y = \frac{1}{4}x^4 - x^2$,

4. If $y = \frac{4}{3}\pi x^3$,

5. If $y = x^4 - 6 + x^{-2} + 7x^{-3}$,

6. If $y = f(x) = c$ for all x, and where c is a constant

7. If $f(x) = 7$, find $f'(x)$

8 If $g(x) = 9x^2 + 4x$, find $g'(x)$

9. If $f(x) = x^4 - 6x + 2$; find $f'(x)$

10. If $f(x) = x^6 + x^2 + 5x$, find $f'(x)$

11. If $f(x) = x^4 + x^2 + 12$, find $f'(x)$

12. If $f(x) = (3x - 1)(x^2 + 4x)$, find $f'(x)$

13. If $y = (5x^2 + 2)^3$, find $\dfrac{dy}{dx}$

14 $g(x) = (2x^3 + 6x^2 + 5)^4$, find $g'(x)$

15. $h(z) = (2z^3 - 6z^2 + 5)^{-4}$, find $h'(z)$

16. If $y = \sqrt{3x}$, find $\dfrac{dy}{dx}$

17.. If $y = \pi^5$, find $\dfrac{dy}{dx}$

18. Find $\dfrac{d}{dx}\left[7x^3\right]$

Answers: 1. $4x^3 - 10x + 6$; **2.** $15x^2 - 4x + 7$; **3.** $x^3 - 2x$; **4.** $4\pi x^2$; **5.** $4x^3 - 2x^{-3} - 21x^{-4}$; **6** 0; **7.** 0; **8** $18x + 4$; **9.** $4x^3 - 6$; **10.** $6x^5 + 2x + 5$; **11.** $4x^3 + 2x$; **12.** $9x^2 + 22x - 4$; **13.** $30x(5x^2 + 2)^2$; **14.** $4(2x^3 + 6x^2 + 5)^3(6x^2 + 12x)$; **15.** $-4(2z^3 - 6z^2 + 5)^{-5}(6z^2 - 12z)$; **16.** $\dfrac{\sqrt{3x}}{2x}$; **17.** 0 ; **18.** $21x^2$

Lesson 10
Implicit Differentiation

Introduction

In Lesson 9, each of the equations we differentiated had been solved for the dependent variable, say y. For example, given $y = (x^2 + 1)^3$ (A)

we found $\dfrac{dy}{dx} = 3(x^2 + 1)^2 \cdot (2x) = 6x(x^2 + 1)^2$ (using the power chain rule).

Observe that equation (A) has been solved for y. Sometimes a given equation might not have been solved for the dependent variable, say y, and solving for y might be tedious or impossible. For such equations we can use a process called **implicit differentiation**, assuming that y is a differentiable function of x. We differentiate first before solving for say, $\dfrac{dy}{dx}$. In the differentiation, we usually apply the **chain rule** to some of the terms of the implicitly defined equations.

For an equation such as $4x^2 + 9y^2 = 36$, which is not a function as is, if we split the graph of this equation into two parts, namely, the upper arc of the ellipse and the lower arc, each arc passes the vertical line test, and thus we can consider two functions.

Notwithstanding the above formal description of implicit functions, for communication purposes, we will further distinguish between how we apply the chain rule, namely the **explicit** application (as on p.99-100), and the **implicit** application. We will also distinguish between some possible cases and some types of terms.

Case 1: y is a function of x; there is an explicit equation relating y and x; the differentiation is with respect to x, and the only variable in the expression to be differentiated is x. In this case, we agree that we are applying the chain rule **explicitly** and we use both Leibniz's notation and the implication that the derivative of a composite function equals the derivative of the outside function multiplied by the derivative of the inside function.

Example 1 Given $y = (x^3 + 2)^4$, find $\dfrac{dy}{dx}$ (**Explicit** application)

Approach 1 Leibniz's notation ("local")	**Approach 2** ("Express"). ("Inside-Outside function" approach
Let $u = x^3 + 2$. Then $y = u^4$; $\dfrac{du}{dx} = 3x^2$	$\dfrac{dy}{dx} = 4(x^3 + 2)^{4-1} \cdot (3x^2)$
$\dfrac{dy}{du} = 4u^3 = 4(x^3 + 2)^3$	$= 12x^2(x^3 + 2)^3$
$y = f(u),\ u = g(x)$	**Note:** $\dfrac{d}{dx}(y) = \dfrac{dy}{dx}$
$\dfrac{dy}{dx} = \dfrac{dy}{du}\dfrac{du}{dx}$, and substituting,	Also, $\dfrac{dy}{dx} = f'[g(x)] \cdot g'(x)$
$\dfrac{dy}{dx} = 4(x^3 + 2)^3 \cdot (3x^2)\ = 12x^2(x^3 + 2)^3$	(Lagrange's notation)

Case 2: z is a function of y, y is a function of x; there is an equation relating z and y; no equation relating y and x; and the differentiation is with respect to x. (Or there is an equation relating y and x; no equation relating z and y; and the differentiation is with respect to y). In this case, we agree that we are applying the chain rule **implicitly**; and we use Leibniz's notation as well as Lagrange's approach. Informally, anytime you differentiate y, multiply by dy/dx.

Example 2 Find $\dfrac{d}{dx}(y^2)$ (**Implicit** application of the chain rule)

Approach 1 Let $z = y^2$; $z = f(y)$	**Approach 2.** ("Express")
$y = g(x)$; Then $\dfrac{dz}{dy} = 2y$	$\dfrac{d}{dx}(y^2) = \dfrac{d}{dx}(y)^2 = 2y\dfrac{dy}{dx}$; where
$\dfrac{dz}{dx} = \dfrac{dz}{dy} \bullet \dfrac{dy}{dx}$	for $(y)^2$, the inside function is y; or
	Approach 3.
$\dfrac{dz}{dx} = 2y\dfrac{dy}{dx}$ (substituting)	$\dfrac{d}{dx}(y^2) = \dfrac{d}{dy}(y^2)\dfrac{dy}{dx} = 2y\dfrac{dy}{dx}$
$\therefore \dfrac{d}{dx}(y^2) = 2y\dfrac{dy}{dx}$	

Example 3 If $y = x^3$, $x = g(t)$, find $\dfrac{dy}{dt}$	**Example 5** If $V = \pi h^3$; $h = g(t)$
$\dfrac{d}{dt}(y) = \dfrac{d}{dt}(x^3)$ [given that $x = g(t)$]	find $\dfrac{dV}{dt}$
$\dfrac{dy}{dt} = 3x^2\dfrac{dx}{dt}$ (Also, $\dfrac{dy}{dt} = \dfrac{dy}{dx} \bullet \dfrac{dx}{dt}$)	$\dfrac{dV}{dt} = \dfrac{d}{dt}(\pi h^3)$
Example 4. $V = r^3$, Find $\dfrac{dV}{dt}$	$= \pi\dfrac{d}{dt}(h^3)$ ($\dfrac{dV}{dt} = \dfrac{dV}{dh}\dfrac{dh}{dt}$)
$\dfrac{d}{dt}(V) = \dfrac{d}{dt}(r^3)$ $V = f(r)$, $r = g(t)$	$\dfrac{dV}{dt} = 3\pi h^2\dfrac{dh}{dt}$.(**Implicit** Appl.)
$\dfrac{dV}{dt} = 3r^2\dfrac{dr}{dt}$ ($\dfrac{dV}{dt} = \dfrac{dV}{dr}\dfrac{dr}{dt}$)	(We have an explicit equation relating V and h, but we do not have an
Equation for V and r; none for r and t.	equation relating h and t)

Case 3 : y is a function of x; the differentiation is with respect to x and the term to be differentiated is a product in terms of x and y, with each variable raised to various powers. In this case, we agree that we are applying the product rule and the chain rule accordingly.

Example 6	**Example 7**
$\dfrac{d}{dx}(xy) = x\dfrac{dy}{dx} + y(1)$	$\dfrac{d}{dx}(x^2 y^3) = x^2\dfrac{d}{dx}(y^3) + y^3\dfrac{d}{dx}(x^2)$
$= x\dfrac{dy}{dx} + y$	$= x^2(3y^2)\dfrac{dy}{dx} + y^3(2x) = 3x^2 y^2\dfrac{dy}{dx} + 2xy^3$

Example 8 Use implicit differentiation to find $\dfrac{dy}{dx}$ for $y^3 + xy - 3 = 0$

We use the chain rule and the product rule accordingly for differentiation.

Step 1: $\dfrac{d}{dx}(y^3) + \dfrac{d}{dx}(xy) - \dfrac{d}{dx}(3) = 0$ Step 4: $\left(3y^2 + x\right)\dfrac{dy}{dx} + y = 0$

Step 2: $3y^2\dfrac{dy}{dx} + x\dfrac{dy}{dx} + y\dfrac{d}{dx}(x) + 0 = 0$ Step 5: $\left(3y^2 + x\right)\dfrac{dy}{dx} = -y$

  (For the second term, use the product rule)

Step 3: $3y^2\dfrac{dy}{dx} + x\dfrac{dy}{dx} + y(1) = 0$ Step 6: $\dfrac{dy}{dx} = -\dfrac{y}{3y^2 + x}$.

Note: Informally, anytime you differentiate y multiply by $\dfrac{dy}{dx}$.

Example 9 Use implicit differentiation to find $\frac{dy}{dx}$ for $\frac{x^2}{9} + \frac{y^2}{4} = 1$

Approach 1	**Approach 2**
$$\frac{x^2}{9} + \frac{y^2}{4} = 1$$	(undoing the denominators first)
$$\frac{1}{9}\frac{d}{dx}(x^2) + \frac{1}{4}\frac{d}{dx}(y^2) = \frac{d}{dx}(1)$$	$$\frac{x^2}{9} + \frac{y^2}{4} = 1$$
$$\frac{1}{9}(2x) + \frac{1}{4}(2y)\frac{dy}{dx} = 0$$	$$\frac{(36)x^2}{9} + \frac{(36)y^2}{4} = 1(36)$$
$$\frac{y}{2}\frac{dy}{dx} = -\frac{2}{9}x$$	$$4x^2 + 9y^2 = 36$$
$$\frac{dy}{dx} = -\frac{2}{9}x \bullet \frac{2}{y}$$	$$4\frac{d}{dx}(x^2) + 9\frac{d}{dx}(y^2) = \frac{d}{dx}(36)$$
$$\frac{dy}{dx} = -\frac{4x}{9y}$$	$$4(2x) + 9(2y)\frac{dy}{dx} = 0$$
	$$8x + 18y\frac{dy}{dx} = 0$$
	$$\frac{dy}{dx} = \frac{-8x}{18y} = -\frac{4x}{9y}$$

You may skip Examples 10-14 on trigonometric, logarithmic and exponential functions, now, since we will cover these basic functions in Chapters 5 and 8. If you do not want to skip them now, the following derivatives would be helpful:

1. $\frac{d}{dx}\sin x = \cos x$; **2.** $\frac{d}{dx}\cos x = -\sin x$; **3.** $\frac{d}{dx}(\ln x) = \frac{1}{x}$. **4.** $\frac{d}{dx}(e^x) = e^x$
(**Implicit** application of the chain rule)

Example 10 Find $\frac{d}{dx}(\sin y)$

Approach 1:
Let $z = \sin y$. (Also, $y = g(x)$)

$$\frac{dz}{dx} = \frac{dz}{dy}\frac{dy}{dx}$$

$$\frac{dz}{dx} = \cos y \frac{dy}{dx} \qquad (\frac{dz}{dy} = \cos y)$$

$$\therefore \frac{d}{dx}(\sin y) = \cos y \frac{dy}{dx}$$

Approach 2:
$$\frac{dz}{dx} = \frac{d}{dx}(\sin y) = \cos y \frac{dy}{dx}$$
(inside function is y)

Example 11 $z = \ln y$, $y = g(x)$

$$\frac{dz}{dx} = \frac{d}{dx}(\ln y) \quad \text{Also, } (\frac{dz}{dx} = \frac{dz}{dy}\frac{dy}{dx})$$

$$= \frac{1}{y}\frac{dy}{dx} \quad \text{(inside function is } y\text{)}$$

Example 12 Find $\frac{d}{dx}(e^y)$; $y = g(x)$

Let $z = e^y$; then $\frac{d}{dx}(e^y) = \frac{dz}{dx} = \frac{dz}{dy}\frac{dy}{dx}$

or simply,

$$\frac{d}{dx}(e^y) = e^y\frac{dy}{dx} \quad \text{(inside function is } y\text{)}$$

Example 13 $z = e^{2y}$. ($y = g(x)$

Find $\frac{dz}{dx}$

Approach 1: $\frac{dz}{dx} = \frac{d}{dx}(e^{2y}) = 2e^{2y}\frac{dy}{dx}$

(inside function is $2y$)

Approach 2 Let $u = 2y$.

Then $z = e^u$

Since $z = f(u)$, $u = g(y)$, $y = h(x)$

$$\frac{dz}{dx} = \frac{dz}{du}\frac{du}{dy}\frac{dy}{dx} \quad \text{(multiple substitution)}$$

$$\frac{dz}{du} = e^u = e^{2y}; \quad \frac{du}{dy} = 2$$

$$\frac{dz}{dx} = e^{2y}(2) \bullet \frac{dy}{dx} = 2e^{2y}\frac{dy}{dx} \quad .$$

Example 14 Use implicit differentiation to find $\dfrac{dy}{dx}$ for $\sin y - \cos xy = 3x + y$

Step 1: $\cos y \dfrac{dy}{dx} - [-\sin xy(x\dfrac{dy}{dx} + y)] = 3 + \dfrac{dy}{dx}$	**Scrapwork** for $\cos xy$ Let $z = \cos xy$
Step 2: $\cos y \dfrac{dy}{dx} - [-x\sin xy \dfrac{dy}{dx} - y\sin xy] = 3 + \dfrac{dy}{dx}$	Let $u = xy$ Then $z = \cos u$
Step 3: $\cos y \dfrac{dy}{dx} + x\sin xy \dfrac{dy}{dx} + y\sin xy = 3 + \dfrac{dy}{dx}$	$\dfrac{dz}{du} = -\sin u = -\sin xy$
Step 4: $\cos y \dfrac{dy}{dx} + x\sin xy \dfrac{dy}{dx} - \dfrac{dy}{dx} = 3 - y\sin xy$	$\dfrac{du}{dx} = x\dfrac{dy}{dx} + y(1)$
Step 5: $\dfrac{dy}{dx}[\cos y + x\sin xy - 1] = 3 - y\sin xy$	$\dfrac{dz}{dx} = \dfrac{dz}{du}\dfrac{du}{dx}$
Step 6: $\dfrac{dy}{dx} = \dfrac{3 - y\sin xy}{\cos y + x\sin xy - 1}$.	$= -\sin xy(x\dfrac{dy}{dx} + y)$

Lesson 10 Exercises A

1. Use implicit differentiation to find $\dfrac{dy}{dx}$ if $y^3 + xy - 3 = 0$

2. Use implicit differentiation to find $\dfrac{dy}{dx}$ if $\dfrac{x^2}{9} + \dfrac{y^2}{4} = 1$

3. Use implicit differentiation to find $\dfrac{dy}{dx}$ if $4y^4 + xy^3 + 3x^2 - 5 = 0$

4. Use implicit differentiation to find $\dfrac{dy}{dx}$ if $x^2 + y^2 = 16$

5. Evaluate the derivative of $3x^2y^3 + xy^4 = -5$ at $(2,1)$

Answers 1. $-\dfrac{y}{3y^2 + x}$; **2.** $-\dfrac{4x}{9y}$; **3.** $-\dfrac{y^3 + 6x}{16y^3 + 3xy^2}$; **4.** $-\dfrac{x}{y}$ **5.** $-\dfrac{13}{44}$

Lesson 10 Exercises B

1-4 Use implicit differentiation to find $\dfrac{dy}{dx}$ for the following::

1. $y^2 + x^3y - 4 = 0$; **2.** $\dfrac{x^2}{25} - \dfrac{y^2}{16} = 1$

3. $5y^3 + xy^2 + 5x^2 - 3 = 0$; **4.** $6y^3 + x^2y^3 + 4y^2 - 2 = 0$

5. Evaluate the derivative of $2x^3y^2 + x^4y = -6$ at $(2,1)$

6. Find $\dfrac{dy}{dx}$: $x + y + x^2 + y^2 + xy + \cos y - 3 = 0$

Answers 1. $-\dfrac{3x^2y}{2y + x^3}$; **2.** $\dfrac{16x}{25y}$; **3.** $-\dfrac{y^2 + 10x}{15y^2 + 2xy}$; **4.** $-\dfrac{2xy^2}{18y + 3x^2y + 8}$; **5.** $-\dfrac{7}{6}$

6. $-\dfrac{2x + y + 1}{x + 2y - \sin y + 1}$.

Lesson 11
Differentiation of Rational Functions

The Power rule for differentiation of polynomials is applicable here.

Power Rule: Given $f(x) = \dfrac{1}{x^n} = x^{-n}$, $f'(x) = -nx^{-n-1}$, where n is positive.

Case 1: For a simple rational function such as $f(x) = \dfrac{1}{x}$, we can write the function in exponential form and then apply the power rule : $f'(x) = nx^{n-1}$ or we can also apply the quotient rule covered below.

Example 1 If $f(x) = \dfrac{1}{x}$, find $f'(x)$.

Solution

$$f(x) = \frac{1}{x}$$
$$= x^{-1}$$
$$f'(x) = -1x^{-1-1}$$
$$= -x^{-2}$$
$$= -\frac{1}{x^2}$$

Case 2: For other rational functions in which the numerator contains a variable, the **Quotient Rule** below, is more efficient even though we can write the function in exponential form and then apply the **Product Rule** for polynomials. We introduce the quotient rule, apply it and then apply the product rule as second method to observe the relative merits of the two approaches.

Quotient Rule: If f and g are differentiable at x and $g(x) \neq 0$, then

$$\left(\frac{f}{g}\right)' = \frac{g \bullet f' - f \bullet g'}{g^2} \quad \text{That is}$$

$$\frac{d}{dx}\left[\frac{f(x)}{g(x)}\right] = \frac{g(x)\frac{d}{dx}[f(x)] - f(x)\frac{d}{dx}[g(x)]}{[g(x)]^2}$$

In words, the derivative of a quotient equals the product of the denominator and the derivative of the numerator **minus** the product of the numerator and the derivative of the denominator, all divided by the square of the denominator.

For curiosity, we could also apply the quotient rule to find the derivative of

$$f(x) = \frac{1}{x} \qquad \text{Applying the Quotient Rule} \quad f'(x) = \frac{x\frac{d}{dx}(1) - 1\frac{d}{dx}(x)}{x^2}$$

$$= \frac{x(0) - 1(1)}{x^2}$$

$$= \frac{0-1}{x^2} = -\frac{1}{x^2}.$$

Example 2 If $y = \dfrac{3x^2 + 1}{x - 2}$, find $\dfrac{dy}{dx}$

Solution

We use three methods: **Quotient** rule; **product** rule; **implicit** differentiation

Method 1: Applying the Quotient Rule

$$\frac{dy}{dx} = \frac{(x - 2)\frac{d}{dx}(3x^2 + 1) - (3x^2 + 1)\frac{d}{dx}(x - 2)}{(x - 2)^2}$$

$$= \frac{(x - 2)(6x) - (3x^2 + 1)(1)}{(x - 2)^2}$$

$$= \frac{6x^2 - 12x - 3x^2 - 1}{(x - 2)^2}$$

$$= \frac{3x^2 - 12x - 1}{(x - 2)^2}$$

Method 2: Rewriting in exponential form and applying the **Product Rule**

$$y = \frac{3x^2 + 1}{x - 2}$$

$$y = (3x^2 + 1)(x - 2)^{-1}$$

Now, applying the product rule

$$\frac{dy}{dx} = (3x^2 + 1)\frac{d(x - 2)^{-1}}{dx} + (x - 2)^{-1}\frac{d(3x^2 + 1)}{dx}$$

$$= (3x^2 + 1)(-1)(1)(x - 2)^{-2} + (x - 2)^{-1}(6x)$$

$$= -(3x^2 + 1)(x - 2)^{-2} + 6x(x - 2)^{-1}$$

$$= \frac{-3x^2 - 1}{(x - 2)^2} + \frac{6x}{x - 2}$$

$$= \frac{-3x^2 - 1}{(x - 2)^2} + \frac{6x(x - 2)}{(x - 2)(x - 2)} \qquad \text{(making the denominators the same)}$$

$$= \frac{-3x^2 - 1 + 6x^2 - 12x}{(x - 2)(x - 2)}$$

$$= \frac{3x^2 - 12x - 1}{(x - 2)(x - 2)}$$

$$= \frac{3x^2 - 12x - 1}{(x - 2)^2} \qquad \text{We obtain the same answer, but the algebra is more involved.}$$

Method 3: Using **Implicit differentiation**

If $y = \dfrac{3x^2 + 1}{x - 2}$, find $\dfrac{dy}{dx}$

Step 1 Multiply both sides of the equation by $x - 2$ or cross-multiply to obtain

$$xy - 2y = 3x^2 + 1$$

Step 2 Differentiate both sides of the equation to obtain

$$x\frac{dy}{dx} + y - 2\frac{dy}{dx} = 6x$$

$$\frac{dy}{dx}(x - 2) = 6x - y$$

Step 3: Solve for $\dfrac{dy}{dx}$

$$\frac{dy}{dx} = \frac{6x - y}{x - 2}$$

Step 4: Replace y on the right-hand side by $y = \dfrac{3x^2 + 1}{x - 2}$ to obtain

$$\frac{dy}{dx} = \frac{6x - \dfrac{(3x^2 + 1)}{x - 2}}{x - 2}$$

$$= \frac{6x(x - 2) - (3x^2 + 1)}{(x - 2)(x - 2)}$$

$$\frac{dy}{dx} = \frac{3x^2 - 12x - 1}{(x - 2)^2} \quad \text{(same result as by Methods 1 and 2)}$$

Now, compare the relative merits of the quotient rule, the power rule and the application of implicit differentiation in Example 2 above. Of course, Method 1, the Quotient Rule takes the least number of steps.

Lesson 11 Exercises A

1. If $f(x) = \dfrac{1}{x}$, find $f'(x)$

2. State the quotient rule.

3. If $y = \dfrac{3x^2 + 1}{x - 2}$, find $\dfrac{dy}{dx}$

4. If $f(x) = \dfrac{3}{5x^2 + 2x - 1}$, find $f'(x)$

5. If $f(x) = (x + 3)^{-2}$ find $f'(x)$

6. If $f(x) = (3 - x)^{-2}$, find $f'(x)$

7. If $y = x^{-2} + \dfrac{1}{x^6}$, find $\dfrac{dy}{dx}$

8. If $y = \dfrac{1}{6x - 5}$, find $\dfrac{dy}{dx}$

9. If $y = \dfrac{x}{x^3 + 5}$, find $\dfrac{dy}{dx}$

10. If $r = \dfrac{t}{t^2 + 5}$, find $\dfrac{dr}{dt}$

11. Find $\dfrac{dy}{dx}$ if $y = \dfrac{3x + 1}{x - 2}$

12. Find $\dfrac{dy}{dx}$ if $y = \dfrac{2}{x^4}$

Answers: **1.** $-\dfrac{1}{x^2}$; **3.** $\dfrac{3x^2 - 12x - 1}{(x - 2)^2}$; **4.** $-\dfrac{3(10x + 2)}{\left(5x^2 + 2x - 1\right)^2}$; **5.** $-\dfrac{2}{(x + 3)^3}$;

6. $\dfrac{2}{(3 - x)^3}$; **7.** $-\dfrac{2}{x^3} - \dfrac{6}{x^7}$; **8.** $-\dfrac{6}{(6x - 5)^2}$; **9.** $-\dfrac{2x^3 - 5}{(x^3 + 5)^2}$

10. $-\dfrac{t^2 - 5}{(t^2 + 5)^2}$; **11.** $-\dfrac{7}{(x - 2)^2}$; **12.** $-\dfrac{8}{x^5}$

Lesson 11 Exercises B

1. If $f(x) = \dfrac{1}{2x}$, find $f'(x)$

2. State the quotient rule.

3. If $y = \dfrac{2x^2 + 1}{x - 3}$, find $\dfrac{dy}{dx}$

4. If $f(x) = \dfrac{4}{3x^2 + x - 1}$, find $f'(x)$

5. If $f(x) = (3x + 2)^{-1}$ find $f'(x)$

6. If $f(x) = (3 - x)^{-1}$, find $f'(x)$

7. If $y = x^{-2} + \dfrac{1}{x^3}$, find $\dfrac{dy}{dx}$

8. If $y = \dfrac{1}{2x - 3}$, find $\dfrac{dy}{dx}$

9. If $y = \dfrac{2x}{x^2 - 3}$, find $\dfrac{dy}{dx}$

10. If $r = \dfrac{t}{t^2 + 3}$, find $\dfrac{dr}{dt}$

11. Find $\dfrac{dy}{dx}$ if $y = \dfrac{2x + 1}{x - 3}$

12. Find $\dfrac{dy}{dx}$ if $y = \dfrac{3}{x^5}$

Answers: **1.** $-\dfrac{1}{2x^2}$; **3.** $\dfrac{2x^2 - 12x - 1}{(x-3)^2}$; **4.** $-\dfrac{24x + 4}{(3x^2 + x - 1)^2}$; **5.** $-\dfrac{3}{(3x + 2)^2}$

6. $\dfrac{1}{(3 - x)^2}$ or $\dfrac{1}{(x - 3)^2}$; **7.** $-\dfrac{2}{x^3} - \dfrac{3}{x^4}$; **8.** $-\dfrac{2}{(2x - 3)^2}$;

9. $-\dfrac{2x^2 + 6}{(x^2 - 3)^2}$; **10.** $-\dfrac{t^2 - 3}{(t^2 + 3)^2}$; **11.** $-\dfrac{7}{(x - 3)^2}$; **12.** $-\dfrac{15}{x^6}$.

Lesson 12
Differentiation of Radical Functions

Example 13: If $y = \sqrt{x-3}$, find $\frac{dy}{dx}$

Method 1 We apply Version 1 of the **chain rule**.

Step 1: Let $u = x - 3$ (A)

Then $y = u^{\frac{1}{2}}$

$$\frac{dy}{du} = \frac{1}{2}u^{-\frac{1}{2}}$$

$$= \frac{1}{2} \cdot \frac{1}{u^{\frac{1}{2}}}$$

$$\frac{dy}{du} = \frac{1}{2\sqrt{x-3}} \quad \text{(B)}$$

Also from (A), $\frac{du}{dx} = 1$ (C)

Step 2: Substitute for $\frac{dy}{du}$ (from B); for $\frac{du}{dx}$ (from C); in $\frac{dy}{dx} = \frac{dy}{du} \cdot \frac{du}{dx}$,

to obtain $\frac{dy}{dx} = \frac{1}{2\sqrt{x-3}}(1)$

$$= \frac{1}{2\sqrt{x-3}}$$

$\therefore \frac{dy}{dx} = \frac{1}{2\sqrt{x-3}}$ (or if we rationalize $\frac{dy}{dx} = \frac{1}{2(\sqrt{x-3})}\frac{\sqrt{x-3}}{(\sqrt{x-3})} = \frac{\sqrt{x-3}}{2x-6}$

Method 2: Rewriting in exponential form and applying
Version 2 of the **chain rule** (power chain Rule)

If $y = \sqrt{x-3}$, find $\frac{dy}{dx}$.

Step 1: Write in exponential form

Then $y = (x-3)^{\frac{1}{2}}$.

Step 2 Differentiate using the power chain rule

Then $\frac{dy}{dx} = \frac{1}{2}(x-3)^{\frac{1}{2}-1}$

$$\frac{dy}{dx} = \frac{1}{2}(x-3)^{-\frac{1}{2}}$$

$$\frac{dy}{dx} = \frac{1}{2} \cdot \frac{1}{(x-3)^{\frac{1}{2}}}$$

$\therefore \frac{dy}{dx} = \frac{1}{2\sqrt{x-3}}$

$\frac{dy}{dx} = \frac{1}{2(\sqrt{x-3})}\frac{\sqrt{x-3}}{(\sqrt{x-3})}$ (rationalizing the denominator)

$$= \frac{\sqrt{x-3}}{2(x-3)} \text{.or} = \frac{\sqrt{x-3}}{2x-6}$$

Method 3: Using implicit differentiation

$$\text{If } y = \sqrt{x-3}, \text{ find } \frac{dy}{dx}.$$

Step 1: Square both sides of the equation to undo the radical sign.

Then $y^2 = x - 3$

Step 2: Differentiate.

$$2y\frac{dy}{dx} = 1$$

$$\frac{dy}{dx} = \frac{1}{2y}$$

$$= \frac{1}{2\sqrt{x-3}} \qquad (y = \sqrt{x-3})$$

$$\therefore \frac{dy}{dx} = \frac{1}{2\sqrt{x-3}}$$

$$\frac{dy}{dx} = \frac{1}{2(\sqrt{x-3})}\frac{\sqrt{x-3}}{(\sqrt{x-3})} \quad ((\text{rationalizing the denominator}))$$

$$= \frac{\sqrt{x-3}}{2(x-3)} \text{ or } = \frac{\sqrt{x-3}}{2x-6}.$$

Now, compare the relative merits of methods 1, 2 and 3. Method 3 seems to be very competitive.

Lesson 12 Exercises A

1. If $y = \sqrt{x-3}$; , find $\dfrac{dy}{dx}$

2.. If $y = \sqrt{\dfrac{2x-3}{3x+2}}$, find $\dfrac{dy}{dx}$

3. If $f(x) = \dfrac{\sqrt{x-1}}{\sqrt{x+1}}$, find $f'(x)$

4. If $g(t) = \dfrac{t^3}{\sqrt[3]{3t^2+1}}$, find $f'(t)$

5.. If $y = \sqrt{5-3x^2}$, find $\dfrac{dy}{dx}$

6. If $y = \dfrac{2}{\sqrt{x}+3}$, find $f'(x)$

7. If $y = (2x-3)^{\frac{2}{3}}$, find $\dfrac{dy}{dx}$

8. If $y = \dfrac{3x}{x+1}$, find $\dfrac{dy}{dx}$

9. $y = \dfrac{x+3}{x^2+x+2}$, find $\dfrac{dy}{dx}$

10. Find $\dfrac{d}{dx}\left[\dfrac{x^2}{7-3x}\right]$

Answers 1. $\dfrac{1}{2\sqrt{x-3}}$; **2.** $\dfrac{13\sqrt{6x^2-5x-6}}{36x^3-6x^2-56x-24}$ **3.** $\dfrac{\sqrt{x^2-1}}{x^3+x^2-x-1}$; **4.**

$\dfrac{7t^4+3t^2}{(3t^2+1)^{\frac{4}{3}}}$ **5.** $-\dfrac{3x}{\sqrt{5-3x^2}}$ or $\dfrac{3x\sqrt{5-3x^2}}{3x^2-5}$; **6.** $-\dfrac{\sqrt{x}}{x(\sqrt{x}+3)^2}$;

7. $\dfrac{4}{3(2x-3)^{\frac{1}{3}}}$; **8.** $\dfrac{3}{(x+1)^2}$ **9.** $-\dfrac{x^2+6x+1}{(x^2+x+2)^2}$; **10.** $-\dfrac{3x^2-14x}{(3x-7)^2}$.

Lesson 12 Exercises B

1. If $y = \sqrt{x-5}$; , find $\dfrac{dy}{dx}$

2.. If $y = \sqrt{\dfrac{3x-2}{2x+3}}$, find $\dfrac{dy}{dx}$

3. If $f(x) = \dfrac{\sqrt{x-2}}{\sqrt{x+2}}$, find $f'(x)$

4. If $g(t) = \dfrac{t^2}{\sqrt[3]{2t+1}}$, find $f'(t)$

5.. If $y = \sqrt{3-2x^2}$, find $\dfrac{dy}{dx}$

6. If $y = 6\sqrt[3]{x^2}$, find $f'(x)$

7. If $y = (3x-2)^{\frac{2}{3}}$, find $\dfrac{dy}{dx}$

8. If $y = \dfrac{2x}{x+1}$, find $\dfrac{dy}{dx}$

9. $y = \dfrac{x+2}{x^2+2x+1}$, find $\dfrac{dy}{dx}$

10. Find $\dfrac{d}{dx}\left[\dfrac{x^3}{7-4x^2}\right]$

Answers: 1. $\dfrac{1}{2\sqrt{x-5}}$ **2.** $\dfrac{13\sqrt{6x^2+5x-6}}{24x^3+56x^2+6x-36}$; **3.** $\dfrac{2\sqrt{x^2-4}}{x^3+2x^2-4x-8}$;

4. $\dfrac{10t^2+6t}{3(2t+1)^{\frac{4}{3}}}$; **5.** $-\dfrac{2x}{\sqrt{3-2x^2}}$ or $\dfrac{2x\sqrt{3-2x^2}}{2x^2-3}$; **6.** $4x^{-\frac{1}{3}}$ **7.** $\dfrac{2}{(3x-2)^{\frac{1}{3}}}$;

8. $\dfrac{2}{(x+1)^2}$; **9.** $-\dfrac{x+3}{(x+1)(x^2+2x+1)}$; **10.** $-\dfrac{4x^4-21x^2}{(4x^2-7)^2}$

CHAPTER 5

Differentiation Techniques II

Lesson 13: **Differentiation of Logarithmic Functions**
Lesson 14: **Differentiation of Exponential Functions**
Lesson 15: **Applications of the Method of Logarithmic Differentiation**

Lesson 13
Differentiation of Logarithmic Functions

Precalculus: Sometimes, we use the natural log base (base e), and sometimes we use other bases, say base a or base b. It is important that we are able to change quickly from one base to another. We want a relationship between the bases e and b ($b > 0$, $b \neq 1$).

Change of base formula for from base a to base b. $\boxed{\log_a x = \dfrac{\log_b x}{\log_b a}}$.

Change of base formula for from base b to base e. the natural log base.

$$\log_b x = \frac{\log_e x}{\log_e b} = \frac{\ln x}{\ln b}$$

$\boxed{\log_b x = \dfrac{\ln x}{\ln b}}$ **Note**: $\log_b e$ and $\ln b$ or $\log_e b$ are reciprocals of each other.

Note also: You can write $\log_e x$ for $\ln x$.(just like $\sqrt[2]{x} = \sqrt{x}$). In fact, at times, indicating the base, e, helps one to write this log in equivalent exponential form.

The **basic rules** for the **differentiation of logarithmic functions** follow.

Approach 1 (From specific to general)	**Approach 2** (From general to specific)
Case 1. Base e	**Case 1.** Base b

Approach 1 (From specific to general)

Case 1. Base e

If $y = \ln x$, $\boxed{\dfrac{dy}{dx} = \dfrac{1}{x}}$ <-(memorize this)
(For the derivation, see p.115, Extra)

Case 2. Base b or any other base

$y = \log_b x$

$= \dfrac{\log_e x}{\log_e b}$ (changing to base e)

$y = \dfrac{\ln x}{\ln b}$

$\dfrac{dy}{dx} = \dfrac{1}{\ln b} \dfrac{d(\ln x)}{dx}$

$= \dfrac{1}{\ln b} \dfrac{1}{x}$ ($\ln b$ is a constant)

$\boxed{\dfrac{dy}{dx} = \dfrac{1}{x \ln b}}$ (base e: $\log_b e = \dfrac{1}{\ln b}$)

Also, $\boxed{\dfrac{dy}{dx} = \dfrac{1}{x} \log_b e}$ (base b)

Approach 2 (From general to specific)

Case 1. Base b
Let y be a differentiable function of x,
*t*hen if $y = \log_b x$, then equivalently

$\boxed{b^y = x}$ ($b > 0$, $b \neq 1$)

$\ln b^y = \ln x$ (taking logs)

$y \ln b = \ln x$

$y = \dfrac{1}{\ln b} \ln x$

$\dfrac{dy}{dx} = \dfrac{1}{\ln b} \bullet \dfrac{1}{x}$ ($\dfrac{1}{\ln b}$ is a constant)

$\dfrac{dy}{dx} = \dfrac{1}{x \ln b}$

Case 2. Base e
If the base $b = e$, (natural log).

$\dfrac{dy}{dx} = \dfrac{1}{x \ln e}$

$\dfrac{dy}{dx} = \dfrac{1}{x}$ (since $\ln e = \log_e e = 1$)

Derivation of the Change of base formula for logarithms   1 1 4

Example: Change $\log_b M$ to base c.

Step 1: Let $\log_b M = y$.

Then, $M = b^y$ (Equivalent exponential definition)

Step 2: Now, take logs of both sides of the equation and introduce the new base, c.

Then, $\log_c M = \log_c b^y$

Step 3: Solve for y:

$$\log_c M = y \log_c b$$

$$\frac{\log_c M}{\log_c b} = y \qquad \text{(Solving for } y)$$

Therefore, $\log_b M = \dfrac{\log_c M}{\log_c b}$ <-----This is known as the change of base formula.

(Here, we changed from base b to base c.)

Differentiation involving natural logarithms

Example 1 If $y = \ln 4x$, find $\dfrac{dy}{dx}$ (Note from above, if $y = \ln x$, $\dfrac{dy}{dx} = \dfrac{1}{x}$)

Solution: **Method 1** We apply the chain rule

Let $u = 4x$. Then $\dfrac{du}{dx} = 4$; Also, $y = \ln u$ and $\dfrac{dy}{du} = \dfrac{1}{u}$

Substituting in $\dfrac{dy}{dx} = \dfrac{dy}{du}\dfrac{du}{dx}$

$$= \frac{1}{u}(4)$$

$$= \frac{1}{4x}(4)$$

$$= \frac{1}{x}$$

Method 2
Since $\ln 4x = \ln x + \ln 4$
$\dfrac{d}{dx}(\ln 4x) = \dfrac{d}{dx}(\ln x) + \dfrac{d}{dx}(\ln 4)$
$= \dfrac{1}{x} + 0$ ($\ln 4$ is a constant)
$= \dfrac{1}{x}$

Therefore, $\dfrac{dy}{dx} = \dfrac{1}{x}$. (By method 2, we predict that $\dfrac{d}{dx}(\ln kx) = \dfrac{1}{x}$ (k a constant))

Example 2 If $y = \ln(4x^2 - 3)$, find $\dfrac{dy}{dx}$

Solution: **Method 1** Leibniz's notation	**Method 2** Using the power chain
Step 1: Let $u = 4x^2 - 3$	chain rule is faster
Then $y = \ln u$	(The "inside-outside function" approach)
$\dfrac{du}{dx} = 8x$	$\dfrac{dy}{dx} = \dfrac{1}{4x^2 - 3}(8x) = \dfrac{8x}{4x^2 - 3}$
$\dfrac{dy}{du} = \dfrac{1}{u}$	**Method 3** (by implicit differentiation)
	If $y = \ln(4x^2 - 3)$
Step 2: Substitute for $\dfrac{dy}{du}$ and $\dfrac{du}{dx}$ in	Then $e^y = 4x^2 - 3$
$\dfrac{dy}{dx} = \dfrac{dy}{du}\dfrac{du}{dx} = \dfrac{1}{u}(8x)$	$e^y \dfrac{dy}{dx} = 8x$ (implicit differentiation)
$= \dfrac{1}{4x^2 - 3}(8x) = \dfrac{8x}{4x^2 - 3}$	$\dfrac{dy}{dx} = \dfrac{8x}{e^y} = \dfrac{8x}{4x^2 - 3}$

Example 3:. If $y = \log_2 x$, find $\dfrac{dy}{dx}$

Method 1 Applying the change of base formula,

$$y = \log_b x = \frac{\log_e x}{\log_e b} = \boxed{\frac{\ln x}{\ln b}}$$

$$y = \log_2 x = \frac{\ln x}{\ln 2} \quad \text{(changing to base e)}$$

$$\frac{dy}{dx} = \frac{1}{\ln 2} \frac{d(\ln x)}{dx} \quad (\log_e x = \ln x)$$

$$= \frac{1}{\ln 2} \frac{1}{x} \quad (\log_e 2 = \ln 2)$$

$$= \frac{1}{x \ln 2} \quad \text{(base e)}$$

$$\frac{dy}{dx} = \frac{1}{x} \log_2 e \quad \text{(base 2)}.$$

Method 2 Starting with the equivalent exponential definition.
Step 1: Express in the equivalent exponential form
$\qquad$ If $y = \log_2 x$, then

$\qquad 2^y = x$
Step 2: Take logs of both sides of the equation and change to base e
$\qquad \ln 2^y = \ln x \qquad (\log_e x = \ln x)$
Step 3: $y \ln 2 = \ln x \qquad (\log_e 2 = \ln 2)$

Step 4:: **Approach 1**	Step 4:: **Approach 2**
Solve for y to obtain $y = \dfrac{\ln x}{\ln 2}$	By implicit differentiation
Differentiate now:	$\ln 2 \dfrac{dy}{dx} = \dfrac{1}{x}$
$\dfrac{dy}{dx} = \dfrac{1}{\ln 2} \bullet \dfrac{1}{x} = \dfrac{1}{x \ln 2}$ (base e)	$\dfrac{dy}{dx} = \dfrac{1}{x \ln 2}$
$\dfrac{dy}{dx} = \dfrac{1}{x} \log_2 e$ (base 2).	

Extra: Show that if $y = \ln x$, then $\dfrac{dy}{dx} = \dfrac{1}{x}$
Method: Apply implicit differentiation)
Step 1: If $y = \ln x \qquad (\ln x = \log_e x)$

$\qquad$ Then $e^y = x \qquad$ (Exponential equivalent definition))
Step 2 Differentiate implicitly with respect to x.

$\qquad$ Then $e^y \dfrac{dy}{dx} = 1$ (implicit differentiation)

$$\frac{dy}{dx} = \frac{1}{e^y}$$

$$\therefore \ \frac{dy}{dx} = \frac{1}{x} \qquad \left(e^y = x\right)$$

Lesson 13 Exercises A

1. If $y = \log_b x$, find $\dfrac{dy}{dx}$

2. If $y = \ln x$ find $\dfrac{dy}{dx}$

3 If $y = \ln 4x$ find $\dfrac{dy}{dx}$

4. If $y = \ln(4x^2 - 3)$, find $\dfrac{dy}{dx}$

5. If $y = \log_2 x$;, find $\dfrac{dy}{dx}$

6. Show that $b^x = e^{x \ln b}$

Answers: 1. $\dfrac{1}{x \ln b}$; **2.** $\dfrac{1}{x}$; **3.** $\dfrac{1}{x}$; **4.** $\dfrac{8x}{4x^2 - 3}$; **5.** $\dfrac{1}{x \log_e 2}$ or $\dfrac{1}{x \ln 2}$

Lesson 13 Exercises B

1. If $y = \log_a x$, find $\dfrac{dy}{dx}$

2. If $y = \log_e x$ find $\dfrac{dy}{dx}$

3 If $y = \ln 9x$ find $\dfrac{dy}{dx}$

4. If $y = \ln(2x^3 - 5)$, find $\dfrac{dy}{dx}$

5. If $y = \log_3 x$;, find $\dfrac{dy}{dx}$

6. Show that $a^x = e^{x \ln a}$

Answers: 1. $\dfrac{1}{x \ln a}$; **2.** $\dfrac{1}{x}$; **3.** $\dfrac{1}{x}$; **4.** $\dfrac{6x^2}{2x^3 - 5}$; **5.** $\dfrac{1}{x \log_e 3}$ or $\dfrac{1}{x \ln 3}$

Lesson 14
Differentiation of Exponential Functions

Differentiation Rules

Approach 1 (From specific to general) **Approach 2** (From general to specific)

Case 1. Base e

$$y = e^x$$

$$\boxed{\dfrac{dy}{dx} = e^x}$$ (memorize this)

Case 2. Base b or any other base

$$y = b^x$$

$$\boxed{b^x = e^{x \ln b}}$$ (Using change of base formula for exponents. Memorize this) (see also p.119)

Let $u = x \ln b$ (and apply the chain rule)

Then $y = e^u = b^x$

$$\frac{dy}{du} = e^u$$

$$\frac{du}{dx} = \ln b$$

$$\frac{dy}{dx} = \frac{dy}{du} \bullet \frac{du}{dx}$$

$$\frac{dy}{dx} = e^u \ln b$$

$$\boxed{\frac{dy}{dx} = b^x \ln b}$$ Note: $e^u = b^x$

Case 1. Base b

If $y = b^x$, then $\boxed{\dfrac{dy}{dx} = b^x \ln b}$

Derivation: We use logarithmic differentiation. See next lesson.

Taking logs of both sides of $y = b^x$

$$\ln y = \ln b^x$$

$$\ln y = x \ln b$$

$$\frac{1}{y}\frac{dy}{dx} = \ln b \quad \text{(applying implicit differentiation)}$$

$$\frac{dy}{dx} = y \ln b$$

$$\frac{dy}{dx} = b^x \ln b \quad (y = b^x)$$

Case 2. Base e (if $b = e$)

$$y = e^x$$

$$\frac{dy}{dx} = e^x \ln e \quad \text{(noting that}$$

$$= e^x(1) \quad \ln e = \log_e e = 1)$$

$$\boxed{\frac{dy}{dx} = e^x}$$

Example 1 If $y = e^{x^2}$, find $\dfrac{dy}{dx}$

Solution: We apply the chain rule

Let $u = x^2$. Then $y = e^u$

$$\frac{du}{dx} = 2x \text{ and } \frac{dy}{du} = e^u$$

Substituting in $\dfrac{dy}{dx} = \dfrac{dy}{du}\dfrac{du}{dx}$

$$\frac{dy}{dx} = e^u(2x)$$

$$= e^{x^2}(2x)$$

$$= 2xe^{x^2}.$$

(Above, the derivative of e^x is e^x;

The derivative of b^x equals b^x times $\ln b$)

Example 2. $f(x) = 2^x$; find $\dfrac{dy}{dx}$

Method 1: Let $y = 2^x$

Then $\dfrac{dy}{dx} = 2^x \ln 2$ (By formula: If $y = b^x$, then $\dfrac{dy}{dx} = b^x \ln b$)

Method 2: By logarithmic differentiation

(Master this method in case you forget the formula)

Take logs of both sides of $y = 2^x$

Then $\ln y = \ln 2^x$

$\ln y = x \ln 2$

$\dfrac{1}{y}\dfrac{dy}{dx} = \ln 2$ (Note: $\ln 2$ is a constant)

$\dfrac{dy}{dx} = y \ln 2$

$\dfrac{dy}{dx} = 2^x \ln 2$ (substituting 2^x for y)

Example 3   Similarly if $y = 4^x$, $\dfrac{dy}{dx} = 4^x \ln 4$. (Using Method 2 above)

Example 4. If $y = 2^{-x}$, find $\dfrac{dy}{dx}$ Take logs of both sides of this equation. Then $\ln y = \ln 2^{-x}$ $\ln y = -x \ln 2$ $\dfrac{1}{y}\dfrac{dy}{dx} = -\ln 2$ $\dfrac{dy}{dx} = -y \ln 2$ $\dfrac{dy}{dx} = -2^{-x} \ln 2$ ($y = 2^{-x}$)	**Example 5.** If $y = 4^{-x}$;, find $\dfrac{dy}{dx}$ Take logs of both sides of this equation Then $\ln y = \ln 4^{-x}$ $\ln y = -x \ln 4$ $\dfrac{1}{y}\dfrac{dy}{dx} = -\ln 4$ $\dfrac{dy}{dx} = -y \ln 4$ $\dfrac{dy}{dx} = -4^{-x} \ln 4$ ($y = 4^{-x}$)

Example 6. Given $y = e^{-x}$

Method 1
(by implicit differentiation)
$\ln y = \ln e^{-x}$; $\ln y = -x \ln e$

$\dfrac{1}{y}\dfrac{dy}{dx} = -1$; $\dfrac{dy}{dx} = -y$ ($\ln e = 1$)

$\phantom{\dfrac{dy}{dx}} = -e^{-x}$ ($y = e^{-x}$)

Method 2 (by chain rule)
Let $u = -x$. Then $y = e^u$

$\dfrac{du}{dx} = -1$; $\dfrac{dy}{du} = e^u = e^{-x}$

$\dfrac{dy}{dx} = \dfrac{dy}{du}\dfrac{du}{dx} = e^{-x}(-1)$

$\phantom{\dfrac{dy}{dx}} = -e^{-x}$

Example 7. If $y = 10^x$; $\dfrac{dy}{dx} = 10^x \ln 10$	**Example 8.** If $y = 10^{-x}$, $\dfrac{dy}{dx} = -10^{-x} \ln 10$.

Extra: Show that if $y = e^x$, $\dfrac{dy}{dx} = e^x$

Take logs and differentiate.

Step 1: $\ln y = \ln e^x$; $\ln y = x \ln e$

Step 2: $\ln y = x$; $\dfrac{1}{y}\dfrac{dy}{dx} = 1$;

Step 3: $\dfrac{dy}{dx} = y$ (**Note:** $\ln e = 1$)

$\phantom{\dfrac{dy}{dx}} = e^x$

Derivation of some formulas

If $y = b^x$, derive $\dfrac{dy}{dx} = b^x \ln b$

Method 1	**Method 2**
Step 1: Take logs of both side of the equation	Step 1: $\quad y = b^x$
$\quad y = b^x$	$\quad\quad b^x = e^{x \ln b}$ (see below)
$\quad \ln y = \ln b^x$	$\quad\quad$ Let $u = x \ln b$
$\quad \ln y = x \ln b$	$\quad\quad$ Then $y = e^u$
Step 2: Differentiate using implicit differentiation (Lesson 10)	Step 2: Differentiate
$\quad \dfrac{1}{y}\dfrac{dy}{dx} = \ln b$	$\quad \dfrac{dy}{du} = e^u$
$\quad \dfrac{dy}{dx} = y \ln b$	$\quad \dfrac{du}{dx} = \ln b$
Step 3: Replace y by b^x	$\quad \dfrac{dy}{dx} = \dfrac{dy}{du} \bullet \dfrac{du}{dx}$
$\quad$ Then $\dfrac{dy}{dx} = b^x \ln b$	$\quad\quad = e^u \ln b = e^{x \ln b} \ln b$
	Step 3: $\dfrac{dy}{dx} = b^x \ln b$ $(e^{x \ln b} = b^x.)$

Application of Method 1 (Repetition of "EXTRA" on the previous page)

If $y = e^x$, find $\dfrac{dy}{dx}$ (This application is to show the power of logarithmic differentiation)

Step 1: Take logs of both side of the equation

$\quad\quad \ln y = \ln e^x \quad (y = e^x)$

$\quad\quad \ln y = x \ln e$

$\quad\quad \ln y = x \quad\quad (\ln e = \log_e e = 1)$

Step 2: Differentiate using implicit differentiation

$\quad\quad \dfrac{1}{y}\dfrac{dy}{dx} = 1$, and from which $\dfrac{dy}{dx} = y$

Step 3: Replace y by e^x. Then $\dfrac{dy}{dx} = e^x$

(From above, we have shown that if $y = e^x$ $\dfrac{dy}{dx} = e^x$.)

Change of base formula for exponents: Change b^x to base a.

Learn this derivation.

Step 1: Let $b^x = a^k$ $\quad\quad$ (1)
Step 2: Take logs (to base a, the new base) of both sides of the equation.

$\quad\quad$ Then $\log_a b^x = \log_a a^k$

$\quad\quad\quad x \log_a b = k \log_a a$

$\quad\quad\quad x \log_a b = k \quad\quad (\log_a a = 1)$

$\quad\quad$ and $\quad k = x \log_a b$

Step 3: Substitute for k in (1).

$\quad\quad$ Then $\boxed{b^x = a^{x \log_a b}}$. If $a = e$. $\boxed{b^x = e^{x \log_e b} = e^{x \ln b}}$ (Memorize)

Lesson 14 Exercises A

1. If $y = b^x$, then $\dfrac{dy}{dx} = ?$

2. If $y = e^x$, then $\dfrac{dy}{dx} = ?$

3 If $y = e^{x^2}$, find $\dfrac{dy}{dx}$

Questions 4-10: Find $\dfrac{dy}{dx}$

4. $f(x) = 2^x$;

5. $f(x) = 2^{-x}$;

6. $f(x) = 4^x$;

7. $f(x) = 4^{-x}$;

8. $f(x) = e^{-x}$;

9. $y = 10^x$;

10. $y = 10^{-x}$;

Answers: 1. $b^x \ln b$; **2.** e^x; **3.** $2xe^{x^2}$; **4.** $2^x \ln 2$; **5.** $-2^{-x} \ln 2$, **6.** $4^x(2\ln 2)$;
7. $-4^{-x}(2\ln 2)$; **8.** $-e^{-x}$; **9.** $10^x \ln 10$; **10.** $-10^{-x} \ln 10$

Lesson 14 Exercises B

1. If $y = a^x$, then $\dfrac{dy}{dx} = ?$

2. If $y = 3e^x$, then $\dfrac{dy}{dx} = ?$

3 If $y = 4e^{x^2}$, find $\dfrac{dy}{dx}$

Questions 4-10:- Find $\dfrac{dy}{dx}$

4. $f(x) = 5^x$;

5. $f(x) = 3^{-x}$;

6. $f(x) = 6^x$;

7. $f(x) = 3^{-2x}$;

8. $y = e^{-3x}$;

9. $y = \pi^{-x}$;

10. $y = 2 \times 10^x$;

Answers: 1. $a^x \ln a$; **2.** $3e^x$; **3.** $8xe^{x^2}$; **4.** $5^x \ln 5$; **5.** $-3^{-x} \ln 3$; **6.** $6^x \ln 6$
7. $-3^{-2x}(2\ln 3)$; **8.** $-3e^{-3x}$; **9.** $-\pi^{-x} \ln \pi$; **10.** $10^x(2\ln 10)$

Memory Reminder

Compare the **change of base formulas** for logarithms and exponents, and observe how for the logarithms, the conversion factor, $\ln b$, divides; but for the exponents, $\ln b$ multiplies the exponent.

For **logarithms:** $\log_b x = \dfrac{\ln x}{\ln b} = \left(\dfrac{\log_e x}{\log_e b} \right)$

For **exponents:** $b^x = e^{x \ln b}$

Lesson 15

Applications of the Method of Logarithmic Differentiation
(Take logs and differentiate)

Logarithmic differentiation can be applied to simplify the differentiation of a differentiable function which is exponential, or the product or quotient of several factors involving exponents. In logarithmic differentiation, we take natural logarithms (logs) before differentiating.

Example 1 Use logarithmic differentiation to find $\dfrac{dy}{dx}$:

(a) $y = a^x$; (b) $y = (x^2 - 4)^5(x - 2)^3$; (c) $y = x^x$

Solution

(a) Step 1: Take logs of both sides of the equation. Then

$y = a^x$ becomes

$\ln y = \ln a^x$

$\ln y = x \ln a$

Step 2: Apply implicit differentiation.

$\dfrac{1}{y}\dfrac{dy}{dx} = \ln a$ ($\ln a$ is a constant)

$\dfrac{dy}{dx} = y \ln a$

$\dfrac{dy}{dx} = a^x \ln a$ $(y = a^x)$

(b) $y = (x^2 - 4)^5(x - 2)^3$

Step 1: Take logs of both sides of the equation.

$\ln|y| = \ln\left|(x^2 - 4)^5(x - 2)^3\right|$

$\ln|y| = \ln\left|(x^2 - 4)^5\right| + \ln\left|(x - 2)^3\right|$ ($\ln AB = \ln A + \ln B$)

$\ln y = 5\ln\left|x^2 - 4\right| + 3\ln|x - 2|$

Step 2: Apply implicit differentiation.

$\dfrac{1}{y}\dfrac{dy}{dx} = \dfrac{10x}{x^2 - 4} + \dfrac{3}{x - 2}$ (see scrapwork for first term)

$\dfrac{dy}{dx} = y\left(\dfrac{10x}{x^2 - 4} + \dfrac{3}{x - 2}\right)$

$= \left[(x^2 - 4)^5(x - 2)^3\right]\left(\dfrac{10x}{x^2 - 4} + \dfrac{3}{x - 2}\right)$

$= \left[(x^2 - 4)^5(x - 2)^3\right]\left(\dfrac{10x}{(x + 2)(x - 2)} + \dfrac{3(x + 2)}{(x - 2)(x + 2)}\right)$

$= \left[(x^2 - 4)^5(x - 2)^3\right]\dfrac{13x + 6}{(x + 2)(x - 2)}$

$= \left[(x^2 - 4)^5(x - 2)^3\right]\dfrac{13x + 6}{x^2 - 4}$

$= \left[(x^2 - 4)^4(x - 2)^3\right](13x + 6)$

$= (13x + 6)\left[(x^2 - 4)^4(x - 2)^3\right]$.

Scrapwork

Let $y_1 = 5\ln(x^2 - 4)$

$u = x^2 - 4$

$5\ln(x^2 - 4) = 5\ln u$

$y_1 = 5\ln u$

$\dfrac{dy_1}{du} = 5\left(\dfrac{1}{u}\right) = \dfrac{5}{x^2 - 4}$

$\dfrac{du}{dx} = 2x$

$\dfrac{dy_1}{dx} = \dfrac{dy_1}{du}\dfrac{du}{dx} = \dfrac{5}{x^2 - 4} \cdot (2x)$

$\dfrac{d}{dx}(5\ln(x^2 - 4)) = \dfrac{10x}{x^2 - 4}$

(c) Find $\dfrac{dy}{dx}$ if $y = x^x$

Step 1: Take logs of both sides of the equation, Then $y = x^x$ becomes

$\ln|y| = \ln\left|x^x\right|$, and from which $\ln|y| = x\ln|x|$

Step 2: Apply implicit differentiation

$\dfrac{1}{y}\dfrac{dy}{dx} = x \bullet \dfrac{1}{x} + \ln|x|(1)$ (apply the product rule: $x\dfrac{d(\ln x)}{dx} + \ln x\dfrac{d(x)}{dx}$)

$\dfrac{1}{y}\dfrac{dy}{dx} = 1 + \ln|x|$

$\dfrac{dy}{dx} = y(1 + \ln|x|)$

$\dfrac{dy}{dx} = x^x(\ln|x| + 1)$. $(y = x^x)$

Example 2 Use logarithmic differentiation to find $\dfrac{dy}{dx}$

$$\text{if } y = \frac{(x^3 - 2x + 1)(5x - 1)^2}{(3x^2 + 2)^3}$$

Solution: $y = \dfrac{(x^3 - 2x + 1)(5x - 1)^2}{(3x^2 + 2)^3}$

$\ln|y| = \ln\left|\dfrac{(x^3 - 2x + 1)(5x - 1)^2}{(3x^2 + 2)^3}\right|$ (Taking logs of both sides of the equation)

$= \ln\left|x^3 - 2x + 1\right| + \ln\left|5x - 1\right|^2 - \ln\left|3x^2 + 2\right|^3$

$\ln|y| = \ln\left|x^3 - 2x + 1\right| + 2\ln\left|5x - 1\right| - 3\ln\left|3x^2 + 2\right|$ $(\log x^n = n\log x)$

$\dfrac{1}{y}\dfrac{dy}{dx} = \dfrac{3x^2 - 2}{x^3 - 2x + 1} + \dfrac{2(5)}{5x - 1} - \dfrac{3(6x)}{3x^2 + 2}$ (differentiating: see scrapwork below)

$\dfrac{dy}{dx} = y\left[\dfrac{3x^2 - 2}{x^3 - 2x + 1} + \dfrac{10}{5x - 1} - \dfrac{18x}{3x^2 + 2}\right]$

$= \dfrac{(x^3 - 2x + 1)(5x - 1)^2}{(3x^2 + 2)^3}\left[\dfrac{3x^2 - 2}{x^3 - 2x + 1} + \dfrac{10}{5x - 1} - \dfrac{18x}{3x^2 + 2}\right]$.

Scrapwork	**Scrapwork**	**Scrapwork**
Let $\left|y_1\right| = \ln\left|x^3 - 2x + 1\right|$	Let $\left|y_2\right| = 2\ln\left|5x - 1\right|$	Let $\left|y_3\right| = 3\ln\left|3x^2 + 2\right|$
Let $u = x^3 - 2x + 1$	$u = 5x - 1$ and $\dfrac{du}{dx} = 5$	$u = 3x^2 + 2$
Then $y_1 = \ln u$		$\dfrac{du}{dx} = 6x$
$\dfrac{dy_1}{du} = \dfrac{1}{u} = \dfrac{1}{x^3 - 2x + 1}$	$\dfrac{dy_2}{du} = \dfrac{2}{u} = \dfrac{2}{5x - 1}$	$y_3 = 3\ln u$
$\dfrac{du}{dx} = 3x^2 - 2$	$\dfrac{dy_2}{dx} = \dfrac{dy_2}{du}\dfrac{du}{dx} = \dfrac{10}{5x - 1}$	$\dfrac{dy_3}{du} = \dfrac{3}{u} = \dfrac{3}{3x^2 + 2}$
$\dfrac{dy_1}{dx} = \dfrac{dy_1}{du}\dfrac{du}{dx}$		$\dfrac{dy_3}{dx} = \dfrac{dy_3}{du}\dfrac{du}{dx} = \dfrac{3(6x)}{3x^2 + 2}$
$= \dfrac{3x^2 - 2}{x^3 - 2x + 1}$		

Lesson 15 Exercises A

1. Use logarithmic differentiation to find $\frac{dy}{dx}$:

(a) $y = a^x$; (b) $y = (x^2 - 4)^5 (x - 2)^3$;

(c) $y = x^x$

2. Use logarithmic differentiation to find $\frac{dy}{dx}$

if $y = \frac{(x^3 - 2x + 1)(5x - 1)^2}{(3x^2 + 2)^3}$

3. If $y^3 = \frac{(x^3 - 2x + 1)(5x - 1)^2}{(3x^2 + 2)^3}$, find $\frac{dy}{dx}$

Answers: 1.a) $a^x \ln a$; (b) $(13x + 6)\left[(x^2 - 4)^4 (x - 2)^3\right]$; (c) $x^x(\ln|x| + 1)$

2. $= \frac{(x^3 - 2x + 1)(5x - 1)^2}{(3x^2 + 2)^3}\left[\frac{3x^2 - 2}{x^3 - 2x + 1} + \frac{10}{5x - 1} - \frac{18x}{3x^2 + 2}\right]$

3. $= \frac{1}{3}\frac{\left[(x^3 - 2x + 1)(5x - 1)^2\right]^{\frac{1}{3}}}{3x^2 + 2}\left[\frac{3x^2 - 2}{x^3 - 2x + 1} + \frac{10}{5x - 1} - \frac{18x}{3x^2 + 2}\right]$

Lesson 15 Exercises B

1 Use logarithmic differentiation to find $\frac{dy}{dx}$:

(a) $y = c^x$; (b) $y = (x^2 - 1)^4 (x - 1)^2$;

(c) $y = x^{x^2}$

2. Use logarithmic differentiation to find $\frac{dy}{dx}$

if $y = \frac{(x^4 - 3x + 1)(4x - 1)^2}{(2x^3 + 1)^3}$

3 If $y^4 = \frac{(x^4 - 3x + 1)(4x - 1)^2}{(2x^3 + 1)^3}$, find $\frac{dy}{dx}$

Answers: 1. (a) $c^x \ln c$; **(b)** $(10x + 2)\left[(x^2 - 1)^3 (x - 1)^2\right]$; **(c)** $x^{x^2 + 1}(2\ln|x| + 1)$

2. $\frac{(x^4 - 3x + 1)(4x - 1)^2}{(2x^3 + 1)^3}\left(\frac{4x^3 - 3}{x^4 - 3x + 1} + \frac{8}{4x - 1} - \frac{18x^2}{2x^3 + 1}\right)$;

3. $\frac{1}{4}\left[\frac{(x^4 - 3x + 1)(4x - 1)^2}{(2x^3 + 1)^3}\right]^{\frac{1}{4}}\left(\frac{4x^3 - 3}{x^4 - 3x + 1} + \frac{8}{4x - 1} - \frac{18x^2}{2x^3 + 1}\right)$

CHAPTER 6
Lesson 16
Rates of Change and Related Rates
Applications of Differentiation I

Rates of Change

A derivative is an instantaneous rate of change. If y is a function of time x, then the rate of change of y with respect to time x is given by $\frac{dy}{dx}$.

Examples: 1. **Speed** is the rate of change of distance s with respect to time t, and would be symbolized by $\frac{ds}{dt}$.

    2.. **Acceleration** is the rate of change of velocity v, with respect to time t, and would be symbolized by $\frac{dv}{dt}$;

    3. The rate of change of **volume**, V, with respect to the radius r, by $\frac{dV}{dr}$.

Example 1 A spherical balloon is being inflated with a gas. At what rate is the volume of the balloon changing instantaneously with respect to the radius of the balloon when the radius is 6.00 cm?

Step 1: Let the volume of the balloon be V, and let its radius be r.

    Then $V = \frac{4}{3}\pi r^3$ (Formula for the volume of a sphere)

    $\frac{dV}{dr} = \left(\frac{4}{3}\pi\right)3r^2 = 4\pi r^2$ (Finding the derivative of V with respect to r)

Step 2: We substitute $r = 6.00$ cm, and $\pi = 3.14$ in the result from step 1.

    Then $\frac{dV}{dr} = 4(3.14)(6)^2$ $\left(\frac{dV}{dr}\Big|_{r=6}\right)$

    $\frac{dV}{dr} = 452$ cubic cm per linear cm of the radius.

The volume is increasing at the rate of 452 cubic cm per linear cm of the radius.

Example 2 If $y = u^2 - 3u + 4$ and $u = 3t^2 + 5t - 6$, find the rate of change of y with respect to t when $t = 3$.

Method 1 $\frac{dy}{du} = 2u - 3$; $\frac{du}{dt} = 6t + 5$	**Method 2** (implicit diff.. w.r.t. time)
$\frac{dy}{dt} = \frac{dy}{du}\frac{du}{dt}$ (explicit chain rule appl.)	$\frac{dy}{dt} = 2u\frac{du}{dt} - 3\frac{du}{dt}$ (Implicit chain rule)
$\frac{dy}{dt} = (2u - 3)(6t + 5)$	$= (2u - 3)\frac{du}{dt}$
$\frac{dy}{dt} = [2(3t^2 + 5t - 6) - 3](6t + 5)$	$\frac{du}{dt} = 6t + 5$
(replacing u by $3t^2 + 5t - 6$)	$\frac{dy}{dt} = (2u - 3)(6t + 5)$
$\frac{dy}{dt} = 36t^3 + 90t^2 - 40t - 75$	$\frac{dy}{dt} = [2(3t^2 + 5t - 6) - 3](6t + 5)$
$\frac{dy}{dt}\Big|_{t=3} = 36(3)^3 + 90(3)^2 - 40(3) - 75$	$= 36t^3 + 90t^2 - 40t - 75$
$= 1587$ (t = 3).	$= 36(3)^3 + 90(3)^2 - 40(3) - 75$
	$= 1587$ (t = 3).

Related Rates Problems

In related rate problems, several variables are related by a single equation and we are interested in the rate of change of one variable with respect to another variable. We observe from the above examples that the rate of change may be specified with respect to any independent variable, even though usually, it is specified with respect to time (time rate of change).

We can obtain a relationship between their respective rates of change by implicitly differentiating both sides of the equation with respect to time. In these problems, the application of calculus (differentiation) is more or less straight forward, and is not the main difficulty you may have; but rather, obtaining the correct relationships (mathematical models) between the variables and quantities is the main difficulty, if any. Sometimes, all the numerical quantities needed to evaluate the required rate of change are given; and sometimes, not all of them are given, and we have to obtain any needed numerical quantity by using the given quantities. (So, remember that when you think that the value of a variable is not given, consider using the given quantities and the proper relationship to obtain this needed value.)

General Procedure for Solving Related Rates Problems

1. As in all word problems, **read. read, and read** the problem, represent the unknowns by letters, and label those quantities that do not change with time as constants and label those that increase as positive and those that decrease as negative (minus sign).

2. Draw a diagram to aid the visualization, especially, if geometry is involved.
3. Write down what is to be found.

4. Write down a relationship between the variables and constants of the problem.
5. Review Lesson 10, and differentiate the equation with respect to the appropriate

variable, and solve for the rate of change (for example, $\frac{dv}{dt}$) asked for in the question.

6. Substitute the given values for the expression on the right-hand-side of the equation, and if all the quantities are not given, check to see if you can calculate them using the given values.
7. Answer the question using a sentence.

Example 3 A spherical balloon is being inflated with a gas at a rate of 3600 cubic cm/minute. At what rate is the radius of the balloon changing instantaneously with respect to time when the radius is 6.00 cm?

Step 1: Let the volume of the balloon be V, and let its radius be r.

Given: $\frac{dV}{dt} = \frac{3600 \text{ cm}^3}{\text{min}}$; $V = \frac{4}{3}\pi r^3$

Required: To find $\frac{dr}{dt}$. $V = f(r)$

$(V = f(r); \ r = g(t))$

Method 1: Differentiate $V = \frac{4}{3}\pi r^3$ implicitly with respect to t. (Lesson 10, Case 2, **chain rule**)

$$\frac{dV}{dt} = 4\pi r^2 \frac{dr}{dt} \quad (1)$$

Step 2: If $r = 6$, $\pi = 3.14$, $\frac{dV}{dt} = 3600$

$3600 = 4(3.14)(6)^2 \frac{dr}{dt}$; and $7.96 = \frac{dr}{dt}$

The radius is increasing at the rate of 7.96 cm / min.

Method 2 (Explicit chain rule appl.)

$$\frac{dV}{dt} = \frac{dV}{dr} \bullet \frac{dr}{dt} \quad (1)$$

From $V = \frac{4}{3}\pi r^3$, $\frac{dV}{dr} = 4\pi r^2$

If $\frac{dV}{dt} = 3600$, $\frac{dV}{dr} = 4\pi r^2$ in (1),

then $3600 = 4\pi r^2 \bullet \frac{dr}{dt}$

$\frac{3600}{4(3.14)(6)^2} = \frac{dr}{dt}$; and

$7.96 = \frac{dr}{dt}$

The radius is increasing at the rate of 7.96 cm / min.

Example 4

Two variables, p and q are related by the equation $2p^2 + q^2 = 6$, and if the rate of change of q with respect to time is 2 cm per sec, find the rate of change of p with respect to time t, if $p = 3$ cm, $q = 18$ cm.

Given $2p^2 + q^2 = 6; \dfrac{dq}{dt} = 2\,\text{cm/sec}; \ p = 3$ cm, $q = 18$ cm.

Required: To find $\dfrac{dp}{dt}$.   ($p = f(t); \ q = g(t)$)

Differentiate implicitly $2p^2 + q^2 = 6$ with respect to time t.

$$4p\frac{dp}{dt} + 2q\frac{dq}{dt} = 0$$

$$4p\frac{dp}{dt} = -2q\frac{dq}{dt}$$

$$\frac{dp}{dt} = -\frac{2q}{4p}\frac{dq}{dt}$$

$$= -\frac{q}{2p}\frac{dq}{dt}$$

$$= -\frac{18}{2(3)}(2) \qquad (\ p = 3 \text{ cm}, q = 18 \text{ cm}, \frac{dq}{dt} = 2 \ cm/\sec)$$

$$= -6 \text{ cm/sec}. \quad (\text{minus sign indicates a decrease.})$$

Therefore, p is decreasing at the rate of 6 cm/sec.

(Here, we are given all the values, and so, to evaluate dp/dt, no other calculation is done.)

Example 5

Three variables x, y and z are related by the Pythagorean theorem, as shown below. As x decreases at the rate of 4 cm per sec, y increases at 2 cm per sec. How fast is z changing when $x = 9$ cm, and $y = 12$ cm?

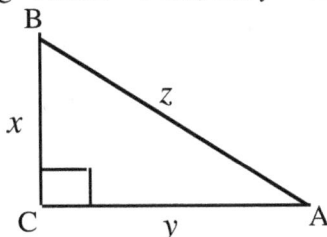

Solution

Given: $z^2 = x^2 + y^2$ (A) , $\frac{dx}{dt} = -4$ cm / sec; (Minus sign because x decreases)

$\frac{dy}{dt} = 2$ cm / sec, $x = 9$, $y = 12$ cm

Required: To find $\frac{dz}{dt}$, when $x = 9$ cm, $y = 12$ cm

Step 1: Differentiate (A) implicitly $z^2 = x^2 + y^2$ with respect to time t .

$$2z\frac{dz}{dt} = 2x\frac{dx}{dt} + 2y\frac{dy}{dt} \qquad (z = f(t); \ x = g(t), \ y = h(t))$$

$$z\frac{dz}{dt} = x\frac{dx}{dt} + y\frac{dy}{dt}$$

$$\frac{dz}{dt} = \frac{x\frac{dx}{dt} + y\frac{dy}{dt}}{z} \qquad \text{(B)}$$

Step 2: Now, we substitute the given numerical values on the right-hand-side of the equation; but we observe that we are not given the value of z. We find it from the Pythagorean theorem: $z^2 = 9^2 + 12^2$ $\quad (81 + 144 = 225)$

$$z = \sqrt{225} = 15$$

Step 3: Now, substitute for z and all the variables (given) on the right-hand side of equation (B). Then,

$$\frac{dz}{dt} = \frac{9(-4) + 12(2)}{15} \quad \text{(We assume the signs of the given numerical values)}$$

$$= \frac{-36 + 24}{15}$$

$$= \frac{-12}{15}$$

$$\frac{dz}{dt} = -\frac{4}{5} \quad \text{(minus sign implies a decrease)}$$

Therefore, z is decreasing at $\frac{4}{5}$ cm/sec.

Example 6 Water is being pumped at the rate of 4 cubic meters per minute into a water tank which is in the shape of a right circular cone. The radius of the base of the cone is 6 meters. and its altitude is 18 meters. How fast is the water level rising when the water level is 3.0 m?

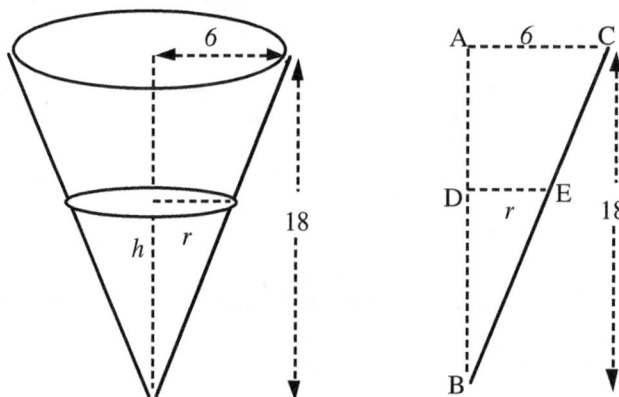

Solution

Let the radius of the surface of the water be r, and let the depth of the water at time t be h. Let the volume of water in the tank at time t be V m^3.

Given: $\frac{dV}{dt} = 4\,\text{m}^3\,/\,\text{min}$, radius of the base of tank = 6 m, altitude = 18 m.

Required: To find $\frac{dh}{dt}$ when $h = 3$ m.   ($V = f(r,h);\ h = g(t)$)

Volume of water in the tank (**a cone**) at anytime is given by $V = \frac{1}{3}\pi r^2 h$ (1)

Since each of V, r, and h is a function of t. and we will take derivatives of all these variables with respect to time, and we are not given any direct information about how r varies with time t, there are two ways to proceed: In method 1, we eliminate r before differentiating; in method 2, we differentiate first and then eliminate r and any derivative of r.

Method 1: We eliminate r by applying similarity of triangles

Since $\triangle ABC$ and $\triangle DBE$ are similar ($\overline{AC}$ is parallel $\overline{DE}$)

$$\frac{r}{6} = \frac{h}{18} \text{ and } r = \frac{h}{3} \quad (r = \frac{6h}{18})$$

$$V = \frac{1}{3}\pi\left(\frac{h}{3}\right)^2 h = \frac{1}{3}\pi\frac{h^2}{9} h \text{ (Substituting } \frac{h}{3} \text{ for } r \text{ in } V = \frac{1}{3}\pi r^2 h)$$

$$V = \frac{\pi h^3}{27}$$

$$\frac{dV}{dt} = \frac{\pi}{27}\cdot 3h^2\frac{dh}{dt} \quad \left(\frac{dV}{dt} = \frac{\pi h^2}{9}\frac{dh}{dt}\right) \text{ (Implicit chain rule . See Lesson 10, Case 2)}$$

$$= \frac{\pi h^2}{9}\frac{dh}{dt}$$

$$4 = \frac{\pi(3)^2}{9}\frac{dh}{dt} \qquad\qquad (\frac{dV}{dt} = 4,\, h = 3)$$

$$\frac{4}{\pi} = \frac{dh}{dt}$$

$$1.3 \approx \frac{dh}{dt}$$

The water level is rising at the rate of 1.3 m/min..

Method 2 We differentiate both sides of $V = \frac{1}{3}\pi r^2 h$ implicitly with respect to

t, followed by differentiating both sides of $r = \frac{h}{3}$ (from $\frac{r}{6} = \frac{h}{18}$)

with respect to t. We want to find $\frac{dh}{dt}$. Review Lesson 10)

Step 1: $V = \frac{1}{3}\pi r^2 h$ (1)

$$\frac{dV}{dt} = \frac{\pi}{3}\frac{d}{dt}(r^2 h)$$

$$\frac{dV}{dt} = \frac{\pi}{3}\left[r^2\frac{dh}{dt} + h(2r)\frac{dr}{dt}\right]$$

(keeping r^2 constant, followed by keeping h constant)

$$\frac{dV}{dt} = \frac{\pi r^2}{3}\frac{dh}{dt} + \frac{2\pi rh}{3}\frac{dr}{dt}$$ (2)

Step 2: Differentiate both sides of $r = \frac{h}{3}$ with respect to t, to obtain

$$\frac{dr}{dt} = \frac{1}{3}\frac{dh}{dt}$$ (3)

Step 3: Substitute for $\frac{dr}{dt}$ from (3) in (2)

Then $\frac{dV}{dt} = \frac{\pi r^2}{3}\frac{dh}{dt} + \frac{2\pi hr}{3} \bullet \frac{1}{3}\frac{dh}{dt}$

$$\frac{dV}{dt} = \left[\frac{\pi r^2}{3} + \frac{2\pi hr}{9}\right]\frac{dh}{dt}$$ (4)

Step 4: Substitute for $\frac{dV}{dt} = 4$, $h = 3$, $r = \frac{h}{3} = \frac{3}{3} = 1$, in (4) to obtain

$$4 = \left[\frac{\pi(1)^2}{3} + \frac{2\pi(3)(1)}{9}\right]\frac{dh}{dt}$$

$$4 = \left[\frac{\pi}{3} + \frac{6\pi}{9}\right]\frac{dh}{dt}$$

$$4 = [\pi]\frac{dh}{dt}$$

$$\frac{4}{\pi} = \frac{dh}{dt}$$

$$\frac{4}{3.14} \approx \frac{dh}{dt}$$

$$1.3 \approx \frac{dh}{dt}$$

$$\frac{dh}{dt} \approx 1.3 \text{ m/minute}$$

The water level is rising at the rate of 1.3 m/minute.

Method 3 Differentiate first, and then substitute for r and its derivatives.

We differentiate both sides of $V = \frac{1}{3}\pi r^2 h$ with respect to t.

$$\frac{dV}{dt} = \frac{\pi}{3}\left[\frac{d}{dt}\left(r^2 h\right)\right]$$

Then $\frac{dV}{dt} = \frac{\pi}{3}\left[r^2 \frac{dh}{dt} + 2hr\frac{dr}{dt}\right]$

$$\frac{3}{\pi} \cdot \frac{dV}{dt} = r^2 \frac{dh}{dt} + 2hr\frac{dr}{dh} \bullet \frac{dh}{dt} \qquad (\frac{dr}{dt} = \frac{dr}{dh} \cdot \frac{dh}{dt})$$

$$= \left(\frac{h}{3}\right)^2 \frac{dh}{dt} + 2h\frac{h}{3} \cdot \frac{1}{3}\frac{dh}{dt} \qquad (\text{From } \frac{r}{6} = \frac{h}{18} \text{ , } r = \frac{h}{3} \text{; and } \frac{dr}{dh} = \frac{1}{3})$$

$$= \left(\frac{h^2}{9} + \frac{2h^2}{9}\right) \cdot \frac{dh}{dt}$$

$$\frac{3}{\pi}\frac{dV}{dt} = \frac{3h^2}{9}\frac{dh}{dt}$$

$$\frac{3}{\pi}(4) = \frac{h^2}{3}\frac{dh}{dt}$$

$$\frac{3}{\pi h^2}(4)(3) = \frac{dh}{dt}$$

$$\frac{36}{\pi(9)} = \frac{dh}{dt} \qquad (h = 3 \text{ cm})$$

$$\frac{4}{\pi} = \frac{dh}{dt}$$

$$\frac{4}{3.14} = \frac{dh}{dt} \text{ m/minute}$$

$$1.3 = \frac{dh}{dt}$$

The water level is rising at the rate of 1 .3 m/minute.

Example 7 Water is poured into a leaky water tank in the shape of a right circular cone at the rate of 12 cubic meters per minute. The tank is 48 meters deep and the radius at the top is 4 meters. Determine how fast the water is leaking away when the water level is 16 meters and the water level is rising at the rate of 3 meters per minute.

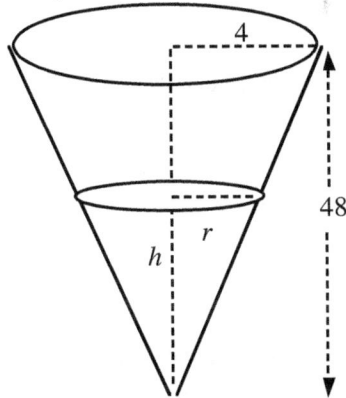

Solution

Let the radius of the surface of the water be r, and let the depth of the water at time t be h. Let the volume of water be V.

Given: $\dfrac{dV_{in}}{dt} = +12$ m³ / min; radius of the conical base of tank = 4 m,

altitude = 48 m.

Required: To find $\dfrac{dV_{\text{leak}}}{dt}$ when $h = 16$ m., and $\dfrac{dh}{dt} = 3\,m/\min$.

The volume of water accumulated in the tank at anytime is given by $V = \frac{1}{3}\pi r^2 h$

Inflow means addition of water; leakage (outflow) means water is being "subtracted".

Method 1: We eliminate r by applying similarity of triangles.

As in the previous example, $\quad \frac{r}{4} = \frac{h}{48}$ and $r = \frac{h}{12}\quad (r = \frac{4h}{48})$

$$V_{in} = \text{Accumulation} + V_{\text{leak}}$$

$$V_{in} = \frac{1}{3}\pi\left(\frac{h}{12}\right)^2 h + V_{\text{leak}} \text{-}$$

$$V_{in} = \frac{1}{3}\pi\frac{h^3}{144} + V_{\text{leak}} \qquad \text{(Substituting } \tfrac{h}{12} \text{ for } r \text{ in } V = \frac{1}{3}\pi r^2 h)$$

$$\frac{d}{dt}(V_{in}) = \frac{1}{3}\pi\frac{d}{dt}\left(\frac{h^3}{144}\right) + \frac{dV_{\text{leak}}}{dt} \qquad \text{(differentiating both sides with respect to } t)$$

$$\frac{dV_{in}}{dt} = \frac{1}{3}\cdot\frac{\pi}{144}\cdot 3h^2\frac{dh}{dt} + \frac{dV_{\text{leak}}}{dt}$$

$$\frac{dV_{in}}{dt} = \frac{\pi h^2}{144}\frac{dh}{dt} + \frac{dV_{\text{leak}}}{dt} \quad \text{(simplifying)}$$

$$12 = \frac{\pi h^2}{144}(3) + \frac{dV_{\text{leak}}}{dt} \quad (\text{ substituting 12 for } \frac{dV_{in}}{dt}, \text{ and 3 for } \frac{dh}{dt})$$

$$12 - \frac{\pi(16)^2}{48} = \frac{dV_{\text{leak}}}{dt}$$

$$12 - \frac{3.14(16)(16)}{48} = \frac{dV_{\text{leak}}}{dt}$$

$$12 - 16.75 = \frac{dV_{\text{leak}}}{dt}$$

$$-4.75 = \frac{dV_{\text{leak}}}{dt}$$

$$\frac{dV_{\text{leak}}}{dt} = -4.75 \ \text{m}^3 \,/\, \text{min} \quad \text{(The minus sign means water is being "subtracted")}$$

The water is leaking away at the rate of $4.75 \ m^3 \,/\, \text{min}$.

Method 2

$$V_{\text{in}} = \text{Accumulation} + V_{\text{leak}}$$

$$V_{\text{in}} = \tfrac{1}{3}\pi r^2 h + V_{\text{leak}}$$

$$\frac{dV_{\text{in}}}{dt} = \frac{\pi}{3}\left[r^2 \frac{dh}{dt} + h(2r)\frac{dr}{dt} \right] + \frac{dV_{\text{leak}}}{dt} \quad \text{(differentiating implicitly with respect to } t)$$

$$= \frac{\pi}{3}\left[r^2 \frac{dh}{dt} + 2hr\frac{dr}{dt} \right] + \frac{dV_{\text{leak}}}{dt}$$

$$= \frac{\pi}{3}\left[r^2 \frac{dh}{dt} + 2hr\frac{dr}{dh}\frac{dh}{dt} \right] + \frac{dV_{\text{leak}}}{dt} \qquad (\frac{dr}{dt} = \frac{dr}{dh}\frac{dh}{dt})$$

$$= \frac{\pi}{3}\left[\left(\frac{h}{12}\right)^2 \frac{dh}{dt} + 2h\frac{h}{12}\left(\frac{h}{12}\right)\frac{dh}{dt} \right] + \frac{dV_{\text{leak}}}{dt} \ (\frac{r}{4} = \frac{h}{48}; \ r = \frac{h}{12}; \frac{dr}{dh} = \frac{1}{12})$$

$$= \frac{\pi}{3}\left[\frac{h^2}{144}\frac{dh}{dt} + \frac{h^2}{72}\frac{dh}{dt} \right] + \frac{dV_{\text{leak}}}{dt}$$

$$= \frac{\pi}{3}\left[\frac{h^2}{144} + \frac{h^2}{72} \right]\frac{dh}{dt} + \frac{dV_{\text{leak}}}{dt}$$

$$= \frac{\pi}{3}\left[\frac{3h^2}{144} \right]\frac{dh}{dt} + \frac{dV_{\text{leak}}}{dt}$$

$$12 = \frac{3.14}{3}\left[\frac{3(16)^2}{144} \right](3) + \frac{dV_{\text{leak}}}{dt} \ (\frac{dh}{dt} = 3; \ h = 16; \ \pi = 3.14; \ \frac{dV_{in}}{dt} = 12)$$

$$12 = 16.75 + \frac{dV_{\text{leak}}}{dt}$$

$$12 - 16.75 = \frac{dV_{\text{leak}}}{dt}$$

$$-4.75 = \frac{dV_{\text{leak}}}{dt}$$

The water is leaking away at the rate of $4.75 \ m^3 \,/\, \text{min}$.

Example 8 A swimming pool is 20 m wide, 30 m long, 8 m deep at the deep end and 3 m deep at the shallow end, so that the bottom of the pool is an inclined plane. Water is being pumped into the pool at the rate of 15 cu. meters per minute. How fast is the water level rising when the level is 3 meters at the deep end of the pool?

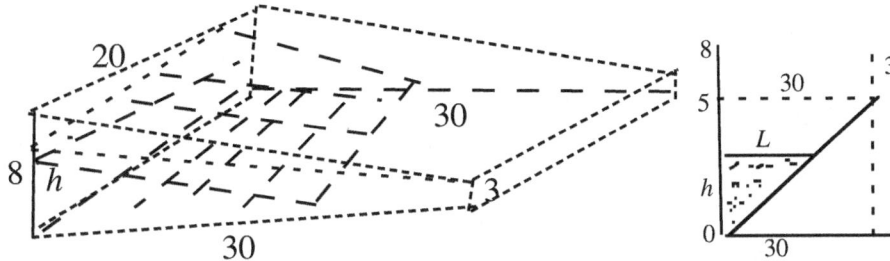

Given: $W = 20$ m, $L = 30$ m, $h_d = 8$ m, $h_s = 3$ m, where W is the width, L is the length, h_d is the height of the deep end, and h_s is the height of the shallow end of a swimming pool. Let h be the depth of the water at the deep end at

time t; $\dfrac{dV_{in}}{dt} = 15$ m^3 / min ,

where $\dfrac{dV_{in}}{dt}$ is the water inflow rate.

Required: To find $\dfrac{dh}{dt}$ when $h = 3$ m.

Solution: Volume, V, of water in pool at depth h is given by

$V = \dfrac{1}{2}WLh$ $V = \dfrac{1}{2}(20)(30 \bullet \dfrac{h}{5})h$ $= 60h^2$	To express L in terms of h : Fill a transparent rectangular container with water, tilt the base and experiment with the water level on one end. Then h is to L as the maximum water level, 5 , is to 30 By similarity of triangles: $\dfrac{h}{L} = \dfrac{5}{30}$ and $L = 6h$ $(0 \leq h \leq 5 = 8 - 3)$

$V = 60h^2$

$\dfrac{dV}{dt} = 60(2h)\dfrac{dh}{dt}$ (differentiating implicitly with respect to t)

$\dfrac{dV}{dt} = 120h\dfrac{dh}{dt}$ (Rate of inflow = Rate of accumulation)

$15 = 120(3)\dfrac{dh}{dt}$ ($\dfrac{dV_{in}}{dt} = 15$ m^3 / min , $h = 3$ m)

$\dfrac{15}{120(3)} = \dfrac{dh}{dt}$

$\dfrac{1}{24} = \dfrac{dh}{dt}$ (or $\dfrac{dh}{dt} = \dfrac{1}{24}$)

The water level is rising at the rate of $\dfrac{1}{24}$ m/min.. (≈ 4cm / min)

Example 9 An airplane flies horizontally at an altitude of 4000 meters and at a speed of 180 kilometers per hour. The airplane passes directly over an observer on the ground. How fast is the airplane approaching the observer when it is 5000 meters away.

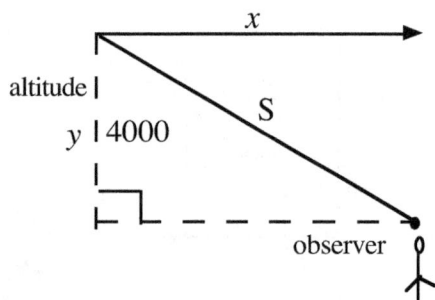

Fig 1

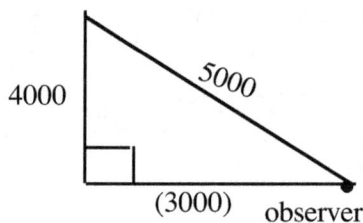

Fig. 2

Given: $\dfrac{dx}{dt} = -180$ km/hr (the minus sign for the approach speed)

Required: To find $\dfrac{dS}{dt}$ when $S = 5000$ m

$S^2 = (4000)^2 + x^2$ (Pythagorean theorem. Fig. 1)

$S^2 = 16 \times 10^6 + x^2$ (Note: x varies, S varies, but $y = 4000$ is constant)

$2S\dfrac{dS}{dt} = 0 + 2x\dfrac{dx}{dt}$ (Differentiating both sides of the equation with respect to t)

$\dfrac{dS}{dt} = \dfrac{2x}{2S}\dfrac{dx}{dt}$

$\dfrac{dS}{dt} = \dfrac{x}{S}\dfrac{dx}{dt}$ (We find x from the Pythagorean theorem, **see scrapwork** below)

$\dfrac{dS}{dt} = \dfrac{3000}{5000}(-180 \text{ km})$ $(S = 5000, y = 4000)$

$\dfrac{dS}{dt} = -108$ km/hr

The airplane is approaching the observer at 108 km/hr.

Scrapwork for (Fig. 2, above)

Finding x using the Pythagorean Theorem.

$(5000)^2 = (4000)^2 + x^2$

$25 \times 10^6 = 16 \times 10^6 + x^2$

$(25 - 16) \times 10^6 = x^2$

$9 \times 10^6 = x^2$

$3 \times 10^3 = x$

$3000 = x$

Lesson 16 Exercises A

1. A spherical balloon is being inflated with a gas at a rate of 3600 cubic cm/minute. At what rate is the volume of the balloon changing instantaneously with respect to radius of the balloon when the radius is 6 cm?

2 If $y = u^2 - 3u + 4$ and $u = 3t^2 + 5t - 6$, find the rate of change of y with respect to t when $t = 3$.

3 A spherical balloon is being inflated with a gas a rate of 3600 cubic cm/minute. At what rate is the radius of the balloon changing instantaneously with respect to time when the radius is 6 cm?

4. Two variables, p and q are related by the equation $2p^2 + q^2 = 6$, and if the rate of change of q with respect to time is 2 cm per sec. Find the rate of change of p with respect to time t , if $p = 3$ cm, $q = 18$ cm.

5. Three variables x, y and z are related by the Pythagorean theorem such that the hypotenuse is z. As x decreases at the rate of 4 cm per sec, y increases at 2 cm per sec. How fast is z changing when $x = 9$ cm, and $y = 12$ cm?

6. Water is being pumped at the rate of 4 cubic meters per minute into a water tank which is in the shape of a right circular cone. The radius of the base of the cone is 6 meters. and its altitude is 18 meters. How fast is the water level rising when the water level is 3 m?

7. Water is poured into a leaky water tank in the shape of a right circular cone at the rate of 12 cubic meters per minute. The tank is 48 meters deep and the radius at the top is 4 meters. Determine how fast the water is leaking away ,when the water level is 16 meters and the water level is rising at the rate of 3 meters per minute.

8 A swimming pool is 20 m wide, 30 long, 8 m deep at the deep end and 3 m deep at the shallow end, so that the bottom of the pool is an inclined plane. Water is being pumped into the pool at the rate of 15 cu. meters per minute. How fast is the water level rising when the level is 3 meters at the deep end of the pool?

9. An airplane flies horizontally at an altitude of 4000 meters and at a speed of 180 kilometers per hour. The airplane passes directly over an observer on the ground. How fast is the airplane approaching the observer when it is 5000 meters away.

Ans.: 1. The volume is increasing at the rate of 452.16 cubic cm per linear cm of the radius

2. When $t = 3$, $\frac{dy}{dt} = 1587$; **3.** The radius is increasing at 7.96 cm / minute.

4. p is decreasing at the rate of 6 cm/sec. **5.** z is decreasing at $\frac{4}{5}$ cm/sec

6. The water level is rising at the rate of 1.3 m/minute
7. The water is leaking away at the rate of 4.75 m/sec

8. The water level is rising at the rate of $\frac{1}{24}$ m/min..

9. The airplane is approaching the observer at 108 km/hr.

Lesson 16 Exercises B

1. A spherical balloon is being inflated with a gas. At what rate is the volume of the balloon changing instantaneously with respect to radius of the balloon when the radius is 4 cm?

2 If $y = u^2 - 2u + 5$ and $u = 4t^2 - 3t - 4$, find the rate of change of y with respect to t when $t = 3$.

3 A spherical balloon is being inflated with a gas a rate of 7200 cubic cm/minute. At what rate is the radius of the balloon changing instantaneously with respect to time when the radius is 12 cm?

4. Two variables, r and s are related by the equation $2r^2 + s^2 = 6$, and if the rate of change of s with respect to time is 2 cm per sec. Find the rate of change of r with respect to time t , if $r = 36$ cm, $s = 48$ cm.

5. Three variables x, y and z are related by the Pythagorean theorem such that the hypotenuse is z. As x decreases at the rate of 6 cm per sec, y increases at 3 cm per sec. How fast is z changing when $x = 9$ cm, and $y = 12$ cm?

6. Oil is being pumped at the rate of 4 cubic meters per minute into an oil tank which is in the shape of a right circular cone. The radius of the base of the cone is 6 meters. and its altitude is 18 meters. How fast is the oil level rising when the oil level is 2 m?

7. Alcohol is poured into a leaky alcohol container in the shape of a right circular cone at the rate of 12 cubic meters per minute. The container is 48 meters deep and the radius at the top is 4 meters. Determine how fast the alcohol is leaking away ,when the alcohol level is 16 meters and the alcohol level is rising at the rate of 3 meters per minute.

8 A swimming pool is 20 m wide, 30 long, 8 m deep at the deep end and 3 m deep at the shallow end, so that the bottom of the pool is an inclined plane. Water is being pumped into the pool at the rate of 15 cu. meters per minute. How fast is the water level rising when the level is 4 meters at the deep end of the pool.

9. An aircraft flies horizontally at an altitude of 4000 meters and at a speed of 200 kilometers per hour. The aircraft passes directly over an observer on the ground. How fast is the aircraft approaching the observer when it is 5000 meters away.

Answers:

1. The volume is increasing at the rate of 201 cm^3 per linear cm of the radius

2. When $t = 3$, $\dfrac{dy}{dt} = 924$

3. The radius is increasing at 3.98 cm/min

4. r is decreasing at the rate of $\dfrac{4}{3}$ cm/sec.

5. z is decreasing at $\dfrac{6}{5}$ cm/sec

6. The water level is rising at the rate of 2.87 m/min

7. The water is leaking away at the rate of 4.75 m/sec

8. The water level is rising at the rate of $\dfrac{1}{32}$ m/min.

9. The airplane is approaching the observer at 120 km/hr.

CHAPTER 7

Lesson 17
Behavior of Functions

Increasing, Decreasing, Positive, and Negative Functions

Introduction

We will need the above terms in describing the behavior of functions. A conceptual understanding of these terms will facilitate a "unified " or " lumped" approach in covering functions. We will make use of the concept of a positive or a negative function in sketching the graphs of polynomial functions, rational functions, exponential functions, logarithmic functions, and trigonometric functions. For example, in order to draw the correct connection (graph) between, say, any two zeros (roots) of a polynomial function, a knowledge of whether the function is positive or negative between the zeros will be invaluable.

Positive and Negative Functions

When we say that a function is **positive** on a certain interval, say , the interval from $x = a$ to $x = b$, we mean that, algebraically, all the y-coordinate values on this interval from a to b are **positive** $(y = f(x) > 0)$ and graphically, the entire curve lies above the x-axis on this interval (Figure 1)

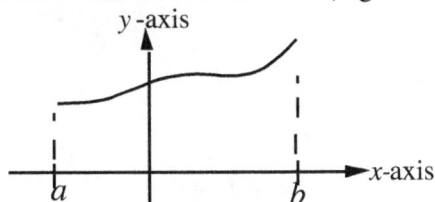

Figure 1

Similarly, when a function is **negative** on the interval from $x = a$ to $x = b$, all the y-coordinate values are negative $(y = f(x) < 0))$ on this interval from a to b, and that, graphically, the entire curve lies below the x-axis (Figure 2).

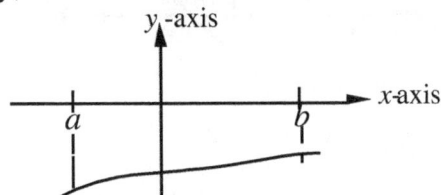

Figure 2

Decreasing and Increasing Functions   138

 We use the above terms in describing critical points, namely, minimum and maximum points, and points of inflection. We will use these terms in describing the behavior of functions as well as in sketching their graphs. For example, we will use the term " concavity " (e.g. concave up or down) in defining critical points (turning points); in describing the behavior of polynomial, rational, trigonometric, exponential and logarithmic functions as well as conic relations, namely the hyperbola and the ellipse.

Increasing Function (strictly monotone increasing function)

We cover this behavior from three points of view, namely the geometric view, the functional view and the derivative (slope) point of view (calculus point of view).

Geometrically, a strictly increasing function is one whose graph (curve or line) rises from left to right, as one reads from left to right (or simply, the curve leans to the right). That is, as we move our eyes horizontally to the right in the coordinate plane, we encounter higher and higher values of y. Thus, any point on the curve is higher than any other point (on the curve) to its left. (Recall from elementary math that a straight line leaning to the right in the page has a positive slope).

More formally, a **function** f is increasing (or strictly increasing) on an interval containing the numbers x_1, x_2 if whenever $x_2 > x_1$, $f(x_2) > f(x_1)$ (Figure 3) (That is, as the x-values increase, the y-values increase.)

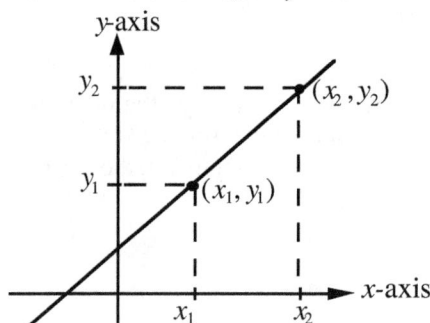

Figure 3: Graph of an increasing function

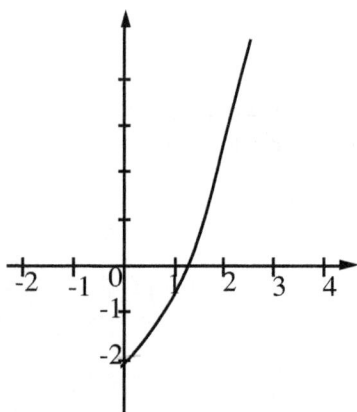

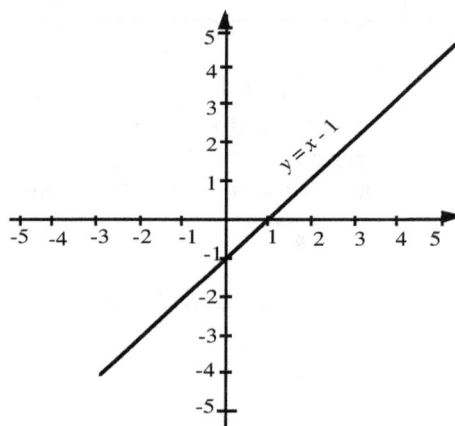

Figure 4: Graph of an increasing function   **Figure 5:** Graph of an increasing function

Example 1 The function given by $f(x) = x - 1$ is an increasing function. (Fig.5)   139

Let us test for $x_1 = 2, x_2 = 5$ (that is $x_2 > x_1$) in $f(x) = x - 1$

Then $f(x_1) = f(2) = 2 - 1 = 1$

$f(x_2) = f(5) = 5 - 1 = 4$

Clearly, $5 > 2$ implies $f(5) > f(2)$ (since $f(5) = 4$ and $f(2) = 1$)

Note: The above is **no** proof, but is only an illustration

Derivative (slope) Point of view (Calculus Point of View)

Theorem

If the function f is differentiable on the open interval (a, b), and if $f'(x) > 0$ on (a, b), then f is increasing on (a, b).

Example 2 Is the function given by $f(x) = x^2$ increasing on $[2, 5]$?

Check: $f'(x) = 2x$

For say $x = 3$, $f'(3) = 2(3) = 6$ (positive)

Since for any number, say 3, in $(2, 5)$ $f'(3) = 2(3) = 6 > 0$,

$f(x) = x^2$ is increasing on $(2, 5)$.

Decreasing Function (strictly monotone decreasing function)

Geometrically, a strictly **decreasing function** is one whose graph (curve or line) falls (drops) from left to right as one reads from the left to the right, (or simply, the curve or line leans to the left). That is, as we move our eyes horizontally from the left to the right in the coordinate plane, we encounter lower and lower values of y. Thus, any point on the curve is lower than any other point (on the curve) to its left (Figure below)

More formally, a **function** f is decreasing (or strictly decreasing) on an interval containing the numbers x_1, x_2 if whenever $x_2 > x_1$, $f(x_2) < f(x_1)$ (That is, as the x-values increase, the y-values decrease) (See Figure 6)

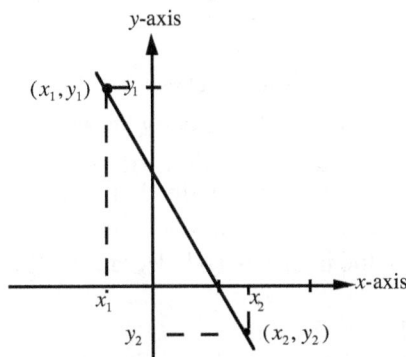

Figure 6: Graph of a decreasing function

Example 3

The function given by $f(x) = -x - 1$ is decreasing on the interval $[-6, -1]$

Let us test for say, $x_1 = -5, x_2 = -2$ (that is, $x_2 > x_1$) in $f(x) = -x - 1$

Then $f(x_1) = f(-5) = -(-5) - 1 = 5 - 1 = 4$

$f(x_2) = f(-2) = -(-2) - 1 = 2 - 1 = 1$. (We could use -6 or -1 for testing)

Clearly, $-2 > -5$ implies $f(-2) < f(-5)$ (since $f(-2) = 1$ and $f(-5) = 4$ and)

Note: The above is **no** proof, but is only an illustration

Derivative (slope) Point of view (Calculus Point of View)
Theorem

If the function f is differentiable on the open interval (a, b), and if $f'(x) < 0$
on (a, b), then f is decreasing on (a, b).

Example

Is the function given by $f(x) = x^2$ decreasing on $(-5, -1)$?

Check: $f'(x) = 2x$

For say $x = -4$, $f'(-4) = 2(-4) = -8$ (negative)

Since for any number, say -4 in $(-5, -1)$, $f'(-4) = 2(-4) = -8 < 0$,

Therefore, $f(x) = x^2$ is decreasing on $(-5, -1)$.

We must note that some functions are neither increasing nor decreasing.
A function which is neither increasing nor decreasing is called a **constant
function.**

Constant function

A function f is **constant** if it is neither increasing nor decreasing.

That is, a function f is constant on an interval containing the numbers x_1, x_2
if whenever $x_2 > x_1$, $f(x_2) = f(x_1)$.

The graph of $f(x) = 5$ in an example of a constant function.

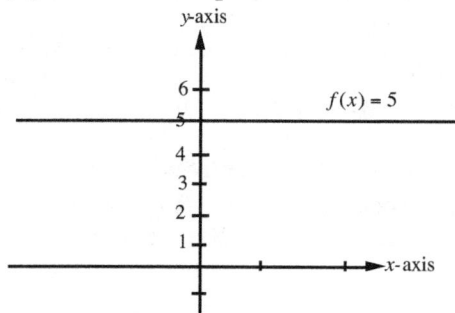

Figure 7: The graph of a constant function: $f(x) = 5$.

About endpoints of increasing intervals and decreasing intervals

In writing the intervals of increasing or decreasing functions from graphs, we
will include the endpoints whenever the endpoints are defined. For example,
with reference to the diagram below, $f(x)$ is increasing on the intervals
$[-2, -1)$, and $[1, 2]$, and decreasing on the intervals $(-1, 1]$, and $[-2, 3]$.

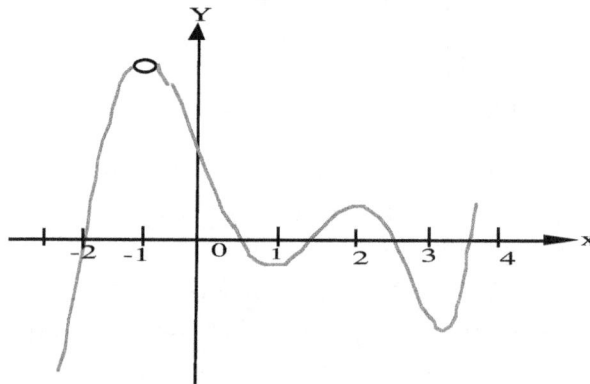

Other Terminology

Nonincreasing function (monotone decreasing function):

A function f is nonincreasing on an interval containing the numbers x_1, x_2

if whenever $x_2 > x_1$. $f(x_2) \leq f(x_1)$

(Note that nonincreasing implies "does not increase" and therefore, it is either constant or decreasing on that interval) Note: Using nonincreasing is consistent with terms such as a "nonpositive" integer which is either zero or a negative integer.

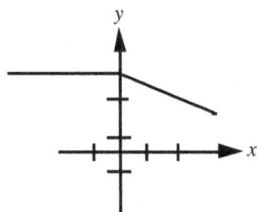

Figure: Graph of a nonincreasing function

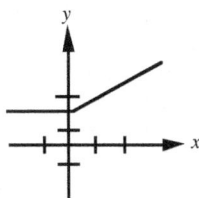

Figure: Graph of a nondecreasing function

Nondecreasing function (monotone increasing function):

A function f is nondecreasing on an interval containing the numbers x_1, x_2

if whenever $x_2 > x_1$, $f(x_2) \geq f(x_1)$

(Note that nondecreasing implies "does not decrease" and therefore, it is either constant or increasing on that interval)

Monotone function A monotone function is either nonincreasing or nondecreasing.

Strictly monotone function A Strictly monotone function is either increasing or decreasing?

Piecewise monotone function

If every finite interval on which a function is defined can be divided into finitely many intervals such that the function is monotone on each of these intervals, then the function is piecewise monotone.

or

A function is piecewise monotone if a given finite interval on which the function is defined can be divided into finitely many intervals such that the function is monotone on each of these intervals.

Example: For $f(x) = x^2$, f is increasing when $x \geq 0$ and decreasing when $x \leq 0$.

Note:
On an interval, a monotone increasing function never decreases.
On an interval, a monotone decreasing function never increases.

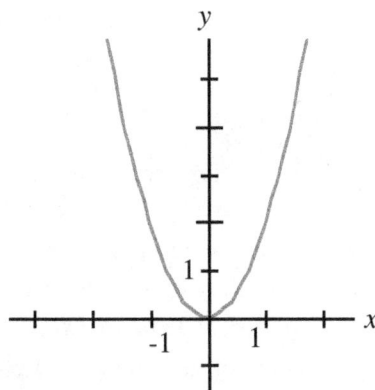

Revisit the next page as you cover subsequent lessons.

About the Next Topic and Lessons 18-21:
Applications of the Derivative: Intervals for Increasing and Decreasing Functions, Minima, Maxima, Concavities and Points of Inflection

Let x_0 be a zero of the first derivative. The slopes of the tangent lines at points immediately before and after the point x_0 are given by the first derivatives.

Uses of the First Derivative Test (We will use a sign diagram)

1. To determine the **intervals** for increasing and decreasing functions.
Here, we find the first derivative, equate it to zero and solve to obtain critical values, which will be used on a sign diagram for testing. (see. p.143)

2. To find the **relative minima and maxima.**
Here also, we find the first derivative, equate it to zero and solve to obtain critical values to be used on a sign diagram for testing.(see p151, 153)

3. To find the **absolute** minima and absolute maxima of functions.(see p.158)
We find the first derivative, set it to zero and solve to obtain critical values which will be used to evaluate the original functional equation. We also evaluate the functional equation for the end points of the domain of the function. The minimum and maximum values may occur at the end points.

4. For **applied minima** and minima problems. (see p.163)

5. To find **inflection points**: Let x_0 be a critical x-value of f. On a sign diagram, if the sign of $f'(x)$ immediately to the left of x_0 is the same as the sign of $f'(x)$ immediately to the right of x_0, then x_0 is at an inflection point.

6. To find **concavity.** (see p.152 for some description of " $f'(x)$ increasing")
If $f'(x)$ is increasing on an interval, then f is concave up on that interval
If $f'(x)$ is decreasing on an interval, then f is concave down on that interval.

Uses of the Second Derivative Test

1. To find the **relative minima** and maxima.
Here, we find the first derivative, set it to zero and solve to obtain critical values. We then find the second derivative and then evaluate the second derivative using the critical values from the first derivative. For a critical x_0,

if $f''(x_0) > 0$ then $f(x_0)$ is a relative minimum; (curve is concave up)

if $f''(x_0) < 0$, $f(x_0)$ is a relative maximum; (curve is concave down) but

if $f''(x_0) = 0$ or is infinite, use the **first derivative** test.

2. To determine **concavities** and points of **inflection**.
Find the first derivative, and find the second derivative. Set the second derivative to zero and solve to obtain values tfor testing intervals for concavity and inflection points, using a sign diagram similar to the one used in the first derivative test. Concavity is determined by the sign of the second derivative.

Let x_{02} be a zero of the second derivative; $x_{02+} =$ to the right of x_{02}; $x_{02-} =$ to its left.

If $f''(x_{02+}) > 0$ in (a, b), f is concave up in (a, b), but concave down if $f''(x_{02+}) < 0$

If $f''(x_{02-}) > 0$ in (a, b), f is concave up in (a, b), but concave down If $f''(x_{02-}) < 0$

3. To determine **inflection points** The point $(x_{02}, f(x_{02}))$ is an inflection point of f if $f''(x_{02}) = 0$ or is undefined and $f''(x)$ changes sign as x increases through x_{02} or $f'''(x_{02}) \neq 0$ if $f'''(x_{02})$ exists. **On a sign diagram, an inflection point is at x_{02} where the second derivative changes sign.**

Application of the derivative to determine intervals for increasing and decreasing functions

Example Determine the intervals on which the function $f(x) = x^3 + 3x^2 - 3$ is increasing or decreasing.

Solution

Step 1: Find the derivative of $f(x) = x^3 + 3x^2 - 3$

Then $f'(x) = 3x^2 + 6x$

Step 2: Equate $f'(x) = 3x^2 + 6x$ to zero and solve for x by factoring or by the quadratic formula: We solve by factoring since the expression is easily factorable.

$3x^2 + 6x = 0$ and factoring $3x(x + 2) = 0$ and solving,

$x = 0$ or $x = -2$

Step 3: There are a number of methods for determining where $3x(x + 2) > 0$ and

where $3x(x + 2) < 0$; however we use the method of sign diagrams.

The product of two factors of the same sign is positive but the product of two factors of different signs is negative. We determine the intervals on which

$3x(x + 2)$ is positive or negative.

Step 4: To find where each factor is positive, assume that each linear factor is positive and solve for x. (Note that saying that $3x$ is positive is the same as saying $3x > 0$). Then for $3x > 0$, we solve to obtain $x > 0$.. Thus, any number greater than 0 makes the linear factor $3x$ positive, and any number less than 0 makes the linear factor $3x$ negative. We show these results on a sign diagram below.

Similarly, for $(x + 2) > 0$, we solve to obtain $x > -2$. Thus, any number greater than -2 makes $(x + 2)$ positive, and any number less than -2 makes $(x + 2)$ negative.

Step 5: Draw a number line and mark the points $x = 0$, $x = -2$

The two points divide the line into three intervals.

Step 6 We write plus signs ("+") over those values of x which make the linear factor positive; and write minus signs ("-") over those values of x which make the linear factor negative

	Factor	Column 1	Column 2	Column 3
		Signs of the intervals		
Row 1	$3x$	$-$	$-$	$+$
Row 2	$x + 2$	$-$	$+$	$+$
Row 3	Product or quotient $(3x)(x + 2)$	$+$	$-$	$+$

Number line: $-\infty$ -2 0 $+\infty$

For $3x$, we write plus signs to the right of the point $x = 0$ and minus signs to its left. For $(x + 2)$, we write plus signs to the right of the point $x = -2$ and minus signs to its left. The linear factors are in the first two rows and the product of the factors is in the third row. The signs in the last row are the results of multiplying the signs in the columns above it. For example, in column 1, the product of the minus signs gives a plus sign in row 3 in that column. For column 2, the product of the minus sign in Row 1 and the plus sign in Row 2 gives a minus sign as a product (column 2, row 3). Finally in column 3, the product of the two plus signs gives a plus sign (column 3, row 3).

Step 7: We inspect the row for the product (row 3) to determine which intervals have plus signs and which intervals have minus signs. The function is increasing on those intervals with plus signs (positive slopes) but decreasing on those intervals with minus signs (negative slopes). Inspection of the columns in row 3, shows that columns 1 and 3 have plus signs and column 2 has a minus sign.

Conclusion: 1. The function is increasing if $x \leq -2$ and also if $x \geq 0$.
        2. The function is decreasing if $-2 \leq x \leq 0$.
        See also the graph of $f(x) = x^3 + 3x^2 - 3$, below.

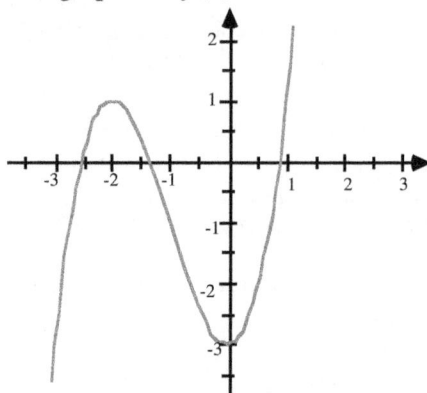

Graph of $f(x) = x^3 + 3x^2 - 3$

Lesson 17 Exercises A

Give both graphical and algebraic answers (with examples) to the following:
1. What is meant by saying that a function is positive on an interval?

2. What is meant by saying that a function is negative on an interval?

3. What is meant by saying that a function is increasing on an interval?
4. What is meant by saying that a function is decreasing on an interval.
5. What is a constant function? Give an example

6. Is the function given by $f(x) = x^2$ increasing on $(2, 5)$?

7.. Is the function given by $f(x) = x^2$ increasing on the interval $(-5, -1)$?
8. Define (a) nonincreasing function; (b) Nondecreasing function
   (c) Monotone function
9. Apply the derivative to determine on which intervals is the following

function is increasing or decreasing: $f(x) = x^3 + 3x^2 - 3$

Answers: 6. Yes; **7.** No; **9.** Increasing if $x \leq -2$ and also if $x \geq 0$, and
        decreasing if $-2 \leq x \leq 0$

Lesson 17 Exercises B

Give both graphical and algebraic answers (with examples) to the following:

1. What is meant by saying that a function is negative on an interval?

2. What is meant by saying that a function is positive on an interval?

3. What is meant by saying that a function is decreasing on an interval?
4. What is meant by saying that a function is increasing on an interval.
5. What is a constant function? Give an example

6. Is the function given by $f(x) = x^2$ increasing on $(3, 6)$?

7.. Is the function given by $f(x) = x^2$ increasing on the interval $(-4, -1)$? .

8. Define (a) nonincreasing function; (b) Nondecreasing function
   (c) Monotone function

9. Apply the derivative to determine on which intervals is the following

   function is increasing or decreasing: $f(x) = x^3 - 3x^2 - 2$

Answers: 6. Yes ; 7. No ; 9. Increasing if $x \le 0$ and also if $x \ge 2$, and
   decreasing if $0 \le x \le 2$.

Lesson 18
Concavities of Curves; Critical Points

Concavities of Curves

As it was in the case of increasing and decreasing behavior, we cover the geometric point of view as well as the derivative (slope) point of view.

Geometric point of view

Sometimes, the curve of a decreasing function will bend (or turn) and begin to increase or the curve of an increasing function will bend and begin to decrease on an interval. When any of this happens, the shape of the curve around the bending or turning point is that of a cup or a bowl (concavity). If the opening of the so formed cup opens upwards, we say that the curve is **concave up** but if the curve opens downwards, we say that the curve is **concave down**. We also refer to whether it opens up or down as the "sense of concavity " or the direction of concavity. When a curve is concave up, the tangent to the curve at the bending point is below the curve, but if the curve is concave down, the tangent to the curve is above the curve.

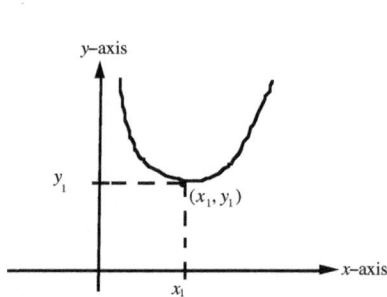

Figure: Curve is concave up
(opens upwards) at (x_1, y_1)

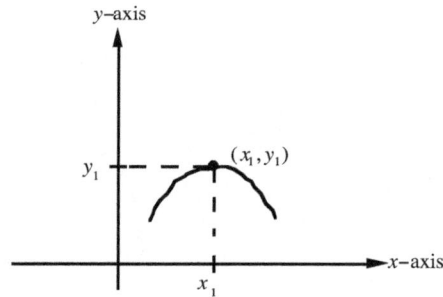

Figure: Curve is concave down
(opens downwards) at (x_1, y_1)

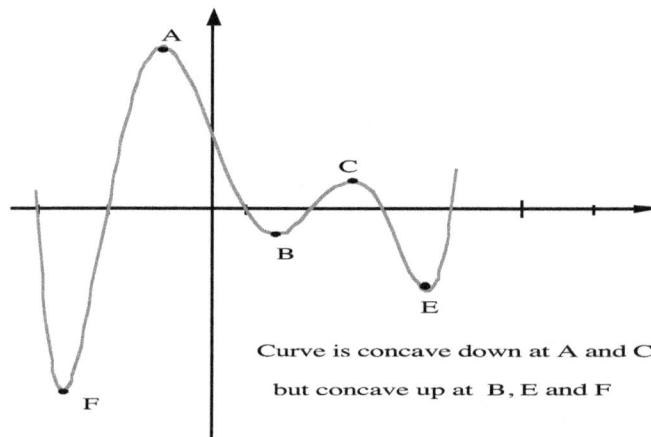

Curve is concave down at A and C;
but concave up at B, E and F

So far as functions and concavity are concerned, only concavity upwards and concavity downwards are of interest.

Derivative (slope) Point of view (Calculus Point of View)

Definition: If f' is increasing (that is, $f''(x) > 0$) on an open interval, then f is concave up on this interval; but if f' is decreasing (that is, $f''(x) < 0$) on an open interval, then f is concave down on this interval. See also p.152.

Theorem (using the second derivative)

If $f''(x) > 0$ on the open interval (a, b), then f is concave up on (a, b); but if $f''(x) < 0$ on the open interval (a, b), then f is concave down on (a, b).

Example 1 Determine concavities of $f(x) = x^3 - x^2 + 5$.(see also p.150)

Solution

Step 1: Find the second derivative:

$$f(x) = x^3 - x^2 + 5$$

$$f'(x) = 3x^2 - 2x$$

$$f''(x) = 6x - 2$$

Step 2: The curve is concave up if

$$6x - 2 > 0$$

$$x > \frac{1}{3};$$

but concave down if

$$6x - 2 < 0$$

$$x < \frac{1}{3}.$$

(We can use this approach for a linear inequality. For a quadratic or higher degree inequality we will use a sign diagram. See p. 150)

Extra: Note below that there is an inflection point at $x = \frac{1}{3}$

The curve is concave up where $x > \frac{1}{3}$ and concave down where $x < \frac{1}{3}$.

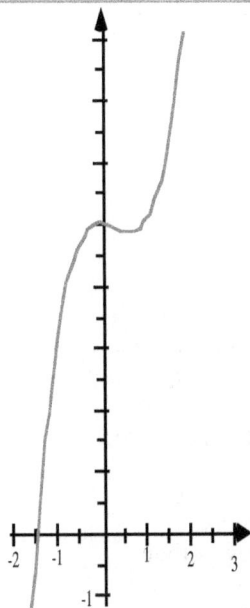

Figure: Graph of $f(x) = x^3 - x^2 + 5$

Some Notables

If $f'(x) > 0$ for all x in (a, b),
then f is increasing in (a, b)

If $f''(x) > 0$ for all x in (a, b),
then f' is increasing in (a, b)

If f' is increasing in (a, b),
then f is concave up in (a, b)

If $f'(x) < 0$ for all x in (a, b),
then f is decreasing in (a, b)

If $f''(x) < 0$ for all x in (a, b),
then f' is decreasing in (a, b)

If f' is decreasing in (a, b),
then f is concave down in (a, b)

Note carefully above, the relationships between f and f' and between f' and f'' .

Critical Points

Turning Points, Minimum Points, Maximum Points, Inflection Points

Pre-Calculus point of view

A critical point is a turning point on a curve. A critical point may be maximum point, a minimum point or an inflection point.

Maximum Point: A given point (turning point) on a curve is a maximum point of the curve if the given point is higher than any other point on the curve in the immediate vicinity (on both sides of the point). The y-coordinate of this point is called a maximum value of the curve. At a maximum point, the curve is concave downwards, and the tangent to the curve at this point lies above the curve. Sometimes, a maximum point of a curve may not be the highest point on the curve. When this happens, this maximum point is higher than only points sufficiently near it and there are other points on the curve which are higher than this point. In this case, we call this maximum point a relative maximum. However, if this maximum point is the highest point on the curve (irrespective of the domain of the function), then we say that this point is the absolute maximum. In the figure below, the points **A** and **C** are maximum points.

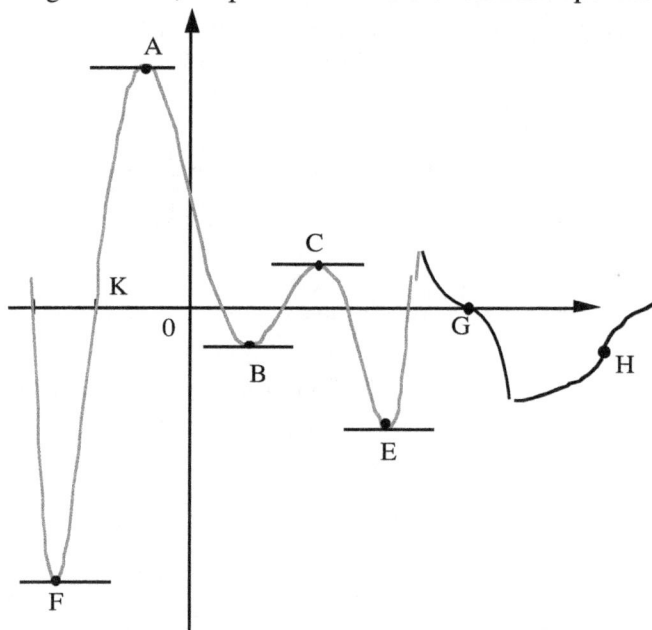

Figure 1: Graph showing critical points

Minimum Point: A given point (turning point) on a curve is a minimum point of the curve if the given point is lower than any other point on the curve in the immediate vicinity (on both sides of the point). At a minimum point, the curve is concave upwards, and the tangent to the curve at this point lies below the curve. The y-coordinate of this point is called a minimum value of the curve. Sometimes, a minimum point of a curve may not be the lowest point on the curve. When this happens, this minimum point is lower than only points sufficiently near it and there are other points on the curve which are lower than this point. In this case, we call this minimum point a relative minimum. However, if this minimum point is the lowest point on the curve (irrespective of the domain of the function), then we say that this point is the absolute minimum. In the above figure, the points **F** and **B** and **E** are minimum points.

Generally, for a polynomial of degree n, there are at most $n-1$ relative minima or maxima. For example, $y = x^2$ has only one minimum point (and this point is an absolute minimum).

Inflection Point: A given point (turning point) on a curve is an inflection point if the curve at this point, changes from being concave up to being concave down or vice versa. At a point of inflection, the curve is either concave down to the left of the point **and** concave up to the right of the point, or it is concave up to the left of the point **and** concave down to the right of the point. Thus, the sense of concavity to the right of the point is opposite to the sense of concavity to the left of the point. Note that a point of inflection is **neither** a minimum point nor a maximum point. Point **G** is an inflection point. At point **G**. the curve is concave up to the left of this point and concave down to the right of this point (Figure 1, above).

Derivative (Slope) Point of View (Calculus Point of View)

A critical point is a point on a curve at which either the first derivative of the function is zero, or the first derivative does not exist. A **critical x-value** for a function f is a number x_0 in the domain of f such that either $f'(x_0) = 0$, or $f'(x_0)$ does not exist but $f(x_0)$ is defined. Note that the slope of the tangent line at a critical value is either zero or undefined. When $f'(x_0) = 0$, x_0 is called a stationary point. A critical point may be a maximum point, a minimum point or an inflection point.

Note: With reference to Figure 1, above, $f'(x_0) = 0$ at A, B, C, E, and F.

Extreme Value Theorem

If f is continuous on the closed interval $[a, b]$ then f attains a maximum at some point in $[a, b]$, and also a minimum at some point in $[a, b]$.

Lesson 18 Exercises

1. Determine the concavities of $f(x) = x^3 - x^2 + 5$

2. State Extreme Value Theorem.

3. Identify the concavities of the following graphs:

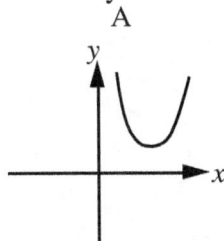

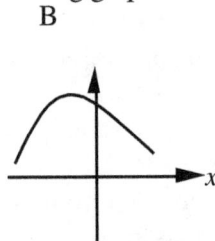

4. Determine the concavities of $f(x) = x^3 - 2x + 5$

Answers: 1. Concave up where $x > \frac{1}{3}$ and concave down when $x < \frac{1}{3}$

3. : **A.** Concave up; **B.** Concave down.

4. Concave up where $x > 0$ and concave down where $x < 0$

More on Concavity and Inflection Points

Example 2   Determine the concavity and inflection points of
$$f(x) = x^4 - 2x^3 - 12x^2 + 12x$$

Step 1: Differentiate $f(x) = x^4 - 2x^3 - 12x^2 + 12x$. to obtain
$$f'(x) = 4x^3 - 6x^2 - 24x + 12$$

Step 2:: Differentiate $f'(x) = 4x^3 - 6x^2 - 24x + 12$ to obtain

the second derivative, $f''(x) = 12x^2 - 12x - 24$

Step 3: Set $f''(x) = 12x^2 - 12x - 24$ to zero. and solve to obtain values for

the sign diagram and for possible points of inflection

Then $12x^2 - 12x - 24 = 0$

$12(x^2 - x - 2) = 0$

$(x + 1)(x - 2) = 0$

$x = -1 \quad x = 2$

(The technique in this example is analogous to the first derivative test for determining minima and maxima.)

Step 3: Determine the sign of $(x + 1)(x - 2)$ within the intervals

$x < -1, \ -1 < x < 2, \ x > 2$ (review p.143)

We draw a sign diagram for $f''(x)$

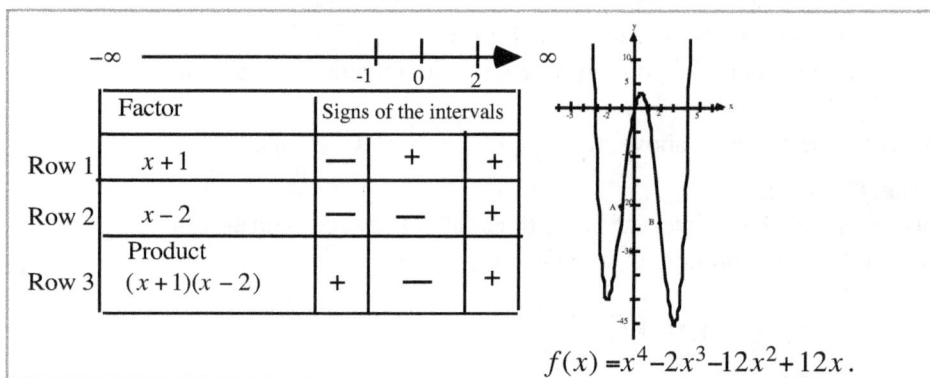

	Factor	Signs of the intervals		
Row 1	$x + 1$	—	+	+
Row 2	$x - 2$	—	—	+
Row 3	Product $(x + 1)(x - 2)$	+	—	+

$f(x) = x^4 - 2x^3 - 12x^2 + 12x$.

Step 4: Determine the signs in the row for $12(x + 1)(x - 2)$ in **Row 3.**

To the left of -1. (column 1) $f''(x)$ is positive (+), That is $f''(x) > 0$

Therefore f is concave up for $x < -1$.

In columns 2, $f''(x)$ is negative (-). That is $f''(x) < 0$

Therefore, f is concave down for $-1 < x < 2$.

In column 3, $f''(x)$ is positive (+). That is $f''(x) > 0$.

Therefore, f is concave up for $x > 2$.

Also, at $x = -1$, $12(x + 1)(x - 2)$ changes sign from "+" to "- " in column 2; and therefore -1 is an inflection point. Also, at $x = 2$, there is a sign change from "-" (in column 2) to "+" (icolumn 3) and therefore, 2 is a point of inflection.

Alternatively,	$f''(-1) = 0$; $f'''(-1) = 12(-2 - 1) = -36 \neq 0$
$f''(x) = 12(x^2 - x - 2)$	$f''(2) = 0$; $f'''(2) = 12(4 - 1) = 36 \neq 0$
$f''' = 12(2x - 1)$	The inflection points are at $x = -1$, and $x = 2$.

Note above that, -1, and 2 are zeros from $f''(x) = 0$.

Lesson 19
Relative Maxima and Minima
First Derivative Test for Relative Maxima and Minima

Let x_0 be the only critical x-value of f in the open interval (a, b), and let f be continuous on the closed interval $[a, b]$. Then

Case 1: If $f'(x) > 0$ for all $x \in (a, x_0)$ and if $f'(x) < 0$ for all $x \in (x_0, b)$, then $f(x_0)$ is a relative maximum. ($f'(x)$ changes sign from $+ve$ to $-ve$).

Case 2: If $f'(x) < 0$ for all $x \in (a, x_0)$ and if $f'(x) > 0$ for all $x \in (x_0, b)$, then $f(x_0)$ is a relative minimum. ($f'(x)$ changes sign from $-ve$ to $+ve$).

Case 3: f may have a relative maximum or a minimum value $f(x_0)$, even though $f'(x)$ does not exist if $f(x_0)$ is **defined.** The x_0 in this case is also considered to be a **critical** x-value. If $f(x_0)$ is not defined, x_0 is not a critical x-value; but we may use x_0 for the purposes of locating intervals. We may also device another method

Case 4: If $f'(x)$ does not change sign, then $f(x_0)$ is neither a relative maximum nor a relative minimum; but $(x_0, f(x_0))$ is an inflection point.

Note above: (Extrema occur at the critical x-values where f' changes signs)

Case 1 implies that as x increases through x_0, if $f'(x)$ changes sign from $+$ to $-$, then $f(x_0)$ is a maximum.

Case 2 implies that as x increases through x_0, if $f'(x)$ changes sign from $-$ to $+$ then $f(x_0)$ is a minimum..

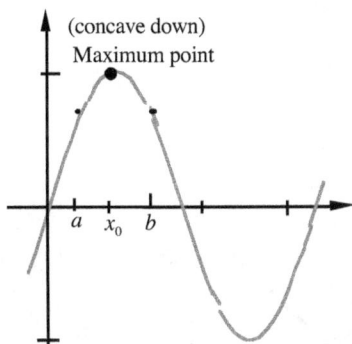

Below: Memory Device

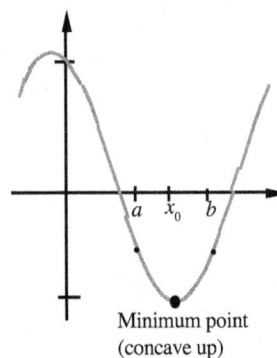

With reference to concavity:

1. If f is concave down at x_0, then $f(x_0)$ is a maximum.

2 If f is concave up at x_0, then $f(x_0)$ is a minimum.

Some meaning of $f'(x)$ increases or decreases See also p. 147.

The first derivative $f'(x)$) of a function at a point on the graph is given by the slope of the tangent line at that point. As the tangent line moves from left to right counterclockwise on the outside of an $\cup$-shaped curve. the slope of the line would be increasing, since the leaning of the line would go from leaning to the left or from a horizontal position to leaning to the right. However, if the tangent line moves from left to right clockwise along the outside of an $\cap$-shaped curve, the slope of the line would be decreasing, since the leaning of the tangent line would go from leaning to the right, or from a horizontal position to leaning to the left. Note that a line leaning to the right in the page has a positive slope and a line leaning to the left in the page has a negative slope.

Memory device: Concavity, minima and maxima (See also p.151)

Let x_{02} be a zero of the second derivative; x_{02+} = to the right of x_{02}; x_{02-} = to its left.

If $f''(x_{02+}) > 0$ in (a, b), f is concave up in (a, b), but concave down if $f''(x_{02+}) < 0$

If $f''(x_{02-}) > 0$ in (a, b), f is concave up in (a, b), but concave down If $f''(x_{02-}) < 0$

Associate "> 0" with concave up; associate "< 0" with concave down; and infer minimum and maximum points from the shapes of "$\cup$ "and "$\cap$".

Some relationships between critical points, relative extrema, inflection points and the derivative

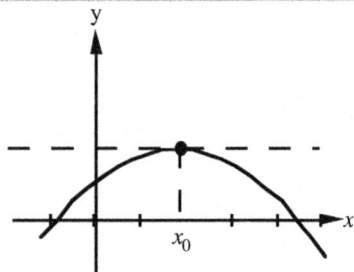

critical point at x_0

1. $f'(x_0) = 0$;

2. Horizontal tangent at x_0

3. Relative maximum

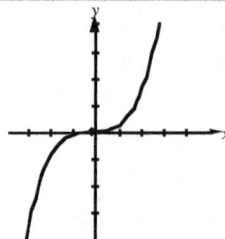

critical point at $x = 0$

1. $f'(0) = 0$;

2. Horizontal tangent

3. Inflection point at $(0,0)$

4. No extremum

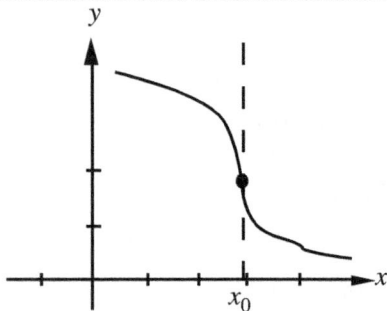

critical point at x_0

No relative extremum

vertical tangent

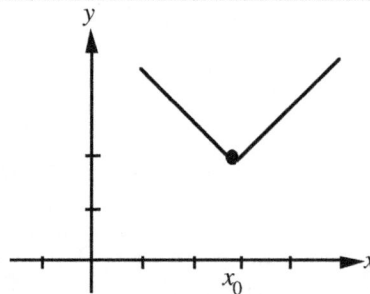

relative minimum at x_0

critical point at x_0

$f'(x_0)$ is undefined.

Note: x_0 means a zero obtained by setting the first derivative to zero. x_{02} is a zero obtained by setting the second derivative to zero.

Second Derivative Test for Relative Maxima and Minima

For a critical x-value x_0, if $f''(x_0) < 0$, $f(x_0)$ is a relative maximum; but
if $f''(x_0) > 0$ then $f(x_0)$ is a relative minimum; Note; x_0 is a zero of Ist derivative.
but if $f''(x_0) = 0$ or is infinite, use the **first derivative** test.

Example 1 Determine the relative maxima and minima for $f(x) = x^4 - 8x^2 + 1$
Method 1 Using the First Derivative Test

Step 1: Differentiate $f(x) = x^4 - 8x^2 + 1$,
Then $f'(x) = 4x^3 - 16x$
Step 2: Equate $f'(x)$ to zero and solve for x

$$4x^3 - 16x = 0$$
$$4x(x^2 - 4) = 0$$
$$4x(x + 2)(x - 2) = 0$$

and $x = 0$, 2, and -2 ; and the critical
x-values are $-2, 0,$ and 2

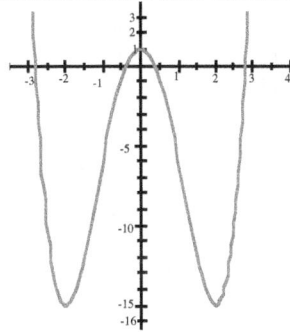

Graph: $f(x) = x^4 - 8x^2 + 1$.

Step 3: Determine the sign of $f'(x) = 4x(x + 2)(x - 2)$ within the intervals
$x < -2, \quad -2 < x < 0, \quad 0 < x < 2, \quad x > 2$ (review p.143)
We use a sign diagram. (You can also use point testing)

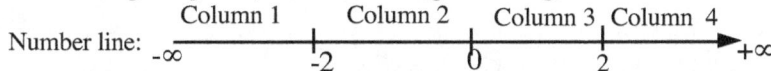

Number line:

		Column 1	Column 2	Column 3	Column 4
	Factor	Signs of the intervals			
Row 1	$4x$	—	—	+	+
Row 2	$x - 2$	—	—	—	+
Row 3	$x + 2$	—	+	+	+
Row 4	Product or quotient $4x(x - 2)(x + 2)$	—	+	—	+

Step 4: Determine the sign changes in the row for $4x(x + 2)(x - 2)$ in **Row 4.**
To the left of -2, (column 1) $f'(x)$ is negative $(-)$, and to the right of -2
(column 2), $f'(x)$ is positive $(+)$. Therefore $f'(x)$ changes sign from
minus to plus and therefore the **critical** x-value -2 is at a relative
minimum point; At $x = 0$, $f'(x)$ changes from $+$ (column 2) to $-$ (column 3),
and the critical x-value 0 is at a relative **maximum** point. At $x = 2$, $f'(x)$
changes from $-$ (column 3) to $+$ (column 4), and therefore 2 is at a relative
minimum point. The minimum values (y-values) are
$f(-2) = (-2)^4 - 8(-2)^2 + 1 = -15$, and $f(2) = (2)^4 - 8(2)^2 + 1 = -15$ (repetition).
The maximum value is $f(0) = (0)^4 - 8(0)^2 + 1 = 1$. The relative minimum points
are at $(-2, -15)$ and $(2, -15)$ and the relative maximum point is at $(0, 1)$.We also
add (From Row 4)**:** f is decreasing between $-\infty$ and -2; increasing between -2
and 0; decreasing between 0 and 2; and increasing between 2 and $+\infty$.

Method 2 Using Second Derivative Test (Steps 1- 2 are as in Method 1)
$$f(x) = x^4 - 8x^2 + 1$$

Step 1: Differentiate $f(x) = x^4 - 8x^2 + 1$. Then we obtain
$$f'(x) = 4x^3 - 16x$$

Step 2: Equate $f'(x)$ to zero and solve for x

$4x^3 - 16x = 0$

$4x(x^2 - 4) = 0$; $4x(x + 2)(x - 2) = 0$ and solving

$x = 0, 2,$ and -2 ; and the critical x-values are $-2, 0,$ and 2.

Step 3: Differentiate $f'(x) = 4x^3 - 16x$ to obtain the second derivative $f''(x)$

Then $f''(x) = 12x^2 - 16$

Step 4: Test the critical values $-2, 0,$ and 2 from step 2 in $f''(x)$

$f''(-2) = 12(-2)^2 - 16 = 32 > 0$ **<---**yields a minimum value, since $f''(-2) > 0$

$f''(0) = 12(0)^2 - 16 = -16 < 0$**<---**yields a maximum value, since $f''(0) < 0$

$f''(2) = 12(2)^2 - 16 = 32 > 0$ **<---**yields a minimum value, since $f''(2) > 0$

The relative minimum values are at $x = -2$ and $x = 2$; and the relative maximum value is at $x = 0$.

As in Method 1, the maximum value is $f(0) = (0)^4 - 8(0)^2 + 1 = 1$; and the minimum value is $f(-2) = (-2)^4 - 8(-2)^2 + 1 = -15$. (repetition)

Example 2 Determine the relative extrema for $f(x) = x^3 - 6x^2 + 12x + 2$

Method 1 Using the first derivative test

Step 1: Differentiate $f(x) = x^3 - 6x^2 + 12x + 2$. Then, we obtain
$$f'(x) = 3x^2 - 12x + 12$$

Step 2: Equate $f'(x)$ to zero and solve for x.

$3x^2 - 12x + 12 = 0$

$3(x^2 - 4x + 4) = 0$

$3(x - 2)^2 = 0$

and $x = 2$ is the only critical x-value
The critical x-value is 2.

Step 3: Determine the sign of $3(x - 2)^2$ immediately before 2 and immediately after 2 (the critical point)

Number line:

	Column 1	Column 2
	$-\infty$ 2	$+\infty$

Factor	Interval Sign	
$x - 2$	—	**+**
$x - 2$	—	**+**
Product $(x - 2)(x - 2)$	**+**	**+**

Step 4: Determine the sign change from the left of 2 to the right of 2

Since $(x - 2)^2$ is always positive there are no sign changes at 2 for $f'(x)$, and f' has no relative minimum nor minimum. ($x = 2$ is at a point of inflection)

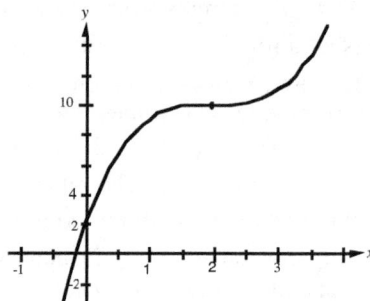

Graph of
$f(x) = x^3 - 6x^2 + 12x + 2$

Method 2 Using the second derivative test
Steps 1-2 are the same as for Method 1

Step 3: Differentiate $f'(x) = 3x^2 - 12x + 12$ to obtain $f''(x) = 6x - 12$

Step 4: Test the critical value $x = 2$ in $f''(x) = 6x - 12$

$f''(2) = 6(2) - 12 = 0$. Since $f''(2) = 0$, the second derivative test fails, and so method 1 above, was the approach to use. Try the first derivative test if f not twice differentiable.

Extra: $x = 2$ is at a point of inflection since $f''(2) = 0$ and $f'''(2) = 6 \neq 0$

Radical Functions
Example 3 Find the relative maximum and minimum values if any for
$$f(x) = \sqrt{x} + 3.$$

Solution   Step 1: $f(x) = \sqrt{x} + 3 = f(x) = x^{\frac{1}{2}} + 3$

Step 2: $f'(x) = \frac{1}{2}x^{-\frac{1}{2}} + 0$

$f'(x) = \frac{1}{2}x^{-\frac{1}{2}}$

Step 3: Set $\frac{1}{2}x^{-\frac{1}{2}} = 0$ and solve for x.

$$x^{-\frac{1}{2}} = 0$$

$$\frac{1}{x^{\frac{1}{2}}} = 0$$

$$\left(\frac{1}{x^{\frac{1}{2}}}\right)^2 = 0^2$$

$$\frac{1}{x} = 0 \qquad \text{(no solution)}$$

Clearly, f' is never zero. Moreover, $f'(0)$ is not defined), but $f(0)$ is defined (since when $x = 0$, $f(x) = \sqrt{x} + 3 = 3$ is defined}. Therefore, even though $f'(x)$ is never zero, we consider 0 as a critical x-value for the purpose of locating the intervals on a number line (see Case 3 of the first derivative test,, p.151). At $x = 0$, the function has a minimum value 3.

Graph of
$f(x) = \sqrt{x} + 3$:

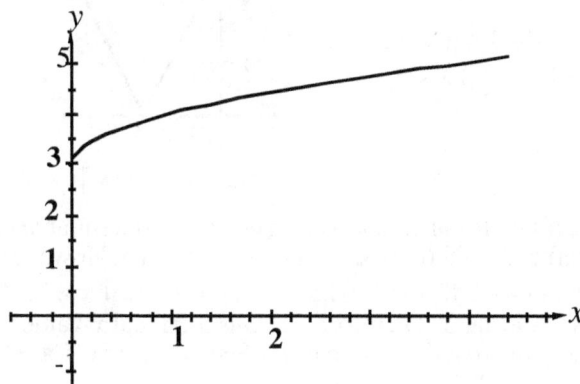

Rational Functions

Example 4 Find the relative maximum and minimum values, if any, for

$$f(x) = \frac{1}{x - 3}$$

Solution We observe that $f(x)$ is not defined when $x = 3$.

Step 1: $f'(x) = -\dfrac{1}{(x - 3)^2}$

Step 2: $-\dfrac{1}{(x - 3)^2} = 0$

In trying to solve for x, we find that the no value of x will satisfy this equation, Therefore, f' is never zero.. Therefore, there is no critical x-value.; and also $f(x)$ is not defined when $x = 3$. There is no maximum nor minimum value.

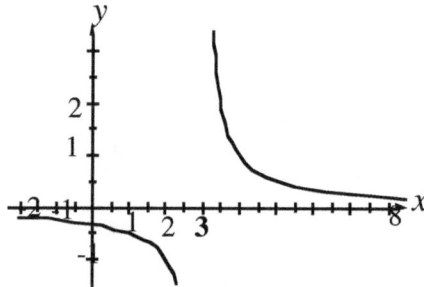

Graph of $f(x) = \dfrac{1}{x - 3}$

Absolute Value Functions

Example 5 Find the relative maximum and minimum values if any for

$$f(x) = |x - 3|$$

Solution $|x - 3| = \begin{cases} x - 3 & \text{if } x - 3 \geq 0, \text{ or } x \geq 3 \\ -(x - 3) & \text{if } x - 3 < 0, \text{ or } x < 3 \end{cases}$

Step 1: We consider the two cases

Case 1: $f(x) = x - 3 \qquad$ if $x \geq 3$
$\qquad\quad f'(x) = 1 \qquad\quad$ if $x \geq 3$

Case 2: $\quad f(x) = -x + 3 \quad$ if $x < 3$
$\qquad\quad f'(x) = -1 \qquad$ if $x < 3$

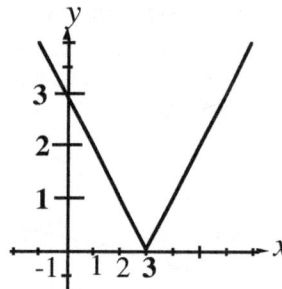

Graph of $f(x) = |x - 3|$

The function is defined for all x and is everywhere differentiable **except** when $x = 3$, since the left and right derivatives are not the same. However, 3 is a critical x-value. Since $f'(x) = -1$ if $x < 3$, while $f'(x) = 1 > 0$ if $x \geq 3$, $f'(x)$ changes sign from minus to plus at 3. Therefore, $x = 3$ is a critical x-value for a minimum value, by the first derivative test. (See also p.151, Case 2.). The minimum value is 0.

Lesson 19 Exercises A

1. Determine the relative maxima and minima for $f(x) = x^4 - 8x^2 + 1$
2. Determine the relative maxima and minima for $f(x) = x^3 - 6x^2 + 12x + 2$
3. Find the relative maximum and minimum values if any for $f(x) = \sqrt{x} + 3$
4. Find the relative maximum and minimum values if any for $f(x) = \dfrac{1}{x-3}$
5. Find the relative maximum and minimum values if any for $f(x) = |x - 3|$

Answers: 1. The minimum value is -15, and the maximum value is 1
Minimum points at $(-2, -15)$ and $(2, -15)$ and maximum point is at $(0, 1)$
2. There is no relative minimum nor minimum.
3. At $x = 0$, the function has a minimum value 3.
4. There is no maximum nor minimum value.
5. $x = 3$ is a critical x-value for a minimum value, 0.

Lesson 19 Exercises B

1. Determine the relative maxima and minima for $f(x) = x^4 - 6x^2 - 2$
2. Determine the relative maxima and minima for $f(x) = 2x^3 - 3x^2 - 12x + 5$
3. Determine the relative maxima and minima for $f(x) = x^2 - 2x - 3$
4. Find the relative maximum and minimum values if any for $f(x) = \sqrt{x} - 2$
5. Find the relative maximum and minimum values if any for $f(x) = \dfrac{1}{x-5}$
6. Find the relative maximum and minimum values if any for $f(x) = |x - 2|$

Answers: 1. The relative maximum value is -2 (when $x = 0$); and the relative
minimum value is -11 (when $\pm\sqrt{3}$).
2. The relative minimum is -15 (when $x = 2$) and the relative maximum is 12
(when $x = -1$)
3. The relative minimum value is -4 (when $x = 1$) ; also absolute minimum.
4. At $x = 0$, the function has a minimum value -2.
5. There is no maximum nor minimum value.
6. $x = 2$ is a critical x-value for a minimum value, 0.

Lesson 20
Maximum Value (Absolute Maximum Value) and Minimum Value (Absolute Minimum Value) of a Function

Definition

Let x_0 be a point on an interval on which f is defined. Then if $f(x_0) \geq f(x)$ for all x on this interval, f has an **absolute maximum** at x_0, and $f(x_0)$ is the absolute maximum of f on this interval; but if $f(x_0) \leq f(x)$ for all x on this interval, then f has an **absolute minimum** at x_0, and $f(x_0)$ is the absolute minimum of f on this interval.

Note: A function defined on an interval. may or may not have a maximum or minimum on this interval. For example, the continuous function $f(x) = x$ has neither a maximum nor a minimum on the open interval $(0,1)$; but the same function has the maximum 1 and the minimum 0 on the closed interval $[0,1]$. The Extreme-Value Theorem, repeated below, clarifies the conditions for a maximum or a minimum.

Extreme Value Theorem

If f is continuous on the closed interval $[a,b]$ then f attains a maximum at some point in $[a,b]$, and also a minimum at some point in $[a,b]$.

That is, a continuous function defined on a closed interval $[a,b]$ has a maximum and a minimum on this interval.

Critical x-value

A **critical x-value** for a function f is a number x_0 in the domain of f such that either $f'(x_0) = 0$, or $f'(x_0)$ does not exist but $f(x_0)$ is defined

Guidelines for Finding the Maximum and Minimum Values of a Continuous Function f on the closed interval $[a,b]$

Given: The function f is continuous on the closed interval $[a,b]$, and differentiable at each interior point of $[a,b]$.

Step 1: Differentiate $f(x)$ to obtain. $f'(x)$.

Step 2:. Set the derivative, $f'(x) = 0$ and solve to obtain the critical x-values, and eliminate those x-values not in the domain of the specified closed interval.

Step 3: Evaluate $f(x)$ for the critical x-values and also for the end-points, a and b (that is find $f(\text{critical } x - \text{value})$, $f(a)$, and $f(b)$).

Step 4: The largest value from Step 3 is the maximum value and the smallest value is the minimum value.

Step 5: Answer the question using a sentence.

Note from above that the maximum or minimum value occurs either at a critical point or at an end point. If the end-points are excluded, then the maximum or minimum value occurs at a critical point

Example 1 Find the maximum and minimum values of the curve given by
$$f(x) = x^4 - 8x^2 + 1 \qquad [-4, 1]$$

Step 1: $f(x) = x^4 - 8x^2 + 1$
$f'(x) = 4x^3 - 16x$ (differentiating)

Step 2: $4x^3 - 16x = 0$ (setting $f'(x) = 0$)
$4x(x + 2)(x - 2) = 0$
$4x = 0$ or $(x + 2)(x - 2) = 0$
Solving, $x = 0$ or $x = 2$, $x = -2$.
The critical x-values are -2, 0 and 2; and we check to see if we should eliminate any of these values. Since the domain of interest is
$[-4, 1]$ (same as $-4 \le x \le 1$), we eliminate 2 which is outside this domain, and keep the critical values -2, and 0 for the next step.

Step 3: Evaluate $f(x)$ for -2, 0,, and for the end points -4 and 1, by substituting these values in $f(x) = x^4 - 8x^2 + 1$ (original functional equation)

$f(-2) = (-2)^4 - 8(-2)^2 + 1 = -15$ (Critical x-value = -2)
$f(0) = (0)^4 - 8(0)^2 + 1 = 1$ (Critical x-value = 0)
$f(-4) = (-4)^4 - 8(-4)^2 + 1 = 129$ (end-point check)
$f(1) = (1)^4 - 8(1)^2 + 1 = -6$ (end-point check)

Step 4. The largest value from Step 3 is the maximum value and the smallest value is the minimum value. The functional values from Step 3 are -15, 1, 129, and -6.
In the closed interval $[-4, 1]$, the maximum value is 129, and the minimum value is -15. The maximum value occurs at $x = -4$, and the minimum value occurs at $x = -2$. See also p.153, Example 1.

Example 2 Find the maximum and minimum values of the curve given by
$$f(x) = x^3 - 6x^2 + 12x + 2 \qquad [-2, 4]$$

Step 1: $f'(x) = 3x^2 - 12x + 12$
$3x^2 - 12x + 12 = 0$
$x^2 - 4x + 4 = 0$ (dividing through by 3)

Step 2: $(x - 2)(x - 2) = 0$
$x = 2$ (a repeated root)

Step 3 Since $x = 2$ is in the given domain $[-2, 4]$, we keep it for the next step.

Step 4: We evaluate $f(x)$ for 2, and the endpoints -2 and 4. by substituting these values in $f(x) = x^3 - 6x^2 + 12x + 2$ (original functional equation)
$f(2) = (2)^3 - 6(2)^2 + 12(2) + 2 = 10$
$f(-2) = (-2)^3 - 6(-2)^2 + 12(-2) + 2 = -54$
$f(4) = (4)^3 - 6(4)^2 + 12(4) + 2 = 18$

Step 4: In the closed interval $[-2, 4]$, the maximum value is 18, and the minimum value is -54 (from step 3).
(The maximum value occurs at $x = 4$, and the minimum value occurs at $x = -2$.)

Example 3 Find the maximum and minimum values of the curve given by

$$f(x) = x^3 + 2x^2 - 8 \qquad [-4, 2]$$

Step 1: $f'(x) = 3x^2 + 4x$

Step 2: $3x^2 + 4x = 0$

Step 3 $x(3x + 4) = 0$

Step 4: $x = 0$, or $(3x + 4) = 0$

$$x = 0, \quad x = -\frac{4}{3}$$

Step 5: Since 0, and $-\frac{4}{3}$ are in the given domain , $[-4, 2]$, we keep them for the next step.

Step 6: We evaluate $f(x)$ for 0, and $-\frac{4}{3}$ and also for the endpoints -4 and 2. by substituting these values in $f(x) = x^3 + 2x^2 - 8$.

$f(0) = (0)^3 + 2(0)^2 - 8 = -8$

$f(-\frac{4}{3}) = (-\frac{4}{3})^3 + 2(-\frac{4}{3})^2 - 8 = -6.8$

$f(-4) = (-4)^3 + 2(-4)^2 - 8 = -40$

$f(2) = (2)^3 + 2(2)^2 - 8 = 8$

Step 7: From step 6, in the closed interval $[-4, 2]$, the maximum value is 8, and the minimum value is -40. The maximum value occurs at $x = 2$, and the minimum value occurs at $x = -4$.

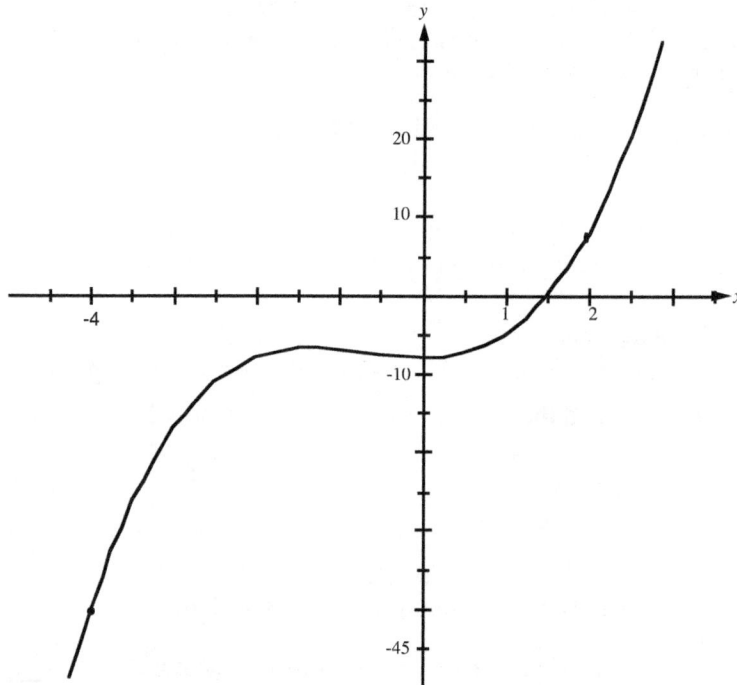

Graph of $f(x) = x^3 + 2x^2 - 8$

Extra: there is an inflection point at $x = -0.7$

Example 4 Find the maximum and minimum values of the curve given by

$$f(x) = \sqrt{x} + 3 \qquad [-1, 1]$$

Step 1: $f(x) = \sqrt{x} + 3$

Graph of $f(x) = \sqrt{x} + 3$

$$= x^{\frac{1}{2}} + 3$$

$$f'(x) = \tfrac{1}{2} x^{\frac{1}{2}-1} + 0$$

$$= \tfrac{1}{2} x^{-\frac{1}{2}}$$

Step 2: $\dfrac{1}{2\sqrt{x}} = 0$

$$\dfrac{1}{\sqrt{x}} = 0 \quad :$$

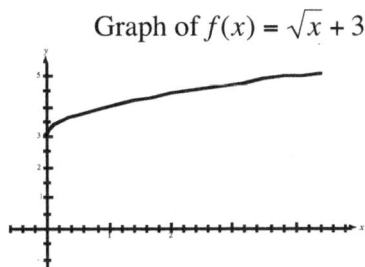

Clearly, f' is never zero. Moreover, $f'(0)$ is not defined. However,

$f(0)$ is defined ($f(0) = \sqrt{0} + 3 = 3$). We consider 0 as critical x-value.
(See Case 3 of the first derivative test,, p.151)

Step 3: We eliminate the end point -1, since $x \geq 0$ (for the square root function).

We evaluate $f(x)$ for 0, and the endpoint 1. by substituting these values

in $f(x) = \sqrt{x} + 3$

$$f(0) = \sqrt{0} + 3 = 3$$

$$f(1) = \sqrt{1} + 3 = 4$$

Note also:: $f(-1) = \sqrt{-1} + 3$ is not real and therefore dropping -1 in Step 3 was justifiable.

Step 4: In the closed interval $[-1, 1]$, the maximum value is 4 and the minimum value is 3. The maximum value occurs at $x = 1$, and the minimum value occurs at $x = 0$. **Note:** We can estimate the location of extremum by graphing.

Lesson 20 Exercises A

Find the maximum and minimum values of the curve given by

$$f(x) = x^4 - 8x^2 + 1 \qquad [-4, 1]$$

2 Find the maximum and minimum values of the curve given by

$$f(x) = x^3 - 6x^2 + 12x + 2 \qquad [-2, 4]$$

3 Find the maximum and minimum values of the curve given by

$$f(x) = x^3 + 2x^2 - 8 \qquad [-4, 2]$$

4 Find the maximum and minimum values of the curve given by

$$f(x) = \sqrt{x} + 3 \qquad [-1, 1]$$

Answers:

1. In the closed interval $[-4, 1]$, the maximum value is 129, and the minimum value is -15. The maximum value occurs at $x = -4$, and the minimum value at $x = -2$.

2. The maximum value is 18, and the minimum value is -54. The maximum value occurs at $x = 4$, and the minimum value at $x = -2$.

3. The maximum value is 8, and the minimum value is -40. The maximum value occurs at $x = 2$, and the minimum value at $x = -4$.

4. The maximum value is 4 and the minimum value is 3. The maximum value occurs at $x = 1$, and the minimum value at $x = 0$.

Lesson 20 Exercises B

Find the maximum and minimum values of the curve given by

$$f(x) = x^4 - 6x^2 + 2 \qquad [-4, 1]$$

2 Find the maximum and minimum values of the curve given by

$$f(x) = x^3 - 6x^2 + 12x + 2 \qquad [-1, 5]$$

3 Find the maximum and minimum values of the curve given by

$$f(x) = x^3 + 2x^2 - 6 \qquad [-2, 4]$$

4 Find the maximum and minimum values of the curve given by

$$f(x) = \sqrt{x} - 2 \qquad [-1, 1]$$

Answers:

1. In the closed interval $[-4, 1]$, the maximum value is 162 (when $x = -4$) , and the minimum value is -7 (when $x = \pm\sqrt{3}$).

2. In the closed interval $[-1, 5]$, the maximum value is 37 (when $x = 5$,) and the minimum value is -17 (when $x = -1$).

3. :In the closed interval $[-2, 4]$, the maximum value is 90 (when $x = 4$) , and he minimum value is -6 (when $x = 0$ or $x = -2$).

4. In the closed interval $[-1, 1]$, the maximum value is -1 (when $x = 1$ and the minimum value is -2 (when $x = 0$)

Lesson 21
Applied Maxima and Minima Problems

As it is in the case of related rates problems, the main difficulty a student may encounter here is not the application of the calculus, but rather being able to obtain the correct relationships between the quantities involved. The calculus part is rather straightforward.

Guidelines for Solving Applied Maxima and Minima Problems

Step **1.** As in all word problems, read and read the problem, represent the unknowns by letters, and write down what is given or known.

Step **2.** Draw a diagram if helpful, and if geometry is involved, draw a diagram and identify the quantity that is to be maximized or minimized.

Step **3.** Write down an equation or equations involving the quantity to be maximized or minimized. Try to eliminate any quantity whose value is not given so as to obtain a single independent variable by substitution.

Step **4.** Write down the interval (based on the physical constraints (restrictions) of the problem.

Step **5.** Differentiate the equation or equations accordingly. In some cases, you may use implicit differentiation.

Step **6.** Set $f'(x) = 0$, and solve to obtain the critical x-values, and eliminate those x-values not satisfying the physical constraints of the problem.

Step **7:** Determine which critical values are for maximum or minimum point (using the **second derivative** test, or otherwise) and reject accordingly any critical values not satisfying the physical constraints of the problem. For example, if you are looking for a maximum value and the second derivative test indicates an x-value for a minimum point, then reject this x-value.

Step **8.** Evaluate $f(x)$ for the critical x-values and also for the end-points.

Step **9.** The largest value from Step 7 is the maximum value and the smallest value is the minimum value.

Step **10.** Answer the question using a sentence.

Example 1: The sum of two numbers is 48. Find these numbers such that their product is a maximum.

Solution

Step 1: Let one of the numbers $= x$.

Then the other number is $48 - x$

The product, p is then $p = x(48 - x)$

We want to maximize $48x - x^2$,

subject to the domain $0 \le x \le 48$

Step 2: Find $\dfrac{dp}{dx}$.

$$\dfrac{dp}{dx} = 48 - 2x \quad (p = 48x - x^2)$$

Step 3: Equate $\dfrac{dp}{dx}$ to zero and solve for x.

$48 - 2x = 0, \ 2x = 48$; and $x = 24$

The critical x-value is 24 which is in the domain $0 \le x \le 48$.

To determine if this value is at maximum or minimum point, we apply the second derivative test.

From $\frac{dp}{dx} = 48 - 2x$, $\frac{d^2p}{dx^2} = -2 < 0$ which means that $x = 24$ is a maximum

x-value. If one number is 24, the other number is $48 - 24 = 24$.
The numbers are 24 and 24; and the maximum product is 576.
(Note: For the end-points, if either number is zero the product is zero)

Example 2 Twice the sum of two numbers is 48. Find these numbers such that their product is a maximum.

Solution
Step 1: Let one of the numbers $= x$.

Let the other number $= y$

Twice their sum is 48 translates to: $2(x + y) = 48$

$x + y = 24$

$y = (24 - x)$

The product, P is then $P = xy$

$$P = x(24 - x)$$

$$= 24x - x^2$$

We want to maximize $P = 24x - x^2$,

subject to the domain $0 \le x \le 24$

Step 2 Find : $\frac{dp}{dx}$

$\frac{dp}{dx} = 24 - 2x$

Step 3: Equate $\frac{dp}{dx}$ to zero and solve for x.

$24 - 2x = 0$, $2x = 24$; and $x = 12$

The critical x-value is 12 which is in the domain $0 \le x \le 24$
To determine if this value is at maximum or minimum point, we apply the second derivative test.

From $\frac{dp}{dx} = 24 - 2x$, $\frac{d^2p}{dx^2} = -2 < 0$ which means that $x = 12$ is a

maximum x-value. When $x = 12, y = 24\text{-}12 = 12$
(If $x = 0, y = 24$, and the product is 0. Similarly if $x = 24, y = 0$, the product is 0.)
The numbers are 12 and 12; and the maximum product is 144.

Example 3 The perimeter of a rectangle is 48. Find the length and the width such that the area is a maximum.

Solution

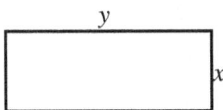

Step 1: Let the width = x; and let the length = y
The perimeter, P is given by
$$P = 2x + 2y = 48 \qquad\qquad \text{(A)}$$
$$x + y = 24$$
$$y = (24 - x)$$
The area A is given by $A = xy$
$$A = x(24 - x) = 24x - x^2$$
We want to maximize $A = 24x - x^2$ (B)
subject to the domain $0 \le x \le 24$

Step 2 Find $\frac{dA}{dx}$: $\frac{dA}{dx} = 24 - 2x$ (differentiating equation (B))

Step 3: Equate $\frac{dA}{dx}$ to zero and solve for x.
$$24 - 2x = 0, \quad 2x = 24; \text{ and } x = 12.$$
The critical x-value is 12 which is in the domain $0 \le x \le 24$
To determine if this value is at maximum or minimum point, we apply the second derivative test.

From $\frac{dA}{dx} = 24 - 2x$, $\frac{d^2A}{dx^2} = -2 < 0$ which means that $x = 12$ is a maximum x-value. When $x = 12$, $y = 24 - 12 = 12$
The length is 12 units; the width is 12 units; and the maximum area is 144 sq. units. (Note: As in Example 2, here, if either side is zero the product is 0.

Example 4 The sum of two numbers is 48. Find these numbers such that the product of the smaller number and the square of the larger number is a maximum.

Solution

Step 1: Let the larger number = x
Then the smaller number = $48 - x$
The square of the larger number = x^2
The product, p is then $p = (48 - x)x^2$
We want to maximize $P = 48x^2 - x^3$,
subject to the domain $0 \le x \le 48$

Step 2 Find $\frac{dp}{dx}$. $\frac{dp}{dx} = 96x - 3x^2$

Step 3: Equate $\frac{dp}{dx}$ to zero and solve for x.
$$96x - 3x^2 = 0, \quad 3x(32 - x); \text{ and } x = 0, \quad x = 32$$
The critical x-values 0 and 32 are in the domain $0 \le x \le 48$
To determine if each value is at a maximum or minimum point, we apply the second derivative test.

From $\dfrac{dp}{dx} = 96x - 3x^2$, $\dfrac{d^2p}{dx^2} = 96 - 6x$

when $x = 0$, $\dfrac{d^2p}{dx^2} = 96 > 0$, which means $x = 0$ is at a minimum point,

and it is rejected, since we are looking for a maximum. See Lesson 19.

When $x = 32$, $\dfrac{d^2p}{dx^2} = 96 - 6(32) = -96 < 0$, which means $x = 32$ is a

maximum x-value (by the second derivative test).

When $x = 32$, $48 - 32 = 16$. The smaller number is 16, and the larger

number is 32. (The maximum product $= (32)^2(16) = 16,384$)

(Note: For the end-points, if either number is zero the product is zero)

Example 5 An open box is to be made from a square sheet of tin. The length of each side of the sheet is L units. Squares of equal size are cut from the four corners of the sheet, and the sides are then bent up to form the box. Find the length of each square cut out such that the volume of the resulting box is a maximum.

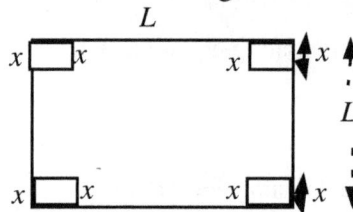

Solution

Step 1: Let the side of each square to be cut out be x units.

Then the length of each side of the finished the box $= L - 2x$ units.

The height of the box $= x$ (The sides are to be bent up to form the box)

The volume, V of the box : $V = (L - 2x)(L - 2x)x$ (L × L × H)

$$V = L^2 x - 4Lx^2 + 4x^3$$

Maximize $V = 4x^3 - 4Lx^2 + L^2 x$ subject to $0 \le x \le \dfrac{L}{2}$

Step 2: $\dfrac{dV}{dx} = 12x^2 - 8Lx + L^2$

Step 3: $12x^2 - 8Lx + L^2 = 0$

$$(2x - L)(6x - L) = 0$$

$$2x - L = 0 \Rightarrow x = \dfrac{L}{2} \quad \text{or} \quad 6x - L = 0 \Rightarrow x = \dfrac{L}{6}$$

The critical x-values are $\dfrac{L}{2}$ and $\dfrac{L}{6}$. The end-points are 0 and L.

Step 4: $V(L/2) = 0$; $V(L/6) = \dfrac{2L^3}{27}$; $V(0) = 0$ $(V(x) = 4x^3 - 4Lx^2 + L^2 x)$

The length of each side of square to be cut out $= \dfrac{L}{6}$ units.

Extra   Second derivative test for critical values. (see also p.153)

$\dfrac{d^2V}{dx^2} = 24x - 8L$; $\left. \dfrac{d^2V}{dx^2} \right|_{x=\frac{L}{2}} = 24(\frac{L}{2}) - 8L = 4L > 0$ ($x = \dfrac{L}{2}$ is a minimum point)

$\left. \dfrac{d^2V}{dx^2} \right|_{x=\frac{L}{6}} = 24(\frac{L}{6}) - 8L = -4L < 0$ ($x = \dfrac{L}{6}$ is a maximum point)

Example 6 An open box is to be made from a square sheet of tin. The length of each side of the sheet is 18 units. Squares of equal size are cut from the four corners of the sheet, and the sides are then bent up to form the box. Find the length of each square cut out such that the volume of the resulting box is a maximum.

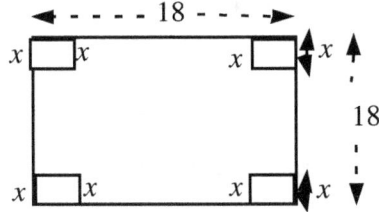

Solution

Step 1: Let the side of each square to be cut out be x units.

Then the length of each side of the finished the box $= 18 - 2x$ units

The height of the box $= x$. (The sides are to be bent up to form the box)

The volume, V of the box : $V = (18 - 2x)(18 - 2x)x$ (L × L × H)

$$V = 324x - 72x^2 + 4x^3$$
$$V = 4x^3 - 72x^2 + 324x \text{ subject to } 0 \le x \le 9 \quad (18/2 = 9)$$

Step 2: $\frac{dV}{dx} = 12x^2 - 144x + 324$

Step 3 $12x^2 - 144x + 324 = 0$

$$12(x - 3)(x - 9) = 0$$
$$(x - 3) = 0 \Rightarrow x = 3 \quad \text{or} \quad (x - 9) = 0 \Rightarrow x = 9$$

The critical x-values are **3** and **9,** The end-points are 0 and 9.

Step 4: $V(3) = 432$; $V(9) = 0$; $V(0) = 0$ $(V(x) = 4x^3 - 72x^2 + 324x)$

The maximum volume 432 occurs when $x = 3$

The length of each side of square to be cut out $= 3$ units.

Extra: Second derivative test for critical values. (see also p.153)

$\frac{d^2V}{dx^2} = -144 + 24x$; $\left.\frac{d^2V}{dx^2}\right|_{x=3} = -144 + 24(3) = -72 < 0$ ($x = 3$ is a maximum point)

$\left.\frac{d^2V}{dx^2}\right|_{x=9} = -144 + 24(9) = 72 > 0$ ($x = 9$ is a minimum point)

Again, the length of each side of square to be cut out $= 3$ units.

Note: Compare the results of Example 6 with those of Example 5.

Example 7 An open box is to be made from a tin sheet of length 15 units and the width 8 units. Squares of equal size are cut from the four corners of the sheet, and the sides are then bent up to form the box. Find the length of each square cut out. such that the volume of the resulting box is a maximum.

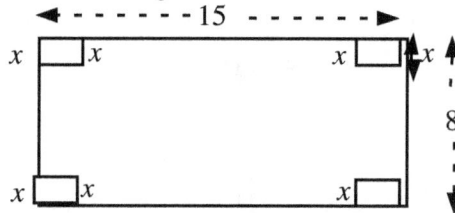

Solution

Step 1: Let the side of each square to be cut out be x units

Then the length of the finished box = $15 - 2x$ units

The width of the finished box = $8 - 2x$ units

The height of the box = x (The sides are to be bent up to form the box)

The volume, V of the box : $V = (15 - 2x)(8 - 2x)x$ (L × W × H)

$\qquad V = 120x - 46x^2 + 4x^3$

$\qquad V = 4x^3 - 46x^2 + 120x$, subject to $0 \le x \le 4$ (8/2 = 4)

Step 2: $\dfrac{dV}{dx} = 12x^2 - 92x + 120$

Step 3 $12x^2 - 92x + 120 = 0$

$\qquad 4(3x - 5)(x - 6) = 0$

$\qquad\qquad (3x - 5) = 0 \Rightarrow x = \dfrac{5}{3}$ or $(x - 6) = 0 \Rightarrow x = 6$

The critical x-values are $\dfrac{5}{3}$ and 6. Since 6 is outside the constraint, $0 \le x \le 4$,

we reject 6 as a critical value; but accept $\dfrac{5}{3}$. The end-points are 0 and 4.

Step 4: $V(5/3) = 90.7$; $V(0) = 0$, $V(4) = 0$.

$\qquad (V(x) = 4x^3 - 46x^2 + 120x)$

The maximum volume 90.7 occurs when $x = \dfrac{5}{3}$

The length of each side of square to be cut out = $\dfrac{5}{3}$ units.

(Note also that $V(6) = -72$ does not make sense in this problem)

Extra: Second derivative test for the critical values, $\dfrac{5}{3}$ and 6.

$\dfrac{d^2V}{dx^2} = 24x - 92$

$\left.\dfrac{d^2V}{dx^2}\right|_{x=\frac{5}{3}} = (24)(\tfrac{5}{3}) - 92 = 40 - 92 = -52 < 0$ ($\dfrac{5}{3}$ is at maximum point)

Example 8: Find the radius and height of the right circular cylinder of maximum volume that can be inscribed in a right-circular cone with radius 3 units and height 5 units.

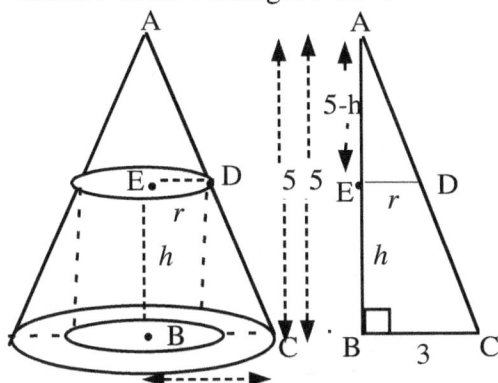

Solution

Step 1: Let the radius of the cylinder = r units.

Let the height of the cylinder = h units

The volume V of the cylinder is given by

$V = \pi r^2 h$ subject to $0 \le r \le 3$, $0 \le h \le 5$

Step 2: We eliminate h using similarity of triangles $\Delta' s\ AED$ and ABC

$\dfrac{ED}{BC} = \dfrac{AE}{AB}$ ($\overline{ED}$ is parallel to $\overline{BC}$)

$\dfrac{r}{3} = \dfrac{5 - h}{5}$

$h = \dfrac{15 - 5r}{3}$ (solving for h)

Approach #1

Step 3: Substitute for h in $V = \pi r^2 h$

Then $V = \pi r^2 \left(\dfrac{15 - 5r}{3} \right)$

$V = \dfrac{\pi}{3} \left(15r^2 - 5r^3 \right)$

Step 4: $\dfrac{dV}{dr} = \dfrac{\pi}{3} \left(30r - 15r^2 \right)$

Step 5: $\dfrac{\pi}{3} \left(30r - 15r^2 \right) = 0$ (Setting $\dfrac{dV}{dr} = 0$)

$30r - 15r^2 = 0$

$15r(2 - r) = 0$

$r = 0,\ \ r = 2.$ (0 and 2 satisfy $0 \le r \le 3$)

The critical r-values are 0 and 2; the endpoints are 0 and 3.

$V(0) = 0;\ V(2) = 20.9$ cubic units $V(3) = 0$

($V(2) = \dfrac{\pi}{3} \left(15(2)^2 - 5(2)^3 \right);\ = \dfrac{20\pi}{3} = 20.9$)   ($V(r) = \dfrac{\pi}{3} \left(15r^2 - 5r^3 \right)$)

The maximum volume 20.9 occurs when $r = 2$; also $h = \dfrac{15 - 5(2)}{3} = \dfrac{5}{3}$

The radius is 2 units, and the height is $\dfrac{5}{3}$ units. ($h = \dfrac{5}{3}$ satisfies $0 \le h \le 5$)

Approach #2

Step 3: $V = \pi r^2 h$ (1)

$$\frac{r}{3} = \frac{5-h}{5} \quad\quad (2)$$

$$\boxed{5r = 15 - 3h} \quad\quad (3)$$

$$3h = 15 - 5r$$

$$h = 5 - \frac{5r}{3}$$

$$\boxed{\frac{dh}{dr} = -\frac{5}{3}}$$

Step 4: Implicitly differentiate V with respect to r $(V = \pi r^2 h)$

$$\frac{dV}{dr} = \pi[r^2 \frac{dh}{dr} + 2hr] \quad\quad \text{(Recall the product rule for differentiation)}$$

$$\frac{dV}{dr} = \pi[r^2(-\tfrac{5}{3}) + 2hr] \quad\quad\quad \left(\frac{dh}{dr} = -\frac{5}{3}\right)$$

$$\pi[-\tfrac{5}{3}r^2 + 2hr] = 0 \quad\quad \text{(setting } \frac{dv}{dr} = 0 \text{ for maximum volume)}$$

$$5r^2 - 6hr = 0$$

$$r(5r - 6h) = 0$$

$$r = 0 \text{ or } 5r - 6h = 0$$

$$\boxed{5r = 6h} \quad\quad\quad (4)$$

Step 6: Solve equation (4), $\boxed{5r = 6h}$, and equation (3), $\boxed{5r = 15 - 3h}$

     simultaneously for r and h

     Equating the right sides of these equations

$$6h = 15 - 3h$$

$$9h = 15$$

$$h = \frac{5}{3}$$

Step 7: $5r = 6(\tfrac{5}{3})$

$$5r = 10$$

$$r = 2$$

The radius is 2 units, and the height is $\frac{5}{3}$ units.

Example 9: Find the radius and height of the right circular cylinder of maximum volume that can be inscribed in a sphere of radius radius 3 units.

Method 1

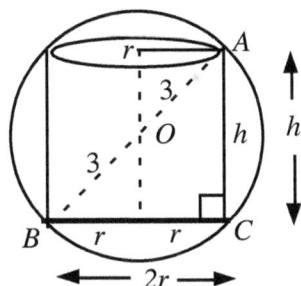

Step 1: Let the radius of the cylinder $= r$ units.
Let the height of the cylinder $= h$ units
The volume V of the cylinder is given by
$V = \pi r^2 h$ subject to $0 \le r \le 3$,

Step 2: We eliminate h applying the Pythagorean theorem to $\triangle ABC$
In $\triangle ABC$, $AB^2 = BC^2 + CA^2$
$6^2 = (2r)^2 + h^2$ and from which $\boxed{h^2 = 36 - 4r^2}$
(This equation connects the cylinder and the sphere)

Approach #1

Step 3: $V = \pi r^2 h$; and $h^2 = 36 - 4r^2$ and $h = (36 - 4r^2)^{\frac{1}{2}}$

Step 4: Substituting for h in $V = \pi r^2 h$, we obtain

$$V = \pi r^2 (36 - 4r^2)^{\frac{1}{2}}$$

$$V = \pi [r^2 (36 - 4r^2)^{\frac{1}{2}}]$$

Step 5 Differentiate V with respect to r (usual product rule differentiation)

$$\frac{dV}{dr} = \pi [r^2 (\tfrac{1}{2})(-8r)(36 - 4r^2)^{-\frac{1}{2}} + (2r)(36 - 4r^2)^{\frac{1}{2}}]$$

$$= \pi \left[\frac{-4r^3}{(36 - 4r^2)^{\frac{1}{2}}} + (2r)(36 - 4r^2)^{\frac{1}{2}} \right]$$

Step 6 Set $\dfrac{dV}{dr}$ to zero and solve for r, (For maximum volume, $\dfrac{dV}{dr} = 0$)

$$\pi \left[\frac{-4r^3}{(36 - 4r^2)^{\frac{1}{2}}} + (2r)(36 - 4r^2)^{\frac{1}{2}} \right] = 0$$

$$\left[\frac{-4r^3}{(36 - 4r^2)^{\frac{1}{2}}} + (2r)(36 - 4r^2)^{\frac{1}{2}} \right] = 0$$

$3r(6 - r^2) = 0$ (simplifying) and $r = 0$, $6 - r^2 = 0$, and

$r = \sqrt{6}$ (we reject $r = -\sqrt{6}$, since the radius cannot be negative)
The critical r-values 0, $\sqrt{6}$ ($\sqrt{6} \approx 2.4$) are in the domain $0 \le r \le 3$; and the end points are 0 and 3. When $r = 0$, $V = 0$, and we ignore $r = 0$ for the next step.

Step 7: Find h. by substituting $r = \sqrt{6}$ in $h = \sqrt{36 - 4r^2}$ (from Step 3)

$$h = \sqrt{36 - 4(\sqrt{6})^2} = \sqrt{36 - 24} = \sqrt{12} = = 2\sqrt{3}$$

$$V = \pi r^2 h; \ V(r) = \pi r^2 (2\sqrt{9 - r^2}) = 2\pi r^2 \sqrt{9 - r^2}$$

$$V(0) = 0; \ V(\sqrt{6}) = 2\pi (\sqrt{6})^2 \sqrt{9 - (\sqrt{6})^2} = 12\pi\sqrt{3}$$

$$V(3) = 2\pi (3)^2 \sqrt{9 - (3)^2} = 2\pi (3)^2 (0) = 0.$$

The maximum value, $12\pi\sqrt{3}$ occurs when $r = \sqrt{6}$.

The radius is $\sqrt{6}$ units and the height is $2\sqrt{3}$ units.

Approach #2

Step 3: $V = \pi r^2 h$ (1)

$\boxed{h^2 = 36 - 4r^2}$ (2) (from Step 2)

Step 4: Implicitly differentiate V with respect to r

$$\frac{dV}{dr} = \pi[r^2 \frac{dh}{dr} + 2hr]$$

$$r^2 \frac{dh}{dr} + 2hr = 0 \quad \text{(setting } dV/dr = 0, \ \pi \neq 0)$$

$$r^2 \frac{dh}{dr} = -2hr$$

$$\frac{dh}{dr} = -\frac{2hr}{r^2}$$

$$\boxed{\frac{dh}{dr} = -\frac{2h}{r}} \qquad (3)$$

Step 5: In equation (2), implicitly differentiate h with respect to r

$$2h\frac{dh}{dr} = -8r$$

$$\frac{dh}{dr} = \frac{-8r}{2h}$$

$$\boxed{\frac{dh}{dr} = \frac{-4r}{h}} \qquad (4)$$

Step 5: From equations (3) and (4)

$$-\frac{2h}{r} = -\frac{4r}{h}$$

$$\boxed{h^2 = 2r^2} \qquad (5)$$

Step 6: Solve equation (2), $\boxed{h^2 = 36 - 4r^2}$, and equation (5) $\boxed{h^2 = 2r^2}$ simultaneously for r and h

Equating the right sides of these equations

$$2r^2 = 36 - 4r^2$$

$6r^2 = 36; \quad r^2 = 6$ and $r = \sqrt{6}$ (we reject $r = -\sqrt{6}$, since the radius cannot be negative)

Step 7: $h^2 = 2(\sqrt{6})^2$ (substituting in equation(5)

$$h^2 = 12; \quad h = 2\sqrt{3}$$

The radius is $\sqrt{6}$ units and the height is $2\sqrt{3}$ units.

Approach #3

Step 3: Squaring both sides of $V = \pi r^2 h$ (to change h to h^2) we obtain
$$V^2 = \pi^2 r^4 h^2$$

Step 4: Substituting for h^2 from Step 2 in $V^2 = \pi^2 r^4 h^2$, we obtain
$$V^2 = \pi^2 r^4 (36 - 4r^2) = \pi^2 (36r^4 - 4r^6)$$

Step 5 Implicitly differentiate V with respect to r.

$$2V \frac{dV}{dr} = \pi^2 (144r^3 - 24r^5) \text{ and from which } \frac{dV}{dr} = \frac{\pi^2}{2V}(144r^3 - 24r^5)$$

Step 6: Equate $\frac{\pi^2}{2V}(144r^3 - 24r^5) = 0$. (For a maximum volume)

$$144r^3 - 24r^5 = 0 \qquad (\text{ Since } V \neq 0 \text{ , } \pi^2 \neq 0 \text{ })$$
$$24r^3(6 - r^2) = 0$$
$$24r^3 = 0 \text{ or } 6 - r^2 = 0\text{; and from which } r = 0 \text{, or } r = \sqrt{6}$$

The critical r-values are 0, $\sqrt{6}$ ($\sqrt{6} \approx 2.4$) are in the domain $0 \leq r \leq 3$; and the end points are 0 and 3. When $r = 0$, $V = 0$, and we ignore $r = 0$ for the next step.

Step 7: From $h^2 = 36 - 4r^2$; $h^2 = 4(9 - r^2)$; and $h = \sqrt{4(9 - r^2)} = 2\sqrt{9 - r^2}$

$$V = \pi r^2 h; \ V(r) = \pi r^2 (2\sqrt{9 - r^2}) = 2\pi r^2 \sqrt{9 - r^2}$$

$$V(0) = 0; \ V(\sqrt{6}) = 2\pi (\sqrt{6})^2 \sqrt{9 - \left(\sqrt{6}\right)^2} = 12\pi\sqrt{3}$$

$$V(3) = 2\pi (3)^2 \sqrt{9 - (3)^2} = 2\pi (3)^2 (0) = 0.$$

The maximum value, $12\pi\sqrt{3}$ occurs when $r = \sqrt{6}$.

Step 8: Find h. by substituting $r = \sqrt{6}$ in $h = 2\sqrt{9 - r^2}$ (from Step 5)
$$h = 2\sqrt{9 - (\sqrt{6})^2} = 2\sqrt{3}$$

The radius is $\sqrt{6}$ units and the height is $2\sqrt{3}$ units.

Method 2

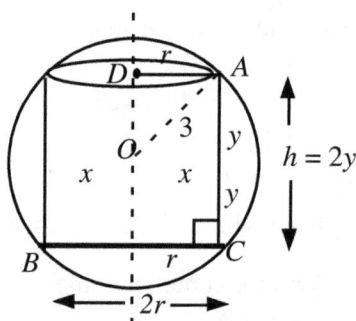

Step 1: Let the radius of the cylinder = r units.
Let the height of the cylinder = h units
The volume V of the cylinder is given by
$\boxed{V = \pi r^2 h}$ subject to $0 \le r \le 3$,

Step 2: We eliminate h applying the Pythagorean theorem to Δ ADO
In Δ ADO, $\quad OA^2 = OD^2 + DA^2$
$3^2 = y^2 + r^2$ and from which $\boxed{y^2 = 9 - r^2}$

Using Approach #1

Step 3; From $y^2 = 9 - r^2$ and $y = (9 - r^2)^{\frac{1}{2}}$; height , h of the cylinder is given
by $h = 2y = 2(9 - r^2)^{\frac{1}{2}}$

Step 4: Substituting for h in $V = \pi r^2 h$, we obtain

$$V = 2\pi r^2 (9 - r^2)^{\frac{1}{2}}$$

$$V = 2\pi \left[r^2 (9 - r^2)^{\frac{1}{2}} \right]$$

Step 5 Differentiate V with respect to r (usual product rule for differentiation)

$$\frac{dV}{dr} = 2\pi \left[r^2 (\tfrac{1}{2})(-2r)(9 - r^2)^{-\frac{1}{2}} + (2r)(9 - r^2)^{\frac{1}{2}} \right]$$

$$= 2\pi \left[\frac{-r^3}{(9 - r^2)^{\frac{1}{2}}} + (2r)(9 - r^2)^{\frac{1}{2}} \right]$$

Step 6 Set $\dfrac{dV}{dr}$ to zero and solve for r, (At maximum volume, $\dfrac{dV}{dr} = 0$)

$$2\pi \left[\frac{-r^3}{(9 - r^2)^{\frac{1}{2}}} + (2r)(9 - r^2)^{\frac{1}{2}} \right] = 0$$

$$\frac{r^3}{(9 - r^2)^{\frac{1}{2}}} = 2r(9 - r^2)^{\frac{1}{2}}$$

(Multiply both sides of the equation by $(9 - r^2)^{\frac{1}{2}}$, or cross-multiply. Then

$$r^3 = 2r(9 - r^2)$$
$$r^3 = 18r - 2r^3$$
$$3r^3 - 18r = 0$$
$$3r(r^2 - 6) = 0$$
$$3r = 0, \text{ or } r^2 - 6 = 0$$
$$r = 0, r = \sqrt{6} \quad \text{(we reject } r = -\sqrt{6}\text{, since the radius cannot be negative)}$$

The critical r-values 0, $\sqrt{6}$ ($\sqrt{6} \approx 2.4$) are in the domain $0 \le r \le 3$; and the end points are 0 and 3. When $r = 0$, $V = 0$, and we ignore $r = 0$ for the next step.

Step 7: Find h by substituting $r = \sqrt{6}$ in $h = 2\sqrt{9 - r^2}$

$$(h = 2y = 2(9 - r^2)^{\frac{1}{2}}, \text{ from Step 3})$$
$$h = 2\sqrt{9 - (\sqrt{6})^2} = 2\sqrt{3}$$
$$V = \pi r^2 h; \ V(r) = \pi r^2 (2\sqrt{9 - r^2}) = 2\pi r^2 \sqrt{9 - r^2}$$
$$V(0) = 0; \ V(\sqrt{6}) = 2\pi (\sqrt{6})^2 \sqrt{9 - (\sqrt{6})^2} = 12\pi\sqrt{3}$$
$$V(3) = 2\pi (3)^2 \sqrt{9 - (3)^2} = 2\pi (3)^2 (0) = 0.$$

The maximum value, $12\pi\sqrt{3}$ occurs when $r = \sqrt{6}$.

The radius is $\sqrt{6}$ units and the height is $2\sqrt{3}$ units.

Observation: In problems such as finding the radius of the largest cylinder which can be inscribed in a sphere, note the following:
1. We write a formula for the cylinder whose volume we want to maximize.
2. We write down a relationship between (we connect) the cylinder and the sphere.

Some Words for Maximum and Minimum Problems
1. greatest -----> Maximum
2. largest ------> Maximum
3. Smallest ----> Minimum
4. Shortest-----> Minimum
5. Nearest------> Minimum
6. Least--------->-Minimum
7. Maximize---> Find the Maximum
8. Minimize --->Find the minimum
9. Closest------> Minimum
10. Farthest ----->Maximum
11. Fewest------> Minimum
12. Highest-----> Maximum.
13. Longest ----> Maximum

Lesson 21 Exercises

1.: The sum of two numbers is 48. Find these numbers such that their product is a maximum.

2 Twice the sum of two numbers is 48. Find these numbers such that their product is a maximum.

3. The perimeter of a rectangle is 48. Find the length and the width such that the area is a maximum.

4 The sum of two numbers is 48. Find these numbers such that the product of one number and the square of the other number is a maximum.

5. An open box is to be made from a square sheet of tin. The length of each side of the sheet is L units. Squares of equal size are cut from the four corners of the sheet, and the sides are then bent up to form the box.. Find the length of each square cut out. such that the volume of the resulting box is a maximum.

6. An open box is to be made from a square sheet of tin. The length of each side of the sheet is 18 units. Squares of equal size are cut from the four corners of the sheet, and the sides are then bent up to form the box.. Find the length of each square cut out. such that the volume of the resulting box is a maximum.

7. An open box is to be made from a square sheet of tin. The length of the sheet is 15 units and the width is 8 units. of each side of the sheet is L units. Squares of equal size are cut from the four corners of the sheet, and the sides are then bent up to form the box.. Find the length of each square cut out. such that the volume of the resulting box is a maximum.

8. Find the radius and height of the right circular cylinder of maximum volume that can be inscribed in a right-circular cone with radius 3 units and height 5 units.

9. Find the radius and height of the right circular cylinder of maximum volume that can be inscribed in a sphere of radius 3 units.

Answers: 1. The numbers are 24 and 24; and the maximum product is 192.

2. The numbers are 12 and 12; and the maximum product is 144.

3. The length is 12 and the width is 12; and the maximum area is 144.

4. The numbers are 32 and 16.

5. The length of each side of square to be cut out $= \frac{L}{6}$ units.

6. The length of each side of square to be cut out $= 3$ units.

7. The length of each side of square to be cut out $= \frac{5}{3}$ units.

8. The radius is 2 units, The height is $\frac{5}{3}$ units.

9. The radius is $\sqrt{6}$ units and the height is $2\sqrt{3}$ units

CHAPTER 8

Lesson 22: **Differentiation of Trigonometric Functions**
Lesson 23: **Differentiation of Inverse Trigonometric Functions**

Lesson 22
Differentiation of Trigonometric Functions

We present the trigonometric functions and their corresponding derivatives.
We will derive some of the basic derivatives later. **Memorize** these basic formulas.

Table of Trigonometric Functions and Corresponding Derivatives

Function: $y = f(x)$	**Derivative:** $\dfrac{dy}{dx} = f'(x)$
1. $y = \sin x$	1. $\cos x$
2. $y = \cos x$	2. $-\sin x$
3. $y = \tan x$	3. $\sec^2 x$
4. $y = \cot x$	4. $-\csc^2 x$
5. $y = \sec x$	5. $\sec x \tan x$
6. $y = \csc x$	6. $-\csc x \cot x$

Mnemonic Help:

1. For the sine: its derivative is cosine (these are the first trig functions you learned)

2. For the cosine: its derivative is minus sine. So note the **minus** sign.

3. For the tangent, cotangent, secant, cosecant, we use a nonsense recitation as follows:

Say aloud the following keeping the layout of the table in mind:

1. tan   sec square.
2. cot   minus cosec square
3. sec   sec tan (note the repetition of sec)
4. cosec minus **cosec cotan** (Repetition of cosec is similar to that of sec x in **3**)

Also, for $\sin, \cos, \tan, \sec, \cot$, and $\csc$, observe the relationships between the terms of the following trigonometric identities. Each derivative is expressed in terms of the other function in each identity.

1. $\sin^2 x + \cos^2 x = 1$;   **2.** $1 + \tan^2 x = \sec^2 x$;   **3.** $1 + \cot^2 x = \csc^2 x$
        (sin goes with cos; tan goes with sec; and cot goes with csc.)

Note also that the derivative of a co-named function (cos, cot, cosec) is minus something; and the derivative of a non co-named function (sine, secant and tangent) is plus something. With the exception of the sine and cosine, each derivative is either the product of two like trigonometric functions or two unlike trigonometric functions.

For example, the derivative $\dfrac{d(\tan x)}{dx} = \sec x \bullet \sec x = \sec^2 x$ (product of two like

functions); while the derivative $\dfrac{d(\sec x)}{dx} = \sec x \bullet \tan x$ (product of two unlike functions)

Say aloud **1-4** over and over and as you say them, look at the table. Take a sheet of paper and try to write the above table, from memory, guided by this mnemonic device.

Repeat the above in the coming days, and thereafter, from time to time, test yourself by writing the contents of the table on paper, being guided by the mnemonic device, and if you do not recall anything, review the table and the device. Note that at least 50% of school and college work is memory work, especially since almost every exam is a closed book exam. Even if an exam were open book, the student who can recall accurately what he/she has learned will work faster than a student who has to look up information from a book. The above suggestion does not mean that you memorize without any understanding. First, understand and then try to find a way to remember what you have understood.

Examples on Differentiation of Trigonometric Functions

We will apply the following theorems in Examples 1 and 2

Theorem 1 $\lim\limits_{x \to 0} \dfrac{\sin x}{x} = 1$ (x in radians). We proved this theorem in Lesson 3)

Theorem 2 $\lim\limits_{x \to 0} \dfrac{1 - \cos x}{x} = 0$

Example 1 Given that $f(x) = \sin x$,

show that $f'(x) = \cos x$.

Solution

$$f'(x) = \lim_{a \to 0} \frac{f(x + a) - f(x)}{a}$$

$$f'(x) = \lim_{a \to 0} \frac{\sin(x + a) - \sin x}{a}$$

$$= \lim_{a \to 0} \frac{\sin x \cos a + \sin a \cos x - \sin x}{a}$$

$$\text{(applying } \sin(A + B) = \sin A \cos B + \sin B \cos A))$$

$$= \lim_{a \to 0} \frac{\sin x \cos a - \sin x + \sin a \cos x}{a} \quad \text{(rearranging the numerator)}$$

$$= \lim_{a \to 0} \frac{\sin x(\cos a - 1) + \sin a \cos x}{a} \quad \text{(factoring out } \sin x)$$

$$= \lim_{a \to 0} \left[\frac{\sin x(\cos a - 1)}{a} + \cos x \frac{\sin a}{a} \right]$$

$$= \lim_{a \to 0} \left[\cos x \frac{\sin a}{a} + \frac{\sin x(\cos a - 1)}{a} \right] \quad \text{(rearranging the terms)}$$

$$= \lim_{a \to 0} \left[\cos x \frac{\sin a}{a} - \frac{\sin x(-\cos a + 1)}{a} \right] \quad \text{(Note } b(c - d) = -b(-c + d))$$

$$= \lim_{a \to 0} \left[\cos x \frac{\sin a}{a} - \frac{\sin x(1 - \cos a)}{a} \right] \quad \text{(Rearranging within parentheses)}$$

$$= \lim_{a \to 0} \left[\cos x \frac{\sin a}{a} - \sin x \frac{(1 - \cos a)}{a} \right]$$

$$= \cos x \lim_{a \to 0} \frac{\sin a}{a} - \sin x \lim_{a \to 0} \frac{(1 - \cos a)}{a}$$

$$= \cos x \bullet (1) - \sin x \bullet (0) \quad (\lim_{a \to 0} \frac{\sin a}{a} = 1; \ \lim_{a \to 0} \frac{1 - \cos a}{a} = 0)$$

$$= \cos x - 0$$

$$\therefore f'(x) = \cos x \ \text{ or } \ \frac{d}{dx}(\sin x) = \cos x.$$

Example 2 Given that $f(x) = \cos x$, show that $f'(x) = -\sin x$.
  Solution

$$f'(x) = \lim_{a \to 0} \frac{f(x + a) - f(x)}{a}$$

$$f'(x) = \lim_{a \to 0} \frac{\cos(x + a) - \cos x}{a}$$

$$= \lim_{a \to 0} \frac{\cos x \cos a - \sin x \sin a - \cos x}{a}$$

$$\text{(Note: } \cos(A + B) = \cos A \cos B - \sin A \sin B)$$

$$= \lim_{a \to 0} \frac{\cos x \cos a - \cos x - \sin x \sin a}{a} \quad \text{(rearranging the numerator)}$$

$$= \lim_{a \to 0} \frac{\cos x (\cos a - 1) - \sin x \sin a}{a} \quad \text{(factoring out } \cos x)$$

$$= \lim_{a \to 0} \left[\frac{\cos x (\cos a - 1)}{a} - \sin x \frac{\sin a}{a} \right] \quad \text{(rewriting as two terms)}$$

$$= \lim_{a \to 0} \left[-\sin x \frac{\sin a}{a} + \frac{\cos x (\cos a - 1)}{a} \right] \quad \text{(rearranging)}$$

$$= \lim_{a \to 0} \left[-\sin x \frac{\sin a}{a} - \frac{\cos x (-\cos a + 1)}{a} \right] \quad \text{Note: } b(c - d) = -b(-c + d)$$

$$= \lim_{a \to 0} \left[-\sin x \frac{\sin a}{a} - \frac{\cos x (1 - \cos a)}{a} \right] \quad \text{(Rearranging within parentheses)}$$

$$= \lim_{a \to 0} \left[-\sin x \frac{\sin a}{a} - \cos x \frac{(1 - \cos a)}{a} \right] \quad \text{(Rearranging)}$$

$$= -\sin x \lim_{a \to 0} \frac{\sin a}{a} - \cos x \lim_{a \to 0} \frac{(1 - \cos a)}{a}$$

$$= -\sin x \cdot (1) - \cos x \cdot (0) \quad (\lim_{a \to 0} \frac{\sin a}{a} = 1; \ \lim_{a \to 0} \frac{1 - \cos a}{a} = 0)$$

$$= -\sin x - 0$$

$$f'(x) = -\sin x \ \text{ or } \ \frac{d}{dx}(\cos x) = -\sin x.$$

Example 3 Given that $y = \tan x$, show that $\frac{d}{dx}(\tan x) = \sec^2 x$.

Solution

$$y = \frac{\sin x}{\cos x} \quad (\tan x = \frac{\sin x}{\cos x})$$

$$\frac{dy}{dx} = \frac{\cos x \frac{d}{dx}[\sin x] - \sin x \frac{d}{dx}[\cos x]}{\cos^2 x}$$

(Using the quotient rule)

$$= \frac{\cos x \cos x - \sin x(-\sin x)}{\cos^2 x}$$

$$= \frac{\cos^2 x + \sin^2 x}{\cos^2 x}$$

$$= \frac{1}{\cos^2 x} \quad (\cos^2 x + \sin^2 x = 1)$$

$$= \sec^2 x \quad (\frac{1}{\cos x} = \sec x)$$

$$\frac{d}{dx}(\tan x) = \sec^2 x.$$

Example 4 Given that $y = \cot x$, show that

$$\frac{d}{dx}(\cot x) = -\csc^2 x.$$

Solution

$$y = \frac{\cos x}{\sin x} \quad (\cot x = \frac{\cos x}{\sin x})$$

$$\frac{dy}{dx} = \frac{\sin x \frac{d}{dx}[\cos x] - \cos x \frac{d}{dx}[\sin x]}{\sin^2 x}$$

(Using the quotient rule)

$$= \frac{\sin x(-\sin x) - \cos x(\cos x)}{\sin^2 x}$$

$$= \frac{-\sin^2 x - \cos^2 x}{\sin^2 x}$$

$$= \frac{-[\sin^2 x + \cos^2 x]}{\sin^2 x}$$

$$(\sin^2 x + \cos^2 x = 1)$$

$$= -\frac{1}{\sin^2 x}$$

$$= -\left(\frac{1}{\sin x}\right)^2 \quad (\frac{1}{\sin x} = \csc x)$$

$$= -\csc^2 x$$

$$\frac{d}{dx}(\cot x) = -\csc^2 x.$$

Example 5 Given that $y = \sec x$, show that $\dfrac{d}{dx}(\sec x) = \sec x \tan x$.

Solution (We apply the chain rule)

$$y = \frac{1}{\cos x} \quad (\sec x = \frac{1}{\cos x})$$

$$y = (\cos x)^{-1}$$

Let $u = \cos x$

Then $y = u^{-1}$

$$\frac{dy}{du} = -u^{-2}$$

$$= -\frac{1}{u^2}$$

$$= -\frac{1}{\cos^2 x} \quad (u = \cos x)$$

$$\frac{du}{dx} = -\sin x \quad (u = \cos x)$$

$$\frac{dy}{dx} = \frac{dy}{du}\frac{du}{dx} \quad \text{(Chain rule)}$$

$$\frac{dy}{dx} = (-\frac{1}{\cos^2 x})(-\sin x)$$

$$\frac{dy}{dx} = \frac{\sin x}{\cos^2 x}$$

$$\frac{dy}{dx} = \frac{\sin x}{\cos x} \cdot \frac{1}{\cos x}$$

$$\frac{dy}{dx} = \tan x \bullet \sec x$$

$$\frac{d}{dx}(\sec x) = \sec x \tan x.$$

Example 6 Given that $y = \csc x$, show that $\dfrac{d}{dx}(\csc x) = -\csc x \cot x$.

Solution (We apply the chain rule)

$$y = \frac{1}{\sin x} \quad (\csc x = \frac{1}{\sin x})$$

$$y = (\sin x)^{-1}$$

Let $u = \sin x$

Then $y = u^{-1}$

$$\frac{dy}{du} = -u^{-2}$$

$$= -\frac{1}{u^2}$$

$$= -\frac{1}{\sin^2 x} \quad (u = \sin x)$$

$$\frac{du}{dx} = \cos x \quad (u = \sin x)$$

$$\frac{dy}{dx} = \frac{dy}{du}\frac{du}{dx} \quad \text{(Chain rule)}$$

$$\frac{dy}{dx} = (-\frac{1}{\sin^2 x})(\cos x)$$

$$\frac{dy}{dx} = -\frac{\cos x}{\sin^2 x}$$

$$\frac{dy}{dx} = -\frac{\cos x}{\sin x} \cdot \frac{1}{\sin x}$$

$$\frac{dy}{dx} = -\cot x \bullet \csc x$$

$$\frac{d}{dx}(\csc x) = -\csc x \cot x.$$

Example 7a If $y = -\frac{1}{3}\cos^3 x$, find $\frac{dy}{dx}$.

Step 1: We will apply the chain rule. (since the given function is composite) Let $u = \cos x$. Then $y = -\frac{1}{3}u^3$ $\frac{du}{dx} = -\sin x \qquad (1)$ $\frac{dy}{du} = -\frac{1}{3}\left(3u^{3-1}\right)$	**Step 2**: $= -\frac{1}{3}\left(3u^2\right)$ $= -u^2$ $\frac{dy}{du} = -\cos^2 x \qquad (2)$ Substituting **(2)** and **(1)** in $\frac{dy}{dx} = \frac{dy}{du}\frac{du}{dx}$ $\frac{dy}{dx} = (-\cos^2)(-\sin x)$ $= \cos^2 x \sin x.$ or $= \sin x \cos^2 x$

Example 7b If $y = \sin^2 x$, find $\frac{dy}{dx}$ **Solution** We apply the chain rule. Let $u = \sin x$ Then $y = u^2$ $\frac{dy}{du} = 2u = 2\sin x$ $\frac{du}{dx} = \cos x$ $\frac{dy}{dx} = \frac{dy}{du}\frac{du}{dx}$ $= 2\sin x \cos x$ or $\sin 2x$	**Example 7c** If $y = \cos^2 x$ find $\frac{dy}{dx}$ **Solution** We can apply the chain rule as in Example 7b, but since we know that $\frac{d(\sin^2 x)}{dx} = 2\sin x \cos x$, we apply the Pythagorean identity From $\sin^2 x + \cos^2 x = 1$ $\cos^2 x = 1 - \sin^2 x$ $\frac{d(\cos^2 x)}{dx} = \frac{d(1)}{dx} - \frac{d(\sin^2 x)}{dx}$ $= 0 \ - 2\sin x \cos x$ $= -2\sin x \cos x$ or $-\sin 2x$ Try the above problem using the chain rule.

Example 7d If $y = \tan^2 x$ Find $\frac{dy}{dx}$ **Solution** We apply the chain rule. Let $u = \tan x$ Then $y = u^2$ $\frac{dy}{du} = 2u = 2\tan x$ $\frac{du}{dx} = \sec^2 x$ $\frac{dy}{dx} = \frac{dy}{du}\frac{du}{dx}$ $= 2\tan x \sec^2 x$	**Example 7e** $y = \sec^2 x$, find $\frac{dy}{dx}$ $\frac{dy}{dx} = \frac{d(\sec^2 x)}{dx}$, $\sec^2 x = \tan^2 x + 1$ $\frac{d(\sec^2 x)}{dx} = \frac{d(\tan^2 x)}{dx} + \frac{d(1)}{dx}$ $= 2\tan x \sec^2 x + 0$ $\frac{dy}{dx} = 2\tan x \sec^2 x$ (From 7d, $\frac{d(\tan^2 x)}{dx} = 2\tan x \sec^2 x$) Also try the above problem using the chain rule

Note above that by applying the following identities, we will save time.(see also p.255)
1. $\cos^2 x = \frac{1}{2}\cos 2x + \frac{1}{2}$; **2.** $\sin^2 x = \frac{1}{2} - \frac{1}{2}\cos 2x$ (From the double angle identities)

Example 8 Application of Implicit Differentiation

$$\text{Find } \frac{dy}{dx} \text{ if } \cos xy = x^2 \sin y \quad \text{(A)}$$

Solution

We differentiate both sides of the equation implicitly with respect to x in steps

Step 1: Differentiate left side of the equation implicitly with respect to x

Let $t = \cos xy$ and

let $u = xy$

Then $t = \cos u$

$$\frac{du}{dx} = x\frac{dy}{dx} + y(1) \quad (du = xdy + ydx; \frac{du}{dx} = x\frac{dy}{dx} + y\frac{dx}{dx}; \frac{du}{dx} = x\frac{dy}{dx} + y)$$

$$\frac{du}{dx} = x\frac{dy}{dx} + y$$

$$\frac{dt}{du} = -\sin u$$

$$= -\sin xy$$

$$\frac{dt}{dx} = \frac{dt}{du}\frac{du}{dx}$$

$$= -\sin xy(x\frac{dy}{dx} + y)$$

$$\therefore \frac{d}{dx}(\cos xy) = -\sin xy(x\frac{dy}{dx} + y) \text{ <--differentiating the left side of equation (A)}$$

Step 2: Differentiate the right side of the equation implicitly with respect to x.

(Keep x^2 constant and differentiate $\sin y$ implicitly followed by keeping

$\sin y$ constant and differentiating x^2.)

$$\frac{d}{dx}(x^2 \sin y) = x^2 \cos y\frac{dy}{dx} + \sin y(2x)$$

$$= x^2 \cos y\frac{dy}{dx} + 2x\sin y \text{ <--differentiating the right side of equation (A)}$$

Step 3: Equate the results from Step 1 and Step 2 to each other. Then

$$-\sin xy(x\frac{dy}{dx} + y) = x^2 \cos y\frac{dy}{dx} + 2x\sin y$$

Note: You could skip showing Steps 1 and 2 and work on both sides of (A) simultaneously.

Step 4: Solve for $\frac{dy}{dx}$: $\quad -x\sin xy\frac{dy}{dx} - y\sin xy = x^2 \cos y\frac{dy}{dx} + 2x\sin y$

$$-x\sin xy\frac{dy}{dx} - x^2 \cos y\frac{dy}{dx} = 2x\sin y + y\sin xy$$

$$-\left(x\sin xy + x^2 \cos y\right)\frac{dy}{dx} = 2x\sin y + y\sin xy$$

$$\frac{dy}{dx} = -\frac{2x\sin y + y\sin xy}{x\sin xy + x^2 \cos y}.$$

Lesson 22 Exercises

Complete the derivative part from memory

Function: $y = f(x)$	**Derivative:** $\dfrac{dy}{dx} = f'(x)$
1. $y = \sin x$	1.
2. $y = \cos x$	2.
3. $y = \tan x$	3.
4. $y = \cot x$	4.
5. $y = \sec x$	5.
6. $y = \csc x$	6.

7. f $y = -\dfrac{1}{3}\cos^3 x$, find $\dfrac{dy}{dx}$

8. Find $\dfrac{dy}{dx}$ if $\cos xy = x^2 \sin y$

9, From questions 1-6, derive $\dfrac{dy}{dx} = f'(x)$

Answers: 7. $\cos^2 x \sin x$; **8.** $\dfrac{dy}{dx} = -\dfrac{2x\sin y + y\sin xy}{x\sin xy + x^2 \cos y}$

Lesson 23

Differentiation of Inverse Trigonometric Functions

Introduction

Given the equation, $y = \sin x$ $\quad -\frac{\pi}{2} \le x \le \frac{\pi}{2}$, what is its inverse?

Solution

The inverse is found (by definition) by interchanging the roles of x and y.

Therefore, the inverse of $y = \sin x$ for $-\frac{\pi}{2} \le x \le \frac{\pi}{2}$, $\quad -1 \le y \le 1$ is

$$x = \sin y \text{ for } -\frac{\pi}{2} \le y \le \frac{\pi}{2}, \quad -1 \le x \le 1.$$

If we solve $x = \sin y$ for y, and use the notation, $\text{Sin}^{-1}x$, we obtain

$y = \text{Sin}^{-1}x$ (Recall previously that the inverse of $f(x)$ was symbolized $f^{-1}(x)$).
Another notation for this inverse is $y = \text{Arcsin } x$.

We may therefore use either $y = \text{Sin}^{-1}x$ or $y = \text{Arcsin } x$ for the inverse of
$y = \sin x$. Similarly, the inverse of $y = \cos x$ is symbolized $y = \text{Cos}^{-1}x$ or
$y = \text{Arccos } x$,; for the inverse of $y = \tan x$, the symbol is $y = \text{Tan}^{-1}x$;
for the inverse of $y = \sec x$, the symbol is $y = \text{Sec}^{-1}x$. for the inverse of
$y = \csc x$, the symbol is $y = \text{Csc}^{-1}x$; for $y = \cot x$, the inverse is $y = \text{Cot}^{-1}x$.

We present the inverse trigonometric functions and their derivatives. We will derive some of
the derivatives later. **Memorize** them. We see them again in Lessons 39, 41, 43 & others.

Table of Inverse trigonometric Functions and Derivatives

Inverse Trig Function $y = f(x)$	**Derivative:** $\frac{dy}{dx} = f'(x)$	Each derivative is $\dfrac{1}{\text{something}}$
1. $y = \text{Sin}^{-1}x$;	$\dfrac{dy}{dx} = \dfrac{1}{\sqrt{1-x^2}}$	Note the $\sqrt{1-x^2}$ for $\text{Sin}^{-1}x$ We will see this again in Lesson 43, under integration by trigonometric substitution.
2. $y = \text{Cos}^{-1}x$;	$\dfrac{dy}{dx} = -\dfrac{1}{\sqrt{1-x^2}}$	
3. $y = \text{Tan}^{-1}x$;	$\dfrac{dy}{dx} = \dfrac{1}{1+x^2}$	Note the $1+x^2$ for $\text{Tan}^{-1}x$ We will see this again in Lesson 41, under integration by trigonometric substitution.
4. $y = \text{Sec}^{-1}x$;	$\dfrac{dy}{dx} = \dfrac{1}{x\sqrt{x^2-1}}$	
5. $y = \text{Csc}^{-1}x$;	$\dfrac{dy}{dx} = -\dfrac{1}{x\sqrt{x^2-1}}$	Note the $x\sqrt{x^2-1}$ for $\text{Sec}^{-1}x$ We will see this again in Lesson 43, under integration by trigonometric substitution.
6. $y = \text{Cot}^{-1}x$;	$\dfrac{dy}{dx} = -\dfrac{1}{1+x^2}$	

Memorize **1, 3** and **4**, and note that **2** is the negative of **1**; **6** is the negative of **3**.
5 is the negative of **4**. The derivatives of the inverse cofunctions **1 & 2**; **3 & 6**;
4 & 5 are negatives of each other. Thus, the derivative of a "co-named" function is
minus something. See also p. 177. **Note** that the derivative of an inverse trigonometric
function is an algebraic function, and **not** a trigonometric function.

Finding the Derivatives of Some Inverse Trigonometric Functions

Example 1 Show that $\dfrac{d}{dx}(\text{Sin}^{-1}x) = \dfrac{1}{\sqrt{1-x^2}}$

By deriving some basic formulas we get insight into the approaches used in finding derivatives

Method 1	Method 2

Method 1

Step 1: Let $\text{Sin}^{-1}x = y$

Then $x = \sin y$ (1)

$\dfrac{dx}{dy} = \cos y$; $\dfrac{dy}{dx} = \dfrac{1}{\dfrac{dx}{dy}}$

$\dfrac{dy}{dx} = \dfrac{1}{\cos y}$ (2)

Step 2: From $\sin^2 y + \cos^2 y = 1$

$\cos y = \sqrt{1 - \sin^2 y}$, and

$\dfrac{dy}{dx} = \dfrac{1}{\sqrt{1 - \sin^2 y}}$ (3)

Step 3: From (1) $x = \sin y$, and

squaring, $x^2 = \sin^2 y$

$\therefore \; \dfrac{d}{dx}(\text{Sin}^{-1}x) = \dfrac{1}{\sqrt{1-x^2}}$

(Replacing $\sin^2 y$ by x^2 in (3)).

Method 2

Step 1: Let $\text{Sin}^{-1}x = y$

Then $\sin y = x$ (1)

Using implicit dfferentiation

$\cos y \dfrac{dy}{dx} = 1$;

and $\dfrac{dy}{dx} = \dfrac{1}{\cos y}$

Step 2: From $\sin^2 y + \cos^2 y = 1$,

$\cos y = \sqrt{1 - \sin^2 y}$, and

$\dfrac{dy}{dx} = \dfrac{1}{\sqrt{1 - \sin^2 y}}$

Step 3: From (1) $x = \sin y$, and

squaring, $x^2 = \sin^2 y$

$\therefore \; \dfrac{d}{dx}(\text{Sin}^{-1}x) = \dfrac{1}{\sqrt{1-x^2}}$

(Replacing $\sin^2 y$ by x^2)

Method 3

Step 1: Let $\text{Sin}^{-1}x = y$

Then $x = \sin y$ (1)

$\dfrac{dx}{dy} = \cos y$;

$\dfrac{dy}{dx} = \dfrac{1}{\dfrac{dx}{dy}} = \dfrac{1}{\cos y}$; and

$\dfrac{dy}{dx} = \dfrac{1}{\cos y}$ (2)

Step 2: Draw a right triangle using

$\sin y = \dfrac{x}{1}$, and calculate the third side.

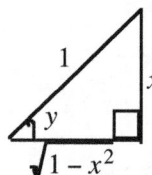

(using Pythagorean theorem).

opposite side $= x$

hypotenuse $= 1$

Step 3: From the right triangle

$\cos y = \dfrac{\sqrt{1-x^2}}{1}$

Substitute for $\cos y$ in (2)

Then $\dfrac{dy}{dx} = 1 \div \dfrac{\sqrt{1-x^2}}{1}$

$\dfrac{dy}{dx} = \dfrac{1}{\sqrt{1-x^2}}$ (reciprocal)

$\therefore \; \dfrac{d}{dx}(\text{Sin}^{-1}x) = \dfrac{1}{\sqrt{1-x^2}}$

Note: If $y = \text{Cos}^{-1}x$, we can similarly show that $\dfrac{dy}{dx} = -\dfrac{1}{\sqrt{1-x^2}}$. Try it.

Example 2 Find $\dfrac{dy}{dx}$ if $y = \text{Sec}^{-1}x$

Step 1: $y = \text{Sec}^{-1}x$ $\qquad$ (A)

is equivalent to

$\sec y = x$ $\qquad$ (B)

Step 2: Perform implicit differentiation with respect to x in equation (B)

Then $\dfrac{d}{dx}(\sec y) = \dfrac{d}{dx}(x)$

$\sec y \bullet \tan y \dfrac{dy}{dx} = 1$ $\qquad$ (C)

Step 3: $\dfrac{dy}{dx} = \dfrac{1}{\sec y \bullet \tan y}$ $\qquad$ (Solving equation (C) for $\dfrac{dy}{dx}$)

Step 4: From the trigonometric identity $1 + \tan^2 y = \sec^2 y$

$\tan y = +\sqrt{\sec^2 y - 1}$,

$\tan y = \sqrt{x^2 - 1}$ (substituting for $\sec y = x$ from (B))

Step 4: $\dfrac{dy}{dx} = \dfrac{1}{x\sqrt{x^2 - 1}}$ (Substitute $\sec y = x$; $\tan y = \sqrt{x^2 - 1}$) in Step 3).

Note above that we could also use the other two methods in Example 1. Try them.

Example 3 Find $\dfrac{dy}{dx}$ if $y = \text{Tan}^{-1}x$

Solution

Step 1: $y = \text{Tan}^{-1}x$ $\qquad$ (A)

is equivalent to

$\tan y = x$ $\qquad$ (B)

Step 2: Perform implicit differentiation with respect x in equation (B)

Then $\sec^2 y \dfrac{dy}{dx} = 1$ $\qquad$ (C)

Step 3: $\dfrac{dy}{dx} = \dfrac{1}{\sec^2 y}$ $\qquad$ (Solving equation (C) for $\dfrac{dy}{dx}$)

Step 4: Using the trigonometric identity $\sec^2 y = \tan^2 y + 1$

$\dfrac{dy}{dx} = \dfrac{1}{\tan^2 y + 1}$, and also, substituting x for $\tan y$ from equation (B)

$\dfrac{dy}{dx} = \dfrac{1}{1 + x^2}$. $(\tan y = x)$

Note above that we could also use the other two methods in Example 1. Try them.

Memory device (I use the following concrete example for recall)

My memory device for remembering for example that $y = \text{Sin}^{-1}x$ is equivalent to $\sin y = x$ is that I recall from the trigonometric tables that $\text{Sin}^{-1}(\tfrac{1}{2}) = \tfrac{\pi}{6} = 30°$ is equivalent to $\sin\tfrac{\pi}{6} = \tfrac{1}{2}$.

Example 4 Find $\dfrac{dy}{dx}$ if $y = Csc^{-1}x$

Solution:

Step 1: $y = Csc^{-1}x$ (A)

is equivalent to

$\csc y = x$ (B)

Step 2: Perform implicit differentiation with respect to x in equation (B).

Then $\dfrac{d}{dx}(\csc y) = \dfrac{d}{dx}(x)$

$-\csc y \bullet \cot y \dfrac{dy}{dx} = 1$ (C)

Step 3: $\dfrac{dy}{dx} = -\dfrac{1}{\csc y \bullet \cot y}$ (D)   (Solving equation (C) for $\dfrac{dy}{dx}$)

Step 4: From the trigonometric identity $1 + \cot^2 y = \csc^2 y$

$\cot y = +\sqrt{\csc^2 y - 1}$,

$\cot y = \sqrt{x^2 - 1}$ (squaring (B) and substituting x^2 for $\csc^2 y$)

Step 5: $\boxed{\dfrac{dy}{dx} = -\dfrac{1}{x\sqrt{x^2 - 1}}}$ (Substitute x for $\csc y$; $\sqrt{x^2 - 1}$ for $\cot y$ in (D.)

Note above that we could also use the other two methods in Example 1.

Applications of Inverse Cofunction Trigonometric Identities

Each of the following identities can be used to find the derivative of an inverse cofunction, knowing the derivative of the other cofunction.

1. $Sin^{-1}x + Cos^{-1}x = \dfrac{\pi}{2}$;   **2**. $Tan^{-1}x + Cot^{-1}x = \dfrac{\pi}{2}$;   **3**. $Sec^{-1}x + Csc^{-1}x = \dfrac{\pi}{2}$

Example 5 Find $\dfrac{d(cot^{-1}x)}{dx}$

Solution

We can do this problem using the previous methods. However as a third method, we apply the inverse cofunction identity

$\tan^{-1}x + \cot^{-1}x = \dfrac{\pi}{2}$,

$\cot^{-1}x = \dfrac{\pi}{2} - \tan^{-1}x$

$\dfrac{d(cot^{-1}x)}{dx} = \dfrac{d}{dx}\left(\dfrac{\pi}{2}\right) - \dfrac{d(tan^{-1}x)}{dx}$

$= 0 - \dfrac{1}{1 + x^2}$

$\boxed{\dfrac{d(Cot^{-1}x)}{dx} = -\dfrac{1}{1 + x^2}}$

(previously, $\dfrac{d(tan^{-1}x)}{dx} = \dfrac{1}{1 + x^2}$)

Example 6 Find $\dfrac{d(cos^{-1}x)}{dx}$

Solution

From $\sin^{-1}x + \cos^{-1}x = \dfrac{\pi}{2}$

$\cos^{-1}x = \dfrac{\pi}{2} - \sin^{-1}x$,

$\dfrac{d(cos^{-1}x)}{dx} = \dfrac{d}{dx}\left(\dfrac{\pi}{2}\right) - \dfrac{d(sin^{-1}x)}{dx}$

$= 0 - \dfrac{1}{\sqrt{1 - x^2}}$

$\boxed{\dfrac{d(cos^{-1}x)}{dx} = -\dfrac{1}{\sqrt{1 - x^2}}}$

(previously, $\dfrac{d(sin^{-1}x)}{dx} = \dfrac{1}{\sqrt{1 - x^2}}$)

Observe also how the following relationships could be obtained from the above inverse cofunction identities because $\dfrac{d}{dx}\left(\dfrac{\pi}{2}\right) = 0$.

1. $\dfrac{d}{dx}(Cos^{-1}x) = -\dfrac{d}{dx}(Sin^{-1}x)$ **2.** $\dfrac{d}{dx}(Cot^{-1}x) = -\dfrac{d}{dx}(Tan^{-1}x)$

3. $\dfrac{d}{dx}(Csc^{-1}x) = -\dfrac{d}{dx}(Sec^{-1}x)$.

Lesson 23 Exercises

A Symbolize the inverse of each of the following:

1. $y = \sin x$, **2.** $y = \cos x$; **3.** $y = \tan x$,; **4.** $y = \sec x$.

B Complete the derivative part from memory.

1. $y = \text{Sin}^{-1}x$; $\dfrac{dy}{dx} =$

2. $y = \text{Cos}^{-1}x$; $\dfrac{dy}{dx} =$

3. $y = \text{Tan}^{-1}x$; $\dfrac{dy}{dx} =$

4. $y = \text{Sec}^{-1}x$; $\dfrac{dy}{dx} =$

5. $y = \text{Csc}^{-1}x$; $\dfrac{dy}{dx} =$

6. $y = \text{Cot}^{-1}x$; $\dfrac{dy}{dx} =$

If $y = \text{Sin}^{-1}x$, derive $\dfrac{dy}{dx} = \dfrac{1}{\sqrt{1-x^2}}$

8. If $y = \text{Sec}^{-1}x$, derive $\dfrac{dy}{dx} = \dfrac{1}{x\sqrt{x^2-1}}$; **9.** If $y = \text{Sin}^{-1}(\sqrt{1-x})$, find $\dfrac{dy}{dx}$

Answers: A: 2. $y = \text{Cos}^{-1}x$ or $y = \text{Arccos } x$; **3.** $y = \text{Tan}^{-1}x$; **4.** $y = \text{Sec}^{-1}x$

9. $-\dfrac{1}{2\sqrt{x}\sqrt{1-x}}$.

CHAPTER 9
Applications of Differentiation B

Lesson: 24: **Equations of Tangent and Normal to a Curve**
Lesson: 25: **Angle of Intersection of Two Curves;**
              Length of Tangent, Normal, Subtangent, and
              Subnormal to a Curve

Lesson 26: **Rectilinear and Circular Motion**

Lesson 27: **Applications of Differentiation to Curvature**

Lesson 24
Equations of Tangent and Normal to a Curve

Perpendicular Lines and their Slopes

Fact 1: If two lines are perpendicular, then their slopes are negative reciprocals
of each other; or simply, if m_1 and m_2 are their slopes, then $m_1 = -\dfrac{1}{m_2}$

or $m_2 = -\dfrac{1}{m_1}$ or $m_1 \cdot m_2 = -1$.

Fact 2: The tangent and the normal to a curve at a given point are perpendicular
to each other (meet at right angles) at the given point.

If the slope of the tangent is m_1, then the slope, m_2, of the normal is $-\dfrac{1}{m_1}$.

Suppose the function $f(x)$ has a finite derivative $f'(x_0)$ at $x = x_0$. Then the
curve has a tangent at the point (x_0, y_0), and the slope, m, of this tangent is given
by $m = f'(x_0)$. From elementary mathematics, the point-slope form of the
equation of the line with slope, m, and passing through the point (x_0, y_0) is
given by $y - y_0 = m(x - x_0)$. By replacing m by $f'(x_0)$, the point-slope form of
the equation of the tangent to the curve $f(x)$ at (x_0, y_0) is given by
$y - y_0 = f'(x_0)(x - x_0)$ and the corresponding equation of the normal at
(x_0, y_0) is given by $y - y_0 = -\dfrac{1}{f'(x_0)}(x - x_0)$.

(Observe that the slopes of the tangent and normal are negative reciprocals of each other.)

Special cases: **Horizontal and vertical tangents and normals**

The special cases are for horizontal and vertical tangents and normals.
Since the tangent and the normal at a point are perpendicular to each other, if the
tangent is horizontal, the normal is vertical, but if the tangent is vertical, the
normal is horizontal. Consequently, if the equation of the horizontal tangent at
(x_0, y_0) is $y = y_0$ (from $y - y_0 = f'(x_0)(x - x_0)$, with $m = f'(x_0) = 0$), the
equation of the normal is $x = x_0$ (vertical line) at this point.

However, if the tangent is vertical, its equation is given by $x = x_0$ and the
equation of the normal is $y = y_0$ (horizontal line).

Example 1

Find equations of the tangent line and the normal line to the curve given by
$f(x) = x^3 - x^2 + 15$ at the point $(-2, 3)$ on the curve.

Solution

For the tangent:

Step 1: Find $f'(x)$, the derivative of $f(x)$.

$$f'(x) = 3x^2 - 2x$$

Step 2: Find $f'(-2)$ (i.e., the derivative of $f(x)$ at $x = -2$)

$$f'(-2) = 3(-2)^2 - 2(-2)$$
$$= 16.$$

The slope of the tangent line at $(-2, 3)$ is 16.

Step 3: Now, with the slope $m = 16$, at $(-2, 3)$, we apply the point-slope form
of the equation of the line (which is given by $y - y_1 = m(x - x_1)$).
Substituting $m = 16$, $x_1 = -2$, $y_1 = 3$, we obtain

$$y - 3 = 16(x - (-2)) \quad \text{<------- point-slope form}$$
$$y - 3 = 16(x + 2)$$
$$y = 16x + 35 \quad \text{<----slope-intercept form.}$$

For the normal

Step 1: Since the tangent and the normal are perpendicular to each other, and the
slope of the tangent is 16, the slope of the normal is $-\frac{1}{16}$.

Step 2: Now, with the slope $m = -\frac{1}{16}$ at $(-2, 3)$, we apply the point-slope
form equation $y - y_1 = m(x - x_1)$ to obtain

$$y - 3 = -\frac{1}{16}(x - (-2)) \quad \text{<---- point-slope form.}$$
$$y - 3 = -\frac{1}{16}(x + 2)$$
$$y = -\frac{1}{16}x + \frac{23}{8} \quad \text{<----slope-intercept form.}$$

Lesson 24 Exercises A

1. If two lines are perpendicular, to each other and the slope of one lines is
   m_1, what is the slope of the other line?

2. At what angle do the tangent and the normal to a curve at a point meet?

3. Two lines are perpendicular to each other, and one of the lines is a vertical
   line, then what type of a line is the other line?

4. Find equations of the tangent line and the normal line to the curve given by
   $f(x) = x^3 - x^2 + 15$ at the point $(-2, 3)$ on the curve.

Answers: **4.** Tangent: $y = 16x + 35$; normal: $y = -\frac{1}{16}x + \frac{23}{8}$.

Lesson 24 Exercises B

1. Find equations of the tangent line and the normal line to the curve given by $y = x^3 - 2x^2 + 4$ at the point $(2, 4)$ on the curve.

2. Find equations of the tangent line and the normal line to the curve given by $f(x) = x^3 - x^2 + 5$ at the point $(-1, 3)$ on the curve.

Answers: **1.** Tangent: $y = 4x - 4$; normal: $y = -\frac{1}{4}x + \frac{9}{2}$

2. Tangent: $y = 5x + 8$; normal: $y = -\frac{1}{5}x + \frac{14}{5}$

Lesson 25

Angle of Intersection of Two Curves; Length of Tangent, Normal, Subtangent, and Subnormal to a Curve

The angle of intersection of two curves c_1, c_2 is defined as the angle between the tangents t_1, t_2 at the point of intersection of c_1 and c_2.

Determining the angle of intersection

Step 1: Determine the points of intersection of the two curves c_1, c_2 by solving the equations of these curves simultaneously.

Step 2: Find the slopes m_1, m_2 of the tangents to the curves at each point of intersection.

Step 3: **Case 1**:(General case)

The angle of intersection, θ is found from the relationship

$$\tan \theta = \frac{m_1 - m_2}{1 + m_1 m_2}, \text{ where } \theta \text{ is the acute angle of intersection when}$$

$\tan \theta > 0$, but if $\tan \theta < 0$, the acute angle of intersection is $180^o - \theta^o$.

Case 2: If $m_1 = -\dfrac{1}{m_2}$, the tangents are perpendicular to each other and the angle of intersection is $90°$.

Case 3: If $m_1 = m_2$, the two tangents are coincident and the angle of intersection is $0°$.

Note above that $\tan \theta = \dfrac{m_1 - m_2}{1 + m_1 m_2}$ is obtained from the trigonometric identity

$$\tan(A - B) = \frac{\tan A - \tan B}{1 + \tan A \tan B}.$$

Example 1 Find the measure of the acute angle of intersection between the curves whose equations are given by $x^2 + y^2 = 4x$ and $x^2 + y^2 = 8$

Step 1: Determine the points of intersection of $x^2 + y^2 = 4x$ and $x^2 + y^2 = 8$ by solving these equations simultaneously.

$$x^2 + y^2 - 4x = 0 \qquad \text{(A)}$$
$$x^2 + y^2 - 8 = 0 \qquad \text{(B)}$$

(A)-(B): $-4x - (-8) = 0$
$$-4x + 8 = 0$$
$$x = 2$$

when $x = 2$ in (A): $2^2 + y^2 - 4(2) = 0$
$$4 + y^2 - 8 = 0$$
$$y^2 = 4$$
$$y = \pm 2 \ \ (y = 2, \text{ or } -2)$$

Thus when $x = 2$, $y = 2$, and also when $x = 2$, $y = -2$

The points of intersection are $P_1(2, 2)$ and $P_2(2, -2)$.

Step 2: Find the slopes m_1, m_2 of the tangents at $P_1(2, 2)$ and at $P_2(2, -2)$, respectively.

At $P_1(2, 2)$: For $x^2 + y^2 - 4x = 0$ (A)

$$2x + 2y\frac{dy}{dx} - 4 = 0 \text{ (differentiating implicitly)}$$

$$x + y\frac{dy}{dx} - 2 = 0 \quad \text{(Dividing through by 2)}$$

$$\frac{dy}{dx} = \frac{2 - x}{y}$$

$$m_1 = \frac{2 - 2}{2} \quad (x = 2, y = 2)$$

$$m_1 = \frac{0}{2}, \text{ and } \boxed{m_1 = 0}$$

At $P_1(2, 2)$: For $x^2 + y^2 - 8 = 0$ (B)

$$2x + 2y\frac{dy}{dx} = 0 \text{ (differentiating implicitly)}$$

$$x + y\frac{dy}{dx} = 0 \quad \text{(Dividing through by 2)}$$

$$\frac{dy}{dx} = -\frac{x}{y}$$

$$m_2 = -\frac{2}{2} = -1 \quad (x = 2, y = 2)$$

$$\boxed{m_2 = -1}$$

Step 3: Find the angle of intersection.

Now, $m_2 = -1$ $m_1 = 0$

Using $\tan\theta = \dfrac{m_2 - m_1}{1 + m_1 m_2}$	**Using** $\tan\theta = \dfrac{m_1 - m_2}{1 + m_1 m_2}$
$\tan\theta = \dfrac{-1 - 0}{1 + 0} = -1$	$\tan\theta = \dfrac{-0 - (-1)}{1 + 0}$
$\tan\theta = -1$	
$\theta_{ref} = 45°$	$\tan\theta = \dfrac{-0 - (-1)}{1 + 0} = \dfrac{1}{1} = 1$
Although $\theta = 135°$ (since the tangent is negative in the second quadrant, the angle of intersection is $180° - 135° = 45°$ (the acute angle)	$\tan\theta = 1$ $\theta_{ref} = 45°$

Step 4: At $P_2(2, -2)$: For $x^2 + y^2 - 4x = 0$ (A)

$$\frac{dy}{dx} = \frac{2 - x}{y}$$

$$m_1 = \frac{2 - 2}{-2} = 0 \quad (x = 2, y = -2)$$

$$m_1 = 0$$

At $P_2(2, -2)$: For $x^2 + y^2 - 8 = 0$ (B)

$$\frac{dy}{dx} = -\frac{x}{y}$$

$$m_2 = -\frac{2}{-2} = 1 \quad (x = 2, y = -2)$$

$$\boxed{m_2 = 1}$$

Step 5: Find the angle of intersection, θ from the relationship.

Now, $m_1 = 0 \quad m_2 = 1$

Using	Using
$\tan \theta = \dfrac{m_2 - m_1}{1 + m_1 m_2}$	$\tan \theta = \dfrac{m_1 - m_2}{1 + m_1 m_2}$
$\tan \theta = \dfrac{1 - 0}{1 + 0} = 1$	$\tan \theta = \dfrac{0 - 1}{1} = -1$
$\tan \theta = 1$	$\tan \theta = -1$
$\theta_{ref} = 45°$	$\theta_{ref} = 45°$
$\theta = 45°$	Although $\theta = 135°$ (since the tangent is negative in the second quadrant, the angle of intersection is $180° - 135° = 45°$ (the acute angle)

The angle of intersection at $P_1(2, 2)$ is $45°$ and at $P_2(2, -2)$ is also $45°$. The angles being the same at the two different points is due to the symmetry of the two curves with respect to the x-axis.

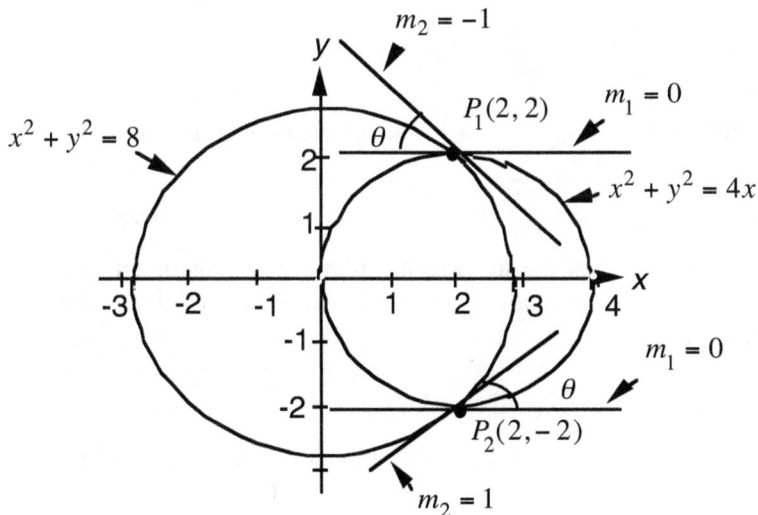

Figure 1

Length of Tangent, Normal, Subtangent, and Subnormal

The length of a **tangent** to a curve at a point on the curve is defined as the length of the segment of the tangent between the point of contact (on the curve) and the x-axis.

The length of the **subtangent** is the length of the **projection** of the tangent (above segment))) **on** the x-axis. The length of the **normal** is defined as the length of the segment of the normal between the contact point of the tangent and the x-axis; and the length of the **projection** of this segment on the x-axis is called the length of the **subnormal**.

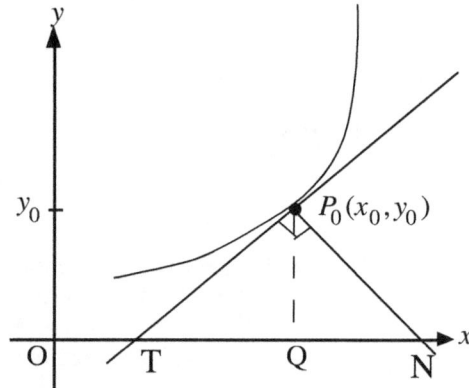

With reference to the above diagram,

Length of subtangent $= \boxed{TQ = \dfrac{y_0}{m}}$: $(m = \text{slope} = \dfrac{y_0}{TQ} = \dfrac{\text{vertical change}}{\text{Horizontal change}})$

$(m$ is the slope of the tangent at $P_0)$

Length of subnormal $= \boxed{QN = my_0}$; $(\dfrac{y_0}{QN} = -\dfrac{1}{m} = \text{slope of } P_0N)$

Length of tangent $= TP_0 = \sqrt{\left(\dfrac{y_0}{m}\right)^2 + \left(y_0\right)^2}$ (Applying the Pythagorean theorem)

Length of normal $= NP_0 = \sqrt{\left(my_0\right)^2 + \left(y_0\right)^2}$ (Applying the Pythagorean theorem)

Example 2 Find the length of the subtangent, subnormal, tangent and normal to the curve whose equation is given by $xy + 4x - y = 16$ at the point $(3, 2)$.

Solution

Step 1: Find the slope of the tangent at $(3, 2)$

$$x\frac{dy}{dx} + y + 4 - \frac{dy}{dx} = 0 \quad \text{(implicitly differentiating with respect to } x)$$

$$x\frac{dy}{dx} - \frac{dy}{dx} + y + 4 = 0$$

$$\frac{dy}{dx}(x - 1) + y + 4 = 0$$

$$\frac{dy}{dx} = \frac{-y - 4}{x - 1} = \frac{-2 - 4}{3 - 1} = -3 \quad (x = 3, y = 2)$$

$m = -3$ (slope of the tangent at $(3, 2)$.

Step 2: Now, $m = -3$, $y_0 = 2$

Length of subtangent $= \left| \dfrac{y_0}{m} \right| = \left| \dfrac{2}{-3} \right| = \dfrac{2}{3}$

Length of subnormal. $\left| my_0 \right| = \left| -3(2) \right| = 6$

Length of tangent $= \sqrt{ \left(\dfrac{y_0}{m} \right)^2 + \left(y_0 \right)^2 }$ (Applying the Pythagorean theorem)

$$= \sqrt{ \left(\dfrac{2}{-3} \right)^2 + (2)^2 } \ = \sqrt{ \dfrac{4}{9} + 4 } = = \ \sqrt{ \dfrac{40}{9} } \ = \dfrac{2\sqrt{10}}{3}$$

Length of normal $= \sqrt{ \left(my_0 \right)^2 + \left(y_0 \right)^2 }$

$$= \sqrt{ ((-3)(2))^2 + (2)^2 } \ = \sqrt{36 + 4} \ = \sqrt{40} \ = 2\sqrt{10} .$$

Lesson 25 Exercises

1. Define the angle of intersection between two curves

2. Find the measure of the acute angle of intersection between the curves whose equations are given by $x^2 + y^2 = 4x$ and $x^2 + y^2 = 8$

3. Define the length of a tangent of a curve at one of its points..

4. Find the length of the subtenant, subnormal, tangent and normal to the curve whose equation is given by $xy + 4x - y = 16$ at the point $(3, 2)$

Answers: **2.** $45°$. **4.** subtangent: $\dfrac{2}{3}$ units; subnormal.: 6 units;

tangent : $\dfrac{2\sqrt{10}}{3}$ units; normal : $2\sqrt{10}$ units.

Lesson 26

Rectilinear and Circular Motion
Rectilinear Motion
(Motion in a Straight Line)

If a body moves along a straight line and travels a distance s, the distance s is a function of the time t taken to travel the distance s, and we can write $s = f(t)$, where $t \geq 0$.

The **linear velocity** v at time t is given by $v = \dfrac{ds}{dt}$.

If $v > 0$, the body is moving in the direction of increasing s.

If $v < 0$, the body is moving in the direction of decreasing s.
If $v = 0$, the body is at rest.

The **linear acceleration** a at time t is given by

$$a = \frac{dv}{dt} = \frac{d^2s}{dt^2}$$

If $a > 0$, v is increasing; but if $a < 0$, v is decreasing.
If v and a have the same sign, the speed of the body is increasing.
If v and a have opposite signs, the speed of the body is decreasing.

Analogy

$v = \dfrac{ds}{dt}$ (First derivative) (velocity: rate of change of distance)	$f' = \dfrac{dy}{dx}$ (First derivative) (slope:: rate of change of y with respect x.)
$a = \dfrac{dv}{dt} = \dfrac{d^2s}{dt^2}$ (second derivative) Rate of change of velocity (acceleration)	$f'' = \dfrac{d}{dx}\left(\dfrac{dy}{dx}\right) = \dfrac{d^2y}{dx^2}$ Rate of change of slope. (second derivative)

Example   A particle is moving along a straight line according to the equation
$s = 2t^3 - 3t^2 + 2t - 4$, where distance s is in ft. and time t is in seconds. Find its velocity and acceleration at the end of 3 seconds.

Solution   Since $s = 2t^3 - 3t^2 + 2t - 4$.
Velocity v, is given by
$v = \dfrac{ds}{dt} = 6t^2 - 6t + 2$
When t = 3, $v = 6(3)^2 - 6(3) + 2$
$v = 38$.
The velocity is $38\, ft/s$.
Acceleration a, is given by
$a = \dfrac{dv}{dt} = 12t - 6$ $(v = 6t^2 - 6t + 2)$
When t = 3, $a = 12(3) - 6$
$a = 30$
The acceleration is $30\, ft/s^2$.

Circular Motion

If a body moves along a circle, its motion is defined in terms of the angular rotation, θ. where θ (in radians) is the central angle subtended by the arc of the circle between the initial position of the body on the circle and the final position on the circle in time t, where $t \geq 0$.

The **angular velocity**, ω, of a body is given by

$\omega = \dfrac{d\theta}{dt}$.

The **angular acceleration,** α, of a body is given by

$\alpha = \dfrac{d\omega}{dt} = \dfrac{d^2\theta}{dt^2}$.

If the body moves with constant angular acceleration, α is a constant.
If the body moves with constant angular velocity (i.e., no acceleration), $\alpha = 0$.

Example   A particle moves in a circle according to the

equation $\theta = t^3 - 20t$, where θ is in radians and t is in seconds. Find its angular displacement, angular velocity and angular acceleration at the end of 5 seconds

Solution Angular displacement $\theta = t^3 - 20t$.

When $t = 5$, $\theta = (5)^3 - 20(5)$
$$= 125 - 100$$
$$= 25$$

Angular displacement = 25 rad.

Angular velocity ω, is given by

$\omega = \dfrac{d\theta}{dt} = 3t^2 - 20$ $(\theta = t^3 - 20t)$

When $t = 5$, $\omega = 3(5)^2 - 20$
$$\omega = 75 - 20$$
$$= 55.$$

The angular velocity after 5 seconds is $55\,\text{rad/s}$.

Angular acceleration α, is given by

$\alpha = \dfrac{d\omega}{dt} = 6t$ $(\omega = 3t^2 - 20)$

When $t = 5$, $\alpha = 6(5)$
$$\alpha = 30$$

The angular acceleration is $30\ rad/s^2$.

Lesson 26 Exercises A

1. A particle is moving along a straight line according to the equation $s = 2t^3 - 3t^2 + 2t - 4$, where distance s is in ft. and time t is in seconds. Find its velocity and acceleration at the end of 3 seconds.

2. A particle moves in a circle according to the equation $\theta = t^3 - 20t$, where θ is in radians and t is in seconds. Find its angular displacement, angular velocity and angular acceleration at the end of 5 seconds.

Answer: 1. velocity $= 38\,ft/s$; acceleration $= 30\,ft/s^2$;

2. displacement $= 25$ rad; angular velocity $55\,rad/s$; acceleration is 30 rad/s^2

Lesson 26 Exercises B

1. A particle is moving along a straight line according to the equation $s = 2t^3 - 4t^2 + 2t - 1$, where distance s is in ft. and time t is in seconds. Find its velocity and acceleration at the end of 3 seconds

2. A particle moves in a circle according to the equation $\theta = t^3 - 12t$, where θ is in radians and t is in seconds. Find its angular displacement, angular velocity and angular acceleration at the end of 4 seconds.

Answer: 1. velocity $= 32\,ft/s$; acceleration $= 28\,ft/s^2$;

2. displacement $= 16$ rad.;. angular velocity $= 36$ rad/s; acceleration $= 24$ rad/s^2

Lesson 27

Applications of Differentiation to Curvature

Derivative of Arc Length (In Lesson 50, will find the arc length)

Let the first derivative of $y = f(x)$ be continuous. Let $P_0(x_0, y_0)$ be a fixed point on the graph of $y = f(x)$ and let $P(x, y)$ be any other point on this curve. If s is the arc length measured from $P_0(x_0, y_0)$ to $P(x, y)$, then the derivative of s is given by

$$\frac{ds}{dx} = \sqrt{1 + \left(\frac{dy}{dx}\right)^2} \text{ , where } s \text{ is increasing with } x \ (y = f(x)).$$

or

$$\frac{ds}{dy} = \sqrt{1 + \left(\frac{dx}{dy}\right)^2} \text{ where } s \text{ is increasing with } y \ (x = g(y)).$$

If Q is located parametrically by $x = f(t)$ and $y = g(t)$, then

$$\frac{ds}{dt} = \sqrt{\left(\frac{dx}{dt}\right)^2 + \left(\frac{dy}{dt}\right)^2} \text{ , where } s \text{ increases with } t$$

All the above formulas can be derived from the Pythagorean relation

$$ds^2 = dx^2 + dy^2$$

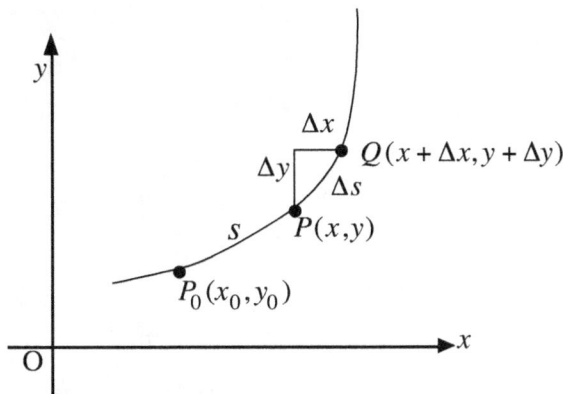

Example 1 Find $\dfrac{ds}{dx}$ at $P(x, y)$ on the parabola $y = 2x^2$

Formula $\dfrac{ds}{dx} = \sqrt{1 + \left(\dfrac{dy}{dx}\right)^2}$

From $y = 2x^2$, $\dfrac{dy}{dx} = 4x$

$$\frac{ds}{dx} = \sqrt{1 + (4x)^2}$$

$$\frac{ds}{dx} = \sqrt{1 + 16x^2}$$

Example 2   Find $\dfrac{ds}{dx}$ if $x = t^3$, $y = t^2$

Solution

Step 1: $\dfrac{ds}{dt} = \sqrt{\left(\dfrac{dx}{dt}\right)^2 + \left(\dfrac{dy}{dt}\right)^2}$

From $x = t^3$, $\dfrac{dx}{dt} = 3t^2$,

From $y = t^2$, $\dfrac{dy}{dt} = 2t$.

Step 2: Substitute $3t^2$ for $\dfrac{dx}{dt}$, and $2t$ for $\dfrac{dy}{dt}$ in $\dfrac{ds}{dt} = \sqrt{\left(\dfrac{dx}{dt}\right)^2 + \left(\dfrac{dy}{dt}\right)^2}$

Then $\dfrac{ds}{dt} = \sqrt{\left(3t^2\right)^2 + (2t)^2}$

$\qquad = \sqrt{9t^4 + 4t^2}$

$\qquad = \sqrt{t^2(9t^2 + 4)}$

$\dfrac{ds}{dt} = t\sqrt{9t^2 + 4}$.

Curvature of a curve

The dictionary meaning of "to curve" is to deviate smoothly (without sharp breaks) from a straight line. The curvature of a curve at a point indicates how fast the tangent line is turning at this point.

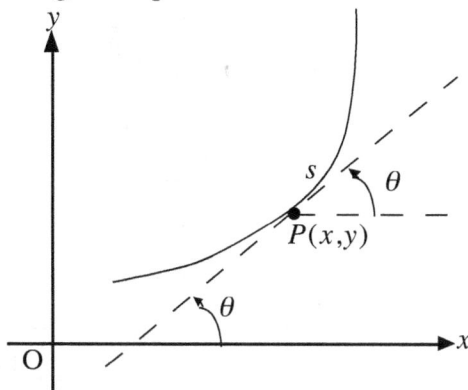

Let the angle of inclination of the tangent line at a point $P(x, y)$ on the curve $y = f(x)$ be θ. Then, the **curvature** K of this curve $y = f(x)$ at $P(x, y)$ is the rate of change of θ with respect to the arc length s.

Thus $K = \dfrac{d\theta}{ds}$, where θ is in radians. and K is positive if θ increases with s.

$$K = \frac{d\theta}{ds} = \lim_{\Delta s \to 0} \frac{\Delta \theta}{\Delta s} = \frac{\dfrac{d^2 y}{dx^2}}{\left[1 + \left(\dfrac{dy}{dx}\right)^2\right]^{\frac{3}{2}}} \qquad (y = f(x))$$

K is positive if the curve is concave up, and negative if concave down.

$$\text{If } x = g(y), \quad K = \frac{-\dfrac{d^2 x}{dy^2}}{\left[1 + \left(\dfrac{dx}{dy}\right)^2\right]^{\frac{3}{2}}}. \qquad (x = g(y))$$

Example Find the curvature of the parabola $y^2 = 9x$ at $(4,6)$

Step 1: The curvature K is given by

$$K = \frac{\dfrac{d^2y}{dx^2}}{\left[1 + \left(\dfrac{dy}{dx}\right)^2\right]^{\frac{3}{2}}}$$

From $y^2 = 9x$

$$2y\frac{dy}{dx} = 9 \,(\text{Implicit differentiation})$$

$$\frac{dy}{dx} = \frac{9}{2y}$$

$$\frac{d^2y}{dx^2} = \frac{9}{2}\frac{d}{dx}(y^{-1})$$

$$= \frac{9}{2}(-1)(y^{-2})\frac{dy}{dx}$$

$$= -\frac{9}{2y^2}\frac{dy}{dx}$$

$$= \left(-\frac{9}{2y^2}\right)\left(\frac{9}{2y}\right)$$

$$\frac{d^2y}{dx^2} = -\frac{81}{4y^3}$$

Step 2: Substitute for $\dfrac{dy}{dx}$ and $\dfrac{d^2y}{dx^2}$

in $K = \dfrac{\dfrac{d^2y}{dx^2}}{\left[1 + \left(\dfrac{dy}{dx}\right)^2\right]^{\frac{3}{2}}}$ to obtain

Step 3:

$$K = \frac{-\dfrac{81}{4y^3}}{\left[1 + \left(\dfrac{9}{2y}\right)^2\right]^{\frac{3}{2}}}$$

$$K = \frac{-\dfrac{81}{4(6^3)}}{\left[1 + \left(\dfrac{9}{2(6)}\right)^2\right]^{\frac{3}{2}}} \qquad (y = 6)$$

$$K = -\frac{3}{4(4)(2)} \bullet \frac{1}{\left[1 + \left(\dfrac{3}{4}\right)^2\right]^{\frac{3}{2}}}$$

$$K = -\frac{3}{32} \bullet \frac{1}{\left[1 + \dfrac{9}{16}\right]^{\frac{3}{2}}}$$

$$K = -\frac{3}{32} \bullet \frac{1}{\left[\dfrac{25}{16}\right]^{\frac{3}{2}}}$$

$$K = -\frac{3}{32} \bullet \frac{1}{\left[\dfrac{5}{4}\right]^{3}}$$

$$K = -\frac{3}{32} \bullet \frac{1}{\dfrac{125}{64}}$$

$$K = -\frac{3}{32} \bullet \frac{64}{125}$$

$$K = -\frac{3}{1} \bullet \frac{2}{125}$$

$$K = -\frac{6}{125} \, \bullet$$

Since K is negative, the curve is concave down.

Lesson 27 Exercises

1. Find $\dfrac{ds}{dx}$ at $P(x,y)$ on the parabola $y = 2x^2$

2 Find $\dfrac{ds}{dx}$ if $x = t^3$, $y = t^2$

3. Find the curvature of the parabola $y^2 = 6x$ at $(3,2)$

Answer: 1. $\sqrt{1 + 16x^2}$; **2.** $t\sqrt{9t^2 + 4}$; **3.** $-\dfrac{6}{125}$

CHAPTER 10
Indefinite Integrals and Antiderivatives

Lesson 28
Introduction to Integration
Inverse of Differentiation

In the past, after covering an operation on numbers, we reversed the steps of that operation to obtain a new operation. After the addition operation, we obtained the subtraction operation; after multiplication, we obtained division; after raising numbers to powers, we extracted the roots of numbers; after finding the logarithms of numbers, we found their antilogarithms. Similarly, after covering trigonometric functions, we covered inverse trigonometric functions. Each of the above operations and its reversed operation are inverses of each other. We also used the inverse operation to check the results of a particular operation, since each operation reverses the action of its inverse, For example, we used addition to check the correctness of a subtraction problem, and multiplication to check a division problem. Each operation and its inverse have practical applications.

So also, it would be quite natural that, in calculus, after learning the differentiation of functions, we reverse the steps to obtain the inverse operation which also quite naturally, we call **antidifferentiation** or simply **integration**.

Let us revisit some previous examples covered under differentiation.

Example 1. If $f(x) = x^4 + x^2 + 12$,

then $f'(x) = 4x^3 + 2x + 0$

$= 4x^3 + 2x$

Example 2 If $f(x) = x^4 + x^2 + 5$, then

$f'(x) = 4x^3 + 2x + 0$

$= 4x^3 + 2x$

Observe above that even though the two given functions are different and differ in the constant terms, the two functions have the same derivative, because the derivative of the constant term in each case is zero.

Let us redo Example 1 and then reverse the steps.

Differentiation

If $f(x) = x^4 + x^2 + 12$

$f'(x) = 4x^3 + 2x$

(Exponent **multiplies** the base and the exponent **decreases** by 1)

Integration (Reverse the steps of differentiation)

$\int (4x^3 + 2x)dx$

$= \dfrac{4x^{3+1}}{3+1} + \dfrac{2x^{1+1}}{1+1} + C$ (Exponent increases by 1,

$= \dfrac{4x^4}{4} + \dfrac{2x^2}{2} + C$ (and new exponent divides.)

$= x^4 + x^2 + C$

Observe that in just reversing the steps in Examples **1** and 2 to obtain the original function, we are unable to obtain the constant terms 12 and 5; and to compensate, we introduce the letter C, which is called the constant of integration. To determine the value of the constant of integration we need more information.

In reversing the steps and applying them to a derivative, we are unable to obtain the exact original function, but rather a **family** of functions which differ from one another by the constant term, the integration constant.

We conclude that by merely reversing the steps of the differentiation process, the result (the integral) we obtain is **not** unique. (More examples: $4x^2 + 4$, $4x^2 + 6$, and $4x^2 - 9$ are antiderivatives of $8x$, since the derivative of each of them is $8x$; the only difference between these integrals is the constant term. We call the "reversed " function the **indefinite integral** of the given function. The indefinite integral is also called the **antiderivative** or the primitive of the given function.

Therefore, given a function $f(x)$, to find the indefinite integral (the antiderivative, or primitive) of $f(x)$ means we are to find another function $g(x) + C$ such that the derivative of $g(x) + C = f(x)$.

Definition: A function $g(x)$ is an antiderivative of a function $f(x)$ if the derivative of $g(x)$ is $f(x)$. (We use $g(x)$ instead of $F(x)$ for pronunciation simplicity)

We symbolize that we are finding the indefinite integral by:

$$\int f(x)\, dx = g(x) + C \qquad\qquad \text{(where } \tfrac{d}{dx}\big[g(x)\big] = f(x))$$

Read " the indefinite integral of $f(x)$ is $g(x) + C$ "

The left side of this equation says " find the indefinite integral of $f(x)$ ", and

this is followed by " is $g(x) + C$. The symbol $\int$ is called an integral sign, $f(x)$ is called the integrand (the expression to be integrated) and the symbol dx indicates that we are to integrate with respect to x. The functions $g(x) + C$ are the antiderivatives of $f(x)$. When the constant of integration, $C = 0$

$$\int f(x)\, dx = g(x)$$

Note also that $\dfrac{d}{dx}\left[\int f(x)\, dx\right] = f(x)$ (differentiation "undoes" integration).

That is, if we integrate a function and differentiate the result, we recover the original function, (This is similar to $(\sqrt[3]{8})^3 = 8$, since finding the cube root followed by cubing recovers the 8). Therefore, after obtaining an antiderivative of a function, we can check this result by differentiating the antiderivative to see if we obtain the original function, the integrand. If we obtain the original integrand, the antiderivative is correct, otherwise it is incorrect.

There is another type of integral called the definite integral, symbolized $\int_a^b f(x)dx$, where a and b are called the limits of integration. If we can find the indefinite integral, finding the definite integral is straightforward and a matter of numerical evaluation. We will not cover the definite integral in this chapter, but we will do so in a different chapter, later.

A simple example on finding the indefinite integral is presented below, followed by examples involving various functions.

Example Find $\int x^3 dx$ **Solution** $\int x^3 dx = \dfrac{x^4}{4} + C.$

Lesson 29
Integration of Polynomial Functions

Case 1: Integration of a Monomial Function

We symbolize the indefinite integral (also called the primitive, or the antiderivatives) by $\int f(x)dx$, where $f(x)$ is called the integrand, and dx identifies the independent variable. We define the **indefinite integral** of a function $f(x)$ as a function $g(x) + C$ such that the derivative of $g(x)$ equals $f(x)$, and where C is an arbitrary constant.

Power Rule for Integration

$\int x^n dx = \frac{x^{n+1}}{n+1} + C$, where C is the integration constant, and $n \neq -1$).

Example 1 Find $\int x^2 dx$

We are to find the family of functions, $g(x) + C$, such that the derivative of each function is x^2. It also means we are to find an antiderivative or a primitive or an indefinite integral of x^2.

Solution $\int x^2 dx = \frac{x^{2+1}}{2+1} + C$ (applying the power rule, $\int x^n dx = \frac{x^{n+1}}{n+1} + C$)

$$= \frac{x^3}{3} + C$$

Let us check the integration by differentiation

$$\frac{d}{dx}\left(\frac{x^3}{3}\right) + \frac{d}{dx}(C)$$

$$= \frac{1}{3}\frac{d}{dx}\left(x^3\right) + \frac{d}{dx}(C)$$

$$= \frac{1}{3}(3x^2) + 0$$

$$= x^2$$

We obtain the original integrand.

Therefore, $\int x^2 dx = \frac{x^3}{3} + C$.

Example 2 Find $\int x^3 dx$
Solution

$$\int x^3 dx = \frac{x^{3+1}}{3+1} + C \quad \text{(applying the power rule, } \int x^n dx = \frac{x^{n+1}}{n+1} + C)$$

$$= \frac{x^4}{4} + C.$$

Case 2: Integration of an integrand containing a constant factor

$$\int af(x)dx = a\int f(x)dx$$

Example Find $\int 5x^3 dx$

$$\int 5x^3 dx = 5\int x^3 dx \qquad \text{(Factor out the constant factor, and integrate.)}$$

$$= 5\left(\frac{x^{3+1}}{3+1}\right) + C$$

$$= \frac{5x^4}{4} + C.$$

Note:

$$\int dx = \int 1 dx$$

$$= x + C$$

$$\int 4 dx = 4x + C \quad \text{(You may view } \int 4 dx \text{ as } \int 4x^0 dx = 4x^{0+1} + C = 4x + C)$$

Case 3: Integration of a Polynomial Function

Approach: Integrate each term of the polynomial.

$$\int [f(x) + \int g(x)] dx = \int f(x)dx + \int g(x)dx$$

Example Find $\int (4x^3 + 2x^2 - 8x + 7)\, dx$

Solution

$$\int (4x^3 + 2x^2 - 8x + 7)\, dx$$

$$= \int 4x^3 dx + \int 2x^2 dx - \int 8x\, dx + \int 7 dx$$

$$= \frac{4x^{3+1}}{3+1} + \frac{2x^{2+1}}{2+1} - \frac{8x^{1+1}}{1+1} + 7x + C$$

$$= \frac{4x^4}{4} + \frac{2x^3}{3} - \frac{8x^2}{2} + 7x + C$$

$$= x^4 + \frac{2x^3}{3} - 4x^2 + 7x + C.$$

Lesson 29A Exercises

1, Complete the power rule for polynomials $\int x^n dx = ?$

2 Find $\int x^2 dx$

3. Find $\int x^3 dx$

4. Find $\int 5x^3 dx$

5. Find $\int (4x^3 + 2x^2 - 8x + 7)\,dx$

6. $\int 4\,dx$

Answers: 1. $\dfrac{x^{n+1}}{n+1} + C$, **2.** $\dfrac{x^3}{3} + C$; **3.** $\dfrac{x^4}{4} + C$; **4.** $\dfrac{5x^4}{4} + C$;

5. $x^4 + \dfrac{2x^3}{3} - 4x^2 + 7x + C$; **6.** $4x + C$

Lesson 29B Exercises

1, Complete the power rule for polynomials $\int x^n dx = ?$

2 . Find $\int x^3 dx$

3. Find $\int x^4 dx$

4. Find $\int 6x^2 dx$

5. Find $\int (2x^4 + 2x^3 - 8x^2 + 2x - 3)\,dx$

6. $\int 5\,dx$

Answers: 1. $\dfrac{x^{n+1}}{n+1} + C$, **2.** $\dfrac{x^4}{4} + C$; **3.** $\dfrac{x^5}{5} + C$; **4.** $2x^3 + C$;

5. $\dfrac{2x^5}{5} + \dfrac{x^4}{2} - \dfrac{8x^3}{3} + x^2 - 3x + C$; **6.** $5x + C$.

Lesson 30

General Substitution Techniques of Integration

An analytical and a guided approach to studying substitution (or change of variable) techniques in integration of functions is introduced in this lesson. A good number of functions can be integrated using simple substitution. Each integrand may be a single composite function, the product of two functions or the quotient of two functions. The basic functions involved include polynomial, rational, radical, exponential, logarithmic, and trigonometric functions. We will cover a guide which will help us to determine quickly if a simple substitution technique will work and also which function to substitute for. Such determination will help reduce or avoid the number of trial-and-error attempts in these techniques.

To facilitate communication, we will classify the substitution (change of a variable) methods as **simple u-substitution** (or t-substitution; any other variable can be used) **or multiple** substitution. We will further divide simple u-substitution, into a **one-step** u-substitution and a **two-step** u-substitution. In multiple substitution, we apply simple substitution more than once using different variables. Another substitution technique called **trigonometric substitution** will **not** be covered in this lesson, but will be covered in a different chapter, later. In multiple substitution, we will apply simple u-substitution followed by trigonometric substitution.

To determine if simple u-substitution will work, the following **guidelines** will be helpful. For communication purposes, we classify simple u-substitution into two main types, namely Type 1 (one-step u-substitution) and Type 2 (two-step u-substitution)

Guidelines for Type 1 (one-step) *u*-substitution

Case 1: Condition for a single composite function

The degree of the "inside" function must be 1 (a linear function).

For example, $\int (2x-1)^4 dx$ satisfies this condition but $\int (2x^2-1)^4 dx$ does **not** satisfy this condition.

We can therefore apply simple u-substitution to find $\int (2x-1)^4 dx$.

Case 2: Conditions for other functions for Type 1 substitution

These two conditions must be satisfied simultaneously.

Condition 1. A given integrand is the product or quotient of functions involving two functions u and u' (read "u-prime") such that the derivative of u equals u' or the derivative of u differs from u' by only a constant factor. For functions whose basic components are polynomials, the degree of u' is 1 less than the degree of u.

Condition 2. The function u' (the derivative) **must** always be in the numerator, but u may be in the numerator or in the denominator

(i.e., u' **cannot** be in the denominator).

Note: The integrands in **A, B** and **E**, below, satisfy the above two conditions, but the integrand in **C** does **not** satisfy the condition guidelines because the derivative part, $4x^2$, (u') is in the denominator. That is, simple u-substitution method will work for **A** and **B** and **E** but not for **C**. (See below.)

A: $\int \dfrac{4x^2}{\sqrt{x^3+1}}\, dx$: **B:** $\int 4x^2 \sqrt{x^3+1}\, dx$: **C:** $\int \dfrac{\sqrt{x^3+1}}{4x^2}\, dx$: **E:** $\int x^2(4x^3+2)^5 dx$

Note also that for functions involving simple **logarithmic functions** such as $\ln x$, if the $\ln x$ part is chosen as u, the derivative $\frac{1}{x}$ will also be in the numerator. However, on simplifying, the "x" will end up in the denominator. Therefore, we can say that the "x" in $\frac{1}{x}$ must be in the denominator. The substitution method will work for **D** and **E** but not for **F**. (See below.)

D: $\int \dfrac{\ln x}{x}\, dx$; **E:** $\int \dfrac{1}{x\ln x}\, dx$; **F:** $\int \dfrac{x}{\ln x}\, dx$: (Note $\dfrac{\frac{1}{x}\ln x}{1} = \dfrac{\ln x}{x}$ but $\dfrac{\ln x}{\frac{1}{x}} = x\ln x$)

Note also: $\mathbf{D} = \int \dfrac{\ln x}{x}\, dx = \int \dfrac{1}{x} \bullet \ln x\, dx$; $\mathbf{E} = \int \dfrac{1}{x\ln x}\, dx = \int \dfrac{\frac{1}{x}}{\ln x}\, dx$

Extra: Simple u-substitution method will **not** work for $\int x\ln x\, dx$. Why?

Guidelines for Type 2 (Two-Step) u-substitution

Let $f(x) = \dfrac{g(x)}{h(x)}$. then a two-step u-substitution will work if the following conditions are satisfied simultaneously..

Condition 1: $h(x)$ is composite with the "inside function" being a linear function (degree 1). Usually $h(x)$ is a power of a linear function.

Condition 2: $g(x)$ is a polynomial function

In the **two-step** u-substitution, the first step expresses the denominator and dx in terms of u.

The second step expresses the rest of the integral in terms of u only.

Examples for a two-step u-substitution are $\int \dfrac{x+6}{(x+4)^3} dx$, $\int \dfrac{x^2+5}{(x+4)^3} dx$ and

$\int \dfrac{x^2+4}{\sqrt{x+4}} dx$. (See Case 3b of Lesson 31; and Example 7 of Lesson 32)

To excuse the author's "dogmatism" in the Type-2 substitution guidelines, it may be added that perhaps, there may be other functions which may fall under Type-2 substitution.

It must also be noted above that the author has attempted to help guide the student with respect to some popular integrals and that in some cases, the student may have to try and err to determine if simple u-substitution will work. The author's philosophy is that some helpful guidelines are better than no guidelines at all.

After millions of years, experience is still the best teacher.

More Representative Examples for Simple U-substitution

Polynomial Functions

1. $\int x^2(4x^3+2)^5\,dx$

Rational Functions

2. $\int \frac{5x}{(x^2+1)^3}\,dx$

3. $\int \frac{x}{x^2+2}\,dx$

4. $\int \frac{x}{(x^2+2)^3}\,dx$

5. $\int \frac{x^2}{(4x^3+2)^5}\,dx$

Integrands Involving Radicals

6. $\int \frac{4x^2}{\sqrt{x^3+1}}\,dx$;

7. $\int 4x^2\sqrt{x^3+1}\,dx$;

8. $\int \frac{x+1}{\sqrt{x^2+2x+2}}\,dx$

9. $\int \frac{\sqrt{x+1}}{1-x}\,dx$

10. $\int \sqrt[3]{2x-3}\,dx$

11. $\int \frac{\sqrt{x}}{\sqrt[3]{x}+1}\,dx$

12. $\int \frac{3x}{\sqrt{1-x}}\,dx$

13. $\int x\sqrt{x+3}\,dx$

Exponential Functions

14. $\int \sqrt{1+e^x}\,dx$;

15. $\int \frac{e^x}{1-e^x}\,dx$;

16. $\int \frac{e^{2x}}{1+e^{2x}}\,dx$

Logarithmic Functions

17. $\int \frac{\ln x}{x}\,dx$

18. $\int \frac{1}{x\ln x}\,dx$

19. $\int \frac{\ln x^2}{x}\,dx$

Trigonometric Functions

20. $\int \sin x\cos x\,dx$

21. $\int \sin^3 x\cos x\,dx$

22. $\int \cos^4 x\sin x\,dx$

23. $\int \cos^2 x\sin x\,dx\,4..$

24. $\int \tan x\sec^2 x\,dx$.

25. $\int \cot x\csc^2 x\,dx$

26. $\int \sec 3x\tan 3x\,dx$

27. $\int \frac{\sin x}{\cos x}\,dx$ ($\tan x$)

28. $\int \frac{\cos x}{\sin x}\,dx$ ($\cot x$)

29. $\int \frac{\sin x}{\cos^2 x}\,dx$

30. $\int \frac{x+6}{(x+4)^3}\,dx$.

31. $\int \frac{x^2+5}{(x+4)^3}\,dx$

32. $\int \frac{x^2+4}{\sqrt{x+4}}\,dx$

Multiple Substitution

33. $\int \sqrt{\frac{1-x}{1+x}}\,dx$

(*u*-sub., plus trig. sub.)

The rest of this lesson will be devoted to applying simple *u*-substitution to polynomial integrands. Thereafter, in subsequent lessons, we will cover more examples involving various functions, some of which are listed in the above table. **Students** should master **simple *u*-substitution** techniques early in the study of integral calculus to the extent that they can readily spot its applicability in unexpected places.

Solved (or Worked) Examples on Simple U-Substitution

We will briefly cover two examples for polynomials. The first example (Method 2) is to show the "power" of simple u-substitution, although this method is an overkill for this example. The second example shows how we can avoid tedious expansion by using simple u-substitution. After these two examples on polynomials, we will continue with the integration of the other functions, and when the need arises, and conditions have been satisfied, we will apply simple u-substitution.

On Polynomials

Example 1: Find $\int x^2(4x^3 + 2)\,dx$

Method 1 (Usual method)

In $\int x^2(4x^3 + 2)\,dx$, we normally multiply to obtain $\boxed{\int(4x^5 + 2x^2)\,dx}$ and then integrate term-by term:

$$\int(4x^5 + 2x^2)\,dx = 4\frac{x^6}{6} + 2\frac{x^3}{3} + c = \frac{2}{3}x^6 + \frac{2}{3}x^3 + c$$

We differentiate to check: $\frac{d}{dx}\left(\frac{2}{3}x^6 + \frac{2}{3}x^3 + c\right) = \frac{2}{3}(6)x^5 + \frac{2}{3}(3)x^2 \boxed{= 4x^5 + 2x^2}$

$$\therefore \int(4x^5 + 2x^2)\,dx = \frac{2}{3}x^6 + \frac{2}{3}x^3 + C$$

Method 2 Simple u-substitution.

Step 1: Let $u = 4x^3 + 2$. Then $\frac{du}{dx} = 12x^2$ and from which $dx = \frac{du}{12x^2}$

Step 2: Substitute u for $4x^3 + 2$; $\frac{du}{12x^2}$ for dx in $\int x^2(4x^3 + 2)\,dx$.

Then $\int x^2 u\frac{du}{12x^2} = \frac{1}{12}\int u\,du = \frac{1}{12}\bullet\frac{u^2}{2} + c = \frac{u^2}{24} + C$

Step 3: Substitute $4x^3 + 2$ for u. Then $\int x^2(4x^3 + 2)dx = \frac{(4x^3 + 2)^2}{24} + C$

$$= \frac{16x^6 + 16x^3 + 4}{24} + C = \frac{2}{3}x^6 + \frac{2}{3}x^3 + \frac{1}{6} + c_1 = \frac{2}{3}x^6 + \frac{2}{3}x^3 + C$$

Example 2. Find $\int x^2(4x^3 + 2)^5\,dx$

Solution Using simple **u-substitution**

In $\int x^2(4x^3 + 2)^5\,dx$, we can expand $x^2(4x^3 + 2)^5$ and then integrate, However we can avoid the tedious expansion and then integrate by using simple u-substitution

Step 1: Let $u = 4x^3 + 2$. Then $\frac{du}{dx} = 12x^2$ and from which $dx = \frac{du}{12x^2}$

Step 2: Substitute u for $4x^3 + 2$; $\frac{du}{12x^2}$ for dx in $\int x^2(4x^3 + 2)^5\,dx$.

Then $\int x^2 u^5 \frac{du}{12x^2} = \frac{1}{12}\int u^5\,du$

$$= \frac{1}{12}\bullet\frac{u^6}{6} + c$$

$$= \frac{u^6}{72} + C.$$

Step 3: Substitute $4x^3 + 2$ for u.

$$\boxed{\text{Then } \int x^2(4x^3+2)^5 = \frac{(4x^3+2)^6}{72} + C}$$

Let us differentiate to check: $\frac{d}{dx}\left[\frac{(4x^3+2)^6}{72} + c\right]$

$$= \frac{1}{72}\left[6(4x^3+2)^5(12x^2)\right] + 0$$

$$= \frac{1}{72}\left[72x^2(4x^3+2)^5\right]$$

$$= x^2(4x^3+2)^5 \text{ <this checks with the original integrand}$$

Now, try by first expanding $x^2(4x^3+2)^5$ and then integrating. (Nice punishment)

Lesson 30 Exercises A

1. Determine by inspection which of the following we can use simple u-substitution to integrate,

a. $\int \frac{4x^2}{\sqrt{x^3+1}}\, dx$; b. $\int 4x^2\sqrt{x^3+1}\, dx$; c. $\int \frac{\sqrt{x^3+1}}{4x^2}\, dx$; d. $\int x^2(4x^3+2)^5 dx$

e. $\int (2x-1)^4 dx$

2. Find $\int x^2(4x^3+2)\, dx$; 3. . Find $\int x^2(4x^3+2)^5 dx$

4. Find $\int \frac{x+1}{\sqrt{x^2+2x+2}}\, dx$

Answers: 1. a, b, d, and e. 2. $\frac{2}{3}x^6 + \frac{2}{3}x^3 + C$; 3. $\frac{(4x^3+2)^6}{72} + C$;

4. $\sqrt{x^2+2x+2} + C$

Lesson 30 Exercises B

1. Determine by inspection which of the following we can use simple u-substitution to integrate,

a. $\int \frac{4x^3}{\sqrt{x^2+1}}\, dx$; b. $\int 4x^3\sqrt{x^4+1}\, dx$; c. $\int \frac{\sqrt{x^2+1}}{4x^3}\, dx$; d. $\int x^3(4x^4+2)^5 dx$

e. $\int (2x-1)^5 dx$; f. $\int x^4(4x^3+2)^5 dx$; g. $\int \frac{4x^2}{\sqrt{x^3+1}}\, dx$

2. Find $\int x^4(4x^5+2)\, dx$; 3. . Find $\int x^3(4x^4+2)^9 dx$

4. Find $\int \frac{x+2}{\sqrt{x^2+4x+2}}\, dx$

Answers: 1. b, d. e, g. 2. $\frac{(4x^5+2)^2}{40} + C$ or $\frac{2}{5}x^{10} + \frac{2}{5}x^5 + C$;

3. $\frac{(4x^4+2)^{10}}{160} + C$; 4. $\sqrt{x^2+4x+2} + C$

Lesson 31
Integration of Rational Functions I
(Integrand is a Rational Expression)

There are a number of methods for integrating rational expressions. Each method depends on the type of rational expression. Below, we briefly cover types of rational expressions.

Definitions

A **rational expression** is an expression which is the ratio of two polynomials, A rational expression may be **proper** or **improper.** In addition, a rational expression may also be **irreducible** or **reducible.**

ln a **proper rational** expression, the degree of the numerator polynomial is less than the degree of the denominator polynomial.

Example $\frac{x+3}{x^2-4}$ (Degree of numerator is 1; degree of denominator is 2.)

In an **improper rational** expression, the degree of the numerator polynomial is greater than or equal to the degree of the denominator polynomial.

Examples   1. $\frac{x^2-9}{x-2}$ (Degree of numerator is 2; degree of denominator is 1.)

2. $\frac{x}{x+2}$ Degree of numerator is 1; Degree of denominator is 1.)

A rational expression is **irreducible** if the numerator and the denominator have **no** common factors other than 1.

A rational expression is **reducible** if the numerator and the denominator have common factors other than 1.

Example 1 The fraction $\frac{x+1}{(x+1)(x-2)}$ is **proper** and **reducible.**

By canceling the common factor $x+1$, we obtain the proper and irreducible fraction $\frac{1}{x-2}$.

Example 2 The fraction $\frac{(x+1)(x+2)}{(x+1)(x-2)}$ is **improper and reducible**.

By canceling the common factor $x+1$, and using long division we obtain $1+\frac{4}{x-2}$, in which the fractional part is **proper** and **irreducible**.

Example 3 The fraction $\frac{x^2+5}{x+1}$ is **improper** and **irreducible**.

Using long division this fraction can be expressed as the sum of a polynomial quotient and a proper, irreducible fraction, Thus $\frac{x^2+5}{x+1}=x-1+\frac{6}{x+1}$.

In working with rational expressions, we shall exclude those values (called **excluded values**) of the variable in the denominator which make the denominator zero. At the excluded values, the rational expressions are undefined. We should note however that some rational expressions are defined for all real values of the independent variable: For .example, $\frac{x}{x^2+1}$ is defined for all real values of x, since the denominator is never zero.

From the above examples, our main concern in the integration of rational functions is to integrate a **proper** and **irreducible** rational fraction. We will assume that the degree of the numerator polynomial is less than the degree of the denominator polynomial and that the numerator and the denominator do not have any common linear or quadratic factors.

Classification of Integration of Rational Functions

We may classify the integration of rational functions by antiderivative type, by integrand type, or by method type.

By **antiderivative type,** when a proper irreducible rational function is integrated, the antiderivative is either a rational function, a logarithmic function, an arctangent function, or the sum of any two or all three types.

The coverage of rational functions in this book is as follows:

Rational Functions I By power rule, by natural log rule;
by u-substitution   See **Lesson 31**

Examples

1. $\int \frac{1}{x^2}dx$; **2.** $\int \frac{1}{x}dx$; **3.** $\int \frac{1}{x-2}dx$ **4.** $\int \frac{x}{x^2+2}dx$; **5.** $\int \frac{x^2}{(4x^3+2)^5}dx$;

6. $\int \frac{5x}{x^2+1}dx$; **7.** $\int \frac{5x}{\left(x^2+1\right)^3}dx$; **8.** $\int \frac{3x}{x^2+4}dx$;

Rational Functions II By Partial Fractions decomposition
The antiderivative type here is a logarithmic function See **Lesson 40.**
Examples

10. $\int \frac{x-5}{x^2+x-2}dx$; **11.** $\int \frac{dx}{a^2-x^2}$; **12.** $\int \frac{x^2+3x+6}{x+2}dx$;

Rational Functions III By Trigonometric Substitution.. See **Lesson 41**
Here the antiderivative type is an arctangent function.

Examples **14.** $\int \frac{dx}{1+x^2}$; **15.** $\int \frac{dx}{a^2+x^2}$; **16,** $\int \frac{1}{1+4x^2}dx$; **17,** $\int \frac{dx}{(1+x^2)^2}$;

18. $\int \frac{dx}{(1+x^2)^3}$; **19.** $\int \frac{x}{x^2-6x+13}dx$; **20.** $\int \frac{1}{4x^2+6x+9}dx$; **21.** $\int \frac{dx}{a^2-x^2}$

Rational Functions I

Basic Formulas

$$\int x^n dx = \begin{cases} \dfrac{x^{n+1}}{n+1} \text{ if } n \neq -1 \\ \\ \ln x \text{ if } n = -1 \end{cases} \qquad \textbf{(A)}$$

(Same as $\int x^n dx = \frac{x^{n+1}}{n+1}$ if $n \neq -1$; but , if $n = -1$, $\int x^n dx = \ln x$)

Below, we cover three main cases.

Case 1: Rational functions such as $\int \frac{1}{x^2}dx$, $\int \frac{1}{x^3}dx$, $\int \frac{1}{x^7}dx$

or $\int \frac{1}{x^n}dx$, $n \neq 1$. Here, we can use the **power rule** $\int x^n dx = \frac{x^{n+1}}{n+1} + C$, the rule we used for polynomials.

Example 1 Find $\int \frac{1}{x^2}dx$

Solution $\int \frac{1}{x^2}dx = \int x^{-2}dx$

$$= \frac{x^{-2+1}}{-2+1} + C$$

$$= \frac{x^{-1}}{-1} + C$$

$$= -\frac{1}{x} + C.$$

Example 2 Find $\int \frac{1}{x^3}dx$

Solution $\int \frac{1}{x^3}dx = \int x^{-3}dx$

$$= \frac{x^{-3+1}}{-3+1} + C$$

$$= \frac{x^{-2}}{-2} + C$$

$$= -\frac{1}{2x^2} + C.$$

Example 3 Find $\int \frac{1}{x^7}dx$

Solution $\int \frac{1}{x^7}dx = \int x^{-7}dx$

$$= \frac{x^{-7+1}}{-7+1} + C$$

$$= \frac{x^{-6}}{-6} + C$$

$$= -\frac{1}{6x^6} + C.$$

Case 2: The simple **rational functions** such as $f(x) = \frac{1}{x}$; $f(x) = \frac{1}{x-2}$

Example 4 Find $\int \frac{1}{x}dx$

We **cannot** use the power rule here since the result would be undefined.

(Note: $\int \frac{1}{x}dx = \int x^{-1}dx = \frac{x^{-1+1}}{-1+1} = \frac{x^0}{0} = \frac{1}{0}$, which is undefined). The impediment here is that $n = -1$ when $f(x)$ is in exponential form. So we resort to the use of logarithms. We use the lower rule in (**A**) above (p.216).

Solution $\int \frac{1}{x}dx = \ln|x| + C$ (which means $\int \frac{1}{x}dx = \ln x + C$ if $x > 0$;

$$\int \frac{1}{x}dx = \ln(-x) + C \text{ if } x < 0.$$

↑

(The negative of or the opposite of x)

Example 5 Find $\int \frac{1}{x-2}dx$

Solution $\int \frac{1}{x-2}dx = \ln|x-2| + C.$

Case 3a: Integrating Rational Functions by Simple U-substitution Method

Rational functions such as $\int \dfrac{5x}{\left(x^2+1\right)^3}dx$, and $\int \dfrac{5x}{x^2+1}dx$ can be found using simple u-substitution.

Example 6 Find $\int \dfrac{5x}{\left(x^2+1\right)^3}dx$ <--This satisfies the condition guidelines in Lesson 30.

Solution   Let $u = x^2+1$

Then $\dfrac{du}{dx} = 2x$ and $dx = \dfrac{du}{2x}$ or $du = 2xdx$

Now, replace dx in $\int \dfrac{5x}{\left(x^2+1\right)^3}dx$ by $\dfrac{du}{2x}$, and x^2+1 by u.

Then $\displaystyle\int \dfrac{5x}{(x^2+1)^3}\,dx = \int \dfrac{5x}{u^3}\cdot\dfrac{du}{2x}$

$$= \int \dfrac{5}{u^3}\cdot\dfrac{du}{2} \text{ (canceling the common } x)$$

$$= \dfrac{5}{2}\int \dfrac{du}{u^3}$$

$$= \dfrac{5}{2}\int u^{-3}\,du$$

$$= \dfrac{5}{2}\cdot\dfrac{u^{-3+1}}{-3+1}+C$$

$$= \dfrac{5}{2}\cdot\dfrac{u^{-2}}{-2}+C$$

$$= -\dfrac{5}{4}\cdot u^{-2}+C$$

$$= -\dfrac{5}{4}\cdot\dfrac{1}{u^2}+C$$

$$= -\dfrac{5}{4(x^2+1)^2}+C \qquad (u = x^2+1)$$

Therefore, $\displaystyle\int \dfrac{5x}{\left(x^2+1\right)^3}dx = -\dfrac{5}{4(x^2+1)^2}+C$.

Example 7

Find $\int \dfrac{x}{x^2+2}\,dx$;

Step 1: Let $u = x^2 + 2$

   Then $\dfrac{du}{dx} = 2x$ and from

   which $dx = \dfrac{du}{2x}$

Step 2: Substitute u for $x^2 + 2$;

   $\dfrac{du}{2x}$ for dx in $\int \dfrac{x}{x^2+2}\,dx$.

   Then $\int \dfrac{x}{u}\dfrac{du}{2x}$

   $= \int \dfrac{1\,du}{2u}$

   $= \dfrac{1}{2}\int \dfrac{du}{u}$

   $= \dfrac{1}{2}\ln|u| + c$

Step 3: Substitute $x^2 + 2$ for u.

   $= \dfrac{1}{2}\ln|x^2+2| + c$

Then $\int \dfrac{x}{x^2+2}\,dx = \dfrac{1}{2}\ln|x^2+2| + c$.

Example 8

Find $\int \dfrac{x}{\left(x^2+2\right)^3}\,dx$

Step 1: Let $u = x^2 + 2$

   Then $\dfrac{du}{dx} = 2x$ and from

   which $dx = \dfrac{du}{2x}$

Step 2: Substitute u for $x^2 + 2$;

   $\dfrac{du}{2x}$ for dx in $\int \dfrac{x}{\left(x^2+2\right)^3}\,dx$.

   Then $\int \dfrac{x}{u^3}\dfrac{du}{2x}$

   $= \int \dfrac{1\,du}{2u}$

   $= \dfrac{1}{2}\int \dfrac{du}{u^3}$

   $= \dfrac{1}{2}\int u^{-3}\,du$

   $= \dfrac{1}{2}\bullet \dfrac{u^{-3+1}}{-3+1} + C$

   $= \dfrac{1}{2}\bullet \dfrac{u^{-2}}{-2} + C$

   $= -\dfrac{1}{4u^2} + C$

Step 3: Substitute $x^2 + 2$ for u.

Then $\int \dfrac{x}{\left(x^2+2\right)^3}\,dx = -\dfrac{1}{4(x^2+2)^2} + C$

Example 9

Find $\int \dfrac{x^2}{\left(4x^3+7\right)^5}\,dx$

Step 1: Let $u=4x^3+7$

Then $\dfrac{du}{dx}=12x^2$ and

from which $dx=\dfrac{du}{12x^2}$

Step 2 Substitute u for

$4x^3+7$; $\dfrac{du}{12x^2}$ for dx in

$\int \dfrac{x^2}{\left(4x^3+2\right)^5}\,dx$ to obtain .

$=\int \dfrac{x^2}{u^5}\bullet\dfrac{du}{12x^2}$

$=\dfrac{1}{12}\int\dfrac{du}{u^5}$

$=\dfrac{1}{12}\int u^{-5}du$

Step 3: Integrate

$=\dfrac{1}{12}\bullet\dfrac{u^{-5+1}}{-5+1}+C$

$=\dfrac{1}{12}\bullet\dfrac{u^{-4}}{-4}+C$

$=-\dfrac{1}{48}\dfrac{1}{u^4}+C$

Step 4: $=-\dfrac{1}{48(4x^3+7)^4}+C$

(Substituting $4x^3+7$ for u)

Example 10

Find $\int \dfrac{3x}{x^2+4}\,dx$

Step 1: Let $u=x^2+4$

Then $\dfrac{du}{dx}=2x$ and from which

$dx=\dfrac{du}{2x}$

Step 2: Substitute u for x^2+4;

$\dfrac{du}{2x}$ for dx in $\int \dfrac{3x}{x^2+4}\,dx$.

Then $\int\dfrac{3x}{u}\dfrac{du}{2x}$

$=\int\dfrac{3du}{2u}$

$=\dfrac{3}{2}\int\dfrac{du}{u}$

$=\dfrac{3}{2}\ln|u|+c$

Step 3: Substitute x^2+4 for u .

Then $\int \dfrac{3x}{x^2+4}\,dx= =\dfrac{3}{2}\ln\left|x^2+4\right|+c$

Note: Simple substitution will not

work in $\int \dfrac{3x+2}{x^2+4}\,dx$. Why?

Extra: Which of the following can be integrated using simple u-substitution ?

$\int \dfrac{3x+2}{4x^2+1}\,dx$ or $\int \dfrac{3x}{4x^2+1}\,dx$?

Example 11

Find $\displaystyle\int \frac{2x^3 + 4x^2 + 11x + 16}{x^2 + 4}\, dx$

Step 1: Using long division

$$\int \frac{2x^3 + 4x^2 + 11x + 16}{x^2 + 4}\, dx$$

$$= \int 2x\, dx + \int 4\, dx + \int \frac{3x}{x^2 + 4}\, dx$$

Step 2: Integrate the first two terms and use the result of Example 10 for

$\left(\displaystyle\int \frac{3x}{x^2 + 4}\, dx = \frac{3}{2}\ln\left|x^2 + 4\right| + c\right)$ to obtain

$$x^2 + 4x + \frac{3}{2}\ln\left|x^2 + 4\right| + c$$

$$\therefore \int \frac{2x^3 + 4x^2 + 11x + 16}{x^2 + 4}\, dx$$

$$= x^2 + 4x + \frac{3}{2}\ln\left|x^2 + 4\right| + c$$

Case 3b: **Rational functions** such as $\int \frac{x+6}{(x+4)^3}dx$,, and $\int \frac{x^2+5}{(x+4)^3}dx$ can be integrated using simple u-substitution, but there is an **extra step** to completely express the integral in terms of u only.

Example 12: Find $\int \frac{x+6}{(x+4)^3}dx$

Step 1: Let $u = x+4$, Then $\frac{du}{dx} = 1$, and $du = dx$. After substituting u for $x+4$, and du for dx in $\int \frac{x+6}{(x+4)^3}dx$, we obtain $\int \frac{x+6}{u^3}du$, and we still have the variable x, which we express in terms of u as follows:

Step 2: From $u = x+4$, $x = u-4$, and substituting for x in $\int \frac{x+6}{u^3}du$, we

obtain $\int \frac{u-4+6}{u^3}du$.

$$= \int \frac{u+2}{u^3}du$$

$$= \int \frac{u}{u^3}du + \int \frac{2}{u^3}du \quad \text{(splitting the numerators)}$$

$$= \int \frac{1}{u^2}du + \int \frac{2}{u^3}du$$

$$= \int u^{-2}du + 2\int u^{-3}du$$

$$= -u^{-1} + \frac{2}{-2u^2} + C$$

$$= -\frac{1}{u} - \frac{1}{u^2} + C$$

$$\int \frac{x+6}{(x+4)^3}dx = -\frac{1}{x+4} - \frac{1}{(x+4)^2} + C.$$

Example 12: Find $\int \frac{x^2+5}{(x+4)^3}dx$.

Step 1: Let $u = x+4$, Then $\frac{du}{dx} = 1$, and $du = dx$. After substituting u for $x+4$, and du for dx in $\int \frac{x^2+5}{(x+4)^3}dx$, we obtain $\int \frac{x^2+5}{u^3}du$, and we still have x^2 which we express in terms of u as follows:

Step 2: From $u = x+4$, $x = u-4$, and squaring, $x^2 = u^2 - 8u + 16$

We now substitute for x^2 in $\int \frac{x^2+5}{u^3}du$, to obtain $\int \frac{u^2-8u+16+5}{u^3}du$

$$\int \frac{u^2-8u+21}{u^3}du = \int \frac{u^2}{u^3}du + \int \frac{-8u}{u^3}du + \int \frac{21}{u^3}du$$

$$= \int \frac{1}{u}du - 8\int \frac{1}{u^2}du + \int \frac{21}{u^3}du$$

Step 3 Integrate $= \ln|u| - \frac{8u^{-1}}{-1}du + \frac{21u^{-2}}{-2} + C$

$$= \ln|u| + \frac{8}{u} - \frac{21}{2u^2} + C$$

$$\int \frac{x^2+5}{(x+4)^3}dx = \ln|x+4| + \frac{8}{x+4} - \frac{21}{2(x+4)^2} + C. \quad (u = x+4)$$

31 Exercises A

1. If $n = -1$ then $\int x^n dx =$

2. If $n \neq -1$, then $\int x^n dx =$

3. Find $\int \frac{1}{x} dx$

4. Why do we write $\ln|x| + C$ instead of $\ln x + C$

5 Find $\int \frac{1}{x-2} dx$

6. Find $\int \frac{1}{x^2} dx$ (Hint: use power rule)

7. Find $\int \frac{1}{x^3} dx$

8. Find $\int \frac{5x}{(x^2+1)^3} dx$

9. Find $\int \frac{x}{x^2+2} dx$;

10. Find $\int \frac{x}{(x^2+2)^3} dx$

11. Find $\int \frac{x^2}{(4x^3+7)^5} dx$

12. Find $\int \frac{3x}{x^2+4} dx$

13, On which of the following can we simple u-substitution to integrate?

$$\int \frac{3x+2}{4x^2+1} dx \text{ or } \int \frac{3x}{4x^2+1} dx$$

14. Find $\int \frac{2x^3+4x^2+11x+16}{x^2+4} dx$

15. Find $\int \frac{x+6}{(x+4)^3} dx$.

16. Find $\int \frac{x^2+5}{(x+4)^3} dx$

Answers: 1. $\ln|x| + C$; **2.** $\frac{x^{n+1}}{n+1} + C$; **3.** $\ln|x| + C$;**5.** $\ln|x-2| + C$;

6. $-\frac{1}{x} + C$; **7.** $-\frac{1}{2x^2} + C$; **8.** $-\frac{5}{4(x^2+1)^2} + C$; **9.** $= \frac{1}{2}\ln|x^2+2| + c$;

10. $-\frac{1}{4(x^2+2)^2} + C$; **11.** $-\frac{1}{48(4x^3+7)^4} + C$; **12.** $\frac{3}{2}\ln(x^2+4) + c$

13. $\int \frac{3x}{4x^2+1} dx$; **14.** $x^2 + 4x + \frac{3}{2}\ln(x^2+4) + c$; **15,** $-\frac{1}{x+4} - \frac{1}{(x+4)^2} + C$

16. $\ln|x+4| + \frac{8}{x+4} - \frac{21}{2(x+4)^2} + C$;

Lesson 31 Exercises B

1. If $n = -1$ then $\int x^n dx =$

2. If $n \neq -1$, then $\int x^n dx =$

3. Find $\int \frac{1}{3x} dx$

4. Why do we write $\ln|x| + C$ instead of $\ln x + C$

5 Find $\int \frac{1}{x-3} dx$

6. Find $\int \frac{1}{x^3} dx$ (Hint: use power rule)

7. Find $\int \frac{1}{x^6} dx$

8. Find $\int \frac{6x}{(x^2-1)^3} dx$

9. Find $\int \frac{x^2}{x^3+2} dx$;

10. Find $\int \frac{x}{(x^2-2)^4} dx$

11. Find $\int \frac{x^3}{(5x^4+2)^4} dx$

12. Find $\int \frac{4x}{x^2+5} dx$

13, On which of the following can we simple u-substitution to integrate?

(a) $\int \frac{5x+2}{4x^2+1} dx$ or (b) $\int \frac{5x}{4x^2+1} dx$

14. Find $\int \frac{x^2+1}{(x-1)^3} dx$

Answers: **3.** $\frac{1}{3}\ln|x| + C$; **5.** $\ln|x-3| + C$ **6.** $-\frac{1}{2x^2} + C$; **7.** $-\frac{1}{5x^5} + C$;

8. $-\frac{3}{2(x^2-1)^2} + C$; **9.** $\frac{1}{3}\ln|x^3+2| + C$; **10.** $-\frac{1}{6(x^2-2)^3} + C$;

11. $-\frac{1}{60(5x^4+2)^3} + C$; **12.** $2\ln(x^2+5) + C$; **13.** b.;

14. $\ln|x-1| - 2(x-1)^{-1} - (x-1)^{-2} + C$. or

$\ln|x-1| - \frac{2}{x-1} - \frac{1}{(x-1)^2} + C$

Lesson 32

Integration of Radical Functions I

Classification of Integration of Radical Functions

We may classify the integration of radical functions by antiderivative type, by integrand type, or by method type. For some functions ,we will use simple u-substitution, and for others we will use trigonometric substitution. In this lesson, we use only simple u-substitution. In Lesson 43, Integration of Radical Functions II, we use trigonometric substitution. Before proceeding, review Lesson 30.

Integration of Radical Functions Using Simple U-Substitution

Examples   **1.** $\int \sqrt{x-5}\, dx$; **2.** $\int \frac{4x^2}{\sqrt{x^3+1}}\, dx$; **3.** $\int 4x^2 \sqrt{x^3+1}\, dx$.

Note: **1** is composite; **2** and **3** contain composite functions.

Example 1 Find $\int \sqrt{x-5}\, dx$

Solution We use simple **u-substitution** with **two approaches**

Approach 1

$\int \sqrt{x-5}\, dx$

Let $u = x - 5$

Then $\frac{du}{dx} = 1$, and $du = dx$

Substituting for $x - 5$ and dx in

$\int \sqrt{x-5}\, dx$, we obtain

$\int \sqrt{u}\, du$

$= \int u^{\frac{1}{2}} du$

$= \frac{u^{\frac{1}{2}+1}}{\frac{1}{2}+1} + C$

$= \frac{2}{3} u^{\frac{3}{2}} + C$

$= \frac{2}{3}(x-5)^{\frac{3}{2}} + C$ (back to x).

$= \frac{2}{3}\left(\sqrt{x-5}\right)^3 + C$ **.** or

$= \frac{2}{3}(x-5)\sqrt{x-5} + C$

Approach 2

Let $u = \sqrt{x-5}$ (A)

Then $u^2 = x - 5$

Using implicit differentiation,

$2u\frac{du}{dx} = 1$, and $2u\, du = dx$

Substituting for $\sqrt{x-5}$ and dx in

$\int \sqrt{x-5}\, dx$, we obtain

$\int u \bullet 2u\, du$

$= \int 2u^2 du$

$= 2 \int u^2 du$

$= 2 \bullet \frac{u^3}{3} + C$

$= \frac{2}{3}\left(\sqrt{x-5}\right)^3 + C$ (back to x). or

$= \frac{2}{3}(x-5)\sqrt{x-5} + C$

Example 2 Find $\int \dfrac{4x^2}{\sqrt{x^3+1}}\,dx$

We use simple **u-substitution** with **two approaches.**

Approach 1	**Approach 2**

Approach 1

Step 1: Let $u = x^3 + 1$.

Then $\dfrac{du}{dx} = 3x^2$ and from which

$$dx = \dfrac{du}{3x^2}$$

Step 2: Now, replace dx in $\int \dfrac{4x^2}{\sqrt{x^3+1}}\,dx$

by $\dfrac{du}{3x^2}$, and $x^3 + 1$ by u. Then ,

$\int \dfrac{4x^2}{\sqrt{x^3+1}}\,dx$

$= \int \dfrac{4x^2}{\sqrt{u}} \cdot \dfrac{du}{3x^2}$

$= \int \dfrac{4}{u^{\frac{1}{2}}} \cdot \dfrac{du}{3}$ (canceling the common x^2)

$= \dfrac{4}{3} \int \dfrac{1}{u^{\frac{1}{2}}}\,du$

$= \dfrac{4}{3} \int u^{-\frac{1}{2}}\,du$

$= \dfrac{4}{3} \cdot \dfrac{u^{-\frac{1}{2}+1}}{-\frac{1}{2}+1} + C$ (using power rule)

$= \dfrac{4}{3} \cdot \dfrac{u^{\frac{1}{2}}}{\frac{1}{2}} + C$

$= \dfrac{8}{3} u^{\frac{1}{2}} + C$

$= \dfrac{8}{3} \sqrt{u} + C$

$= \dfrac{8}{3} \sqrt{x^3+1} + C$ (replacing u by $x^3 + 1$

$\int \dfrac{4x^2}{\sqrt{x^3+1}}\,dx = \dfrac{8}{3}\sqrt{x^3+1} + C$.

Note above that $\sqrt{x^3+1}$, is composite.

Approach 2

Step 1: **Rationalize the integrand**

Let $u = \sqrt{x^3+1}$.

Then $u^2 = x^3 + 1$

Using implicit differentiation,

$2u\dfrac{du}{dx} = 3x^2$, and from which

$$dx = \dfrac{2u\,du}{3x^2}$$

Step 2: Substitute u for $\sqrt{x^3+1}$;

$\dfrac{2u\,du}{3x^2}$ for dx in $\int \dfrac{4x^2}{\sqrt{x^3+1}}\,dx$ to

obtain

$\int \dfrac{4x^2}{u} \bullet \dfrac{2u\,du}{3x^2}$

$= \dfrac{8}{3} \int 1\,du$

$= \dfrac{8}{3} u + c$

$= \dfrac{8}{3} \sqrt{x^3+1} + c$ $(u = \sqrt{x^3+1})$.

Example 3. Find $\int 4x^2 \sqrt{x^3 + 1}\, dx$

 Solution : We use simple **u-substitution** with **two approaches**

Approach 1 .

Step 1 Let $u = x^3 + 1$.

Then $\dfrac{du}{dx} = 3x^2$ and from which $dx = \dfrac{du}{3x^2}$

Step 2: Substitute u for $x^3 + 1$; and $\dfrac{du}{3x^2}$

for dx in $\int 4x^2 \sqrt{x^3 + 1}\, dx$. Then

$$\int 4x^2 \sqrt{x^3 + 1}\, dx = \int 4x^2 u^{\frac{1}{2}} \cdot \frac{du}{3x^2}$$

$$= \frac{4}{3} \int u^{\frac{1}{2}} du$$

$$= \frac{4}{3} \int \frac{u^{\frac{3}{2}}}{\frac{3}{2}}\, du$$

$$= \frac{4}{3} \left(\frac{2}{3} u^{\frac{3}{2}} \right) + C$$

$$= \frac{8}{9} \left(u^{\frac{3}{2}} \right) + C$$

$$= \frac{8}{9} (x^3 + 1)^{\frac{3}{2}} + C$$

$$= \frac{8}{9} \left(\sqrt{x^3 + 1} \right)^3 + C$$

$$\int 4x^2 \sqrt{x^3 + 1}\, dx = \frac{8}{9} \left(\sqrt{x^3 + 1} \right)^3 + C \,.$$

$$= \frac{8}{9} (x^3 + 1)\sqrt{x^3 + 1} + C$$

Note above that $\sqrt{x^3 + 1}$, is composite
with the "inside" function being $x^3 + 1$.

Approach 2

Step 1: Rationalize the integrand

 Let $u = \sqrt{x^3 + 1}$.

 Then $u^2 = x^3 + 1$

Then $2u \dfrac{du}{dx} = 3x^2$ and from which

$$dx = \frac{2u\, du}{3x^2}$$

Step 2: Substitute u for $\sqrt{x^3 + 1}$;

$\dfrac{2u\, du}{3x^2}$ for dx in $\int 4x^2 \sqrt{x^3 + 1}\, dx$ to
obtain

$$\int 4x^2 u \bullet \frac{2u\, du}{3x^2}$$

$$= \frac{8}{3} \int u^2 du$$

$$= \frac{8}{3} \bullet \frac{u^3}{3} + c$$

$$= \frac{8}{9} u^3 + c$$

$$= \frac{8}{9} \left(\sqrt{x^3 + 1} \right)^3 + c \quad (u = \sqrt{x^3 + 1}).$$

or $= \dfrac{8}{9} (x^3 + 1)\sqrt{x^3 + 1} + C$

Example 4 Find $\int \dfrac{dx}{1+\sqrt{x}}$

Solution: We use simple **u-substitution** with **two approaches**

Both approaches rationalize the integrand.

Approach 1

$\int \dfrac{dx}{1+\sqrt{x}}$

Let $\sqrt{x} = u$

Then $x = u^2$

$\dfrac{dx}{du} = 2u$, or $2u\dfrac{du}{dx} = 1$

and from either,

$dx = 2u\,du$

Substitute accordingly.

Then $\int \dfrac{dx}{1+\sqrt{x}} = \int \dfrac{2u\,du}{1+u}$

$= 2\int \dfrac{u}{1+u}\,du$

$= 2\int\left(1 - \dfrac{1}{1+u}\right)du$

$= 2\left(\int 1\,du - \int \dfrac{1}{1+u}\,du\right)$

$= 2[u - \ln(u+1)] + C$

$= 2u - 2\ln(u+1)] + C$

$= 2\sqrt{x} - 2\ln(\sqrt{x}+1) + C$

$\qquad (\sqrt{x} = u)$

Approach 2

$\int \dfrac{dx}{1+\sqrt{x}}$

Step 1: Let $\sqrt{x} = u^2$

Then $x = u^4$

$4u^3\dfrac{du}{dx} = 1$, and $dx = 4u^3\,du$

Substitute accordingly.

Then $\int \dfrac{dx}{1+\sqrt{x}}$

$= \int \dfrac{4u^3\,du}{1+u^2}$

$= 4\int \dfrac{u^3}{1+u^2}\,du$

Step 2: $= 4\int\left(u - \dfrac{u}{1+u^2}\right)du$ (by long division)

$= 4\left[\int u\,du - \int \dfrac{u}{1+u^2}\,du\right]$ (2nd term by simple-subst.)

$= 4\left[\dfrac{u^2}{2} - \dfrac{1}{2}\ln(u^2+1)\right] + C$ (by simple-substitution).

$= 2u^2 - 2\ln(u^2+1)] + C$

$= 2\sqrt{x} - 2\ln(\sqrt{x}+1) + C \qquad (\sqrt{x} = u^2)$

Note: In Approach 2, Step 2:

We used simple t-substitution to find $\int \dfrac{u}{1+u^2}\,du$ as follows: **:** Let $t = u^2+1$,

then $\dfrac{dt}{du} = 2u$, and $du = \dfrac{dt}{2u}$

$\int \dfrac{u\,dt}{2ut} = \dfrac{1}{2}\int \dfrac{dt}{t} = \dfrac{1}{2}\ln t = \dfrac{1}{2}\ln(1+u^2)$.

Example 5 Find $\int \dfrac{\sqrt{1+x}}{1-x}\,dx$

Solution We use simple u-substitution

Step 1: Rationalize the integrand

Let $u = \sqrt{1+x}$

Then $u^2 = 1 + x$ (A)

Now, we find an expression for $1 - x$ in terms of u.

From (A), $-u^2 = -(1+x)$ (First multiply both sides of (A) by -1

     $2 - u^2 = -1 - x + 2$ (Add 2 to both sides of the equation)

     $2 - u^2 = 1 - x$

     $x = u^2 - 1$ (solving for x)

     $\dfrac{dx}{du} = 2u$ or implicitly $, -2u\dfrac{du}{dx} = -1$

     $dx = 2u\,du$

Step 2: Substitute u for $\sqrt{1+x}$; $2u\,du$ for dx in $\int \dfrac{\sqrt{1+x}}{1-x}\,dx$

$$\int \frac{\sqrt{1+x}}{1-x}\,dx = \int \frac{u}{2-u^2} \cdot \frac{2u\,du}{1}$$

$$= 2\int \frac{u^2\,du}{2-u^2}$$

$$= -2\int \frac{u^2\,du}{u^2-2}$$

$$= -2\left(\int 1\,du + \int \frac{2\,du}{u^2-2} \right) \qquad \text{(by division)}$$

$$= -2\int du - 4\int \frac{du}{u^2-2}$$

$$= -2u - 4\int \frac{du}{u^2-2}$$

$$= -2u - 4\left[\int -\frac{\sqrt{2}}{4} \cdot \frac{1\,du}{u+\sqrt{2}} + \int \frac{\sqrt{2}}{4} \cdot \frac{1\,du}{u-\sqrt{2}} \right]$$

$$= -2u + \sqrt{2}\int \frac{1\,du}{u+\sqrt{2}} - \sqrt{2}\int \frac{1\,du}{u-\sqrt{2}}$$

$$= -2u + \sqrt{2}\ln|u+\sqrt{2}| - \sqrt{2}\ln|u-\sqrt{2}| + C$$

For $\int \dfrac{1}{u^2-2}\,du$:

We use partial fraction decomposition to integrate. See Rational Functions II (Lesson 40) to review partial fraction decomposition from Pre-Calculus)

$$= -2\sqrt{1+x} + \sqrt{2}\ln\left|\sqrt{1+x}+\sqrt{2}\right| - \sqrt{2}\ln\left|\sqrt{1+x}-\sqrt{2}\right| + C$$

$$= -2\sqrt{1+x} + \sqrt{2}\ln\left|\frac{\sqrt{1+x}+\sqrt{2}}{\sqrt{1+x}-\sqrt{2}}\right| + C .$$

Example 6 Find $\int \sqrt[3]{2x-3}\, x\, dx$

Method 1: Use simple u-substitution.
Step 1: Rationalize the integrand.

Let $u = \sqrt[3]{2x-3}$ or $u = (2x-3)^{\frac{1}{3}}$

$$u^3 = 2x - 3$$

$3u^2 \dfrac{du}{dx} = 2$ and $dx = \dfrac{3u^2}{2}\, du$ (by implicit differentiation)

Also, $x = (u^3 + 3)/2$ (for the factor

"x" in $\int \sqrt[3]{2x-3}\, x\, dx$

Step 2: Substitute u for $\sqrt[3]{2x-3}$;

$\dfrac{3u^2}{2} du$ for dx in $\int \sqrt[3]{2x-3}\, x\, dx$.

Then $\int \sqrt[3]{2x-3}\, x\, dx$

$$= \int \frac{u(u^3+3)}{2} \bullet \frac{3u^2}{2}\, du$$

$$= \frac{3}{4}\left(\int u^3(u^3+3) \right) du$$

$$= \frac{3}{4}\left(\int u^6\, du + \int 3u^3\, du \right)$$

$$= \frac{3}{4}\left(\frac{u^7}{7} + \frac{3u^4}{4} \right) + C = \frac{3u^7}{28} + \frac{9u^4}{16} + C$$

Step 3: We change back to the
variable x. $(u = \sqrt[3]{2x-3})$

$$= \frac{3(\sqrt[3]{2x-3})^7}{28} + \frac{9(\sqrt[3]{2x-3})^4}{16} + C$$

$$= \frac{3(2x-3)^{\frac{7}{3}}}{28} + \frac{9(2x-3)^{\frac{4}{3}}}{16} + C$$

$$= (2x-3)^{\frac{4}{3}}[\frac{3(2x-3)}{28} + \frac{9}{16}] + C$$

$$= (2x-3)^{\frac{4}{3}}[\frac{6x-9}{28} + \frac{9}{16}] + C$$

$$= (2x-3)^{\frac{4}{3}} \frac{4(6x-9) + 7(9)}{112} + C$$

$$= (2x-3)^{\frac{4}{3}}[\frac{24x+27}{112}] + C$$

$$= \frac{3(8x+9)}{112}(2x-3)^{\frac{4}{3}} + C$$

$$= \tfrac{3}{112}(8x+9)(2x-3)^{\frac{4}{3}} + C$$

$$= \tfrac{3}{112}(8x+9)(2x-3)(\sqrt[3]{2x-3}) + C$$

$$= \tfrac{3}{112}(16x^2 - 6x - 27)(\sqrt[3]{2x-3}) + C$$

Method 2: Use simple u-substitution
without rationalizing the integrand.
Step 1: Let $u = 2x - 3$.

Then $\dfrac{du}{dx} = 2$ and $dx = \dfrac{du}{2}$

Step 2: Substitute u for $2x - 3$,

$\dfrac{du}{2}$ for dx in $\int \sqrt[3]{2x-3}\, x\, dx$ to

obtain $\int u^{\frac{1}{3}} x \bullet \dfrac{du}{2}$. We will express x

in terms of u. From $u = 2x - 3$.

$x = \dfrac{u+3}{2}$. Substituting for x in

$$= \int u^{\frac{1}{3}} x \bullet \frac{du}{2}, \text{ we obtain}$$

$$= \int \frac{u^{\frac{1}{3}}(u+3)}{2} \bullet \frac{du}{2}$$

$$= \frac{1}{4}\int u^{\frac{1}{3}}(u+3)\, du$$

$$= \frac{1}{4}\int \left(u^{\frac{4}{3}} + 3u^{\frac{1}{3}} \right) du$$

$$= \frac{1}{4}(\frac{3u^{\frac{7}{3}}}{7} + \frac{9u^{\frac{4}{3}}}{4}) + C$$

$$= \frac{3}{4}(\frac{u^{\frac{7}{3}}}{7} + \frac{3u^{\frac{4}{3}}}{4}) + C$$

$$= \frac{3}{4}(\frac{4u^{\frac{7}{3}} + 21u^{\frac{4}{3}}}{28}) + C$$

$$= \frac{3}{112}(4u^{\frac{7}{3}} + 21u^{\frac{4}{3}}) + C$$

$$= \frac{3}{112}[4u^2 u^{\frac{1}{3}} + 21u u^{\frac{1}{3}}] + C$$

$$= \frac{3u^{\frac{1}{3}}}{112}[4u^2 + 21u] + C$$

$$= \tfrac{3}{112}\sqrt[3]{2x-3}\left[4(2x-3)^2 + 21(2x-3)\right] + C$$

$$= \tfrac{3}{112}\sqrt[3]{2x-3}\left[4(4x^2-12x+9)+42x-63\right] + C$$

$$= \tfrac{3}{112}\sqrt[3]{2x-3}\left[16x^2-48x+36+42x-63\right] + C$$

$$= \frac{3}{112}\sqrt[3]{2x-3}\left[16x^2 - 6x - 27\right] + C.$$

$$= \frac{3}{112}\left[16x^2 - 6x - 27\right]\sqrt[3]{2x-3} + C$$

Example 7 Find $\int \frac{x^2+4}{\sqrt{x+4}}dx$ (This example is similar to Case 3b of Lesson 31).

Step 1 Let $u = x+4$, Then $\frac{du}{dx} = 1$, and $du = dx$. After substituting u for

$x+4$, and du for dx in $\int \frac{x^2+4}{\sqrt{x+4}}dx$, we obtain $\int \frac{x^2+4}{u^{\frac{1}{2}}}du$, and we still

have x^2 which we express in terms of u as follows:

Step 2: From $u = x+4$, $x = u-4$, and squaring,

$\quad x^2 = u^2 - 8u + 16$

We now substitute for x^2 in $\int \frac{x^2+4}{u^{\frac{1}{2}}}du$, to obtain

$$\int \frac{u^2 - 8u + 16 + 4}{u^{\frac{1}{2}}}\,du$$

$$= \int \frac{u^2}{u^{\frac{1}{2}}}\,du + \int \frac{-8u}{u^{\frac{1}{2}}} + \int \frac{20}{u^{\frac{1}{2}}}$$

$$= \int u^{\frac{3}{2}}\,du - \int 8u^{\frac{1}{2}}\,du + \int 20u^{-\frac{1}{2}}\,du$$

Step 3: Integrate now.

$$= \tfrac{2}{5}u^{\frac{5}{2}} - \tfrac{2}{3}\left(8u^{\frac{3}{2}}\right) + \tfrac{2}{1}\left(20u^{\frac{1}{2}}\right) + C$$

$$= u^{\frac{1}{2}}\left(\frac{2}{5}u^2 - \frac{16}{3}u + 40\right) + C$$

$$= (x+4)^{\frac{1}{2}}\left[\tfrac{2}{5}(x+4)^2 - \tfrac{16}{3}(x+4) + 40\right] + C$$

$$= (x+4)^{\frac{1}{2}}\left[\frac{6x^2 + 48x + 96}{15} - \frac{80(x+4)}{15} + \frac{600}{15}\right] + C$$

$$\int \frac{x^2+4}{\sqrt{x+4}}dx = \tfrac{2}{15}\sqrt{x+4}\left[3x^2 - 16x + 188\right] + C.$$

Lesson 32 Exercises A

1. Find $\displaystyle\int \sqrt{x-5}\,dx$

2. Find the $\displaystyle\int \frac{4x^2}{\sqrt{x^3+1}}\,dx$

3. Find $\displaystyle\int 4x^2\sqrt{x^3+1}\,dx$

4. Find $\displaystyle\int \frac{dx}{1+\sqrt{x}}$

5. Find $\displaystyle\int \sqrt[3]{2x-3}\,x\,dx$

6. Find $\displaystyle\int \frac{\sqrt{1+x}}{1-x}\,dx$

7. Find $\displaystyle\int \frac{x^2+4}{\sqrt{x+4}}\,dx$

Answers: 1. $\dfrac{2}{3}\left(\sqrt{x-5}\right)^3+C$ or $\dfrac{2}{3}(x-5)\sqrt{x-5}+C$; **2.** $\dfrac{8}{3}\sqrt{x^3+1}+C$;

3. $\dfrac{8}{9}\left(\sqrt{x^3+1}\right)^3+c$; **4.** $2\sqrt{x}-2\ln(\sqrt{x}+1)+C$;

5. $\dfrac{3}{112}(16x^2-6x-27)(\sqrt[3]{2x-3})$; **6.** $-2\sqrt{1+x}+\sqrt{2}\ln\left|\dfrac{\sqrt{1+x}+\sqrt{2}}{\sqrt{1+x}-\sqrt{2}}\right|+C$;

7. $\dfrac{2}{15}\sqrt{x+4}\left[3x^2-16x+188\right]+C$

Lesson 32 Exercises B

1. Find $\displaystyle\int \sqrt{x-3}\,dx$

2. Find the $\displaystyle\int \frac{4x^3}{\sqrt{x^4+2}}\,dx$

3. Find $\displaystyle\int 4x^4\sqrt{x^5+3}\,dx$

4. Find $\displaystyle\int \frac{dx}{1+\sqrt{x}}$

5. Find $\displaystyle\int \sqrt[3]{3x-2}\,x\,dx$

6. Find $\displaystyle\int \frac{\sqrt{2+x}}{2-x}\,dx$

7. Find $\displaystyle\int \frac{\sqrt{x}}{\sqrt[3]{x}+2}\,dx$

Answers: 1. $\dfrac{2}{3}\left(\sqrt{x-3}\right)^3+C$ or $\dfrac{2}{3}(x-3)\sqrt{x-3}+C$; **2.** $2\sqrt{x^4+2}$;

3. $\dfrac{8}{15}\left(\sqrt{x^5+3}\right)^3+C$ or $\dfrac{8}{15}(x^5+3)\sqrt{x^5+3}+C$

4. $2\sqrt{x}-2\ln(\sqrt{x}+1)+C$;

5. $\dfrac{1}{14}(2x+1)\left(\sqrt[3]{3x-2}\right)^4+C$; or $\dfrac{1}{14}\left(6x^2-x-2\right)(\sqrt[3]{3x-2})+C$

6. $-2\sqrt{2+x}+2\ln\left|\dfrac{2+\sqrt{2+x}}{-2+\sqrt{2+x}}\right|$

7. $\dfrac{6}{7}x^{\frac{7}{6}}-\dfrac{12}{5}x^{\frac{5}{6}}+8\sqrt{x}-48x^{\frac{1}{6}}+48\sqrt{2}\,\mathrm{Tan}^{-1}(\tfrac{1}{2}x^{\frac{1}{6}}\sqrt{2})+C.$

CHAPTER 11

Lesson 33

Integration of Exponential Functions

Given a function $f(x)$, to find the indefinite integral (the antiderivative, or the primitive) of $f(x)$ means we are to find another function $g(x) + C$ such that the derivative of $g(x) + C$ is equal to $f(x)$.

Rules for Integrating Exponential Functions

Case 1. Base e	**Case 2.** Any base b
$\int e^x dx = e^x + C$	$\int b^x dx = \dfrac{b^x}{\ln b} + C \qquad (b > 0, b \neq 1)$
	Note: $\dfrac{b^x}{\ln b} + C = \dfrac{b^x}{\log_e b} + C$

For Case 2: Let us show that $\int b^x dx = \dfrac{b^x}{\ln b} + C$.

Step 1: $b^x = e^{x \ln b}$ (using the change of base formula for exponents)

Then $\int b^x dx = \int e^{x \ln b} dx$

Let $u = x \ln b$. Then $\dfrac{du}{dx} = \ln b$, and $\dfrac{du}{\ln b} = dx$ or $du = \ln b\, dx$.

Step 2: Substitute $\dfrac{du}{\ln b}$ for dx, u for $x \ln b$ in $\int e^{x \ln b} dx$ to obtain

$$\int b^x dx = \int \frac{1}{\ln b} e^u du = \frac{1}{\ln b} \int e^u du$$

$$= \frac{1}{\ln b} e^u + C = \frac{1}{\ln b} e^{x \ln b} + C \quad (u = x \ln b)$$

$\int b^x dx = \dfrac{b^x}{\ln b} + C$ (substituting b^x for $e^{x \ln b}$)

Memory device for the derivative and integral of b^x:

Derivative: $\ln b$ multiplies b^x: $\dfrac{d}{dx}(b^x) = b^x \ln b$

Integral: $\ln b$ divides b^x: $\int b^x dx = \dfrac{b^x}{\ln b} + C$ (Note the reversal of the role of $\ln b$)

For Case 2, let $b = e$ in $\int b^x dx = \dfrac{b^x}{\ln b} + C$. Then $\int e^x dx = \dfrac{e^x}{\ln e} + C = e^x + C$

$$(\ln e = \log_e e = 1)$$

Example 1 Find $\int \dfrac{e^x}{1-e^x}\,dx$

Solution We use simple u-substitution.

Let $u = 1 - e^x$

Then $\dfrac{du}{dx} = -e^x$, and $dx = -\dfrac{du}{e^x}$ or $-du = e^x dx$

Substitute u for $1 - e^x$, $-\dfrac{du}{e^x}$ for dx in $\int \dfrac{e^x}{1-e^{2x}}\,dx$

Then $\int \dfrac{e^x}{u} \cdot \dfrac{du}{(-e^x)}$ becomes

$$-\int \frac{1}{u}\,du$$

$$-\ln|u| + c$$

$$-\ln\left|1 - e^x\right| + c$$

Example 2 Find $\int \dfrac{e^{2x}}{1+e^{2x}}\,dx$

Solution We use simple u-substitution

Let $u = 1 + e^{2x}$

Then $\dfrac{du}{dx} = 2e^{2x}$, and $dx = \dfrac{du}{2e^{2x}}$

Substitute u for $1 + e^{2x}$, $\dfrac{du}{2e^{2x}}$ for dx in $\int \dfrac{e^{2x}}{1+e^{2x}}\,dx$

Then $\int \dfrac{e^{2x}}{u} \cdot \dfrac{du}{2e^{2x}}$ becomes

$$= \int \frac{1}{2u}\,du$$

$$= \frac{1}{2}\int \frac{1}{u}\,du$$

$$= \frac{1}{2}\ln|u| + c$$

$$= \frac{1}{2}\ln\left|1 + e^{2x}\right| + c$$

Lesson 33 Exercises A

1. $\int b^x dx = ?$

2. $\int e^x dx = ?$

3. Find $\int \dfrac{e^x}{1 - e^x} dx$

4. Find $\int \dfrac{e^{2x}}{1 + e^{2x}} dx$

5. $\int e^{2x} dx$

6. $\int e^{-2x} dx$

Answers: 3. $-\ln\left|1 - e^x\right| + C$; **4.** $\frac{1}{2}\ln\left|1 + e^{2x}\right| + C$. **5.** $\frac{1}{2}e^{2x} + C$;

6. $-\frac{1}{2}e^{-2x} + C$

Lesson 33 Exercises B

1. $\int c^x dx = ?$

2. $\int e^x dx = ?$

3. Find $\int \dfrac{e^{2x}}{e^x + 2} dx$

4. Find $\int \dfrac{e^x}{2 - e^x} dx$

5. Find $\int \dfrac{e^{3x}}{1 + e^{3x}} dx$

6. $\int e^{3x} dx$

7. $\int e^{-3x} dx$

7.

Answers: 1. $\dfrac{c^x}{\ln c} + C$; **3.** $e^x - 2\ln\left|e^x + 2\right| + C$; **4.** $-\ln\left|2 - e^x\right| + C$

5. $\frac{1}{3}\ln\left|1 + e^{3x}\right| + C$; **6.** $\frac{1}{3}e^{3x} + C$, **7.** $-\frac{1}{3}e^{-3x} + C$.

Lesson 34
Integration of Trigonometric Functions

We present the trigonometric functions and their corresponding integrals. Use the trigonometric derivatives, learned previously, to help you recall some of the entries in the table. Memorize these basic formulas. Later, we will derive some of them.

Trigonometric Functions and Corresponding Antiderivatives

Function: $f(x)$	**Antiderivative** (Indefinite Integral) $\int f(x)dx$
1. $y = \sin x$	1. $-\cos x + C$
2. $y = \cos x$	2. $\sin x + C$
	$\qquad\qquad$ A $\qquad\qquad\qquad\qquad$ B
3. $y = \tan x$	3. $\ln\lvert\sec\rvert + C \qquad$ or $\qquad -\ln\lvert\cos x\rvert + C$
4. $y = \cot x$	4. $-\ln\lvert\csc x\rvert + C \qquad$ or $\qquad \ln\lvert\sin x\rvert + C$
5. $y = \sec x$	5. $\ln\lvert\sec x + \tan x\rvert + C$
6. $y = \csc x$	6. $\ln\lvert\csc x - \cot x\rvert + C$
7. $y = \sec^2 x$	7. $\tan x + C$ (from the derivative table of Lesson 22)
8. $y = \csc^2 x$	8. $-\cot x + C$ (from the derivative table of Lesson 22)
9. $y = \sec x \tan x$	9. $\sec x + C$ (from the derivative table of Lesson 22)
10. $y = \csc x \cot x$	10. $-\csc x + C$ (from the derivative table of Lesson 22)

Mnemonic Help:
The rules of signs of the integrals for the first four basic trigonometric functions are the opposites of those for the derivatives we covered under differentiation previously (Lesson 22). However, for $\tan x$ and $\cot x$, column A is easier to recall since the basic names of the integrals are the same as those of the derivatives learned in Lesson 22 (p.177), which you have already memorized.

Examples: (Ignoring the integration constant for simplicity)
The integral of $\sin x = -\cos x$ ($\sin x$ is non co-named)
The integral of $\cos x = \sin x$ ($\cos x$ is co-named)

1. For the first two entries (sin to cos): use the first two entries of the differentiation tables. For the next four entries (tan to csc), we can modify the memory device for the differentiation to obtain the integrals.
(a) First note the natural log (ln-**ell-en**) and the absolute value symbols. Each of them uses the "natural log of the absolute value of something".
(b) We use the mnemonic device we used for memorizing the derivatives with some minor modification.

2. For the last four entries in the table (from $\sec^2 x$ to $\csc x \cot x$), interchange the entries in the table for the derivatives (Lesson 22), noting their signs.
Say aloud the following, keeping the layout of the table in mind (entries **7-10**):

1. tan (ell-en) sec (you may skip the **"ell-en"** to make the recitation simple)
2. cot minus (ell-en) cosec (You may also use **cot** (ell-en) **sin**
3. sec (ell-en) sec plus tan (plus sign as with the derivative of **sec**).
4. cosec (ell-en) cosec minus **cotan** (minus sign as with the derivative of **csc**)
Note above that each of the integrals that cannot be obtained by mere interchange of the derivatives (in Lesson 22) begins with natural log of the absolute value

More on the Memory Device

Note the similarities between the derivatives and the integrals. Note the natural log and the absolute value bars of the integrals and also, how the plus and minus signs are carried over from the derivatives to the integrals.

Differentiation (Lesson 22)	**Integration** (Present Lesson)
1. tan sec square.	**1. tan** ln|sec| **or** $-\ln\|\cos x\|$
2. cot minus cosec square	**2. cot** $-\ln\|\text{cosec}\|$ or **cot** ln|sin|
3. sec sec tan	
(note the repetition of sec)	**3. sec** ln|sec+ tan|
4 cosec minus cosec cotan	(note the repetition of sec)
(Repetition of cosec is similar to that of sec x in **3**)	**4 cosec** ln|cosec − cotan|
	(Repetition of cosec is similar to sec x .)

Say aloud **1-4** over and over and as you say them, look at the table. Take a sheet of paper and try to write the above table, from memory, guided by this mnemonic device.

Repeat the above in the coming days, and thereafter, from time to time test yourself by writing the contents of the table on paper, being guided by the mnemonic device, and if you do not recall anything, review the table and the device. Note that at least 50% of school and college work is memory work, especially since almost every exam is a closed book exam. Even if an exam were open book, the student who can recall accurately what he/she has learned will work faster than a student who has to look up information from a book. The above suggestion does not mean that you memorize without any understanding. First, understand and then try to find a way to remember what you have understood.

About the use of absolute values above: $\int \tan x\, dx = \ln |\sec x| + C$ implies

(a) $\int \tan x\, dx = \ln \sec x + C$, for all x such that $\sec x \geq 1$, or

(b) $\int \tan x\, dx = \ln(-\sec x) + C$, for all x such that $x \leq -1$.

Derivation of some basic integration formulas

Example 1 Show that $\int \tan x\, dx = \ln |\sec x| + C$ (We apply simple u-substitution)

Step 1: $\int \tan x\, dx = \int \frac{\sin x}{\cos x}\, dx$

Let $u = \cos x$

Then $\frac{du}{dx} = -\sin x$, and

$-\frac{du}{\sin x} = dx$ or $-du = \sin x\, dx$

Step 2: Substitute u for $\cos x$, and

$-\frac{du}{\sin x}$ for dx or $-du = \sin x\, dx$

Step 3: Then $\int \frac{\sin x}{u}\frac{(-du)}{\sin x} = -\int \frac{du}{u}$

$= -\ln|u| + C$

$= -\ln|\cos x| + C$ $(u = \cos x)$

$= \ln\left|(\cos x)^{-1}\right| + C$

$\int \tan x\, dx = \ln|\sec x| + C$

$(\cos x)^{-1} = \frac{1}{\cos x} = \sec x)$

Example 2 Show that $\int \cot x\, dx = \ln|\sin x| + C = -\ln|\csc x| + C$.

Solution   We apply simple u-substitution

Step 1: $\int \cot x\, dx = \int \frac{\cos x}{\sin x}\, dx$

Let $u = \sin x$

Then $\frac{du}{dx} = \cos x$, and

$\frac{du}{\cos x} = dx$ or $du = \cos x\, dx$

Step 2: Substitute u for $\sin x$, and $\frac{du}{\cos x}$

for dx (That is, $du = \cos x\, dx$)

Then $\int \frac{\cos x}{u}\frac{du}{\cos x} = \int \frac{du}{u} = \ln|u| + C$

$\therefore$ $\int \cot x\, dx = \ln|\sin x| + C$

$\left(= \ln\left|\frac{1}{\csc x}\right| = \ln\left|(\csc x)^{-1}\right| = -\ln|\csc x|\right)$

Example 3 Show that $\int \sec x \, dx = \ln|\sec x + \tan x| + C$

Step 1: $\int \sec x \, dx = \int \frac{\sec x (\sec x + \tan x)}{\sec x + \tan x} \, dx$

(multiplying both numerator and denominator by $\sec x + \tan x$)

$$\int \frac{\sec^2 x + \sec x \tan x}{\sec x + \tan x} \, dx$$

Let $u = \sec x + \tan x$

$\frac{du}{dx} = \sec x \tan x + \sec^2 x$ $(\frac{d}{dx} \sec x = \sec x \tan x; \frac{d}{dx}(\tan x) = \sec^2 x)$

$dx = \frac{du}{\sec x \tan x + \sec^2 x}$ or

$du = (\sec x \tan x + \sec^2 x) dx$

Step 2: Substitute u for $\sec x + \tan x$, and $\frac{du}{\sec x \tan x + \sec^2 x}$ for dx

(That is, $du = (\sec x \tan x + \sec^2 x) dx$)

Then $\int \sec x \, dx = \int \frac{\sec^2 x + \sec x \tan x}{u} \cdot \frac{du}{\sec^2 x + \sec x \tan x}$

$$= \int \frac{du}{u}$$

$$= \ln|u| + C$$

$$= \ln|\sec x + \tan x| + C \qquad (u = \sec x + \tan x)$$

Therefore, $\int \sec x \, dx = \ln|\sec x + \tan x| + C$ (See page 301-302 for another method)

Example 4 Show that $\int \csc x \, dx = \ln|\csc x - \cot x| + C$

Step 1: $\int \csc x \, dx = \int \frac{\csc x (\csc x - \cot x)}{\csc x - \cot x} \, dx$

(multiplying both numerator and denominator by $\csc x - \cot x$)

$$= \int \frac{\csc^2 x - \csc x \cot x}{\csc x - \cot x} \, dx$$

Let $u = \csc x - \cot x$. Then $\frac{du}{dx} = -\csc x \cot x - (-\csc^2 x)$

$(\frac{d}{dx} \csc x = -\csc x \cot x; \frac{d}{dx} \cot x = -\csc^2 x)$

$dx = \frac{du}{\csc^2 x - \csc x \cot x}$, or $du = (\csc^2 x - \csc x \cot x) dx$

Step 2: Substitute u for $\csc x - \cot x$, and $\frac{du}{\csc^2 x - \csc x \cot x}$ for dx.

Then $\int \csc x \, dx = \int \frac{\csc^2 x - \csc x \cot x}{u} \cdot \frac{du}{\csc^2 x - \csc x \cot x}$

$$= \int \frac{du}{u}$$

$$= \ln|u| + C$$

$$= \ln|\csc x - \cot x| + C \qquad (u = \csc x - \cot x)$$

Therefore, $\int \csc x \, dx = \ln|\csc x - \cot x| + C$.

More Examples on Integration of Trigonometric Functions

Example 5. Find $\int 5\sin x\, dx$

Solution

$$\int 5\sin x\, dx = 5\int \sin x\, dx$$
$$= 5(-\cos x) + C$$
$$= -5\cos x + C$$

Example 6. Find $\int \dfrac{\sin x}{\cos^2 x}\, dx$

Solution
Method 1: Since the numerator is the **derivative** of $\cos x$ (in the denominator) we can use the u-substitution method.

Step 1: Let $u = \cos x$, then $\cos^2 x = u^2$,

and $\dfrac{du}{dx} = -\sin x$ and from which $dx = -\dfrac{du}{\sin x}$

Step 2: Substitute $-\dfrac{du}{\sin x}$ for dx and u^2 for $\cos^2 x$ in $\int \dfrac{\sin x}{\cos^2 x}\, dx$ to obtain

$$\int \frac{\sin x}{u^2}(-\frac{du}{\sin x})$$
$$= \int -\frac{1}{u^2}\, du$$
$$= -\int u^{-2}\, du$$
$$= -\frac{u^{-2+1}}{-2+1} + C$$
$$= -\frac{u^{-1}}{-1} + C$$
$$= \frac{1}{u} + C$$

$$\int \frac{\sin x}{\cos^2 x}\, dx = \frac{1}{\cos x} + C \quad \text{(replacing } u \text{ by } \cos x)$$
$$= \sec x + C.$$

Method 2

Step 1: $\dfrac{\sin x}{\cos^2 x} = \dfrac{1}{\cos x}\cdot\dfrac{\sin x}{\cos x}$
$$= \sec x \tan x$$

Step 2: Since the derivative of $\sec x$ is $\sec x \tan x$, the antiderivative of $\sec x \tan x$ is $\sec x + C$.

Therefore, $\int \dfrac{\sin x}{\cos^2 x}\, dx = \sec x + C.$

Example 7. Find $\int \cos^2 x \sin x \, dx$ (Also, see Lesson 37)

Solution   We apply simple u-substitution

Step 1: Since $\sin x$ is the derivative of $\cos x$,
let $u = \cos x$

Then $\dfrac{du}{dx} = -\sin x$, and from which $dx = -\dfrac{du}{\sin x}$

Step 2: Substitute $-\dfrac{du}{\sin x}$ for dx and u^2 for $\cos^2 x$ in $\int \cos^2 x \sin x \, dx$ to
   obtain

$$\int u^2 \sin x \left(-\frac{du}{\sin x}\right)$$

$$= \int -u^2 \, du$$

$$= -\frac{u^3}{3} + C$$

$$= -\frac{1}{3} \cos^3 x + C$$

Example 8. Find $\int \sec 3x \tan 3x \, dx$

Solution

Step 1: Let $u = \sec 3x$.   (since the derivative of $\sec x$ is $\sec x \tan x$)

Then $\dfrac{du}{dx} = 3\sec 3x \tan 3x$ and from which $dx = \dfrac{du}{3\sec 3x \tan 3x}$

Step 2: Substitute $\dfrac{du}{3\sec 3x \tan 3x}$ for dx in $\int \sec 3x \tan 3x \, dx$ to obtain

$$\int \frac{\sec 3x \tan 3x \, du}{3\sec 3x \tan 3x}$$

$$= \int \frac{du}{3} \quad \text{(sec } 3x \tan 3x \text{ in the numerator and denominator cancel out)}$$

$$= \frac{1}{3}u + C$$

$$= \frac{1}{3}\sec 3x + C$$

$\int \sec 3x \tan 3x \, dx = \frac{1}{3}\sec 3x + C$.

Extra: Show $-\ln(\csc x + \cot x) = \ln(\csc x - \cot x)$

$$-\ln(\csc x + \cot x) = \ln(\csc x + \cot x)^{-1}$$

$$= \ln\left(\frac{1}{\csc x + \cot x}\right)$$

$$= \ln\left(\frac{1 \bullet (\csc x - \cot x)}{(\csc x + \cot x)(\csc x - \cot x)}\right)$$

$$= \ln\left(\frac{\csc x - \cot x}{\csc^2 x - \cot^2 x}\right)$$

$$= \ln\left(\frac{\csc x - \cot x}{1}\right) \text{ Note: } \csc^2 x - \cot^2 x = 1$$

$$= \ln(\csc x - \cot x)$$

Lesson 34 Exercises

In 1-10, complete the antiderivative part from memory

Function: $f(x)$ $\int f(x)dx + C$ (Antiderivative)

1. $y = \sin x$ 1.

2. $y = \cos x$ 2.

3. $y = \sec^2 x$ 3.

4. $y = \csc^2 x$ 4.

5. $y = \sec x \tan x$ 5.

6. $y = \csc x \cot x$ 6.

7. $y = \tan x$ 7.

8. $y = \cot x$ 8.

9. $y = \sec x$ 9.

10. $y = \csc x$ 10.

11.. Find $\int 5\sin x\, dx$

12. Find $\int \dfrac{\sin x}{\cos^2 x} dx$

13. Find $\int \cos^2 x \sin x\, dx$

14. Find $\int \sec 3x \tan 3x\, dx$

We can use Simple U-substitution for 15-27. Do it

15. Find $\int \dfrac{\sin x}{\cos x}\, dx$

16. Find $\int \sin x \cos x\, dx$

17, Find $\int \sin^5 x \cos x\, dx$

18. Find $\int \cos^5 x \sin x\, dx$

19. $\int \sin^n x \cos x\, dx$

20. $\int \cos^n \sin x\, dx$

21. $\int \dfrac{\cos x}{1 + \sin x}\, dx$

22. $\int \dfrac{\sin x}{1 + \cos x}\, dx$

23. $\int \dfrac{\sin x}{\cos^2 x}\, dx$

24. $\int \sec 3x \tan 3x\, dx$

25. $\int \dfrac{\cos x}{\sin x}\, dx$

26. $\int \dfrac{\sin 2x}{1 - \cos 2x} dx$

27. $\int \cot 2x\, dx$

Answers: 11 $-5\sin x + C$; **12.** $\sec x + C$; **13.** $-\frac{1}{3}\cos^3 x + C$; **14** $\frac{1}{3}\sec 3x + C$

15. $-\ln|\cos x| + C$ or $\ln|\sec x| + C$; **16.** $\frac{1}{2}\sin^2 x + C$; **17.** $\frac{1}{6}\sin^6 x + C$;

18. $-\frac{1}{6}\cos^6 x + C$; **19.** $\dfrac{\sin^{n+1} x}{n+1} + C$; **20.** $-\dfrac{\cos^{n+1} x}{n+1} + C$; **21.** $\ln|\sin x + 1| + C$

22. $-\ln|\cos x + 1| + C$ **23.** $\sec x + C$; **24.** $\frac{1}{3}\sec 3x + C$; **25.** $\ln|\sin x| + C$

26. $\frac{1}{2}\ln|1 - \cos 2x| + C$; **27.** $\frac{1}{2}\ln|\sin 2x| + C$.

Lesson 35

Integration of Algebraic Functions whose Antiderivatives are Inverse Trigonometric Functions
(Antiderivatives of the derivatives of Inverse Trigonometric Functions.)

Since in **Lesson 23**, we learned the derivatives of the inverse trigonometric functions, there is little new here to memorize.

The derivative of an inverse trigonometric function is an **algebraic function**. For example, the derivative of the inverse trigonometric function $\arcsin x$ or

$Sin^{-1}x$ is $\dfrac{1}{\sqrt{1-x^2}}$ $\quad\quad$ ($\dfrac{1}{\sqrt{1-x^2}}$ is purely algebraic,)

Symbolically, $\dfrac{d}{dx}(Sin^{-1}x) = \dfrac{1}{\sqrt{1-x^2}}$. When we reverse the process, we would be

integrating $\dfrac{1}{\sqrt{1-x^2}}$, which is algebraic. There are therefore, some algebraic

functions whose antiderivatives are inverse trigonometric functions.

Note: $\dfrac{d}{dx}(Sin^{-1}x) = \dfrac{1}{\sqrt{1-x^2}}$, but $\displaystyle\int \dfrac{1}{\sqrt{1-x^2}}\,dx = Sin^{-1}x + C$

Analogy: The derivative of x^3 is $3x^2$; but the antiderivative of $3x^2$ is $x^3 + C$.

Algebraic Integrand	Antiderivative	Inverse Trig Function	Derivative
1. $\dfrac{1}{\sqrt{1-x^2}}$	$Sin^{-1}x + C$	1. $y = Sin^{-1}x;$	$\dfrac{dy}{dx} = \dfrac{1}{\sqrt{1-x^2}}$
2. $-\dfrac{1}{\sqrt{1-x^2}};$	$Cos^{-1}x + C$	2. $y = Cos^{-1}x;$	$\dfrac{dy}{dx} = -\dfrac{1}{\sqrt{1-x^2}}$
3. $\dfrac{1}{1+x^2};$	$Tan^{-1}x + C$	3. $y = Tan^{-1}x;$	$\dfrac{dy}{dx} = \dfrac{1}{1+x^2}$
4. $\dfrac{1}{x\sqrt{x^2-1}};$	$Sec^{-1}x + C$	4. $y = Sec^{-1}x;$	$\dfrac{dy}{dx} = \dfrac{1}{x\sqrt{x^2-1}}$
5. $-\dfrac{1}{x\sqrt{x^2-1}};$	$Csc^{-1}x + C$	5. $y = Csc^{-1}x;$	$\dfrac{dy}{dx} = -\dfrac{1}{x\sqrt{x^2-1}}$
6. $-\dfrac{1}{1+x^2};$	$Cot^{-1}x + C$	6. $y = Cot^{-1}x;$	$\dfrac{dy}{dx} = -\dfrac{1}{1+x^2}$

Note: It is technically **incorrect** to say that we are integrating inverse trigonometric functions. (We integrate inverse trigonometric functions in **Lesson 39**.)What is **correct** to say is that the integrals of some algebraic functions are inverse trigonometric functions, A correct topic would be "the antiderivatives of the derivatives of inverse trigonometric functions". An **incorrect** topic would be ' antiderivatives of inverse trigonometric functions". Note above , for example that one approach for justifying that the antiderivative

of $\dfrac{1}{1+x^2}$ is $Tan^{-1}x + C$ is that the derivative of $Tan^{-1}x$ is $\dfrac{1}{1+x^2}$. Another

approach is by trigonometric substitution (see lesson **41**)

Lesson 35 Exercises

Complete the antiderivative part from memory

Algebraic Integrand	Indefinite Integral
$f(x)$	$\int f(x)dx + C$

1. $\dfrac{1}{\sqrt{1-x^2}}$ 1.

2. $-\dfrac{1}{\sqrt{1-x^2}}$ 2.

3. $\dfrac{1}{1+x^2}$ 3.

4. $\dfrac{1}{x\sqrt{x^2-1}}$ 4.

5. $\dfrac{1}{x\sqrt{x^2-1}}$ 5.

6. $-\dfrac{1}{1+x^2}$ 6.

7. Show that $\dfrac{d}{dx}(\operatorname{Sin}^{-1}x) = \dfrac{1}{\sqrt{1-x^2}}$

Lesson 36
Integration-by-Parts

Integration-by-parts is often used when the integrand is the product of two functions as well as when the integrand is a single function such as the logarithmic function or an inverse trigonometric function or to find an integral such as $\int \sin^2 x\,dx$. A guided as well as a classified approach is used to cover two versions of the integration by parts **formula**, namely the American version and the British version.

American Version (Version 1)

$$\int u\,dv = uv - \int v\,du$$

In words, the integral = Product of u and the integral of the dv part minus the integral of the product of the derivative of u and the integral of the dv part. Generally, we choose the dv-part as the part that is readily integrable and the other part is the u-part. **See** next page for a guide.

Derivation of Version 1 formula:
Let u and v be two functions of x.
Then $d(uv) = u\,dv + v\,du$.   (1)

$u\,dv = d(uv) - v\,du$ (2)

$\int u\,dv = uv - \int v\,du$ (3)

British Version (Version 2)

$$\int [uv]\,dx = u\int v\,dx - \int\left(\frac{du}{dx} \cdot \int v\,dx\right)dx$$

You may use the memory device:

$$\int (uv)\,dx = u\int v - \int\left(\frac{du}{dx} \cdot \int v\right)dx$$

In words, the integral of the product uv of two functions is equal to the product of u and the integral of v, minus the integral of the product of the derivative of u and the integral of v. Choose the v-part so that v is readily integrable, and the other function will be the u-part. (or let the part that is difficult to integrate be the u-part); but there are some qualifications as we shall see in the examples that will follow.
See next page for a guide for choosing the u-part.

It does not matter which version we use, since the subsequent corresponding statements would all be identical. In practice, the author finds Version 2 more straightforward.

Derivation of Version 2 formula:

Approach #1
Let u and t be two functions of x.

Then $\dfrac{d}{dx}(ut) = u\dfrac{dt}{dx} + t\dfrac{du}{dx}$.   (1)

$u\dfrac{dt}{dx} = \dfrac{d}{dx}(ut) - t\dfrac{du}{dx}$ (2)

$\int u\dfrac{dt}{dx} \bullet dx = ut - \int t\dfrac{du}{dx} \bullet dx$ (3)

Let $\dfrac{dt}{dx} = v$ (4)

Then $t = \int v\,dx$ (5)

Substitute for $\dfrac{dt}{dx}$ from (4), for t from (5) in (3). Then

$\int (uv)\,dx = u\int v\,dx - \int\left\{\dfrac{du}{dx} \bullet \int v\,dx\right\}dx$

Approach #2
Let r and s be two functions of x.
Then $d(rs) = r\,ds + s\,dr$.   (1)

$r\,ds = d(rs) - s\,dr$ (2)

$\int r\,ds = rs - \int s\,dr$ (**Version 1**)   (3)

Let $r = u$ (4)

Then $\dfrac{dr}{dx} = \dfrac{du}{dx}$ and $dr = \dfrac{du}{dx} \bullet dx$ (5)

Let $\dfrac{ds}{dx} = v$ Then $ds = v\,dx$ (6)

and $s = \int v\,dx$ (7)

Now, substitute for r from (4), for dr from (5), for ds from (6), for s from (7) in (**3**). Then

$\int (uv)\,dx = u\int v\,dx - \int\left\{\dfrac{du}{dx} \cdot \int v\,dx\right\}dx$

(Version 2)

Classifying Integration-by-Parts by Process Types

There are three main process types in the application of integration-by-parts.

Type **1**: The process terminates after a single application of the formula.
Type **2**: The process terminates after two or more applications of the formula.
Type **3**: The process does not terminate but the original integral is reproduced.
Note: For integrating a **single** function, u = function and $dv = 1$ for Version 1
(or $v = 1$ for Version 2)
Below, we will classify the examples according to function types but in addition, we will also mention the process type. Also, for some examples, whenever the simple substitution technique (learned previously) will work, we will cover this substitution technique as another method.

Choosing the u-part

Choose the u-part according to the following order, from left to right:
Logarithmic Functions (L); Inverse Trigonometric Functions (I); Algebraic Functions (A); Trigonometric Functions (T); Exponential functions (E)..
However, there is flexibility in the above order. Memory device: **L I A T E**

Classifying Integration-by-Parts
(by Function Types)

Case 1: Product of a polynomial and a trigonometric function
Example: Find $\int x^2 \sin x \, dx$

Since in the above order (LIATE), from left to right, the algebraic function (A) comes before the trigonometric function (T), we let the u-part $= x^2$ and let $\sin x$ be the dv-part (v-part for Version 2) The application of the integration formula terminates after two applications.. After two differentiations, the polynomial part would be reduced to a constant and only the trigonometric function remains and which we can integrate readily. If we otherwise choose $\sin x$ as the u-part and x^2 as the dv-part and differentiate and integrate accordingly, we will obtain cosines and sines, while the degree of the monomial part would increase and we will not get a readily integrable function as the process proceeds. Note above that both x^2 and $\sin x$ are readily integrable. The process type here is **Type 2**.
Note that all the steps are the same irrespective of which Version we use.

Starting with Version 1	Starting with Version 2
Step 1: Let $u = x^2$ $\quad dv = \sin x \, dx$	**Step 1:** Let $u(x) = x^2$ $\quad v(x) = \sin x$
$du = 2x\,dx$ $\quad\quad v = \int \sin x\,dx$ $\quad\quad\quad\quad\quad = -\cos x$	$\quad \dfrac{du}{dx} = 2x \quad\quad \int v\,dx = \int \sin x\,dx$ $\quad\quad\quad\quad\quad\quad\quad = -\cos x$
Now, substitute in	Now, substitute in
$\int u\,dv = uv - \int v\,du$ (Formula). Then	$\int [uv]\,dx = u\int v\,dx - \int\left(\dfrac{du}{dx}\cdot\int v\,dx\right)dx$
$\int x^2 \sin x \, dx$	The rest of the steps are the same as those in Version 1 on the left.
$= -x^2 \cos x - \int 2x(-\cos x)dx$	-------------------------------------
$= -x^2 \cos x + 2\int x \cos x\,dx$	Explanations for the left side
Step 2:	**<--Step 2: (Version 1)**
$= -x^2 \cos x + 2[x\sin x - \int \sin x\,dx]$	$\quad u = x;\ du/dx = 1$ or $du = dx$
$= -x^2 \cos x + 2[x\sin x - (-\cos x)] + C$	$dv = \sin x\,dx;\ v = \int \sin x\,dx = -\cos x$
$= -x^2 \cos x + 2[x\sin x + \cos x] + C$	**<--(Version 2)** $u = x;\ du/dx = 1$
$= -x^2 \cos x + 2x\sin x + 2\cos x + C$	$v = \sin x;\ \int v\,dx = \int \sin x\,dx = -\cos x$.

Case 2: Product of a polynomial and an exponential function

Example: Find $\int x^2 e^x dx$ (From the order **LIATE**, from left to right, the algebraic function (A) comes before the exponential function (E).)

This case is like Case 1, and we let the u-part $= x^2$. The process type is **Type 2**

Starting with Version 1	Starting with Version 2
Step 1 Let $u = x^2$ and $dv = e^x\,dx$	Let $u(x) = x^2$ and $v(x) = e^x$
$v = \int e^x\,dx = e^x$; $\dfrac{du}{dx} = 2x$ and	$\dfrac{du}{dx} = 2x$; $\int v\,dx = \int e^x dx = e^x$
$du = 2x\,dx$. Substituting in	$\int [uv]dx = u\int v\,dx - \int\left(\dfrac{du}{dx}\cdot\int v\,dx\right)dx$
$\int u\,dv = uv - \int v\,du$	The rest of the steps are the same as
$\int x^2 e^x dx$	those in Version 1 on the left.
$= x^2 e^x - \int 2x e^x\,dx$	
Step 2: $= x^2 e^x - 2[\int x e^x\,dx]$	Explanations for the left side
$= x^2 e^x - 2[x e^x - \int (1)e^x dx]$	$\leftarrow u = x$, $\dfrac{du}{dx} = 1$ and $du = dx$; for
$= x^2 e^x - 2[x e^x - e^x] + C$	version 1.
$= x^2 e^x - 2x e^x + 2e^x + C$	For version 2: $\int v\,dx = \int e^x dx = e^x$
$= e^x(x^2 - 2x + 2) + C$.	

Case 2+: Product of a polynomial function and an exponential function with base other than e.

Example: Find $\int x^2 3^x\,dx$. The process type is **Type 2,** (Two applications)

Starting with Version 1	Starting with Version 2
Step 1: Let $u = x^2$ $\bigg\vert$ $dv = 3^x\,dx$	**Step 1:** Let $u(x) = x^2$ $v(x) = 3^x$
$du = 2x\,dx$ $\bigg\vert$ $v = \int 3^x\,dx = \dfrac{3^x}{\ln 3}$	$\dfrac{du}{dx} = 2x$ $\int v(x)dx = \dfrac{3^x}{\ln 3}$
Now, substitute in $\int u\,dv = uv - \int v\,du$	Now, substitute in
$\int x^2 3^x dx = x^2 \dfrac{3^x}{\ln 3} - \dfrac{1}{\ln 3}\int 2x \cdot 3^x\,dx$	$\int [uv]dx = u\int v\,dx - \int\left(\dfrac{du}{dx}\cdot\int v\,dx\right)dx$
$= x^2 \dfrac{3^x}{\ln 3} - \dfrac{2}{\ln 3}\left[x\cdot\dfrac{3^x}{\ln 3} - \int (1)\dfrac{3^x}{\ln 3}dx\right]$	The rest of the steps are the same as those in Version 1 on the left.
Step 2 $= x^2 \dfrac{3^x}{\ln 3} - \dfrac{2}{\ln 3}\left[\dfrac{3^x x}{\ln 3} - \int (1)\dfrac{3^x}{\ln 3}dx\right]$	Explanations for the left side
$= x^2 \dfrac{3^x}{\ln 3} - \dfrac{2}{\ln 3}\left[\dfrac{3^x x}{\ln 3} - \dfrac{1}{\ln 3}\int 3^x dx\right]$	$\leftarrow u = x$; $\dfrac{du}{dx} = 1$; $du = dx$
$= x^2 \dfrac{3^x}{\ln 3} - \dfrac{2}{\ln 3}\left[\dfrac{3^x x}{\ln 3} - \dfrac{1}{\ln 3}\left(\dfrac{3^x}{\ln 3}\right)\right] + C$	$v = \int 3^x\,dx = \dfrac{3^x}{\ln 3}$ (version 1)
$= 3^x\left(\dfrac{x^2(\ln 3)^2 - 2x\ln 3 + 2}{(\ln 3)^3}\right) + C$	$\int v(x)dx = \dfrac{3^x}{\ln 3}$ (version 2)

Extra $\int x^3 e^{x^2}\,dx = \int x^2 \cdot x e^{x^2}\,dx = \frac{1}{2}e^{x^2}(x^2 - 1) + C$; ($u = x^2$; $dv = x e^{x^2}\,dx$)

We integrated the second term of the formula by substitution.

Case 3: Product of a polynomial and a logarithmic function

Example 1 Find $\int x \ln x \, dx$. (From the order **LIATE** the logarithmic function (L) comes before the algebraic function (A).)

The process type here is **Type 1**

Starting with Version 1	**Starting with Version 2**
Step 1:	**Step 1:**
Let $u = \ln x$ $dv = x\,dx$ $\frac{du}{dx} = \frac{1}{x}$; $du = \frac{1}{x}dx$ $v = \int x\,dx = \frac{x^2}{2}$	Let $u = \ln x$ $v = x$ $\frac{du}{dx} = \frac{1}{x}$ $\int v\,dx = \int x\,dx = \frac{x^2}{2}$

Substituting in $\int u\,dv = uv - \int v\,du$ $\int x \ln x\,dx = \ln x\left(\frac{x^2}{2}\right) - \int\left(\frac{1}{x}\cdot\frac{x^2}{2}\right)dx$ $\qquad = \ln x\left(\frac{x^2}{2}\right) - \frac{1}{2}\int x\,dx$ **Step 2** $= \ln x\left(\frac{x^2}{2}\right) - \frac{1}{2}\cdot\frac{x^2}{2} + C$ $\qquad = \ln x\left(\frac{x^2}{2}\right) - \frac{x^2}{4} + C$ $\int x \ln x\,dx = \frac{x^2}{2}\ln x - \frac{x^2}{4} + C.$	Now, substitute in $\int [uv]dx = u\int v\,dx - \int\left(\frac{du}{dx}\cdot\int v\,dx\right)dx$ The rest of the steps are the same as those in Version 1 on the left.

Example 2: Find $\int x^2 \ln x \, dx$ (From the order **LIATE** the logarithmic function (L) comes before the algebraic function (A)

Also,, x^2 is easily integrable, but $\ln x$ is **not** easily integrable.

We let the u-part $= \ln x$ .and the dv-part $= x^2$. The process type here is **Type 1**

Starting with Version 1	**Starting with Version 2 2**
Step 1:	**Step 1:**
Let $u = \ln x$ $dv = x^2\,dx$ $\frac{du}{dx} = \frac{1}{x}$; $du = \frac{1}{x}dx$ $v = \int x^2\,dx = \frac{x^3}{3}$	Let $u = \ln x$ $v = x^2$ $\frac{du}{dx} = \frac{1}{x}$ $\int v\,dx = \int x^2\,dx = \frac{x^3}{3}$

Substituting in $\int u\,dv = uv - \int v\,du$ $\int x^2 \ln x\,dx = \ln x \bullet \frac{x^3}{3} - \int\left(\frac{1}{x}\bullet\frac{x^3}{3}\right)dx$ $\qquad = \frac{x^3}{3}\ln x - \frac{1}{3}\int x^2\,dx$ $\qquad = \frac{x^3}{3}\ln x - \frac{1}{3}\bullet\frac{x^3}{3} + C$ $\qquad = \frac{x^3}{3}\ln x - \frac{x^3}{9} + C.$	Now, substitute in $\int [uv]dx = u\int v\,dx - \int\left(\frac{du}{dx}\bullet\int v\,dx\right)dx$ The rest of the steps are the same as those in Version 1 on the left.

Case 3b: Product of a log function and a rational function

Example 1: Find $\int \frac{\ln x}{x}\, dx$ $(= \int x^{-1} \ln x\, dx)$ (From the order **LIATE**, the logarithmic function (L) comes before the algebraic function (A)

Here, $\frac{1}{x} = x^{-1}$ is easily integrable, but $\ln x$ is **not** easily integrable.
The process type here is **Type 3**

Starting with Version 1		Starting with Version 2	
Step 1:		**Step 1:**	
Let $u = \ln x$	$dv = x^{-1}dx$ or $\frac{1}{x}dx$	Let $u = \ln x$	$v = x^{-1}$
$\frac{du}{dx} = \frac{1}{x}$; $du = \frac{1}{x}dx$	$v = \int x^{-1}dx = \ln x$	$\frac{du}{dx} = \frac{1}{x}$	$\int v\,dx = \int x^{-1}dx = \ln x$

Substituting in $\int u\,dv = uv - \int v\,du$	Now, substitute in
$\int \frac{\ln x}{x}\, dx = \ln x \bullet \ln x - \int\left(\frac{1}{x} \bullet \ln x\right)dx$	$\int[uv]dx = u\int v\,dx - \int\left(\frac{du}{dx}\cdot \int v\,dx\right)dx$
$\int \frac{1}{x}\ln x\,dx = (\ln x)^2 - \int \frac{1}{x}\ln x\,dx$	The rest of the steps are the same as Version 1 on the left.
$2\int \frac{1}{x}\ln x\,dx = (\ln x)^2$ (add $\int\frac{1}{x}\ln x\,dx$ to sides)	---
	-Method 2: We can also use U-substitution. Let $u = \ln x$
$\int \frac{1}{x}\ln x\,dx = \frac{1}{2}(\ln x)^2 + C$ (divide sides by 2)	$dx = x\,du$; $\int u\,du \gg . \frac{u^2}{2} = \frac{(\ln x)^2}{2}$

Case 4: Product of an exponential and a trigonometric function

Example: Find $\int e^x \sin x\, dx$. (There is flexibility in the order **LIATE**)

Here, it does not matter how many times we apply the formula, we will always have a product of an exponential function and a trigonometric function , and the application of the formula would not terminate; however the original integral would be reproduced on the right-hand side, and when this happens, we stop the application of the formula and transpose accordingly this reproduction to the left-side of the equation and then solve for the original integral. Also it does not matter which part is the u-part, We can let the u-part $= e^x$ or $\sin x$.
The process type here is **Type 3**.

Starting with Version 1

Find $\int e^x \sin x\, dx$. Let $u = e^x$ and $dv = \sin x\, dx$; $du = e^x dx$; $v = -\cos x$
Substitute in $\int u\,dv = uv - \int v\,du$

Step 1: $\int e^x \sin x\, dx = e^x(-\cos x) - \int\left(e^x(-\cos x)\right)dx$

$= -e^x \cos x + \int e^x \cos x\, dx$

Step 2 : $= -e^x \cos x + e^x \sin x - \int e^x \sin x\, dx$ ($\int e^x \sin x\, dx$ reproduced)

$\int e^x \sin x\, dx + \int e^x \sin x\, dx = -e^x \cos x + e^x \sin x$ (Add $\int e^x \sin x\, dx$ to both sides)

$2\int e^x \sin x\, dx = -e^x \cos x + e^x \sin x$

$\int e^x \sin x\, dx = -\frac{1}{2}e^x \cos x + \frac{1}{2}e^x \sin x + C$

Case 4 Example   Find $\int e^x \sin x\, dx$.   **Starting with Version 2**

Let $u(x) = e^x$ and $v(x) = \sin x$ $\int v\,dx = -\cos x$; $du/dx = e^x$;

$\int[uv]dx = u\int v\,dx - \int\left(\frac{du}{dx}\cdot \int v\,dx\right)dx$ The rest of the steps are the same as above

Case 5: Product of a polynomial and an inverse trigonometric function

Example: Find $\int x \tan^{-1}(x)\, dx$.

From the order **LIATE**, the inverse trigonometric function (I) comes before the algebraic function (A).

Here, we let $u = \tan^{-1}(x)$

The process type here is **Type 1.** However, we also use trigonometric substitution to complete the integration process.

Starting with Version 1	**Starting with Version 2**
Step 1: Let $u = \tan^{-1}(x)$ and $dv = xdx$	**Step 1:** Let $u = \tan^{-1}(x)$ and $v = x$

Starting with Version 1

Step 1: Let $u = \tan^{-1}(x)$ and
$$dv = xdx$$
$$\frac{du}{dx} = \frac{1}{1+x^2} \text{ and}$$
$$du = \frac{1}{1+x^2}dx$$
$$v = \int dv = \int xdx = \frac{x^2}{2}$$

Step 2: Substitute the above in
$\int u\,dv = uv - \int v\,du$ to obtain
$\int x \tan^{-1}(x)\, dx$

$$= \tan^{-1}(x) \bullet \frac{x^2}{2} - \int \frac{x^2}{2} \bullet \frac{1}{1+x^2}dx$$

$$= \frac{x^2}{2}\tan^{-1}(x) - \frac{1}{2}\int \frac{x^2}{1+x^2}dx$$

$$= \frac{x^2}{2}\tan^{-1}(x) - \frac{1}{2}\left[\int 1dx - \int \frac{1}{1+x^2}dx\right]$$

$$= \frac{x^2}{2}\tan^{-1}(x) - \frac{1}{2}\left[x - \tan^{-1}(x)\right] + C$$

$$= \frac{x^2}{2}\tan^{-1}(x) - \frac{1}{2}x + \frac{1}{2}\tan^{-1}(x) + C$$

Starting with Version 2

Step 1: Let $u = \tan^{-1}(x)$ and
$$v = x$$
$$\frac{du}{dx} = \frac{1}{1+x^2}$$
$$\int vdx = \int xdx = \frac{x^2}{2}$$

Step 2: Substitute the above in

$$\int [uv]dx = u\int vdx - \int \left(\frac{du}{dx} \bullet \int vdx\right)dx$$

The rest of the steps are the same as those in Version 1 on the left.

Explanations for the left side

$\longleftarrow$ $\int \frac{1}{1+x^2}dx = \tan^{-1}(x)$, since

$$\frac{d}{dx}[\tan^{-1}(x)] = \frac{1}{1+x^2}$$

Case 6: Product of a polynomial function and a radical function

Example 1: Find $\int x\sqrt{x+3}\,dx$ (We use two methods)

Method 1 (by parts: using **Version 1**)

Step 1: We write the radical part in exponential form and then apply integration by parts

For $\int x(x+3)^{\frac{1}{2}}\,dx$ let $u = x$ and let
$$dv = (x+3)^{\frac{1}{2}}\,dx$$

Step 2: Find v

Let $w = x + 3$

Then $\dfrac{dw}{dx} = 1$ or $dx = dw$

$$v = \int (x+3)^{\frac{1}{2}}\,dx$$
$$= \int w^{\frac{1}{2}}\,dw$$
$$= \frac{2}{3}w^{\frac{3}{2}} = \frac{2}{3}(x+3)^{\frac{3}{2}}$$
$$v = \frac{2}{3}(x+3)^{\frac{3}{2}}$$

Step 3: Find:. $\dfrac{du}{dx} = 1$ or $du = dx$
    (from $u = x$)

Step 4: Substitute the above accordingly in a
$$\int u\,dv = uv - \int v\,du.$$

Then $\int x(x+3)^{\frac{1}{2}}\,dx$

$$= x \bullet \frac{2}{3}(x+3)^{\frac{3}{2}} - \int (1)\frac{2}{3}(x+3)^{\frac{3}{2}}\,dx$$
$$= \frac{2}{3}x(x+3)^{\frac{3}{2}} - \frac{2}{3}\int (x+3)^{\frac{3}{2}}\,dx$$
$$= \frac{2}{3}x(x+3)^{\frac{3}{2}} - \frac{2}{3} \bullet \frac{2}{5}(x+3)^{\frac{5}{2}} + C$$
$$= \frac{2}{3}x(x+3)^{\frac{3}{2}} - \frac{4}{15}(x+3)^{\frac{5}{2}} + C$$
$$= \frac{2}{3}x\sqrt{(x+3)^3} - \frac{4}{15}\sqrt{(x+3)^5} + C$$
$$= \left(\frac{2}{3}x(x+3) - \frac{4}{15}(x+3)^2\right)\sqrt{x+3} + c$$
$$= \frac{2}{5}(x^2 + x - 6)\sqrt{x+3} + c.$$

Note: By differentiating the results of each integration method, we would obtain the original integrand.

Method 2 (by substitution)

Step 1: Let $u = \sqrt{x+3}$

Then $u^2 = x + 3$

$x = u^2 - 3$

$\dfrac{2u\,du}{dx} = 1$ and $dx = 2u\,du$

Step 2: Substitute the above in
$$\int x\sqrt{x+3}\,dx \text{ to obtain}$$
$$= \int (u^2 - 3)u \bullet 2u\,du$$
$$= 2\int (u^4 - 3u^2)\,du$$
$$= 2[\frac{u^5}{5} - \frac{3u^3}{3}] + C$$
$$= 2\left[\frac{u^5}{5} - u^3\right] + C$$
$$= 2\left[\frac{\sqrt{(x+3)^5}}{5} - \sqrt{(x+3)^3}\right] + C$$
$$= \frac{2}{5}\sqrt{(x+3)^5} - 2\sqrt{(x+3)^3} + C$$
$$= \frac{2}{5}(x+3)^{\frac{5}{2}} - 2(x+3)^{\frac{3}{2}} + C$$
$$= \frac{2}{5}(x+3)^2\sqrt{x+3} - 2(x+3)\sqrt{x+3} + c$$
$$= \frac{2}{5}(x^2 + x - 6)\sqrt{x+3} + c$$

Approach 3 (by substitution)

Let $u = x + 3$

Then $du/dx = 1$ and $du = dx$

$\int x\sqrt{x+3}\,dx = \int x u^{\frac{1}{2}}\,du$. To express the integral completely in terms u only we apply $x = u - 3$

Then $\int x u^{\frac{1}{2}}\,du$ becomes
$$\int (u-3)u^{\frac{1}{2}}\,du = \int u^{\frac{3}{2}}\,du - 3\int u^{\frac{1}{2}}\,du$$
$$= \frac{2}{5}u^{\frac{5}{2}} - 3 \bullet \frac{2}{3}u^{\frac{3}{2}} + C$$
$$= \frac{2}{5}(x+3)^{\frac{5}{2}} - 3 \bullet \frac{2}{3}(x+3)^{\frac{3}{2}} + C$$
$$= \frac{2}{5}(x+3)^{\frac{5}{2}} - 2(x+3)^{\frac{3}{2}} + C$$

Observe above that the substitution method is less involved than integration by parts

Also, by converting each result to simplified radical form, the results would be identical.

Example 2: Find $\int \dfrac{3x}{\sqrt{1-x}}\,dx$ **Solution:** We use two methods

Method 1 (by parts using Version 2)
Type 1 process
Step 1: We write this as product and then apply integration by parts

$$\int \frac{3x}{\sqrt{1-x}}\,dx = 3x(1-x)^{-\frac{1}{2}}$$

Let $u = 3x$ and let $v = (1-x)^{-\frac{1}{2}}$

Step 2: $\int v(x)dx = \int(1-x)^{-\frac{1}{2}}dx$.

     Let $w = (1-x)$.

     Then $\dfrac{dw}{dx} = -1$ and $dx = -dw$

$\int v(x)dx = -\int w^{-\frac{1}{2}}dw = -2(1-x)^{\frac{1}{2}}$
Substitute the above accordingly in

$$\int[uv]dx = u\int vdx - \int\left(\frac{du}{dx}\cdot\int vdx\right)dx$$

Step 3: $\int \dfrac{3x}{\sqrt{1-x}}\,dx$

$= 3x(-2)(1-x)^{\frac{1}{2}} - \int[3(-2)(1-x)^{\frac{1}{2}}]dx$

$= -6x\sqrt{1-x} + 6\int(1-x)^{\frac{1}{2}}dx$

$= -6x\sqrt{1-x} - \dfrac{2}{3}\cdot 6(1-x)^{\frac{3}{2}} + C$

$= -6x\sqrt{1-x} - 4\sqrt{(1-x)^3} + C$

$= -4\sqrt{(1-x)^3} - 6x\sqrt{1-x} + C$

Step 4 Simplifying:

$= -4\sqrt{(1-x)^2}\sqrt{1-x} - 6x\sqrt{1-x} + C$

$= -4(1-x)\sqrt{1-x} - 6x\sqrt{1-x} + C$

$= \sqrt{1-x}(-4(1-x) - 6x) + C$

$= \sqrt{1-x}(-4-2x) + C$

$= (-4-2x)\sqrt{1-x} + C$ --

$= -2(x+2)\sqrt{1-x} + C$.

Method 2 (by substitution)
Step 1: Let $u = \sqrt{1-x}$

     Then $u^2 = 1-x$

       $x = 1 - u^2$

   $\dfrac{dx}{du} = -2u$ and $dx = -2udu$

Step 2:

$\int \dfrac{3x}{\sqrt{1-x}}\,dx = \int \dfrac{3(1-u^2)(-2udu)}{u}$

$= \int 3(1-u^2)(-2du)$

$= \int 6u^2 du - \int 6du$

$= 6\dfrac{u^3}{3} - 6u + C$

$= 2u^3 - 6u + C$

$= 2\sqrt{(1-x)^3} - 6\sqrt{1-x} + C$

Step 3 simplifying

$= 2\sqrt{(1-x)^2}\sqrt{(1-x)} - 6\sqrt{1-x} + C$

$= 2(1-x)\sqrt{(1-x)} - 6\sqrt{1-x} + C$

$= \sqrt{(1-x)}[2(1-x) - 6] + C$

$= \sqrt{(1-x)}(-2x-4) + C$

$= (-2x-4)\sqrt{(1-x)} + C$

$= -2(x+2)\sqrt{(1-x)} + C$

Approach 3: we can also use approach 3 of the previous example. Try it.

Observe above that the substitution method is less involved than integration by-parts.

Comment: From the results of the above two methods, we can say that when a method for finding an integral is not specified and we use any method, we may obtain an answer which is equivalent to the answer in the textbook or on a class exam; but which seems to differ from the answer in the textbook. There are two approaches to check your result: One approach is to differentiate each answer to see if you obtain the original integrand. The other approach is to simplify each answer as much as possible to see if they are identical. Instructors should be careful when grading such problems, since a student might have obtained a correct answer which seems to differ from instructor's answer.

Case 7: Composite Function (of trig function with log function)

Example: Find $\int \sin(\ln x)dx$ (Simple u-substitution plus integration-by-parts)
The process type here is **Type 3** (from Step 4 below)

Step 1: Write the integrand as a product in terms of u.

$$\text{Let } u = \ln x \qquad\qquad (1)$$
$$\frac{du}{dx} = \frac{1}{x} \text{ and } \boxed{dx = x\,du}; \; du = \frac{dx}{x}$$

Step 2: Substitute $u = \ln x$, $dx = x\,du$ in $\int \sin(\ln x)dx$ to obtain

$$\int \sin u \bullet x\,du \qquad\qquad (2)$$

We need to write x in terms of u.

From (1) $u = \log_e x$ and (**Note** this technique for the future)

$$e^u = x \quad \text{(equivalently)} \qquad \textbf{Note:} \;\; \ln x = \log_e x$$

Step 3: Now replace x by e^u in

$\int \sin(\ln x)dx$ to obtain

$\int \sin u \bullet e^u du$ or $\int e^u \sin u\,du$

Step 4: **Integrate by parts**. (see **Case 4** example) (u-part $= e^u$; $dv = \sin x$)

$$\int e^u \sin u\,du = -e^u \cos u - \int e^u(-\cos u)du$$
$$= -e^u \cos u + \int e^u(\cos u)du$$
$$= -e^u \cos u + e^u \sin u - \int e^u \sin u\,du$$
$$\int e^u \sin u\,du + \int e^u \sin u\,du = -e^u \cos u + e^u \sin u$$
$$2\int e^u \sin u\,du = -e^u \cos u + e^u \sin u$$
$$\int e^u \sin u\,du = -\frac{1}{2}e^u \cos u + \frac{1}{2}e^u \sin u + C$$
$$\int x \sin(\ln x)\frac{dx}{x} = -\frac{1}{2}x\cos(\ln x) + \frac{1}{2}x\sin(\ln x) + C$$
$$\int \sin(\ln x)dx = -\frac{1}{2}x\cos(\ln x) + \frac{1}{2}x\sin(\ln x) + C$$

Summary of Guidelines for applying the integration-by-parts formula
Choosing the u-part

Choose the u-part according to the **L I A T E** order (p.245) covered previously; but note that there is flexibility in the **L I A T E** order.

1a. If one of the functions is $\ln x$, let, $u = \ln x$ and let the other function be the dv–part. (v-part for version 2), because $\int \ln x\,dx = x \ln x - x + c$ is found, coincidentally, using the integration-by-parts formula.. Note also that because $\frac{d}{dx}(\ln x) = \frac{1}{x}$, a rational function, if the other function is a polynomial (e.g., x^2) the x in $\frac{1}{x}$ would be a factor of the polynomial, and the process is of **Type 1**, and terminates quickly.. See Case 3, Example 2.

1b. If one of the functions is an inverse trigonometric function such as $\tan^{-1} x$, let, $u =$ the inverse trigonometric function (e.g., $u = \tan^{-1} x$) and let the other function be the dv–part. (v-part for version 2), because for example,

$\int \tan^{-1} x\, dx = x\tan^{-1} - \frac{1}{2}\ln(x^2 + 1) + c$ is found , coincidentally,, using the integration-by-parts formula.. Note also that because the derivative of an inverse trigonometric function is an algebraic function, rational function,

if the other function is a polynomial (e.g., x^2) the second term of the formula would be a polynomial or a rational function, which would be readily integrable by previous methods.. See Case 3, Example 2.

2. If one of the functions is a polynomial and the other function is a trigonometric function (e.g., $\sin x$) or an exponential function (e.g., e^x), let u = the polynomial part and let the other function be the dv–part. (or v-part for Version 2). This choice is appropriate since after one or more differentiations, the polynomial part would be reduced to a constant, and only trigonometric or exponential function remains, and which can be integrated readily. See Cases 1 & 2.

3. For the product of an exponential and a trigonometric function (e.g., $e^x \sin x$), it does not matter which function is the u–part . Let $u = e^x$, or let $u = \sin x$ .The process type here is **Type 3**. See Case 4.

Lesson 36 Exercises A

1. Explain in words to a class mate the two equivalent versions of the integration-by -parts formula. What are the relative merits of each version?

a. $\int u\, dv = uv - \int v\, du$;   **b.** $\int [uv] dx = u\int v\, dx - \int \left(\frac{du}{dx} \cdot \int v\, dx\right) dx$

2. Name the types of functions which lend themselves to integration by parts.

3. Name three main process types you may encounter when you attempt to integrate by parts.

4. Find $\int x \ln x\, dx$;   **5.** Find $\int \frac{3x}{\sqrt{1-x}}\, dx$ (try two methods)

6. Find $\int x^2 \sin x\, dx$;   **7.** Find $\int x^2 3^x\, dx$

8, $\int x^2 \ln x\, dx$;   **9..** Find $\int e^x \sin x\, dx$

10. Find $\int x\tan^{-1}(x)\, dx$

11. Find $\int x\sqrt{x+3}\, dx$ (Try by parts and by substitution. Which method is faster?)

12. Find $\int \sin(\ln x) dx$

Answers: **4.** $\frac{x^2}{2}\ln x - \frac{x^2}{4} + C$;   **5.** $-2(x+2)\sqrt{1-x} + C$;

6. $-x^2 \cos x + 2\cos x + 2x\sin x + C$; **7.** $3^x\left(\frac{x^2(\ln 3)^2 - 2x\ln 3 + 2}{(\ln 3)^3}\right) + C$;

8. $\frac{x^3}{3}\ln x - \frac{x^3}{9} + C$; **9.** $-\frac{1}{2}e^x \cos x + \frac{1}{2}e^x \sin x + C$;

10. $\frac{x^2}{2}\tan^{-1}(x) - \frac{1}{2}x + \frac{1}{2}\tan^{-1}(x) + C$; **11.** $= \frac{2}{5}(x^2 + x - 6)\sqrt{x+3} + c$;

12. $-\frac{1}{2}x\cos(\ln x) + \frac{1}{2}x\sin(\ln x) + C$.

Lesson 36 Exercises B

1. Explain in words to a class mate the two equivalent versions of the integration-by -parts formula. What are the relative merits of each version?

a. $\int u\,dv = uv - \int v\,du$;   **b.** $\int [uv]dx = u\int v\,dx - \int \left(\frac{du}{dx} \cdot \int v\,dx\right)dx$

2. Name the types of functions which lend themselves to integration by parts.

3. Name three main process types you may encounter when you attempt to integrate by parts.

4. Find $\int x^2 \ln x\,dx$;   **5.** Find $\int \frac{3x}{\sqrt{2-x}}\,dx$ (try two methods)

6. Find $\int x\sin x\,dx$;   **7.** Find $\int x^2 4^x\,dx$; 8, $\int x^3 \ln x\,dx$;

9. Find $\int e^x \cos x\,dx$;   **10.** Find $\int x^2 \tan^{-1}(x)\,dx$

11. Find $\int x\sqrt{x-3}\,dx$ (Try by parts and by substitution. Which method is faster?)

12. Find $\int \cos(\ln x)dx$

Answers: 4. $\frac{x^3}{3}\ln x - \frac{x^3}{9} + C$;   **5.** $-2(x+4)\sqrt{2-x} + C$;

6. $\sin x - x\cos x$;   **7.** $\dfrac{4^x(2x^2(\ln 2)^2 - 2x\ln 2 + 1)}{4(\ln 2)^3}$;

8. $\frac{x^4}{4}\ln x - \frac{x^4}{16} + C$;   **9.** $\frac{1}{2}e^x \cos x + \frac{1}{2}e^x \sin x + C$;

10. $\frac{x^3}{3}\tan^{-1}(x) - \frac{1}{6}x^2 + \frac{1}{6}\ln(x^2 + 1)$;   **11.** $= \frac{2}{5}(x^2 - x - 6)\sqrt{x+3} + c$;

12. $\frac{1}{2}x\cos(\ln x) + \frac{1}{2}x\sin(\ln x) + C$.

EXTRA: See Appendix A, p. 436 for series representations of functions

Case 8: Product of a logarithmic function and an exponential function

Consider $\int e^x \ln x\,dx$. If $u = \ln x$, $\frac{d}{dx}(\ln x) = \frac{1}{x}$:

$\int e^x \ln x\,dx = e^x \ln x - \int e^x \frac{1}{x}\,dx$ (Integrating by parts)

$\int e^x \frac{1}{x}\,dx$ is not integrable by parts applying previous methods. However, we can obtain an approximation for this integral by applying the power series expansion of e^x given by $e^x = 1 + x + \frac{x^2}{2!} + \frac{x^3}{3!} + ...$

Then $\int e^x \ln x\,dx = e^x \ln x - \int \frac{1}{x}(1 + x + \frac{x^2}{2!} + \frac{x^3}{3!} + ...)\,dx$

$= e^x \ln x - \int(\frac{1}{x} + 1 + \frac{x}{2} + \frac{x^2}{6} + ...)\,dx$

$\int e^x \ln x\,dx = e^x \ln x - (\ln x + x + \frac{x^2}{4} + \frac{x^3}{18} + ...) + C$

Better approximation can be obtained by adding more terms to the series part.

Case 9: Product of a logarithmic function and a trigonometric function

Consider $\int \cos x \ln x\,dx = \sin x \ln x - \int \frac{1}{x}\sin x\,dx$ .An approximation to

$\int \frac{1}{x}\sin x\,dx$ can be found by applying $\sin x = x - \frac{x^3}{3!} + \frac{x^5}{5!} - \frac{x^7}{7!} + \frac{x^9}{9!} + ...$

Lesson 37
Integration of Powers and More Products of Trigonometric Functions

Some of the functions covered in this lesson can be handled using general simple u-substitution, while for others, we need the application of trigonometric identities and occasionally, we will apply integration by parts. We will cover a number of cases.

Tool box: **1.** $\sin^2 x + \cos^2 x = 1$; **2.** $1 + \tan^2 x = \sec^2$; **3.** $1 + \cot^2 x = \csc^2 x$

Case 1 Product of **any power of sine** and **first power of cosine** or the product of **any power cosine** and **first power of sine**.
This case can be handled using simple *u*-substitution.

Examples (a) Find $\int \sin x \cos x \, dx$ (Here, $u = \sin x$, or $u = \cos x$

Answer: $\frac{1}{2} \sin^2 x + C$ or $-\frac{1}{2} \cos^2 x + C$

(b) $\int \sin^3 x \cos x \, dx$ (Here, $u = \sin x$); Answer: $\frac{1}{4} \sin^4 x + C$

(c) $\int \sin x \cos^3 x \, dx$ (Here, $u = \cos x$); Answer: $-\frac{1}{4} \cos^4 x + C$

(d) $\int \sin^5 x \cos x \, dx$ (Here, $u = \sin x$); Answer: $\frac{1}{6} \sin^6 x + C$

(e) $\int \sin x \cos^6 x \, dx$ (Here, $u = \cos x$). Answer: $-\frac{1}{7} \cos^7 x + C$

Case 2 Second **power of sine** and **second power of cosine**

That is, **(a)** Find $\int \sin^2 x \, dx$; **(b)** Find $\int \cos^2 x \, dx$
This case can be handled using the trigonometric identities in the tool box as well as using integration by parts.

Tool box

1. $\cos 2x = \cos^2 x - \sin^2 x$;

2. $\cos 2x = 2\cos^2 x - 1$; and from this $\boxed{\cos^2 x = \frac{1}{2}\cos 2x + \frac{1}{2}}$

3. $\cos 2x = 1 - 2\sin^2 x$; and from this $\boxed{\sin^2 x = \frac{1}{2} - \frac{1}{2}\cos 2x}$

The problem here may be remembering the **boxed** identities.
Let us find a way to recall these on exams:

First memorize identity **1** ($\cos 2x = \cos^2 x - \sin^2 x$) since this form is easy to remember; and then using this identity in conjunction with the basic identity

$\sin^2 x + \cos^2 x = 1$, obtain $\boxed{\cos^2 x = \frac{1}{2}\cos 2x + \frac{1}{2}}$ and $\boxed{\sin^2 x = \frac{1}{2} - \frac{1}{2}\cos 2x}$

Solution
Method 1 Using the trigonometric identities in the tool box.

(a) Find $\int \sin^2 x \, dx$

Applying $\boxed{\sin^2 x = \frac{1}{2} - \frac{1}{2}\cos 2x}$

$$\int \sin^2 x \, dx = \int (\frac{1}{2} - \frac{1}{2}\cos 2x) \, dx$$
$$= \frac{1}{2}x - \frac{1}{2}(\frac{1}{2})\sin 2x + C$$
$$= \frac{1}{2}x - \frac{1}{4}\sin 2x + C$$
$$= -\frac{1}{2}\sin x \cos x + \frac{1}{2}x + C$$

(b) Find $\int \cos^2 x \, dx$

Applying $\boxed{\cos^2 x = \frac{1}{2}\cos 2x + \frac{1}{2}}$

$$\int \cos^2 x \, dx = \int (\frac{1}{2}\cos 2x + \frac{1}{2}) \, dx$$
$$= \frac{1}{2}(\frac{1}{2})\sin 2x) + \frac{1}{2}x + C$$
$$= \frac{1}{4}\sin 2x + \frac{1}{2}x + C$$
$$= \frac{1}{2}\sin x \cos x + \frac{1}{2}x + C$$

Note above: Use simple u-substitution to find $\int \cos 2x \, dx$, with $u = 2x$.

Note also: $\sin 2x = 2\sin x \cos x$

Method 2 Finding $\int \sin^2 x \, dx$ using integration by parts, type 3.

$$\int \sin^2 x \, dx = \int \sin x \sin x \, dx \qquad (u = \sin x, \, dv = \sin x \, (\text{or } v = \sin x \text{ for version 2})$$
$$= \sin x(-\cos x) - \int \cos x(-\cos x) \, dx$$
$$= -\sin x \cos x + \int \cos^2 x \, dx$$
$$= -\sin x \cos x + \int (1 - \sin^2 x) \, dx \qquad (\text{From } \sin^2 x + \cos^2 x = 1)$$
$$= -\sin x \cos x + \int 1 \, dx - \int \sin^2 x \, dx$$

$2\int \sin^2 x \, dx = -\sin x \cos x + \int dx \qquad (\text{adding } \int \sin^2 x \, dx \text{ to both sides})$

$2\int \sin^2 x \, dx = -\sin x \cos x + x + C$

$\int \sin^2 x \, dx = -\frac{1}{2}\sin x \cos x + \frac{1}{2}x + C$

Method 2 Finding $\int \cos^2 x \, dx$ using integration by parts, type 3.

$$\int \cos^2 x \, dx = \int \cos x \cos x \, dx$$
$(u = \cos x, \, dv = \cos x \, (\text{or } v = \cos x$ for version 2)

$$= \cos x \sin x - \int (-\sin x)(\sin x) \, dx$$
$$= \cos x \sin x + \int \sin^2 x \, dx$$
$$= \cos x \sin x + \int (1 - \cos^2 x) \, dx$$
$$\qquad (\text{From } \sin^2 x + \cos^2 x = 1)$$
$$= \cos x \sin x + \int 1 \, dx - \int \cos^2 x \, dx$$

$2\int \cos^2 x \, dx = \cos x \sin x + \int 1 \, dx$

$(\text{adding } \int \cos^2 x \, dx \text{ to both sides})$

$2\int \cos^2 x \, dx = \cos x \sin x + x + C$

$\int \cos^2 x \, dx = \frac{1}{2}\sin x \cos x + \frac{1}{2}x + C$

Method 3 Find $\int \cos^2 x \, dx$
From above, previously

$\int \sin^2 x \, dx = -\frac{1}{2}\sin x \cos x + \frac{1}{2}x + C$ (A)

Also, $\cos^2 x = 1 - \sin^2 x$ (Pythagorean)

$\int \cos^2 x \, dx = \int 1 \, dx - \int \sin^2 x \, dx$
$$= x - (-\frac{1}{2}\sin x \cos x + \frac{1}{2}x) + C$$
$$= x + \frac{1}{2}\sin x \cos x - \frac{1}{2}x + C$$
$$= \frac{1}{2}x + \frac{1}{2}\sin x \cos x + C$$

Relative Merits of Methods 1, 2 & 3
In Method 1, you need to recall two identities:

$\sin^2 x = \frac{1}{2} - \frac{1}{2}\cos 2x$ (A); and

$\cos^2 x = \frac{1}{2}\cos 2x + \frac{1}{2}$ (B)

In Method 3, we need to recall (A) and (C); in Method 2, you need to recall a single identity, namely, identity

$\qquad \sin^2 x + \cos^2 x = 1$ (C)

(easy to recall, but more steps)

Case 3: Other powers of sine and cosine

Example (a) Find $\int \sin^4 x\, dx$; (b) $\int \cos^4 x\, dx$; (c) $\int \sin^3 x\, dx$; (d) $\int \cos^5 x\, dx$

Here, we write each power greater than 2 in terms of powers of the second power, since the highest power in the trigonometric identities is 2.

Examples: (a) $\int \sin^4 x\, dx = \int (\sin^2 x)^2\, dx$; (b) $\int \cos^4 x\, dx = \int (\cos^2 x)^2\, dx$

(c) $\int \sin^3 x\, dx = \int \sin^2 x \bullet \sin x\, dx$ (d) $\int \sin^5 x = \int (\sin^2 x)^2 \bullet \sin x\, dx$

(a) Finding $\int \sin^4 x\, dx$

(Apply $\sin^2 x = \frac{1}{2} - \frac{1}{2}\cos 2x$

followed by

$\cos^2 2x = \frac{1}{2}\cos 4x + \frac{1}{2}$)

$\int \sin^4 x\, dx$

$= \int (\sin^2 x)^2\, dx$

$= \int \left(\frac{1}{2} - \frac{1}{2}\cos 2x\right)\left(\frac{1}{2} - \frac{1}{2}\cos 2x\right) dx$

$= \int \left(\frac{1}{4} - (2)\frac{1}{4}\bullet \cos 2x + \frac{1}{4}\cos^2 2x\right) dx$

$= \int \left(\frac{1}{4} - \frac{1}{2}\cos 2x + \frac{1}{4}\cos^2 2x\right) dx$

$= \int \left(\frac{1}{4} - \frac{1}{2}\cos 2x + \frac{1}{4}\{\frac{1}{2}\cos 4x + \frac{1}{2}\}\right) dx$

$= \int \left(\frac{1}{4} - \frac{1}{2}\cos 2x + \frac{1}{8}\cos 4x + \frac{1}{8}\right) dx$

$= \int \left(\frac{2}{8} - \frac{1}{2}\cos 2x + \frac{1}{8}\cos 4x + \frac{1}{8}\right) dx$

$= \int \left(\frac{3}{8} - \frac{1}{2}\cos 2x + \frac{1}{8}\cos 4x\right) dx$

We integrate now:

$= \frac{3}{8}x - \frac{1}{2}\bullet\frac{1}{2}\sin 2x + \frac{1}{8}\bullet\frac{1}{4}\sin 4x + C$

$= \frac{3}{8}x - \frac{1}{4}\sin 2x + \frac{1}{32}\sin 4x + C$

(b) Finding $\int \cos^4 x\, dx$

(Imitate (a) above and apply

$\cos^2 x = \frac{1}{2}\cos 2x + \frac{1}{2}$.

followed by

$\cos 2x = \frac{1}{2}\cos 4x + \frac{1}{2}$)

$\int \cos^4 x\, dx$

$= \int (\cos^2 x)^2\, dx$

$= \int \left(\frac{1}{2}\cos 2x + \frac{1}{2}\right)\left(\frac{1}{2}\cos 2x + \frac{1}{2}\right) dx$

(c) Finding $\int \sin^3 x\, dx$

$= \int \sin^2 x \bullet \sin x\, dx$

Method 1: (Apply $\sin^2 x = 1 - \cos^2 x$)

$= \int (1 - \cos^2 x)\bullet \sin x\, dx$

$= \int \sin x\, dx - \int \cos^2 x \sin x\, dx$

$= -\cos x - (-\int u^2\, du)$ (Let $u = \cos x$.
Complete the integr. using u-subst.)

$= -\cos x + \frac{u^3}{3} + C$

$= -\cos x + \frac{\cos^3 x}{3} + C$

Method 2: If we begin by applying $\sin^2 x = \frac{1}{2} - \frac{1}{2}\cos 2x$, we will subsequently have to replace $\cos 2x$ by $2\cos^2 x - 1$, before using simple u-substitution. Therefore, Method 1 is the shorter approach.

(d) $\int \sin^5 x\, dx$

$= \int (\sin^2 x)^2 \bullet \sin x\, dx$

Apply $\sin^2 x = 1 - \cos^2 x$

$= \int (1 - \cos^2 x)(1 - \cos^2 x)\sin x\, dx$

$= \int (1 - 2\cos^2 x + \cos^4 x)\sin x\, dx$

$= -\int (1 - 2u^2 + u^4)\, du$ $(u = \cos x)$

$= -u + \frac{2u^3}{3} - \frac{u^5}{5} + C$

$= -\cos x + \frac{2\cos^3}{3} - \frac{\cos^5 x}{5} + C.$

Extra. Finding $\int \cos^5 x\, dx$: imitate (d) above.

Note: From $\sin^2 x = \frac{1}{2} - \frac{1}{2}\cos 2x$ to $\sin^2 2x = \frac{1}{2} - \frac{1}{2}\cos 4x$: multiply the angle on the left by 2 and multiply the angle on the right by 2.

Note also: For $\int \sin^3 x\, dx$ and $\int \cos^3 x\, dx$, if we wish to memorize more identities, we can also apply the following:

1. $\sin^3 x = \frac{1}{4}(3\sin x - \sin 3x)$ (from $\sin 3x = 3\sin x - 4\sin^3 x$)

2. $\cos^3 x = \frac{1}{4}(3\cos x + \cos 3x)$ (from $\cos 3x = 4\cos^3 x - 3\cos x$)

For more curiosity, we can also use **integration by parts** to find $\int \sin^3 x\, dx$ just as we did for $\int \sin^2 x\, dx$ and $\int \cos^2 x\, dx$ in Case 2, above.

Finding $\int \sin^3 x\, dx$ using integration-by-parts

$\int \sin^3 x\, dx$

$= \int \sin^2 x \sin x\, dx$

$= \sin^2 x(-\cos x) - \int(2\sin x\cos x)(-\cos x)dx$

$= -\sin^2 x\cos x + 2\int\cos^2 x\sin x\, dx$

$= -\sin^2 x\cos x - \dfrac{2\cos^3 x}{3} + C$ which is

equivalent to $= -\cos x + \dfrac{\cos^3 x}{3} + C$ as shown below:

Scrapwork

1. $\frac{d}{dx}(\sin^2 x) = 2\sin x\cos x$

2, integrate

$\boxed{\int\cos^2 x\sin x\, dx}$ by u-subst.

Let $t = \cos x$, Then

$\boxed{t^2 = \cos^2 x}$; $\dfrac{dt}{dx} = -\sin x$

$dx = -\dfrac{dt}{\sin x}$ or $dt = -\sin x\, dx$

$-\int t^2 dt = -\dfrac{t^3}{3} = -\dfrac{\cos^3 x}{3} + C$

Showing equivalence

Step 1 $-\sin^2 x\cos x - \dfrac{2\cos^3 x}{3} + C$

$= -\cos x(1 - \cos^2 x) - \dfrac{2\cos^3 x}{3} + C$

$= -\cos x + \cos^3 x - \dfrac{2\cos^3 x}{3} + C$

Step 2

$= -\cos x + \dfrac{3\cos^3 x}{3} - \dfrac{2\cos^3 x}{3} + C$

$= -\cos x + \dfrac{\cos^3 x}{3} + C$

Observe that Method 1 of Case 3 is shorter than by-parts method.

Extra

By the way, we can also find $\int\sin^4 x\, dx$ using integration by parts.

$\int\sin^4 x\, dx = \int\sin^3 x\sin x\, dx$ $(u = \sin^3 x;\ dv = \sin x\, dx)$

$= \sin^3 x(-\cos x) - \int 3\sin^2 x\cos x(-\cos x)dx$

$= -\sin^3 x\cos x + 3\int\sin^2 x\cos^2 x\, dx$

$= -\sin^3 x\cos x + 3\int\sin^2 x(1 - \sin^2 x)dx$

$\int\sin^4 x\, dx = -\sin^3 x\cos x + 3\int\sin^2 x\, dx - 3\int\sin^4 x\, dx$

$4\int\sin^4 x\, dx = -\sin^3 x\cos x + 3\int\sin^2 x\, dx$

(adding $3\int\sin^4 x\, dx$ to both sides)

$4\int\sin^4 x\, dx = -\sin^3 x\cos x + 3\left[-\frac{1}{2}\sin x\cos x + \frac{1}{2}x\right] + C$

$4\int\sin^4 x\, dx = -\sin^3 x\cos x - \frac{3}{2}\sin x\cos x + \frac{3}{2}x + C$

$\int\sin^4 x\, dx = -\frac{1}{4}\sin^3 x\cos x - \frac{3}{8}\sin x\cos x + \frac{3}{8}x + C$

Scrapwork

$\frac{d}{dx}(\sin^3 x)$

$= 3\sin^2 x\cos x$

using the chain rule.

<-Equiv., to previous result

Case 4 Product of second power of sine and second power of cosine

That is, find $\int \sin^2 x \cos^2 x \, dx$

Here, apply $\sin^2 x = \frac{1}{2} - \frac{1}{2}\cos 2x$ and $\cos^2 x = \frac{1}{2}\cos 2x + \frac{1}{2}$

$\int \sin^2 x \cos^2 x \, dx$

$= \int \left(\frac{1}{2} - \frac{1}{2}\cos 2x\right)\left(\frac{1}{2}\cos 2x + \frac{1}{2}\right) dx$

$= \int \left(\frac{1}{4}\cos 2x + \frac{1}{4} - \frac{1}{4}\cos^2 2x - \frac{1}{4}\cos 2x\right) dx$

$= \int \left(\frac{1}{4} - \frac{1}{4}\cos^2 2x\right) dx$ <-- difference between two squares

$= \int \left(\frac{1}{4} - \frac{1}{4}[\frac{1}{2}\cos 4x + \frac{1}{2}]\right) dx$ $(\cos^2 2x = \frac{1}{2}\cos 4x + \frac{1}{2}; \cos^2 x = \frac{1}{2}\cos 2x + \frac{1}{2})$

$= \int \left(\frac{1}{4} - \frac{1}{8}\cos 4x - \frac{1}{8}\right) dx$

$= \int \left(\frac{1}{8} - \frac{1}{8}\cos 4x\right) dx$

$= \frac{1}{8}x - \frac{1}{8} \cdot \frac{1}{4}\sin 4x + C$ (integrating)

$= \frac{1}{8}x - \frac{1}{32}\sin 4x + C.$

Case 5 Products of **powers of sine** and **of cosine greater than 2**

Examples: (a). Find $\int \sin^3 x \cos^2 x \, dx$; **(b)** Find $\int \sin^2 x \cos^3 x \, dx$

The first step here is to reduce the odd power to an even power and then apply the identity $\sin^2 x + \cos^2 x = 1$.

(a) $\int \sin^3 x \cos^2 x \, dx$

$= \int \sin^2 x \sin x \cos^2 x \, dx$

$= \int \cos^2 x \sin x \sin^2 x \, dx$ (Rearranging so that a product is amenable to u-substitution)

(observe that if we had only $\int \cos^2 x \sin x \, dx$, we could readily use simple u-substitution,

with $u = \cos x$. but in this case, we have to express the factor $\sin^2 x$ in terms of $\cos x$,

using $\sin^2 x = 1 - \cos^2 x$ (From $\sin^2 x + \cos^2 x = 1$)

$= \int \cos^2 x \sin x \sin^2 x \, dx$

$= \int \cos^2 x \sin^2 x \sin x \, dx$

$= \int (\cos^2 x)(1 - \cos^2 x)\sin x \, dx$

$= \int u^2 (1 - u^2)(-du)$ $(u = \cos x, \frac{du}{dx} = -\sin x, -\frac{du}{\sin x} = dx$ or $du = -\sin x \, dx)$

$= -\int (u^2 - u^4) \, du$ (We appreciate the power of simple u-substitution. **Lesson 30**)

$= -\left[\frac{u^3}{3} - \frac{u^5}{5}\right] + C$

$= -\frac{u^3}{3} + \frac{u^5}{5} + C$

$= -\frac{\cos^3 x}{3} + \frac{\cos^5 x}{5} + C.$

(b) Find $\int \sin^2 x \cos^3 x \, dx$

Solution: Imitate (a) above with $u = \sin x$ and $\cos^2 x = 1 - \sin^2 x$

Ans: $= \dfrac{\sin^3 x}{3} - \dfrac{\sin^5 x}{5} + C$

Extra: Find $\int \sin^5 x \cos^3 x \, dx$

Solution

$\int \sin^5 x \cos^3 x \, dx$

$= \int (\sin^2 x)^2 \sin x \cos^2 x \cos x \, dx$

$= \int (\sin^2 x)^2 \sin x (1 - \sin^2 x) \cos x \, dx \qquad \left(\cos^2 x = 1 - \sin^2 x \right)$

$= \int (u^2)^2 u (1 - u^2) \, du \quad (u = \sin x, \; \dfrac{du}{dx} = \cos x, \; \dfrac{du}{\cos x} = dx, \text{ or } du = \cos x \, dx)$

$= \int (u^5 - u^7) \, du$

$= \dfrac{u^6}{6} - \dfrac{u^8}{8} + C$

$= \dfrac{\sin^6 x}{6} - \dfrac{\sin^8 x}{8} + C.$

Case 6 Powers of secant and tangent

Example 1 Find $\int \sec^2 x \, dx$

Solution $\int \sec^2 x \, dx = \tan x + C$ (Note that the derivative of $\tan x = \sec^2 x$.)

Example 2 Find $\int \sec^3 x \, dx$ (For Method 2, see appendix)

$\int \sec^3 x \, dx = \int \sec^2 x \sec x \, dx$

$\qquad\qquad = \int (1 + \tan^2 x) \sec x \, dx$

$\qquad\qquad = \int \sec x \, dx + \int \tan^2 x \sec x \, dx$

$\qquad\qquad = \int \sec x \, dx + \int \tan x \bullet \tan x \sec x \, dx$ (Integrate the second term by-parts)

$\qquad\qquad = \ln|\sec x + \tan x| + \tan x \sec x - \int \sec^2 x \sec x \, dx$

$\int \sec^3 x \, dx = \ln|\sec x + \tan x| + \tan x \sec x - \int \sec^3 x \, dx$

$2\int \sec^3 x \, dx = \ln|\sec x + \tan x| + \tan x \sec x$ (Type 3 integration-by-parts)

$\int \sec^3 x \, dx = \tfrac{1}{2} \ln|\sec x + \tan x| + \tfrac{1}{2} \tan x \sec x + C.$

Example 4 Find $\int \sec^4 x \, dx$

$\int \sec^4 x \, dx$

$= \int (\sec^2 x)(\sec^2 x) \, dx$

$= \int (1 + \tan^2 x) \sec^2 \, dx \quad (\sec^2 x = 1 + \tan^2 x)$

$= \int (1 + u^2) \, du \quad (\text{Let } u = \tan x . \dfrac{du}{dx} = \sec^2 x, dx = \dfrac{du}{\sec^2 x}, \text{ or } du = \sec^2 x \, dx)$

$= u + \dfrac{u^3}{3} + C$

$= \tan x + \dfrac{\tan^3 x}{3} + C.$

Example 4 Find $\int \sec^5 x \, dx$

Step 1: $\int \sec^5 x \, dx = \int (\sec^2 x)^2 \sec x \, dx$

$$= \int (\tan^2 x + 1)^2 \sec x \, dx$$

$$= \int (\tan^4 x + 2\tan^2 x + 1)\sec x \, dx$$

$$= \int \tan^4 x \sec x \, dx + 2\int \tan^2 x \sec x \, dx + \int \sec x \, dx$$

Step 2: Find the first two integrals using integration-by-parts.

$$= \int \tan^3 x \bullet \tan x \sec x \, dx + 2\int \tan x \bullet \tan x \sec x \, dx + \int \sec x \, dx$$

$(\,\textbf{1.}\ u = \tan^3 x \,;\ dv = \tan x \sec x \, dx \,;\ \textbf{2.}\ u = \tan x \ \ dv = \tan x \sec x \, dx\,)$

$$= \tan^3 x \sec x - 3\int \tan^2 x \sec^3 x \, dx + 2\tan x \sec x - 2\int \sec^3 x \, dx + \int \sec x \, dx$$

Step 3: Let $\tan^2 x = \sec^2 x - 1$ in the first integral..

$$= \tan^3 x \sec x - 3\int (\sec^2 x - 1)\sec^3 x \, dx + 2\tan x \sec x - 2\int \sec^3 x \, dx + \int \sec x \, dx$$

$$= \tan^3 x \sec x - 3\int \sec^5 x + 3\int \sec^3 x \, dx + 2\tan x \sec x - 2\int \sec^3 x \, dx + \int \sec x \, dx$$

$$\int \sec^5 x \, dx = \tan^3 x \sec x - 3\int \sec^5 x \, dx + \int \sec^3 x \, dx + 2\tan x \sec x + \int \sec x \, dx$$

$$\text{(adding the } \int \sec^3 x \, dx \text{ terms)}$$

Step 4: Add $3\int \sec^5 x \, dx$ to both sides of the equation; and

$$4\int \sec^5 x \, dx = \tan^3 x \sec x + \int \sec^3 x \, dx + 2\tan x \sec x + \int \sec x \, dx \quad \text{(Type 3 Int.)}$$

Step 5: Obtain $\int \sec^3 x \, dx$ from Example 2, Case 6 and find $\int \sec x \, dx$.

$$4\int \sec^5 x = \tan^3 x \sec x + \tfrac{1}{2}\ln|\sec x + \tan x| + \tfrac{1}{2}\tan x \sec x + 2\tan x \sec x +$$

$$\ln|\sec x + \tan x|$$

$$4\int \sec^5 x = \tan^3 x \sec x + \tfrac{3}{2}\ln|\sec x + \tan x| + \tfrac{5}{2}\tan x \sec x$$

$$\int \sec^5 x = \frac{\tan^3 x \sec x}{4} + \tfrac{3}{8}\ln|\sec x + \tan x| + \tfrac{5}{8}\tan x \sec x$$

$$\int \sec^5 x = \frac{\tan^3 x \sec x}{4} + \tfrac{5}{8}\tan x \sec x + \tfrac{3}{8}\ln|\sec x + \tan x| + C$$

Example 5 Find $\int \sec^6 x \, dx$

$$\int \sec^6 x \, dx$$

$$= \int (\sec^4 x)(\sec^2 x)\, dx$$

$$= \int (\sec^2 x)^2 (\sec^2 x)\, dx$$

$$= \int (\tan^2 x + 1)^2 (\sec^2 x)\, dx \qquad (\sec^2 x = \tan^2 x + 1)$$

$$= \int (u^2 + 1)(u^2 + 1)\, dx \ \ (u = \tan x \,.\ \frac{du}{dx} = \sec^2 x,,\ dx = \frac{du}{\sec^2 x},\ \text{or}\ du = \sec^2 x \, dx)$$

$$= \int (u^4 + 2u^2 + 1)\, du$$

$$= \frac{u^5}{5} + \frac{2u^3}{3} + u + C$$

$$= \frac{\tan^5 x}{5} + \frac{2\tan^3 x}{3} + \tan x + C \,.$$

Example 6 Find $\int \tan^2 x \, dx$
Solution **Method 1**

$\int \tan^2 x \, dx$

$= \int (\sec^2 - 1) \, dx$

$= \tan x - x + C$

Example 6 **Method 2** **Find** $\int \tan^2 x \, dx$

Previously, $\int \sec^2 x \, dx = \tan x + c$

$\tan^2 x = \sec^2 x - 1$, (Pythagorean Identity)

$\int \tan^2 x \, dx = = \int \sec^2 x \, dx - \int 1 \, dx$

$= \tan x - x + c$

Example 7 Find $\int \tan^3 x \, dx$

$\int \tan^3 x \, dx$

$= \int \tan^2 x \tan x \, dx$

$= \int (\sec^2 x - 1) \tan x \, dx$

$= \int \tan x \sec^2 x \, dx - \int \tan x \, dx$ (First integral: $u = \tan x$, $du = \sec^2 x \, dx$)

$= \dfrac{\tan^2 x}{2} - \ln|\sec x| + C$.

Example 8 Find $\int \tan^4 x \, dx$
Solution

$\int \tan^4 x \, dx = \int \tan^2 x \tan^2 x \, dx$

$\qquad = \int (\sec^2 x - 1) \tan^2 x \, dx$

$\qquad = \int \tan^2 x \sec^2 x - \int \tan^2 x \, dx$

$\qquad = \int \tan^2 x \sec^2 x - \int (\sec^2 x - 1) \, dx$

$\qquad = \int \tan^2 x \sec^2 x - \int \sec^2 x \, dx + \int 1 \, dx$

$\qquad = \int u^2 \, du - \int \sec^2 x \, dx + \int 1 \, dx$ (Ist integral: $u = \tan x$, $du = \sec^2 x \, dx$)

$\qquad = \dfrac{\tan^3 x}{3} - \tan x + x + C$.

(See also Example 4)

Example 9 Find $\int \tan^5 x \, dx$

$\int \tan^5 x \, dx = \int (\tan^3 x)(\tan^2 x) \, dx$

$\qquad = \int (\tan^3 x)(\sec^2 x - 1) \, dx$

$\qquad = \int (\tan^3 x)(\sec^2 x) \, dx - \int \tan^3 x \, dx$

$\qquad = \int (\tan^3 x)(\sec^2 x) \, dx - \int (\tan^2 x)(\tan x) \, dx$

$\qquad = \int (\tan^3 x)(\sec^2 x) \, dx - \int (\sec^2 x - 1)(\tan x) \, dx$

$\qquad = \int (\tan^3 x)(\sec^2 x) \, dx - \int (\tan x \sec^2 x - \tan x) \, dx$

$\qquad = \int (\tan^3 x)(\sec^2 x) \, dx - \int \tan x \sec^2 x \, dx + \int \tan x) \, dx$

$\qquad = \int u^3 \, du - \int u \, du + \ln|\sec x|$ ($u = \tan x$, $du = \sec^2 x \, dx$)

$\qquad = \dfrac{u^4}{4} - \dfrac{u^2}{2} + \ln|\sec x| + C$

$\qquad = \dfrac{\tan^4 x}{4} - \dfrac{\tan^2 x}{2} + \ln|\sec x| + C$.

Case 7 Products of powers of tangent and of secant

Example 1. Find $\int \sec x \tan x \, dx$

Solution: We know that the derivative of $\sec x$ is $\sec x \tan x$, and therefore an antiderivative of $\sec x \tan x$ is $\sec x + C$

Therefore, $\int \sec x \tan x \, dx = \sec x + C$

For curiosity, let us subject $\int \sec x \tan x \, dx$ to simple u-substitution method.

Let $u = \sec x$. Then $\frac{du}{dx} = \sec x \tan x$ and $\frac{du}{\sec x \tan x} = dx$ or $du = \sec x \tan x \, dx$

Substituting for $\sec x$ and dx in $\int \sec x \tan x \, dx$.

$\int \sec x \tan x \, dx$

$= \int du$ $\left(\int \sec x \tan x \, dx = \int u \tan x \cdot \dfrac{du}{\sec x \tan x} = \int \dfrac{u^1}{1^u} du = \int du \right)$

$= u + C$

$= \sec x + C$ (same as from memory).

Example 2 Find $\int \sec^2 x \tan x \, dx$

Note that the derivative of $\tan x = \sec^2 x$, we can use simple u-substitution

Let $u = \tan x$, Then $\frac{du}{dx} = \sec^2 x$, and $dx = \frac{du}{\sec^2 x}$ or $du = \sec^2 x \, dx$

Substituting in $\int \sec^2 x \tan x \, dx$, we obtain

$\int u \, du$

$= \dfrac{u^2}{2} + C$

$= \dfrac{\tan^2 x}{2} + C$.

Example 3 Find $\int \sec^2 x \tan^2 x \, dx$

Since the derivative of $\tan x = \sec^2 x$, we use simple u-substitution ($u = \tan x$)

$\int \tan^2 x \sec^2 x \, dx$ (Rewriting)

$= \int u^2 \, du$ $\left(\dfrac{du}{dx} = \sec^2 x, \text{ and } dx = \dfrac{du}{\sec^2 x} \text{ or } du = \sec^2 x \, dx \right.$.

$= \dfrac{u^3}{3} + C$

$= \dfrac{\tan^3 x}{3} + C$.

Note in the above problem that if we let $u = \sec x$, simple u-substitution will not work.

Example 4 Find $\int \tan^3 x \sec^2 x \, dx$, Simple u-substitution ($u = \tan x$), will work for any product $\tan^n x \sec^2$, where n is a positive integer.

Solution

$\int \tan^3 x \sec^2 x \, dx$

$= \int u^3 \, du$

$= \dfrac{u^4}{4} + C$

$= \dfrac{\tan^4 x}{4} + C$.

Note above that if we express the integrand in terms of $\tan x$, we obtain

$\int \tan^3 x(\tan^2 x + 1)\,dx$

$= \int \tan^5 x\,dx + \int \tan^3 x\,dx$, and since we have already found these integrals previously, we could add the results. However, simple u-substitution was very efficient that we would not think of using this "$\tan x$ approach" normally .

Example 5: Find $\int \tan^2 x \sec^3 x\,dx$.

Here, express the integrand in terms of $\sec x$, using $\tan^2 x = \sec^2 x - 1$

$\int \tan^2 x \sec^3 x\,dx$

$= \int (\sec^2 x - 1)\sec^3 x\,dx \qquad (\tan^2 x = \sec^2 x - 1)$

$= \int \sec^5 x - \int \sec^3 x\,dx$

From Case 6 Examples 4 and 2: (see these examples)

$\int \sec^5 x = \dfrac{\tan^3 x \sec x}{4} + \frac{3}{8}\ln|\sec x + \tan x| + \frac{5}{8}\tan x \sec x + C$, and

$\int \sec^3 x\,dx = \frac{1}{2}\ln|\sec x + \tan x| + \frac{1}{2}\tan x \sec x + C$

$\int \tan^2 x \sec^3 x\,dx = \int \sec^5 x - \int \sec^3 x\,dx$

$\qquad = \dfrac{\tan^3 x \sec x}{4} + \frac{3}{8}\ln|\sec x + \tan x| + \frac{5}{8}\tan x \sec x -$

$\qquad\qquad\qquad \left(\frac{1}{2}\ln|\sec x + \tan x| + \frac{1}{2}\tan x \sec x\right) + C$

$\int \tan^2 x \sec^3 x\,dx = \dfrac{\tan^3 x \sec x}{4} - \frac{1}{8}\ln|\sec x + \tan x| + \frac{1}{8}\tan x \sec x + C$.

Case 8 Powers of cosecant and cotangent

Tool box: $1 + \cot^2 x = \csc^2$

Example 1 Find $\int \csc^2 x\,dx$

Solution $= -\cot x + C$ (since the derivative of $\cot x = -\csc^2 x$).

Example 2 Find $\int \csc^3 x\,dx$

Step 1 $\int \csc^3 x\,dx$

$\qquad = \int \csc^2 x \csc x\,dx$

$\qquad = \int (1 + \cot^2 x)\csc x\,dx$

$\qquad = \int \csc x\,dx + \int \cot^2 x \csc x\,dx$

$\qquad = \int \csc x\,dx + \int \cot x \bullet \csc x \cot x\,dx$

Step 2: For the second term, integrate-by-parts: $u = \cot x$, $dv = \csc x \cot x\,dx$;

$\qquad\qquad v = -\csc x$, $\dfrac{du}{dx} = -\csc^2 x$; $du = -\csc^2 x\,dx$

$\qquad \int \csc^3 x\,dx = \ln|\csc x - \cot x| + \cot x(-\csc x) - \int(-\csc^2 x)(-\csc x)$

$\qquad \int \csc^3 x\,dx = \ln|\csc x - \cot x| - \cot x \csc x - \int \csc^3 x$

$\qquad 2\int \csc^3 x\,dx = \ln|\csc x - \cot x| - \cot x \csc x$ (Type 3 Integration-by-parts)

$\qquad \int \csc^3 x\,dx = \frac{1}{2}\ln|\csc x - \cot x| - \frac{1}{2}\cot x \csc x + C$.

Example 3 Find $\int \csc^4 x\, dx$

$\int \csc^4 x\, dx$

$= \int \csc^2 x \bullet \csc^2 x\, dx$

$= \int (1 + \cot^2 x) \bullet \csc^2 x\, dx$

$= -\int (1 + u^2)\, du$ (let $u = \cot x$, then $\frac{du}{dx} = -\csc^2 x$, and $du = -\csc^2 x\, dx$)

$= -u - \frac{u^3}{3} + C$

$= -\cot x - \frac{\cot^3 x}{3} + C$.

Example 4 Find $\int \csc^5 x\, dx$

Step 1: $\int \csc^5 x\, dx$

$= \int (\csc^2 x)^2 \csc x\, dx$

$= \int (\cot^2 x + 1)^2 \csc x\, dx$

$= \int (\cot^4 x + 2\cot^2 x + 1)\csc x\, dx$

$= \int \cot^4 x \csc x\, dx + 2\int \cot^2 x \csc x\, dx + \int \csc x\, dx$.

Step 2: Find the first two integrals using integration-by-parts.

$= \int \cot^3 x \bullet \cot x \csc x\, dx + 2\int \cot x \bullet \cot x \csc x\, dx + \int \csc x\, dx$

(For first integral: $u = \cot^3 x$, $dv = \csc x \cot x\, dx$; For 2nd integral: $u = \cot x$)

$= -\cot^3 x \csc x - 3\int \cot^2 x \csc^3 x\, dx - 2\cot x \csc x - 2\int \csc^3 x\, dx + \int \csc x\, dx$

Step 3: Let $\cot^2 x = \csc^2 x - 1$ in the first integral..

$= -\cot^3 x \csc x - 3\int (\csc^2 x - 1)\csc^3 x\, dx - 2\cot x \csc x - 2\int \csc^3 x\, dx + \int \csc x\, dx$

$= -\cot^3 x \csc x - 3\int \csc^5 x + 3\int \csc^3 x\, dx - 2\cot x \csc x - 2\int \csc^3 x\, dx + \int \csc x\, dx$

$\int \csc^5 x = -\cot^3 x \csc x - 3\int \csc^5 x + \int \csc^3 x\, dx - 2\cot x \csc x + \int \csc x\, dx$

(adding the $\int \csc^3 x\, dx$ terms.

Step 4: Add $3\int \csc^5 x\, dx$ to both sides of the equation. (Type 3 integr. by parts)

$4\int \csc^5 x = -\cot^3 x \csc x + \int \csc^3 x\, dx - 2\cot x \csc x + \int \csc x\, dx$

Step 5: Obtain : $\int \csc^3 x\, dx$ from Example 2, Case 8; and find $\int \csc x\, dx$.

$4\int \csc^5 x = -\cot^3 x \csc x + \frac{1}{2}\ln|\csc x - \cot x| - \frac{1}{2}\cot x \csc x - 2\cot x \csc x +$
$$\ln|\csc x - \cot x|$$

$\int \csc^5 x = -\frac{\cot^3 x \csc x}{4} + \frac{3}{8}\ln|\csc x - \cot x| - \frac{5}{8}\cot x \csc x + C$

Example 5 Find $\int \cot^2 x\, dx$

$\int \cot^2 x\, dx$

$= \int (\csc^2 x - 1)\, dx$

$= \int \csc^2 x\, dx - \int 1\, dx$

$= -\cot x - x + C$. (Note that the derivative of $\cot x = -\csc^2 x$)

Example 6 Find $\int \cot^3 x\, dx$

$\int \cot^3 x\, dx$

$= \int \cot^2 x \cot x\, dx$

$= \int (\csc^2 - 1) \cot x\, dx$

$= \int \csc^2 \cot x\, dx - \int \cot x\, dx$

$= \int \cot x \csc^2 dx - \int \cot x\, dx$

$= -\int u\, du - \int \cot x\, dx$ (Let $u = \cot x$, $\frac{du}{dx} = -\csc^2 x$ and $du = -\csc^2 x\, dx$)

$= -\frac{\cot^2 x}{2} - \ln|\sin x| + C.$

Example 7 Find $\int \cot^4 x\, dx$

$\int \cot^4 x\, dx$

$= \int \cot^2 x \cot^2 dx$

$= \int (\csc^2 x - 1) \cot^2 dx$

$= \int \csc^2 x \cot^2 dx - \int \cot^2 x\, dx$

$= \int \csc^2 x \cot^2 x\, dx - \int (\csc^2 x - 1) dx$

$= \int \cot^2 x \csc^2 x\, dx - \int \csc^2 dx + \int 1 dx$

$= -\int u^2 du - \int \csc^2 dx + \int 1 dx$ ($u = \cot x$, $\frac{du}{dx} = -\csc^2 x$ and $du = -\csc^2 x\, dx$)

$= -\frac{u^3}{3} - (-\cot x) + x + C$ (Note that the derivative of $\cot x = -\csc^2 x$)

$= -\frac{\cot^3 x}{3} + \cot x + x + C.$

Example 8 Find $\int \cot^5 x\, dx$

$\int \cot^5 x\, dx = \int (\cot^3 x)(\cot^2 x) dx$

$= \int (\cot^3 x)(\csc^2 x - 1) dx$

$= \int (\cot^3 x)(\csc^2 x) dx - \int \cot^3 x\, dx$

$= \int (\cot^3 x)(\csc^2 x) dx - \int (\cot^2 x)(\cot x) dx$

$= \int (\cot^3 x)(\csc^2 x) dx - \int (\csc^2 x - 1)(\cot x) dx$

$= \int (\cot^3 x)(\csc^2 x) dx - \int (\cot x \csc^2 x - \cot x) dx$

$= \int (\cot^3 x)(\csc^2 x) dx - \int \cot x \csc^2 x\, dx + \int \cot x\, dx$

$= -\int u^3 du + \int u\, du + \ln|\sin x|$

$= -\frac{u^4}{4} + \frac{u^2}{2} + \ln|\sec x| + C$

$= -\frac{\cot^4 x}{4} + \frac{\cot^2 x}{2} + \ln|\sin x| + C.$

Example 9 Find $\int \csc^2 x \cot^2 x dx$

$\int \csc^2 x \cot^2 x dx$

$= \int \cot^2 x \csc^2 x \, dx$ (rewriting)

$= -\int u^2 du$ (Let $u = \cot x$, $\dfrac{du}{dx} = -\csc^2 x$ and $du = -\csc^2 x \, dx$ or $dx = -\dfrac{du}{\csc^2 x}$

$= -\dfrac{u^3}{3} + C$

$= -\dfrac{\cot^3}{3} + C$.

Example 10 Find $\int \csc^2 \cot^3 x dx$

$= -\int u^3 du$ (Let $u = \cot x$, $\dfrac{du}{dx} = -\csc^2 x$ and $du = -\csc^2 x \, dx$)

$= -\dfrac{u^4}{4} + C$

$= -\dfrac{\cot^4 x}{4} + C$.

Example 11 Find $\int \csc^3 x \cot^2 x dx$

Here, express the integrand in terms of $\csc x$, using $\cot^2 x = \csc^2 x - 1$

$\int \cot^2 x \csc^3 x dx$

$= \int (\csc^2 x - 1)\csc^3 x \, dx$ $(\cot^2 x = \csc^2 x - 1)$

$= \int \csc^5 x - \int \csc^3 x \, dx$

From Case 8, Examples 4 and 2: (see these examples)

$\int \csc^5 x = -\dfrac{\cot^3 x \csc x}{4} + \dfrac{3}{8}\ln|\csc x - \cot x| - \dfrac{5}{8}\cot x \csc x + C$, and

$\int \csc^3 x dx = \dfrac{1}{2}\ln|\csc x - \cot x| - \dfrac{1}{2}\cot x \csc x + C$

$\int \cot^2 x \csc^3 x dx = \int \csc^5 x - \int \csc^3 x \, dx$

$$= -\dfrac{\cot^3 x \csc x}{4} + \dfrac{3}{8}\ln|\csc x - \cot x| - \dfrac{5}{8}\cot x \csc x -$$

$$(\dfrac{1}{2}\ln|\csc x - \cot x| - \dfrac{1}{2}\cot x \csc x) + C$$

$\int \csc^3 x \cot^2 x dx = -\dfrac{\cot^3 x \csc x}{4} - \dfrac{1}{8}\ln|\csc x - \cot x| - \dfrac{1}{8}\cot x \csc x + C$.

Case 9: Integrals of the forms

(a) $\int \sin mx \sin nx \, dx$; (b) $\int \cos mx \cos nx \, dx$; (c) $\int \sin mx \cos nx \, dx$

For the above integrals, we can apply the following identities .

	Signs on the right tside
A: $\sin mx \sin nx = \frac{1}{2}[\cos(m-n)x - \cos(m+n)x]$ or $2\sin mx \sin nx = \cos(m-n)x - \cos(m+n)x$	$-\ -\ +$
B: $\cos mx \cos nx = \frac{1}{2}[\cos(m+n)x + \cos(m-n)x]$ or $2\cos mx \cos nx = \cos(m+n)x + \cos(m-n)x$	$+\ +\ -$
C: $\sin mx \cos nx = \frac{1}{2}[\sin(m+n)x + \sin(m-n)x]$ or $2\sin mx \cos nx = \sin(m+n)x + \sin(m-n)x$	$+\ +\ -$

A **memory** device to help recall the **above** identities follows.

1. For **A** and **B** (sin sin or cos cos, the right hand side of each identity is in terms of **cos** function only.
   Note in **B** that in some books, the order of the right side may be reversed
2. For **C**, (sin cos,) the right hand side is in terms of **sin** function only,
   Note above also that when the left side is in terms of the same function , the right side is in terms of **cos** only, while if the left side is in terms of cos and sin, the right side is in terms of **sin** only.

Example 1 Find $\int \sin 3x \sin 2x \, dx$
Solution This is of form identity **A**.

Apply $\sin mx \sin nx = \frac{1}{2}[\cos(m-n)x - \cos(m+n)x]$

$\int \sin 3x \sin 2x \, dx = \int \frac{1}{2}[\cos(3-2)x - \cos(3+2)x \, dx]$

$\int \sin 3x \sin 2x \, dx = \int \frac{1}{2}[\cos x - \cos 5x \, dx]$

$\qquad\qquad = \frac{1}{2}\int \cos x \, dx - \int \cos 5x \, dx$

$\int \sin 3x \sin 2x \, dx = \frac{1}{2}\sin x - \frac{1}{2} \cdot \frac{1}{5}\sin 5x + C$

$\int \sin 3x \sin 2x \, dx = \frac{1}{2}\sin x - \frac{1}{10}\sin 5x + C$.

Example 2 Find $\int \cos 3x \cos 2x \, dx$
Solution This is of form identity **B**.

Apply $\int \cos mx \cos nx \, dx = \int \frac{1}{2}[\cos(m+n)x + \cos(m-n)x] dx$

$\int \cos 3x \cos 2x \, dx = \frac{1}{2}\int [\cos(3+2)x + \cos(3-2)x] dx$

$\qquad\qquad = \frac{1}{2}\int [\cos 5x + \cos x] dx$

$\qquad\qquad = \frac{1}{2}[\int \cos 5x \, dx + \int \cos x \, dx]$

$\qquad\qquad = \frac{1}{2}\left[\frac{1}{5}\sin 5x + \sin x\right] + C$

$\qquad\qquad = \frac{1}{10}\sin 5x + \frac{1}{2}\sin x + C$.

Example 3   Find $\int \sin 2x \cos 3x\, dx$
Solution   This is of form identity **C**.

Apply $\sin mx \cos nx = \frac{1}{2}[\sin(m+n)x + \sin(m-n)x]$

$\int \sin mx \cos nx\, dx = \int \frac{1}{2}[\sin(m+n)x + \sin(m-n)x]dx$

$\int \sin 2x \cos 3x\, dx = \int \frac{1}{2}[\sin(2+3)x + \sin(2-3)x]dx$

$$= \int \frac{1}{2}[\sin 5x + \sin(-x)]dx$$

$$= \int \frac{1}{2}[\sin 5x - \sin x]dx$$

$$= \frac{1}{2}\left[\int \sin 5x - \int \sin x\, dx\right]$$

$$= \frac{1}{2}\left[\frac{1}{5}(-\cos 5x) - (-\cos x)\right] + C$$

$$= -\frac{1}{10}\cos 5x + \frac{1}{2}\cos x + C.$$

Note: We can also use identity **A** in Case 9 to derive the identity

$\sin^2 x = \sin x \sin x = \frac{1}{2}\left[\cos(1-1)x - \cos(1+1)x\right] = \frac{1}{2}\left[\cos 0 - \cos 2x\right]$

$$= \frac{1}{2} - \frac{1}{2}\cos 2x$$

Lesson 37 Exercises

1.

(a) Find $\int \sin x \cos x\, dx$

(b) $\int \sin^3 x \cos x\, dx$;

(c) $\int \sin x \cos x^3\, dx$;

(d) $\int \sin^5 x \cos x\, dx$;

(e) $\int \sin x \cos^6 x\, dx$;

Ans: (a) $\frac{1}{2}\sin^2 x + C$; (b) $\frac{1}{4}\sin^4 x + C$; (c) $-\frac{1}{4}\cos^4 x + C$; (d) : $\frac{1}{6}\sin^6 x + C$

(e) $-\frac{1}{7}\cos^7 x + C$

---

2. (a) Find $\int \sin^2 x\, dx$; | (b) Find $\int \cos^2 x\, dx$

Ans (a) $= -\frac{1}{2}\sin x \cos x + \frac{1}{2}x + C$; (b) $= \frac{1}{2}\sin x \cos x + \frac{1}{2}x + C$

---

3. (a) Find $\int \sin^4 x\, dx$; | (c) $\int \sin^3 x\, dx$;

(b) Find $\int \cos^4 x\, dx$ | (d) $\int \sin^5 x\, dx$

Ans (a) $\frac{3}{8}x - \frac{1}{4}\sin 2x + \frac{1}{32}\cos 4x + C$; **(b)** $\frac{1}{4}\sin 2x + \frac{1}{32}\sin 4x + \frac{3}{8}x + C$

(c) $= -\cos x + \frac{\cos^3 x}{3} + C$; **(d)** $= -\cos x + \frac{2\cos^3}{3} - \frac{\cos^5 x}{5} + C$

---

4. (a) Find $\int \sin^2 x \cos^2 x\, dx$ | (c) Find $\int \sin^2 x \cos^3 x\, dx$

(b) Find $\int \sin^3 x \cos^2 x\, dx$

Ans (a) $\frac{1}{8}x - \frac{1}{32}\sin 4x + C$; **(b)** $= -\frac{\cos^3 x}{3} + \frac{\cos^5 x}{5} + C$;

(c) $= \frac{\sin^3 x}{3} - \frac{\sin^5 x}{5} + C$

5. Find $\int \sec^2 x\,dx$	**11.** Find $\int \tan^3 x\,dx$
6. Find $\int \sec^3 x\,dx$	**12.** Find $\int \tan^4 x\,dx$
7. Find $\int \sec^4 x\,dx$	**13.** Find $\int \tan^5 x\,dx$
8. Find $\int \sec^5 x\,dx$	**14.** Find $\int \sec x \tan x\,dx$
9. Find $\int \sec^6 x\,dx$	**15.** Find $\int \sec^2 x \tan x\,dx$
10. Find $\int \tan^2 x\,dx$	**16.** Find $\int \sec^2 x \tan^2 x\,dx$

Ans: 5. $\int \sec^2 x\,dx = \tan x + C$; **6.** $\frac{1}{2}\ln|\sec x + \tan x| + \frac{1}{2}\tan x \sec x + C$

$\quad$ **7.** $= \tan x + \frac{\tan^3 x}{3} C$; **8.** $\frac{\tan^3 x \sec x}{4} + \frac{3}{8}\ln|\sec x + \tan x| + \frac{5}{8}\tan x \sec x + C$

9. $\frac{\tan^5 x}{5} + \frac{2\tan^3 x}{3} + \tan x + C$; **10.** $\tan x - x + C$; **11.** $\frac{\tan^2 x}{2} - \ln|\sec x| + C$

12. $\frac{\tan^3 x}{3} - \tan x + x + C$ $\quad$ **13.** $\frac{\tan^4 x}{4} - \frac{\tan^2 x}{2} + \ln|\sec x| + C$; **14.** $\sec x + C$

15. $\frac{\tan^2 x}{2} + C$; **16.** $\frac{\tan^3 x}{3} + C$

17. Find $\int \tan^3 x \sec^2 x\,dx$	**20.** Find $\int \csc^3 x\,dx$
18. Find $\int \tan^2 x \sec^3 x\,dx$	**21.** Find $\int \csc^4 x\,dx$
19. Find $\int \csc^2 x\,dx$	

Ans: 17. $\frac{\tan^4 x}{4} + C$; **18.** $\frac{\tan^3 x \sec x}{4} - \frac{1}{8}\ln|\sec x + \tan x| + \frac{1}{8}\tan x \sec x + C$

19. $-\cot x + C$; **20.** $\frac{1}{2}\ln|\csc x - \cot x| - \frac{1}{2}\cot x \csc x + C$;

21. $-\cot x - \frac{\cot x^3}{3} + C$;

22. Find $\int \csc^5 x\,dx$	**27.** Find $\int \csc^2 x \cot^2 dx$
23. Find $\int \cot^2 x\,dx$	
24 Find $\int \cot^3 x\,dx$	**28.** Find $\int \cot^3 x \csc^2 x\,dx$
25 Find $\int \cot^4 x\,dx$	
26. Find $\int \cot^5 x\,dx$	**29.** Find $\int \cot^2 x \csc^3 x\,dx$

Ans: 22. $-\frac{\cot^3 x \csc x}{4} + \frac{3}{8}\ln|\csc x - \cot x| - \frac{5}{8}\cot x \csc x + C$

23. $-\cot x - x + C$; **24.** $-\frac{\cot^2 x}{2} - \ln|\sin x| + C$; **25** $-\frac{\cot^3 x}{3} + \cot x + x + C$

26. $-\frac{\cot^4 x}{4} + \frac{\cot^2 x}{2} + \ln|\sin x| + C$; **27.** $-\frac{\cot^3}{3} + C$; **28** $-\frac{\cot^4 x}{4} + C$

29. $-\frac{\cot^3 x \csc x}{4} - \frac{1}{8}\ln|\csc x - \cot x| - \frac{1}{8}\cot x \csc x + C$.

Lesson 37 Case 8 Exercises

1. Find $\int \sin 3x \sin 2x \, dx$

2 Find $\int \cos 3x \cos 2x \, dx$

3. Find $\int \sin 2x \cos 3x \, dx$

4. Find $\int \sin 5x \cos 3x \, dx$

Ans: 1. $\frac{1}{2}\sin x - \frac{1}{10}\sin 5x + C$; **2** $\frac{1}{10}\sin 5x + \frac{1}{2}\sin x + C$

3. $-\frac{1}{10}\cos 5x + \frac{1}{2}\cos x + C$; **4..** $-\frac{1}{16}\cos 8x - \frac{1}{4}\cos 2x + C$

Lesson 38
Integration of Logarithmic Functions

Case 1: Base e

(Integrate by parts, with $u = \ln x$ and $dv = 1dx$.(version 1) or $v = 1$ (version 2)
see p.244 for integration-by-parts formula)

$\int \ln x\, dx = x \ln x - \int \frac{1}{x} x\, dx$

$\qquad = x \ln x - \int 1\, dx$

$\qquad = x \ln x - x + c$

$\boxed{\int \ln x\, dx = x \ln x - x + c}$

Note: Do not confuse the integration of logarithmic functions with

$\int \frac{1}{x}\, dx = \ln x$, which is the integration of a rational function.

Case 2: Any base b

Method 1

Step 1: change from base b to base e (the natural log base)

By the change of base formula: $\log_b x = \frac{\ln x}{\ln b}$ ($\log_b x = \frac{\log_e x}{\log_e b}$)

$\int \log_b x\, dx = \frac{1}{\ln b} \int \ln x\, dx$ ($\frac{1}{\ln b}$ is a constant)

$\qquad = \frac{1}{\ln b}(x \ln x - x) + C$ (Using the result from Case 1, above)

$\qquad = x \frac{\ln x}{\ln b} - \frac{x}{\ln b} + C$

$\qquad = x \log_b x - \frac{x}{\ln b} + C$ ($\log_b x = \frac{\ln x}{\ln b}$)

$\boxed{\int \log_b x\, dx = x \log_b x - \frac{x}{\ln b} + C}$

Method 2

Integrate by parts, with $u = \log_b x$ and $dv = 1$ (for version 1) or $v = 1$ for version 2)

$\int \log_b x\, dx = x \log_b x - \int \frac{1}{\ln b} \frac{1}{x} x\, dx$ ($\frac{d}{dx} \log_b x = \frac{1}{\ln b} \frac{d}{dx} \ln x = \frac{1}{\ln b} \cdot \frac{1}{x}$)

$\qquad = x \log_b x - \frac{1}{\ln b} \int 1dx$

$\qquad = x \log_b x - \frac{x}{\ln b} + C$

$\boxed{\int \log_b x\, dx = x \log_b x - \frac{x}{\ln b} + C}$.

Note the following distinctions

1. Derivative $\boxed{\frac{d}{dx}\left(\frac{1}{x}\right) = -\frac{1}{x^2}}$

$\left(y = \frac{1}{x} = x^{-1};\ \frac{dy}{dx} = -1x^{-2} = -\frac{1}{x^2} \right)$

3. Integral: $\boxed{\int \frac{1}{x} dx = \ln x + C}$

2. Derivative: $\frac{d}{dx}(\ln x) = \frac{1}{x}$

$\left(y = \ln x;\ \frac{dy}{dx} = \frac{1}{x} \right)$

4. Integral: $\int \ln x\, dx = x \ln x - x + C$
(Integrating by parts)

Extra Find $\int (\ln x)^2 \, dx$

Solution

Method 1 (*u*-subst. plus Int-by-parts)

Let $u = \ln x$

Then $\dfrac{du}{dx} = \dfrac{1}{x}$

$dx = x \, du$ (Now, express x in terms of u)

$e^u = x$ ($u = \log_e x = \ln x$)

$dx = e^u \, du$ (substituting e^u for x)

Substitute $u = \ln x$; $dx = e^u \, du$ in .

$\int (\ln x)^2 \, dx$ to obtain

$\int u^2 e^u \, du$

Using the result of Case 2 Example, p.246

$\int u^2 e^u \, du = e^u (u^2 - 2u + 2) + C_1$

$= x[(\ln x)^2 - 2\ln x + 2)] + C$ ($e^u = x$)

$= x(\ln x)^2 - 2x \ln x + 2x + C$ ($u = \ln x$)

Method 2 By-parts only knowing that

$\int \ln x = x \ln x - x + C$

$\int (\ln x)^2 \, dx = \int (\ln x)(\ln x) dx$

Let the first $\ln x = u$ and let the second $\ln x = v$ or dv
Integrate by parts.

$\int (\ln x)^2 \, dx =$

$\ln x \int \ln x \, dx - \int [\frac{1}{x} \bullet (\int \ln x \, dx)] dx$

$\ln x [x \ln x - x] - \int \frac{1}{x} \bullet (x \ln x - x) dx$

$\ln x [x \ln x - x] - \int (\ln x - 1) dx]$

$\ln x [x \ln x - x] - [x \ln x - x - x] + C$

$x(\ln x)^2 - x \ln x - x \ln x + 2x + C$

$= x(\ln x)^2 - 2x \ln x + 2x + C$

Lesson 38 Exercises A

1. Determine by inspection which of the following we can use simple *u*-substitution to integrate,

a. $\int \frac{\ln x}{x} \, dx$;   **b.** $\int \frac{1}{x \ln x} \, dx$;   **c.** $\int \frac{x}{\ln x} \, dx$

2. $\int \ln x \, dx = ?$

3. $\int \log_b x \, dx = ?$

4. Find $\int \frac{\ln x}{x} \, dx$

5. Find $\int \frac{1}{x \ln x} \, dx$

Answers: **1.** a, b,; **2.** $x \ln|x| - x + c$; **3.** $\frac{x \ln x}{\ln b} - \frac{x}{\ln b} + C$;   **4.** $\frac{1}{2}(\ln x)^2$;
     5. $\ln(\ln(x))$.

Lesson 38 Exercises B

1. $\int \ln(3x + 2) dx$;

2. $\int \ln x \, dx = ?$

3. $\int \log_a x \, dx = ?$

4. Find $\int x \ln x \, dx$

5. $\int x^2 \ln x \, dx$

6. $\int \frac{\ln x}{x^2} \, dx$

7. $\int \frac{\ln x^2}{x} \, dx$

Answers: **1.** $\frac{1}{3}(3x + 2)\ln(3x + 2) - x + C$ **4.** $\frac{1}{2}x^2 \ln x - \frac{1}{4}x^2 + C$;

    5.. $\frac{1}{3}x^3 \ln x - \frac{1}{9}x^3 + C$;   **6.** $-\frac{\ln x}{x} - \frac{1}{x} + C$;   **7.** $(\ln x)^2 + C$ (from $\frac{[\ln(x^2)]^2}{4}$)

Lesson 39
Integration of Inverse Trigonometric Functions
(Integrate-by-parts)

$$\int \text{Sin}^{-1}x\,dx, \int \text{Cos}^{-1}x\,dx, \int \text{Tan}^{-1}x\,dx, \int \text{Sec}^{-1}x\,dx, \int \text{Csc}^{-1}x\,dx, \int \text{Cot}^{-1}x\,dx$$

We apply integration by parts (Lesson 36) to integrate inverse trigonometric functions.

Example 1: Find $\int \text{Sin}^{-1}x\,dx$ (Integrating by parts; process Type 1)
Step 1:
Let $u = \text{Sin}^{-1}x$, and let $dv = 1dx$ (version 1) or let $v = 1$ (for version 2)

Then $\int \text{Sin}^{-1}x\,dx = x\text{Sin}^{-1}x - \int \dfrac{x}{\sqrt{1-x^2}}\,dx$

Step 2: For $\int \dfrac{x}{\sqrt{1-x^2}}\,dx$

Let $t = 1 - x^2$ $\dfrac{dt}{dx} = -2x$, $dx = -\dfrac{dt}{2x}$

substituting in $\int \dfrac{x}{\sqrt{1-x^2}}$ we obtain

$$\int \frac{x}{t^{\frac{1}{2}}} \bullet \left(\frac{-dt}{2x}\right)$$

Now, putting everything together,

$\int \text{Sin}^{-1}x\,dx = x\text{Sin}^{-1}x - \int \dfrac{x}{t^{\frac{1}{2}}} \bullet \left(\dfrac{-dt}{2x}\right)$

$\qquad = x\text{Sin}^{-1}x + \dfrac{1}{2}\int t^{-\frac{1}{2}}dt$

$\qquad = x\,\text{Sin}^{-1}x + \dfrac{1}{2} \bullet \dfrac{t^{-\frac{1}{2}+1}}{-\frac{1}{2}+1} + C$

$\qquad = x\,\text{Sin}^{-1}x + \dfrac{1}{2} \bullet \dfrac{t^{\frac{1}{2}}}{\frac{1}{2}} + C$

$\qquad = x\,\text{Sin}^{-1}x + t^{\frac{1}{2}} + C$

$\qquad = x\,\text{Sin}^{-1}x + (1-x^2)^{\frac{1}{2}} + C$

$\int \text{Sin}^{-1}x\,dx = x\,\text{Sin}^{-1}x + \sqrt{1-x^2} + C.$

Integration by parts Formula
(Version 1)

$$\int u\,dv = uv - \int v\,du$$

$$dv = dx$$
$$v = x$$

$$du = \frac{1}{\sqrt{1-x^2}}\,dx$$

Version 2

$$\int (uv)\,dx = u\int v - \int\left(\frac{du}{dx} \bullet \int v\right)dx$$

$$\frac{du}{dx} = \frac{1}{\sqrt{1-x^2}}$$

Note above that the derivative of $\text{Sin}^{-1}x$ is $\dfrac{1}{\sqrt{1-x^2}}$; and the integral

$\int \dfrac{1}{\sqrt{1-x^2}}\,dx = \text{Sin}^{-1}x + C$; but $\int \text{Sin}^{-1}x\,dx = x\,\text{Sin}^{-1}x + \sqrt{1-x^2} + C$.

Inverse Cofunction Identities: Can be used to find inverse cofunction integrals.
1. $\text{Sin}^{-1}x + \text{Cos}^{-1}x = \frac{\pi}{2}$; **2.** $\text{Tan}^{-1}x + \text{Cot}^{-1}x = \frac{\pi}{2}$; **3.** $\text{Csc}^{-1}x + \text{Sec}^{-1}x = \frac{\pi}{2}$,

Knowing $\int \text{Sin}^{-1}x\,dx$, we can readily find $\int \text{Cos}^{-1}x\,dx$. See Example 5 below.

Example 2: Find $\int \text{Tan}^{-1} x\, dx$

Step 1:

Let $u = \text{Tan}^{-1} x$, and let $dv = 1 dx$ (version 1) or let $v = 1$ (for version 2)

Then $\int \text{Tan}^{-1} x\, dx = x \text{Tan}^{-1} x - \int \dfrac{x}{1+x^2}\, dx$

Step 2: For $\int \dfrac{x}{1+x^2}\, dx$ (integrate by u-subst.)

Let $t = 1 + x^2$. $\dfrac{dt}{dx} = 2x$, $dx = \dfrac{dt}{2x}$.

Substituting in $\int \dfrac{x}{1+x^2}$ we obtain

$$\int \dfrac{x}{t} \cdot \left(\dfrac{dt}{2x}\right) = \dfrac{1}{2}\int \dfrac{dt}{t}$$

Now, putting everything together,

$\int \text{Tan}^{-1} x\, dx = x \text{Tan}^{-1} x - \dfrac{1}{2}\int \dfrac{dt}{t}$

$= x \text{Tan}^{-1} x - \dfrac{1}{2}\ln t + C$

$\int \text{Tan}^{-1} x\, dx = x \text{Tan}^{-1} x - \dfrac{1}{2}\ln(1+x^2) + C$

> Integration by parts Formula
> (**Version 1**)
> $$\int u\, dv = uv - \int v\, du$$
> $dv = dx$
> $v = x$
> $du = \dfrac{1}{1+x^2}\, dx$
>
> **Version 2**
> $$\int (uv) dx = u \int v - \int \left(\dfrac{du}{dx} \cdot \int v\right) dx$$
> $v = 1$
> $\int v\, dx = \int 1 dx = x$
> $\dfrac{du}{dx} = \dfrac{1}{1+x^2}$ (version 2)

Note above that the derivative of $\text{Tan}^{-1} x$ is $\dfrac{1}{1+x^2}$; and the integral

$\int \dfrac{1}{1+x^2}\, dx = \text{Tan}^{-1} x + C$; but $\int \text{Tan}^{-1} x\, dx = x \text{Tan}^{-1} x - \dfrac{1}{2}\ln(1+x^2) + C$.

Example 3: Find $\int \text{Cos}^{-1} x\, dx$

Step 1:

Let $u = \text{Cos}^{-1} x$, and let $dv = 1 dx$ (version 1) or let $v = 1$ (for version 2)

$\int \text{Cos}^{-1} x\, dx = x \text{Cos}^{-1} x - \int \left(-\dfrac{x}{\sqrt{1-x^2}}\right) dx$

$\int \text{Cos}^{-1} x\, dx = x \text{Cos}^{-1} x + \int \dfrac{x}{\sqrt{1-x^2}}\, dx$

Step 2: For $\int \dfrac{x}{\sqrt{1-x^2}}\, dx$

Let $t = 1 - x^2$; $\dfrac{dt}{dx} = -2x$,

$dx = -\dfrac{dt}{2x}$

Substitute in $\int \dfrac{x}{\sqrt{1-x^2}}\, dx$ to obtain

$$\int \dfrac{x}{t^{\frac{1}{2}}} \cdot \left(\dfrac{-dt}{2x}\right)$$

Now, putting everything together,

$\int \text{Cos}^{-1} x\, dx = x \text{Cos}^{-1} x + \int \dfrac{x}{t^{\frac{1}{2}}} \cdot \left(\dfrac{-dt}{2x}\right)$

$\left(\dfrac{du}{dx} = -\dfrac{1}{\sqrt{1-x^2}}\right.$ (version 2))

Step 3:

$= x \text{Cos}^{-1} x - \dfrac{1}{2} \cdot \dfrac{t^{-\frac{1}{2}+1}}{-\frac{1}{2}+1} + C$

$= x \text{Cos}^{-1} x - \dfrac{1}{2}\int t^{-\frac{1}{2}} dt$

$= x \text{Cos}^{-1} x - \dfrac{1}{2} \cdot \dfrac{t^{\frac{1}{2}}}{\frac{1}{2}} + C$

$= x \text{Cos}^{-1} x - t^{\frac{1}{2}} + C$

$= x \text{Cos}^{-1} x - (1-x^2)^{\frac{1}{2}} + C$

$\int \text{Cos}^{-1} x\, dx$

$= x \text{Cos}^{-1} x - \sqrt{1-x^2} + C$

Example 4: Find $\int \operatorname{Sec}^{-1}x\, dx$
Step 1:
Let $u = \operatorname{Sec}^{-1}x$, and let $dv = 1dx$ (version 1) or let $v = 1$ (for version 2)

Then $\int \operatorname{Sec}^{-1}x\, dx = x\operatorname{Sec}^{-1}x - \int \dfrac{x}{x\sqrt{x^2-1}}\, dx$

$\int \operatorname{Sec}^{-1}x\, dx = x\operatorname{Sec}^{-1}x - \int \dfrac{1}{\sqrt{x^2-1}}\, dx$

$\left(\dfrac{du}{dx} = \dfrac{1}{x\sqrt{x^2-1}} \text{ (version 2)}\right)$

Step 2: For $\int \dfrac{1}{\sqrt{x^2-1}}\, dx$.   (A)

Let $x = \sec\theta$; (trigonometric substitution)

Then $\dfrac{dx}{d\theta} = \sec\theta\tan\theta$, and

$dx = \sec\theta\tan\theta\, d\theta$

Substitute in $\int \dfrac{1}{\sqrt{x^2-1}}\, dx$, (A) ,to obtain

$\int \dfrac{1}{\sqrt{\sec^2\theta-1}}\sec\theta\tan\theta d\theta$

$= \int \dfrac{1}{\sqrt{\tan^2\theta}}\sec\theta\tan\theta d\theta$

$= \int \dfrac{1}{\tan\theta}\sec\theta\tan\theta d\theta$

$= \int \sec\theta d\theta$

$\int \dfrac{1}{\sqrt{x^2-1}}\, dx = \ln|\sec\theta + \tan\theta| + C$

$(\sec^2\theta - 1 = \tan^2\theta)$

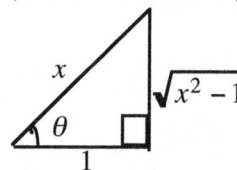

$\sec\theta = x$
$\tan\theta = \sqrt{x^2-1}$

Now, putting everything together,

$\int \operatorname{Sec}^{-1}x\, dx = x\operatorname{Sec}^{-1}x - \ln\left|x + \sqrt{x^2-1}\right| + C .$

To summarize:

1. $\int \operatorname{Sin}^{-1}x\, dx = x\operatorname{Sin}^{-1}x + \sqrt{1-x^2} + C$

2. $\int \operatorname{Cos}^{-1}x\, dx = x\operatorname{Cos}^{-1}x - \sqrt{1-x^2} + C$

3. $\int \operatorname{Tan}^{-1}x\, dx = x\operatorname{Tan}^{-1}x - \frac{1}{2}\ln(1+x^2) + C$

4. $\int \operatorname{Sec}^{-1}x\, dx = x\operatorname{Sec}^{-1}x - \ln\left|x + \sqrt{x^2-1}\right| + C$

5. $\int \operatorname{Csc}^{-1}x\, dx = x\operatorname{Csc}^{-1}x + \ln\left|x + \sqrt{x^2-1}\right| + C$

6. $\int \operatorname{Cot}^{-1}x\, dx == x\operatorname{Cot}^{-1}x + \frac{1}{2}\ln(1+x^2) + C$

EXTRA **Applications of Inverse Cofunction Identities:**

Example 5: We redo Example 3, p.275

Find $\int \text{Cos}^{-1}x\,dx$

Method 2

We use the inverse cofunction identity $\sin^{-1}x + \cos^{-1}x = \frac{\pi}{2}$, and

$\int \text{Sin}^{-1}x\,dx =$
$$x\text{Sin}^{-1}x + \sqrt{1-x^2} + C;$$

to find $\int \text{Cos}^{-1}x\,dx$

Solution: $\text{Cos}^{-1}x = \frac{\pi}{2} - \text{Sin}^{-1}x$

$\int \text{Cos}^{-1}x\,dx = \int \frac{\pi}{2}dx - \int \text{Sin}^{-1}x\,dx$

$\quad = \frac{\pi}{2}x - (x\text{Sin}^{-1}x + \sqrt{1-x^2}) + C$

$\quad = \frac{\pi}{2}x - x\text{Sin}^{-1}x - \sqrt{1-x^2} + C$

$\quad = x(\frac{\pi}{2} - \text{Sin}^{-1}x) - \sqrt{1-x^2} + C$

$\quad = x\text{Cos}^{-1}x - \sqrt{1-x^2} + C$

(**Note:** $\text{Cos}^{-1}x = \frac{\pi}{2} - \text{Sin}^{-1}x$)

Example 6: Find $\int \text{Csc}^{-1}x\,dx$

Given: 1. $\int \text{Sec}^{-1}x\,dx =$
$$x\text{Sec}^{-1}x - \ln\left|x + \sqrt{x^2-1}\right| + C; \text{ and}$$
$\quad\quad$ 2. $\text{Csc}^{-1}x + \text{Sec}^{-1}x = \frac{\pi}{2}$,

$\quad\quad$ find $\int \text{Csc}^{-1}x\,dx$.

Solution $\text{Csc}^{-1}x = \frac{\pi}{2} - \text{Sec}^{-1}x$

$\int \text{Csc}^{-1}x\,dx = \int \frac{\pi}{2}dx - \int \text{Sec}^{-1}x\,dx$

$= \frac{\pi}{2}x - (x\text{Sec}^{-1}x - \ln\left|x + \sqrt{x^2-1}\right|) + C$

$= \frac{\pi}{2}x - x\text{Sec}^{-1}x + \ln\left|x + \sqrt{x^2-1}\right| + C$

$= x(\frac{\pi}{2} - \text{Sec}^{-1}x) + \ln\left|x + \sqrt{x^2-1}\right| + C$

$= x\text{Csc}^{-1}x + \ln\left|x + \sqrt{x^2-1}\right| + C$

(**Note:** $\frac{\pi}{2} - \text{Sec}^{-1}x = \text{Csc}^{-1}x$)

Lesson 39 Exercises

1. $\int \text{Sin}^{-1}x\,dx$;

2. $\int \text{Cos}^{-1}x\,dx$;

3. $\int \text{Tan}^{-1}x\,dx$

4. $\int \text{Sec}^{-1}x\,dx$

5. $\int \text{Csc}^{-1}x\,dx$

6. Find $\int \text{Cot}^{-1}x\,dx$

7. $\int x\text{Tan}^{-1}x\,dx$

8. $\int x\text{Sin}^{-1}x\,dx$

9. $\int \text{Sin}^{-1}\left(\frac{x}{2}\right)dx$

Ans:

7. $\frac{x^2}{2}\text{Tan}^{-1}x - \frac{1}{2}x + \frac{1}{2}\text{Tan}^{-1}x + C$;

8. $\frac{x^2}{2}\text{Sin}^{-1}x + \frac{1}{4}x\sqrt{1-x^2} - \frac{1}{4}\text{Sin}^{-1}x + C$;

9. $x\text{Sin}^{-1}\left(\frac{x}{2}\right) + \sqrt{4-x^2} + C$ or $x\text{Sin}^{-1}\left(\frac{x}{2}\right) + 2\sqrt{1 - \frac{x^2}{4}} + C$

CHAPTER 12

Lesson 40
Introduction to Partial Fraction Decomposition
Integration of Rational Functions II

Introduction to Partial Fraction Decomposition

Consider the addition operations: $\frac{2}{5} + \frac{1}{4} = \frac{13}{20}$

$$\frac{3}{x+2} + \frac{2}{x-1} = \frac{5x+1}{(x+2)(x-1)}$$

In each of the above equations, the terms of left-hand side are the **partial fractions** of the right-hand sides. Note that each left-hand side consists of terms whereas each right-hand side is a single fraction. In integral calculus, there are occasions when we like to break up a single fraction into its equivalent sum of partial fractions. The procedure for breaking up a given single fraction into a sum of terms is called the **partial fraction decomposition** of the given fraction. We shall assume that the rational functions to be decomposed are proper rational functions.

Note that a rational function is **proper** if the degree of the numerator polynomial is less than the degree of the denominator polynomial. If the given fraction is not proper, we will use long division to obtain a polynomial and a proper rational fraction, and then decompose the proper fraction part.

The form of the decomposition depends on the form of the denominators. We will cover the various cases.

Case 1. The denominator has unrepeated (distinct) linear factors of form

$ax + b$. In this case, there is one partial fraction of form $\dfrac{A}{ax+b}$ for

each linear factor, where A is a constant to be determined.

Example: $\dfrac{1}{(x-1)(x+2)} = \dfrac{A}{x-1} + \dfrac{B}{x+2}$, where A and B are
constants to be determined.

Case 2 The denominator has repeated linear factors, each linear factor $(ax + b)$
repeated n times. In this case, there will be n partial fractions of form

$$\frac{A_1}{ax+b} + \frac{A_2}{(ax+b)^2} + \frac{A_3}{(ax+b)^3} + \dots \frac{A_n}{(ax+b)^n}$$

Example 1: $\dfrac{x^2+5}{(x+4)^3} = \dfrac{A_1}{(x+4)} + \dfrac{A_2}{(x+4)^2} + \dfrac{A_3}{(x+4)^3}$

Example 2: $\dfrac{4x+1}{x^2} = \dfrac{A}{x} + \dfrac{B}{x^2}$; **Example 3:** $\dfrac{3x}{x^3} = \dfrac{A}{x} + \dfrac{B}{x^2} + \dfrac{C}{x^3}$

Case 3 The denominator has one unrepeated (distinct) and irreducible quadratic factor ($ax^2 + bx + c$). In this case, the partial fraction decomposition is of the form

$\dfrac{Ax + b}{ax^2 + bx + c}$ where A and B are constants to be determined.

Note: If $b^2 - 4ac$ is negative, then the quadratic factor $ax^2 + bx + c$ is **irreducible**.

For example, $x^2 + 1$ is irreducible since $b^2 - 4ac = 0^2 - 4(1)(1) = -4$ Another way to determine the above irreducibility is if the quadratic factor cannot be decomposed into linear factors without introducing imaginary numbers, then it is irreducible.

Example $\dfrac{2x}{(x^2 + 3)(x^2 + 2)} = \dfrac{Ax + B}{x^2 + 3} + \dfrac{Cx + D}{x^2 + 2}$

Case 4 The denominator has repeated irreducible quadratic factors repeated n times. For each quadratic factor, there will be n corresponding quadratic factors of form

$$\dfrac{(A_1 x + B_1)}{ax^2 + bx + c} + \dfrac{A_2 x + B_2}{(ax^2 + bx + c)^2} + \ldots + \dfrac{A_n x + B_n}{(ax^2 + bx + c)^n}$$

Example $\dfrac{x + 6}{(x^2 + 4)^3} = \dfrac{(A_1 x + B_1)}{x^2 + 4} + \dfrac{A_2 x + B_2}{(x^2 + 4)^2} + \dfrac{A_3 x + B_3}{(x^2 + 4)^3}$

If we can recall the various forms of the partial fraction, then the only other work we have to do is to determine the constants A, B, C, D, etc., in the numerators. Below, we present the process of determining the constants and finding the partial fraction decomposition using examples.

General Procedure:

Step 1: Set up the proper partial fraction sums involving the constants to be determined as illustrated above.

Step 2: Add the partial fractions to obtain a single fraction.

Step 3: Equate the numerator of the partial fraction sum to the numerator of the original fraction.

Step 4: Determine the constants. There are two methods for determining the constants, namely, a general method and a substitution method. In the general method, we equate the coefficients of like powers of the variable on both sides of the equation from Step 3, and solve simultaneous equations. In the substitution method, we properly choose, in turns, values for the variable so as to eliminate some of the constants and determine the other constants.

From the next page, we illustrate the application of partial fraction decomposition in **integrating** some rational functions.

Integration of Rational Functions II
by Partial Fraction Decomposition

Example 1 Find $\int \dfrac{x-5}{x^2+x-2}\,dx = \int \dfrac{x-5}{(x-1)(x+2)}\,dx$

Step 1: (precalculus part)

Decompose the fraction into partial fractions. The partial fraction is of the form below:

$$\frac{x-5}{(x-1)(x+2)} = \frac{A}{x-1} + \frac{B}{x+2}$$

$$\frac{x-5}{(x-1)(x+2)} = \frac{A(x+2) + B(x-1)}{(x-1)(x+2)} \quad (2)$$

Since the denominators are identical, we equate the numerators. (or multiply both sides by $(x-1)(x+2)$)

Then, $x - 5 = A(x+2) + B(x-1)$ (3)

Step 2: Determine the values of A and B.

Method 1: **General method**

Expand the right-hand side of equation (3) and factor out the x.

$$x - 5 = Ax + 2A + Bx - B$$
$$x - 5 = (A+B)x + 2A - B$$

Equate the coefficients of the like powers of x on the left-hand side to the coefficients of like powers of x on the right hand side. Then for the coefficients,

$$1 = A + B \qquad\qquad (5)$$
$$-5 = 2A - B \qquad\qquad (6)$$

Note: We could write -5 as $-5x^0$ and $(2A - B)$ as $(2A - B)x^0$

We now solve equations (5) and (6) simultaneously for A and B.

Solving, $A = -\dfrac{4}{3}$, and $B = \dfrac{7}{3}$

Substituting $A = -\dfrac{4}{3}$, $B = \dfrac{7}{3}$ in eqn. **(1)**

$$\frac{x-5}{(x-1)(x-2)} = \frac{-\frac{4}{3}}{x-1} + \frac{\frac{7}{3}}{x+2}$$

$$\frac{x-5}{(x-1)(x-2)} = -\frac{4}{3}\left(\frac{1}{x-1}\right) + \frac{7}{3}\left(\frac{1}{x+2}\right)$$

Alternative Step 2:

Method 2: Substitution method

Determine the values of A and B. This method involves properly choosing values for x so as to eliminate some of the constants, and then solving the resulting equations accordingly. However, this method may not work if all the unknowns drop out (this may occur if there are repeated factors). We now apply this method to equation (3).

Rewriting equation (3):

$$x - 5 = A(x+2) + B(x-1) \quad (3)$$

If we let $x = -2$ in equation (3) so that the A-term drops out, then

$$-2 - 5 = A(-2+2) + B(-2-1)$$
$$-7 = A(0) + B(-3)$$
$$-7 = 0 - 3B,\text{ and from which}$$
$$\frac{7}{3} = B.$$

If we let $x = 1$ in equation (3) so that the B term drops out, then

$$1 - 5 = A(1+2) + B(1-1)$$
$$-4 = 3A + 0 \text{ and } -\frac{4}{3} = A$$

$$\therefore A = -\frac{4}{3}, \text{ and } B = \frac{7}{3}.\text{Again, we}$$

obtain the same results as by the general method, and

$$\frac{x-5}{(x-1)(x+2)} =$$
$$-\frac{4}{3}\left(\frac{1}{x-1}\right) + \frac{7}{3}\left(\frac{1}{x+2}\right)$$

Step 3: (**The calculus part**) We integrate now.

$$\int \frac{x-5}{(x-1)(x+2)}\,dx = \int\left[-\frac{4}{3}\left(\frac{1}{x-1}\right) + \frac{7}{3}\left(\frac{1}{x+2}\right)\right]dx$$

$$= -\frac{4}{3}\ln|x-1| + \frac{7}{3}\ln|x+2| + C$$

Note above: In the determination involving quadratic factors, we may use a combination of the general method and the substitution method. We should also note that if the given rational expression is not proper (i.e. degree of the numerator polynomial is greater than or equal to that of the denominator polynomial), then we shall use the long division process to reduce the improper rational expression to a polynomial and a proper rational expression (preferably in its lowest terms), and then proceed to apply the decomposition process to the fractional part.

Example 2 Find $\int \dfrac{dx}{a^2 - x^2}$, (Denominator is the difference of two squares)

Method 1: Using Partial Fractions
(In method 2, p.297, we also use trigonometric substitution)

Step 1::(Precalculus) Decompose the fraction into partial fractions.

$$\frac{1}{a^2 - x^2} = \frac{1}{(a + x)(a - x)} \quad (1)$$

$$\frac{1}{(a + x)(a - x)} = \frac{B}{a + x} + \frac{C}{a - x} \quad (2)$$

$$\frac{1}{(a + x)(a - x)} = \frac{B(a - x) + C(a + x)}{(a + x)(a - x)}$$

Since the denominators are the same, equate the numerators to each other.

Thus, $B(a - x) + C(a + x) = 1 \quad (3)$

Step 2: Find the coefficients B and C.
Method 1: General method

$$Ba + Ca + Cx - Bx = 1$$
$$aB + aC + (C - B)x = 1 \quad \text{(Factoring)}$$
$$Ba + Ca = 1 \quad \text{(Equating the constants)}$$
$$C - B = 0 \quad \text{(Equating the coefficient of the } x\text{-terms)}$$

We solve $Ba + Ca = 1$ and $C - B = 0$ simultaneously for B and C..

$$Ba + Ca = 1$$
$$C = B$$
$$Ba + Ba = 1$$

$2Ba = 1$ and $B = \dfrac{1}{2a}$; $C = \dfrac{1}{2a}$

Alternative Step 2
Method 2: Substitution method
Find the coefficients B and C. This method involves properly choosing values for x so as to eliminate some of the constants, and then solving the resulting equations accordingly. However, this method may not work if all the unknowns drop out (this may occur if there are repeated factors)

$B(a - x) + C(a + x) = 1 \quad (3)$
If we let $x = a$ in equation (3) so that the B–term drops out, then

$$B(a - a) + C(a + a) = 1$$

$2aC = 1$ and $C = \dfrac{1}{2a}$

If we let $x = -a$ in equation (3) so that the C–term drops out, then

$$B(a - (-a)) + C(a + (-a)) = 1$$

$2aB = 1$ and $B = \dfrac{1}{2a}$.

Therefore. $B = \dfrac{1}{2a}$ and $C = \dfrac{1}{2a}$.

Step 3: Integrate now.(the calculus part)

$$\int \frac{dx}{a^2 - x^2} = \frac{1}{2a} \int \frac{dx}{a + x} + \frac{1}{2a} \int \frac{dx}{a - x}$$

$$\int \frac{dx}{a^2 - x^2} = \frac{1}{2a} \ln(a + x) - \frac{1}{2a} \ln(a - x) + C.$$

$$\int \frac{dx}{a^2 - x^2} = \frac{1}{2a} \ln\left|\frac{a + x}{a - x}\right| + C \quad \text{or} \quad = \frac{1}{a} \ln\left|\frac{a + x}{\sqrt{a^2 - x^2}}\right| + C \quad \text{<---see also p. 298.}$$

(**Note:** For $\int \dfrac{dx}{a - x}$, $u = a - x$, $-du = dx$)

Example 3 Find $\int \frac{x^2 + 3x + 6}{x + 2} \, dx$

Step 1: (**Precalculus part**)

Find the partial fraction decomposition of $\frac{x^2 + 3x + 6}{x + 2}$

Since the expression is improper, we use the long division process.

$$
\begin{array}{r}
x + 1 \\
x + 2 \overline{\smash{)}\, x^2 + 3x + 6} \\
\underline{x^2 + 2x} \\
x + 6 \\
\underline{x + 2} \\
4
\end{array}
$$

Thus $\frac{x^2 + 3x + 6}{x + 2} = x + 1 + \frac{4}{x + 2}$

Step 2: (**calculus part**) We integrate now.

$$\int \frac{x^2 + 3x + 6}{x + 2} \, dx = \int \left(x + 1 + \frac{4}{x + 2} \right) dx$$

$$= \int x \, dx + \int 1 \, dx + \int \frac{4}{x + 2} \, dx$$

$$= \int x \, dx + \int 1 \, dx 1 + \int \frac{4}{x + 2} \, dx$$

$$= \frac{1}{2} x^2 + x + 4 \ln |x + 2| + C$$

(In the above case, we do not decompose the fractional part any more. The next example requires further decomposition.)

Example 4 Find $\int \frac{x^3 + 2x^2 + 3x + 4}{x^2 - 4} \, dx$

Step 1: (**The precalculus part**) Find the partial fraction decomposition of

$$\frac{x^3 + 2x^2 + 3x + 4}{x^2 - 4}$$

By the long division process we obtain the following;

$$\frac{x^3 + 2x^2 + 3x + 4}{x^2 - 4} = x + 2 + \frac{7x + 12}{x^2 - 4}$$

We now apply the decomposition process to $\frac{7x + 12}{x^2 - 4}$

$$\frac{7x + 12}{x^2 - 4} = \frac{7x + 12}{(x + 2)(x - 2)}$$

Let $\frac{7x + 12}{(x + 2)(x - 2)} = \frac{A}{x + 2} + \frac{B}{x - 2}$ \qquad (1)

$$\frac{7x + 12}{(x + 2)(x - 2)} = \frac{A(x - 2) + B(x + 2)}{(x - 2)(x + 2)}$$

Equating the numerators to each other,

$$7x + 12 = A(x - 2) + B(x + 2) \qquad (2)$$

We now proceed to determine the constants A and B, say ,by the **substitution method.**

Letting $x = 2$ in equation (2), the A-term drops out.

$$7(2) + 12 = A(2 - 2) + B(2 + 2)$$

$26 = A(0) + 4B$ and from which $\frac{13}{2} = B$

Letting $x = -2$ in equation (2) the B-term drops out.

Therefore, $A = \frac{1}{2}$, and $B = \frac{13}{2}$.

We now substitute $A = \frac{1}{2}$, and $B = \frac{13}{2}$ in equation (1), to obtain

$$\frac{7x + 12}{(x + 2)(x - 2)} = \frac{1}{2}\left(\frac{1}{x + 2}\right) + \frac{13}{2(x - 2)}$$

Substituting in the original given rational expression,

$$\frac{x^3 + 2x^2 + 3x + 4}{x^2 - 4} = x + 2 + \frac{1}{2}\left(\frac{1}{x + 2}\right) + \frac{1}{2}\left(\frac{13}{x - 2}\right).$$

Step 2: (**the calculus part**) We integrate now.

$$\int \frac{x^3 + 2x^2 + 3x + 4}{x^2 - 4}\,dx = \int x\,dx + 2\,dx + \frac{1}{2}\int \frac{1}{x + 2}\,dx + \frac{1}{2}\int \frac{13}{x - 2}\,dx$$

$$= \frac{x^2}{2} + 2x + \frac{1}{2}\ln|x + 2| + \frac{13}{2}\ln|x - 2| + C.$$

Lesson 40 Exercises-A Partial Fractions

1. Write $\frac{13}{20}$ as the indicated sum of two common fractions in lowest terms and with different denominators

2. Find $\int \frac{x - 5}{(x - 1)(x + 2)}\,dx$

3. Find $\int \frac{x^2 + 3x + 6}{x + 2}\,dx$

4. Find $\int \frac{x^3 + 2x^2 + 3x + 4}{x^2 - 4}\,dx$

5. Find $\int \frac{1}{x^2 + x - 2}\,dx$

6. Find $\int \frac{1}{x^4 - 16}\,dx$

Answers: 2. $-\frac{4}{3}\ln|x - 1| + \frac{7}{3}\ln|x + 2| + C$; **3.** $\frac{1}{2}x^2 + x + 4\ln|x + 2| + C$

4. $\frac{x^2}{2} + 2x + \frac{1}{2}\ln|x + 2| + \frac{13}{2}\ln|x - 2| + C$;

5. $\frac{1}{3}\ln\left|\frac{x - 1}{x + 2}\right| + C$;, or $\frac{1}{3}\ln|x - 1| - \frac{1}{3}\ln|x + 2| + C$;

6. $\frac{1}{32}\ln\left|\frac{x - 2}{x + 2}\right| - \frac{1}{16}\text{Tan}^{-1}\left(\frac{x}{2}\right) + C$

Lesson 40 Exercises-B Partial Fractions   284

Find the following:

1. $\int \dfrac{2}{(x-2)(x+1)}\, dx$;   **2.** $\int \dfrac{x-3}{(x-1)(x+2)}\, dx$;   **3.** $\int \dfrac{3x+11}{x^2-4}\, dx$;

4. $\int \dfrac{7x-21}{x^2-6x+8}\, dx$ **5.** $\int \dfrac{2x^3-x^2+3x-1}{x^2-25}\, dx$; **6.** $\int \dfrac{2+x^2}{x(x-1)(x+3)}\, dx$

7. $\int \dfrac{x^2-2}{x^3-4x}\, dx$

Answers:

1. $\dfrac{2}{3}\ln|x-2| - \dfrac{2}{3}\ln|x+1| + C$; **2.** $\dfrac{5}{3}\ln|x+2| - \dfrac{2}{3}\ln|x-1| + C$;

3. $\dfrac{17}{4}\ln|x-2| - \dfrac{5}{4}|x+2| + C$; **4.** $\dfrac{7}{2}\ln|x-4| + \dfrac{7}{2}\ln|x-2| + C$;

5. $x^2 - x + \dfrac{291}{10}\ln|x+5| + \dfrac{239}{10}\ln|x-5| + C$;

6. $-\dfrac{2}{3}\ln|x| + \dfrac{3}{4}\ln|x-1| + \dfrac{11}{12}\ln|x+3| + C$; **7.** $\dfrac{1}{4}\ln|x^2-4| + \dfrac{1}{2}\ln|x| + C$

Lesson 41
Introduction to Trigonometric Substitution
Integration of Rational Functions III

Introduction to Trigonometric Substitution

So far, we have used simple u-substitution and partial fraction decomposition to find the integrals of various functions. However, there are some integrals which simple u-substitution and partial fraction decomposition cannot handle.

These integrands involve quadratic expressions such as $\frac{1}{1+x^2}$, $\frac{1}{a^2+x^2}$

$\sqrt{a^2-x^2}$, $x\sqrt{x^2-a^2}$, $\sqrt{1-x^2}$, and $x\sqrt{x^2-1}$. To handle such expressions, we resort to a technique called **trigonometric substitution**. In trigonometric substitution, we use a trigonometric function to represent a variable. For example, let $x = \sin\theta$, or let $x = \sin t$. In addition to the above examples, trigonometric substitution can handle other expressions such as $\sqrt{a^2+x^2}$.

Previously (Lesson 23), we learned that the derivative, $\frac{d}{dx}\operatorname{Sin}^{-1}x = \frac{1}{\sqrt{1-x^2}}$, and

from which we can deduce that the antiderivative, $\int \frac{1}{\sqrt{1-x^2}}\,dx$ is $\operatorname{Sin}^{-1}x + C$.

However, we could still use trigonometric substitution to find $\int \frac{1}{\sqrt{1-x^2}}\,dx$,

even if we did not know that the derivative of $\operatorname{Sin}^{-1}x$ is $\frac{1}{\sqrt{1-x^2}}$. See p.308

We will show for examples that: **1.** $\int \frac{1}{a^2+x^2}\,dx = \frac{1}{a}\operatorname{Tan}^{-1}\left(\frac{x}{a}\right) + C$;

2. $\int \frac{1}{1+x^2}\,dx = \operatorname{Tan}^{-1}x + C$; **3.** $\int \frac{1}{x\sqrt{x^2-a^2}}\,dx = \frac{1}{a}\operatorname{Sec}^{-1}\left(\frac{x}{a}\right) + C$.

Basis of the Trigonometric Substitution Technique

Trigonometric substitution is based on the two Pythagorean identities:

$\qquad \sin^2\theta + \cos^2\theta = 1$ $\qquad\qquad$ (A)

$\qquad \tan^2\theta + 1 = \sec^2\theta$ $\qquad\qquad$ (B)

Observe that from (A), $1 - \sin^2\theta = \cos^2\theta$ $\qquad$ (C)

$\qquad\qquad\qquad 1 - \cos^2\theta = \sin^2\theta$ $\qquad$ (D)

Observe also that in (B), $\tan^2\theta + 1 = \sec^2\theta$ $\qquad$ (E)

and from (B) also: $\sec^2\theta - 1 = \tan^2\theta$ $\qquad$ (F)

In integration by trigonometric substitution, we try to replace each of the two-term expressions on the left side of the equations in (C), (D), (E), and (F), by a one-term square on the right side of these equations so that on simplifying, the variable part does not involve a radical sign. The trigonometric function to use will be determined by the forms of these above identities.

Note also that if convenient, we can also use the third Pythagorean identity

$1 + \cot^2\theta = \csc^2\theta$. (In fact, we have used this identity on p. 290, **Extra**.)

We cover examples according to types of functions: rational functions, then radical functions.

Integration of Rational Functions III
(by Trigonometric Substitution)

Case 1: Denominator of integrand is the sum of two squares

Using $x = \tan\theta$

Example 1: Show that $\int \dfrac{dx}{1+x^2} = \text{Tan}^{-1}x + C$ .or $\arctan x + C$

(Even though from Lessons 23 and 35, we know that the derivative of $\text{Tan}^{-1}x$

is $\dfrac{1}{1+x^2}$, and from which we can deduce that $\int \dfrac{dx}{1+x^2} = \text{Tan}^{-1}x + C$, for
completeness , in this example, we use a new technique called trigonometric
substitution. For this type of the denominator of the integrand , we substitute the
tangent function for the variable x)

Step 1: $\int \dfrac{dx}{1+x^2}$ (A)

Let $x = \tan\theta$ $-\frac{\pi}{2} \le \theta \le \frac{\pi}{2}$ (B)

and from (B) $\tan\theta = \frac{x}{1}$ <------ we use this to draw a right triangle.

Also from (B), $\theta = \text{Tan}^{-1}x$ <-------- we will use this in the last step (C)

From (B) $\dfrac{dx}{d\theta} = \sec^2\theta$ and from which

$dx = \sec^2\theta \, d\theta$ (D)

Also, $x^2 = \tan^2\theta$ (E) (squaring B)

Step 2: Now, substitute for dx from
(D), and for x^2 from (E) in (A) to

$\tan\theta = \dfrac{x}{1}$

$= \dfrac{\text{opposite side}}{\text{adjacent side}}$

obtain $\int \dfrac{\sec^2\theta \, d\theta}{1+\tan^2\theta} = \int \dfrac{\sec^2\theta \, d\theta}{\sec^2\theta}$ <-- $(1 + \tan^2\theta = \sec^2\theta)$

$= \int d\theta = \theta + C = \text{Tan}^{-1}x + C$

$\int \dfrac{dx}{1+x^2} = \text{Tan}^{-1}x + C$ (From (C), $\theta = \text{Tan}^{-1}x$)

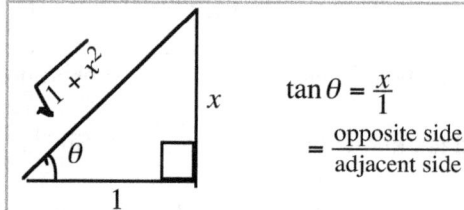

Some Helpful Tips

(a) For problems involving the **sum of two squares** as denominators,
try the **tangent** function for substitution.

Sample candidates are $\dfrac{1}{a^2+x^2}, \dfrac{1}{1+x^2}, \dfrac{1}{\sqrt{a^2+x^2}}, \dfrac{1}{\sqrt{1+x^2}}$

(b) For problems involving the **square root** of the **difference of two
squares**, if the **minus sign** precedes the x^2-term, try the
sine function . Sample candidates are $\sqrt{a^2-x^2} , \dfrac{1}{\sqrt{1-x^2}}, \sqrt{1-x^2}$

(C) For the **square root** of the **difference of two squares**, if the
minus sign precedes the **constant term**, try the **secant** function.

Candidates are: $\sqrt{x^2-1}, x\sqrt{x^2-a^2} , \dfrac{1}{x\sqrt{x^2-1}}, \dfrac{1}{\sqrt{x^2-1}}$ and $\dfrac{\sqrt{x^2-16}}{x}$

Example 2: Show that $\int \dfrac{dx}{a^2 + x^2} = \dfrac{1}{a}\text{Tan}^{-1}\dfrac{x}{a} + C$.

$$\int \dfrac{dx}{a^2 + x^2} \qquad\qquad\qquad\qquad \text{(A)}$$

Step 1: Let $x = a\tan\theta$ $\qquad -\dfrac{\pi}{2} \le \theta \le \dfrac{\pi}{2}$ $\qquad$ (B)

$\qquad \tan\theta = \dfrac{x}{a}$ <----- We use this to draw a right triangle.

$\qquad$ Then $\theta = \text{Tan}^{-1}\dfrac{x}{a}$ <-------- we will use this in the last step $\quad$ (C)

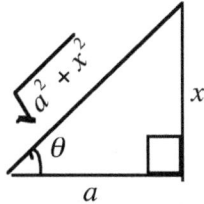

$$\tan\theta = \dfrac{x}{a} = \dfrac{\text{opposite side}}{\text{adjacent side}}$$

$\qquad$ From (B) $\dfrac{dx}{d\theta} = a\sec^2\theta$ and from which

$\qquad dx = a\sec^2\theta\, d\theta.$ $\qquad\qquad\qquad$ (D)

$\qquad$ Also $x^2 = a^2\tan^2\theta$ $\qquad\qquad\qquad$ (E)

Step 2: Substitute for dx from (D), and for x^2 from (E) in (A) to obtain

$$\int \dfrac{a\sec^2\theta\, d\theta}{a^2 + a^2\tan^2\theta} = \int \dfrac{a\sec^2\theta\, d\theta}{a^2(1 + \tan^2\theta)}$$

$$= \int \dfrac{a\sec^2\theta\, d\theta}{a^2\sec^2\theta} \qquad (1 + \tan^2\theta = \sec^2\theta)$$

$$= \int \dfrac{d\theta}{a}$$

$$= \dfrac{1}{a}\int d\theta$$

$$= \dfrac{1}{a}\theta + C$$

$$= \dfrac{1}{a}\text{Tan}^{-1}\dfrac{x}{a} + C \quad (\text{From (C), } \theta = \text{Tan}^{-1}\dfrac{x}{a})$$

$$\int \dfrac{dx}{a^2 + x^2} = \dfrac{1}{a}\text{Tan}^{-1}\dfrac{x}{a} + C.$$

When $a = 1$, we obtain $\int \dfrac{dx}{1 + x^2} = \text{Tan}^{-1}x + C.$ (as in Example 1)

Example 3: Find $\int \dfrac{1}{1+4x^2}\,dx$.   (A)

Method 1

Step 1: Let us write $\int \dfrac{1}{1+4x^2}\,dx$ in

the form $\int \dfrac{dx}{1+x^2}$ whose

antiderivative is $\text{Tan}^{-1}x$ to obtain

$\int \dfrac{dx}{1+(2x)^2}$

Let $u = 2x$. Then $\dfrac{du}{dx} = 2$ and

$dx = \dfrac{du}{2}$.

Substitute accordingly in (A).

Then $\dfrac{1}{2}\int \dfrac{1}{1+u^2}\,du$ (B)

Step 2: Let $u = \tan\theta$,

Then $u^2 = \tan^2\theta$ and (B) becomes

$\dfrac{1}{2}\int \dfrac{1}{1+\tan^2\theta}\,d\theta$ (C)

Also, $\dfrac{du}{d\theta} = \sec^2\theta$, and $du = \sec^2\theta\,d\theta$.

Substituting, we obtain

$\dfrac{1}{2}\int \dfrac{1}{\sec^2\theta}\,\sec^2\theta\,d\theta$

$= \dfrac{1}{2}\int d\theta$

Step 3: $= \dfrac{1}{2}\theta + C$ (integrating)

$= \dfrac{1}{2}\text{Tan}^{-1}(u) + C$

$\int \dfrac{1}{1+4x^2}\,dx = \dfrac{1}{2}\text{Tan}^{-1}2x + C$.

$(u = 2x)$

Draw a right triangle using $\tan\theta = \dfrac{u}{1}$

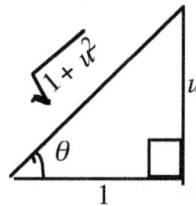

Method 2   289

Step 1: Let us write $\int \dfrac{1}{1+4x^2} \, dx$ in

the form $\int \dfrac{dx}{a^2+x^2}$

$\int \dfrac{1}{1+4x^2} \, dx = \int \dfrac{1}{4(\frac{1}{4}+x^2)} \, dx$

$= \dfrac{1}{4} \int \dfrac{1}{\frac{1}{4}+x^2} \, dx$

$= \dfrac{1}{4} \int \dfrac{1}{\left(\frac{1}{2}\right)^2 + x^2} \, dx$. (A)

Step 2. Now, $a = \dfrac{1}{2}$

Let $x = a \tan\theta$ $\qquad -\dfrac{\pi}{2} \le \theta \le \dfrac{\pi}{2}$ (B)

$x = \dfrac{1}{2} \tan\theta$

$\tan\theta = 2x$. Then $\theta = \text{Tan}^{-1} 2x$
(We will use this in the last step)

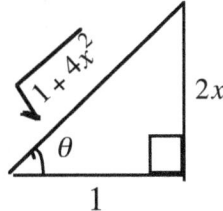

From (B), $\dfrac{dx}{d\theta} = \dfrac{1}{2} \sec^2\theta$ and from

which $dx = \dfrac{1}{2} \sec^2\theta \, d\theta$ $\qquad$ (D)

Also $x^2 = \dfrac{1}{4} \tan^2\theta$ (E)

Step 3: Now, substitute for dx and for x from (D) and (E) respectively in (A)
to obtain

$= \dfrac{1}{4} \int \dfrac{\frac{1}{2}\sec^2\theta \, d\theta}{\frac{1}{4} + \left(\frac{1}{2}\right)^2 \tan^2\theta}$ $\qquad$ $\left(dx = \dfrac{1}{2}\sec^2\theta \, d\theta \, ; \ x^2 = \dfrac{1}{4}\tan^2\theta \right)$

$= \dfrac{1}{4} \int \dfrac{\frac{1}{2}\sec^2\theta \, d\theta}{\frac{1}{4} + \frac{1}{4}\tan^2\theta}$

$= \dfrac{1}{4} \int \dfrac{\frac{1}{2}\sec^2\theta \, d\theta}{\frac{1}{4}(1 + \tan^2\theta)}$

$= \dfrac{1}{4} \int \dfrac{\frac{1}{2}\sec^2\theta \, d\theta}{\frac{1}{4}\sec^2\theta}$ $\qquad$ $(1 + \tan^2\theta = \sec^2\theta)$

$= \dfrac{1}{2} \int d\theta$

$= \dfrac{1}{2}\theta + C$

$= \dfrac{1}{2}\text{Tan}^{-1} 2x + C$ $\qquad$ $(\theta = \text{Tan}^{-1} 2x \text{ from Step 2})$

Therefore, $\int \dfrac{1}{1+4x^2} \, dx = \dfrac{1}{2}\text{Tan}^{-1} 2x + C$.

Method 3

Step 1: (same as Step 1 of Method 1, and we repeat it below)

Let us write $\int \dfrac{1}{1+4x^2}\,dx$ in

the form $\int \dfrac{dx}{a^2+x^2}$

$\int \dfrac{1}{4(\frac{1}{4}+x^2)}\,dx$

$= \int \dfrac{1}{4(\frac{1}{4}+x^2)}\,dx$

$= \dfrac{1}{4}\int \dfrac{1}{\frac{1}{4}+x^2}\,dx$

$= \dfrac{1}{4}\int \dfrac{1}{\left(\frac{1}{2}\right)^2+x^2}\,dx$ (B).

Step 2: From memory,

$\int \dfrac{dx}{a^2+x^2} = \dfrac{1}{a}\,\text{Tan}^{-1}\dfrac{x}{a}+C$ (C)

comparing (B) and (C), $a=\frac{1}{2}$

$\dfrac{1}{4}\int \dfrac{1}{\left(\frac{1}{2}\right)^2+x^2}\,dx = \dfrac{1}{4}\left[\dfrac{1}{\frac{1}{2}}T^{-1}\left(\dfrac{x}{\frac{1}{2}}\right)\right]+C$

$= \dfrac{1}{2}\,Tan^{-1}(2x)+C$

$\therefore \int \dfrac{1}{1+4x^2}\,dx = \dfrac{1}{2}\,Tan^{-1}(2x)+C.$

Extra: Show that $\int -\dfrac{dx}{1+x^2} = \text{Cot}^{-1}x+C$ or $\text{arc}\cot x + C$.

Step 1: $\int -\dfrac{dx}{1+x^2}$ (A)

Let $x=\cot\theta$ $0 \le \theta \le \pi$ (B)

and from (B) $\cot\theta = \frac{x}{1}$ <------ we use this to draw a right triangle.

Also from (B), $\theta = \text{Cot}^{-1}x$ <-------- we will use this in the last step.   (C)

From (B) $\dfrac{dx}{d\theta} = -\csc^2\theta$ and from which

$dx = -\csc^2\theta\,d\theta$.   (D)

Also $x^2 = \cot^2\theta$ (E)   (squaring (B)

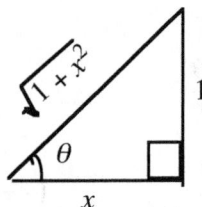

Step 2: Now, substitute for dx from (D), and for x^2 from (E) in (A) to obtain

$\int -\dfrac{(-\csc^2\theta\,d\theta)}{1+\cot^2\theta} = \int \dfrac{\csc^2\theta\,d\theta}{\csc^2\theta}$ $(1+\cot^2\theta = \csc^2\theta)$.

$= \int d\theta$

$= \theta + C$

$= \text{Cot}^{-1}x + C$ (From (C), $\theta = \text{Cot}^{-1}x$)

$\int -\dfrac{dx}{1+x^2} = \text{Cot}^{-1}x + C.$

Exception
The following example is not a rational function but rather a **radical function.**
By simple u-substitution in Steps 1 and 2, we integrate a **rational function** in
Step 3. See also Lesson 43.

Example 4 Find $\int \dfrac{\sqrt{x}}{\sqrt[3]{x}+1}\,dx$ Note: This is not a rational function.

We begin with u-substitution, and in Step 3, we apply trigonometric substitution
to one of the integrals.

Step 1: Rationalize the integral. Since there are two different roots, the square
root , with the radical index 2, and the cube root with the index 3, find the
LCM of 2 and 3 which is 6.; and we

let $u^6 = x$.

From $u^6 = x$

$\dfrac{dx}{du} = 6u^5$

$dx = 6u^5 du$ and

$\sqrt{x} = u^3$

$\sqrt[3]{x} = u^2$

	Scrapwork	
	1. If $u^6 = x$;	**2.** If $u^6 = x$;
	$\left(u^6\right)^{\frac{1}{2}} = x^{\frac{1}{2}}$	$\left(u^6\right)^{\frac{1}{3}} = x^{\frac{1}{3}}$
	$u^3 = x^{\frac{1}{2}} = \sqrt{x}$	$u^2 = x^{\frac{1}{3}} = \sqrt[3]{x}$

Step 2: Substitute u^3 for $\sqrt{x}$; u^2 for $\sqrt[3]{x}$; $6u^5 du$ for dx in $\int \dfrac{\sqrt{x}}{\sqrt[3]{x}+1}\,dx$. Then

$\int \dfrac{\sqrt{x}}{\sqrt[3]{x}+1}\,dx = \int \dfrac{u^3}{u^2+1} \bullet \dfrac{6u^5 du}{1}$

$= 6\int \dfrac{u^8}{u^2+1}\,du$

$= 6\int \left(u^6 - u^4 + u^2 - 1 + \dfrac{1}{u^2+1}\right)du$ (using long division)

Step 3: $= 6\left(\dfrac{u^7}{7} - \dfrac{u^5}{5} + \dfrac{u^3}{3} - u + \text{Tan}^{-1}(u)\right) + C$ (4th integral by trig. substitution)

$= \dfrac{6}{7}u^7 - \dfrac{6}{5}u^5 + 2u^3 - 6u + 6\text{Tan}^{-1}(u)) + C$

$\int \dfrac{\sqrt{x}}{\sqrt[3]{x}+1}\,dx = \dfrac{6}{7}x^{\frac{7}{6}} - \dfrac{6}{5}x^{\frac{5}{6}} + 2x^{\frac{1}{2}} - 6x^{\frac{1}{6}} + 6\text{Tan}^{-1}(x^{\frac{1}{6}})) + C$.

Note: $u = x^{\frac{1}{6}}$, $u^7 = x^{\frac{7}{6}}$, $u^5 = x^{\frac{5}{6}}$, $u^3 = x^{\frac{3}{6}} = x^{\frac{1}{2}}$.

Case 2:   **Denominator of the Integrand is a**
power of the sum of two squares

Using $x = \tan\theta$

Example 5: Find $\int \dfrac{dx}{(1+x^2)^2}$.   ,

Here, as in Case 1, Example 1, we substitute the tangent function for the variable x

Step 1: $\int \dfrac{dx}{(1+x^2)^2}$ (A)

Let $x = \tan\theta$ $-\frac{\pi}{2} \le \theta \le \frac{\pi}{2}$ (B)

and from (B) $\tan\theta = \frac{x}{1}$ <------ we use this to draw a right triangle.

Also from (B), $\theta = \text{Tan}^{-1}x$ <--------we will use this in the last step   (C)

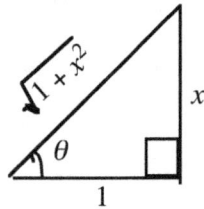

$\sin\theta = \dfrac{x}{\sqrt{1+x^2}}$

$\cos\theta = \dfrac{1}{\sqrt{1+x^2}}$

From (B) $\dfrac{dx}{d\theta} = \sec^2\theta$ and from which

$dx = \sec^2\theta\, d\theta$.   (D)

Also $x^2 = \tan^2\theta$ (E)   (squaring (B))

Step 2: Now, substitute for dx from (D), and for x^2 from (E) in (A) to obtain

$\int \dfrac{\sec^2\theta\, d\theta}{(1+\tan^2\theta)^2} = \int \dfrac{\sec^2\theta\, d\theta}{(\sec^2\theta)^2}$

$= \int \dfrac{d\theta}{\sec^2\theta}$

$= \int \cos^2\theta\, d\theta$

$= \int \frac{1}{2}\cos 2\theta\, d\theta + \int \frac{1}{2}\, d\theta$

$= \left(\frac{1}{2}\right)\left(\frac{1}{2}\right)\sin 2\theta + \frac{1}{2}\theta + C$

$= \frac{1}{4}(2)\sin\theta\cos\theta + \frac{1}{2}\theta + C$

$= \frac{1}{2}\sin\theta\cos\theta + \frac{1}{2}\theta + C$

$= \frac{1}{2}\cdot\dfrac{x}{\sqrt{1+x^2}}\cdot\dfrac{1}{\sqrt{1+x^2}} + \frac{1}{2}\arctan x + C$

$\int \dfrac{dx}{(1+x^2)^2} = \dfrac{x}{2(1+x^2)} + \frac{1}{2}\arctan x + C$.

Scrapwork

⇓

$\left(\cos^2\theta = \frac{1}{2}\cos 2\theta + \frac{1}{2}\right)$ <-trig id.

$\left(\int \cos\theta\, d\theta = \sin\theta + C\right)$ and

$\int \frac{1}{2}\cos 2\theta\, d\theta = \left(\frac{1}{2}\right)\frac{1}{2}\sin 2\theta + C$

using simple u-substitution.

$\sin 2\theta = 2\sin\theta\cos\theta$

$\left(\sqrt{1+x^2}\right)\left(\sqrt{1+x^2}\right) = 1+x^2$

Example 6: Find $\int \dfrac{dx}{(1+x^2)^3}$. $\left(\int \dfrac{dx}{(1+x^2)^n}, \quad n=3 \right)$

For this example, we use trigonometric substitution
where, we substitute the tangent function for the variable x

Step 1: $\int \dfrac{dx}{(1+x^2)^3}$ (A)

Let $x = \tan\theta$ $\quad -\frac{\pi}{2} \le \theta \le \frac{\pi}{2}$ (B)

and from (B) $\tan\theta = \frac{x}{1}$ <------ we use this to draw a right triangle.

Also from (B), $\theta = \text{Tan}^{-1}x$ <-------we will use this in the last step (C)

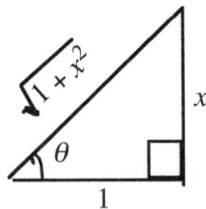

$\sin\theta = \dfrac{x}{\sqrt{1+x^2}}$

$\cos\theta = \dfrac{1}{\sqrt{1+x^2}}$

The following does not refer to this triangle;
but will be used in step 4 below

$\sin 4\theta = 4\sin\theta\cos\theta\cos 2\theta$

From (B) $\dfrac{dx}{d\theta} = \sec^2\theta$ and from which $dx = \sec^2\theta\, d\theta$. (D)

Also $x^2 = \tan^2\theta$ (E) (squaring (B))

Step 2: Now, substitute for dx from (D), and for x^2 from (E) in (A) to obtain

Step 2: $\int \dfrac{\sec^2\theta\, d\theta}{(1+\tan^2\theta)^3}$

$= \int \dfrac{\sec^2\theta\, d\theta}{(\sec^2\theta)^3}$

$= \int \dfrac{\sec^2\theta\, d\theta}{(\sec^6\theta)}$

$= \int \dfrac{d\theta}{\sec^4\theta}$

$= \int \cos^4\theta\, d\theta$

$= \int (\cos^2\theta)^2\, d\theta$

$= \int \left(\frac{1}{2}\cos 2\theta + \frac{1}{2}\right)^2 d\theta$

Step 3: $= \int \left(\frac{1}{4}\cos^2 2\theta + 2\cdot\frac{1}{4}\cos 2\theta + \frac{1}{4}\right)d\theta$

$\left(\cos^2\theta = \frac{1}{2}\cos 2\theta + \frac{1}{2}\right)$

$= \int \frac{1}{4}\left\{\left(\frac{1}{2}\cos 4\theta + \frac{1}{2}\right) + \frac{1}{2}\cos 2\theta + \frac{1}{4}\right\}d\theta$

$= \int \left(\frac{1}{8}\cos 4\theta + \frac{1}{8} + \frac{1}{2}\cos 2\theta + \frac{1}{4}\right)d\theta$

$= \int \left(\frac{1}{8}\cos 4\theta + \frac{1}{2}\cos 2\theta + \frac{3}{8}\right)d\theta$

$= \left(\frac{1}{8}\right)\frac{1}{4}\sin 4\theta + \frac{1}{2}\left(\frac{1}{2}\right)\sin 2\theta + \frac{3}{8}\theta + C$

$= \frac{1}{32}\sin 4\theta + \frac{1}{4}\sin 2\theta + \frac{3}{8}\theta + C$

Step 4: $= \dfrac{1}{32}\left(4\sin\theta\cos\theta\right)\left(\cos^2\theta - \sin^2\theta\right) + \dfrac{1}{4}\left(2\sin\cos\theta\right) + \dfrac{3}{8}\theta + C$

$= \frac{1}{8}\left(\sin\theta\cos\theta\right)\left(\cos^2\theta - \sin^2\theta\right) + \frac{1}{2}\left(\sin\theta\cos\theta\right) + \frac{3}{8}\theta + C$

$= \frac{1}{8}\left(\dfrac{x}{\sqrt{1+x^2}}\dfrac{1}{\sqrt{1+x^2}}\right)\left(\dfrac{1}{1+x^2} - \dfrac{x^2}{1+x^2}\right) + \frac{1}{2}\left(\dfrac{x}{1+x^2}\cdot\dfrac{1}{1+x^2}\right) + \frac{3}{8}\arctan x + C$

$= \frac{1}{8}\left(\dfrac{x}{1+x^2}\right)\left(\dfrac{1-x^2}{1+x^2}\right) + \frac{1}{2}\left(\dfrac{x}{1+x^2}\right) + \frac{3}{8}\arctan x + C$

$= \dfrac{3x^3 + 5x}{8(1+x^2)^2} + \frac{3}{8}\arctan x + C.$

Case 3: Integration of a Rational Function with a Quadratic Trinomial Denominator

Complete the square, apply simple u-substitution, followed by trigonometric substitution.

Example 7 : Find $\int \dfrac{x}{x^2 - 6x + 13}\,dx$ (Note: In Step 2 $\tan^2\theta + 1 = \sec^2 x$)

Step 1: $\int \dfrac{x}{x^2 - 6x + 13}\,dx = \int \dfrac{x}{x^2 - 6x + (-3)^2 - (-3)^2 + 13}\,dx$

$$= \int \dfrac{x}{(x-3)^2 - 9 + 13}\,dx \quad \text{(completing the square)}$$

$$= \int \dfrac{x}{(x-3)^2 + 4}\,dx \qquad \begin{array}{l} \text{Let } u = x - 3 \text{ . Then} \\ du = dx \text{, and } x = u + 3 \end{array}$$

$$\int \dfrac{x}{(x-3)^2 + 4}\,dx = \int \dfrac{u+3}{u^2 + 4}\,du$$

$$\int \dfrac{u+3}{u^2+4}\,du = \int \dfrac{u}{u^2+4}\,du + \int \dfrac{3}{u^2+4}\,du \quad \text{(Splitting the numerators)}$$

Step 2a: For $\int \dfrac{u}{u^2+4}\,du$ (first integral)

Let $t = u^2 + 4$. Then $\dfrac{dt}{du} = 2u$

Substituting accordingly,

$$\int \dfrac{u}{u^2+4}\,du = \int \dfrac{u}{t} \cdot \dfrac{dt}{2u}\,du \text{ (see Case 4)}$$

$$= \dfrac{1}{2}\int \dfrac{1}{t}\,dt$$

$$= \dfrac{1}{2}\ln|t| + C_1$$

$$= \dfrac{1}{2}\ln|u^2 + 4| + C_1$$

$$= \dfrac{1}{2}\ln|(x-3)^2 + 4| + C_1$$

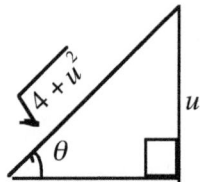

Figure

Step 2b: For $\int \dfrac{3}{u^2+4}\,du$ (second integral)

Let $u = 2\tan\theta$ (then $\tan\theta = \dfrac{u}{2}$)

$$u^2 = 4\tan^2\theta$$

$$\dfrac{du}{d\theta} = 2\sec^2\theta \text{ and } du = 2\sec^2\theta\,d\theta$$

$$\int \dfrac{3}{u^2+4}\,du = 3\int \dfrac{2\sec^2\theta\,d\theta}{4\tan^2\theta + 4}$$

$$= 3\int \dfrac{2\sec^2\theta\,d\theta}{4(\tan^2\theta + 1)}$$

$$= 3\int \dfrac{2\sec^2\theta\,d\theta}{4\sec^2\theta}$$

$$= \dfrac{3}{2}\int d\theta$$

$$= \dfrac{3}{2}\theta + C_2$$

$$= \dfrac{3}{2}\tan^{-1}\left(\dfrac{u}{2}\right) + C_2 \quad (\theta = \tan^{-1}\left(\dfrac{u}{2}\right))$$

$$= \dfrac{3}{2}\tan^{-1}\left(\dfrac{x-3}{2}\right) + C_2 \quad \text{(see Figure)}$$

Step 3: Above, we find the two integrals in Steps 2a and 2b

$$\int \dfrac{u}{u^2+4}\,du + \int \dfrac{3}{u^2+4}\,du = \dfrac{1}{2}\ln|(x-3)^2 + 4| + C_1 + \dfrac{3}{2}\tan^{-1}\left(\dfrac{x-3}{2}\right) + C_2$$

$$= \dfrac{1}{2}\ln|(x-3)^2 + 4| + \dfrac{3}{2}\tan^{-1}\left(\dfrac{x-3}{2}\right) + C_1 + C_2$$

Therefore, $\int \dfrac{x}{x^2 - 6x + 13}\,dx = \dfrac{1}{2}\ln|(x-3)^2 + 4| + \dfrac{3}{2}\tan^{-1}\left(\dfrac{x-3}{2}\right) + C$

Example 8a: Find $\int \frac{1}{4x^2 + 6x + 9} dx$

Step 1: $\int \frac{1}{4x^2 + 6x + 9} dx = \int \frac{1}{4[x^2 + \frac{3}{2}x + \frac{9}{4}]} dx$

Step 2: Write denominator in the form

$$a^2 + 1$$

$$\int \frac{dx}{4[x^2 + \frac{3}{2}x + \left(\frac{3}{4}\right)^2 - \left(\frac{3}{4}\right)^2 + \frac{9}{4}]}$$

$$= \frac{1}{4} \int \frac{dx}{\left(x + \frac{3}{4}\right)^2 - \frac{9}{16} + \frac{9}{4}}$$

$$= \frac{1}{4} \int \frac{dx}{\left(x + \frac{3}{4}\right)^2 + \frac{36-9}{16}}$$

$$= \frac{1}{4} \int \frac{dx}{\left(x + \frac{3}{4}\right)^2 + \frac{27}{16}}$$

$$= \frac{1}{4} \int \frac{\left(\frac{16}{27}\right)}{\left[\left(x + \frac{3}{4}\right)^2 + \frac{27}{16}\right]\frac{16}{27}} dx$$

$$= \frac{1}{4} \cdot \frac{16}{27} \int \frac{dx}{\left(x + \frac{3}{4}\right)^2 \cdot \frac{16}{27} + 1}$$

$$= \frac{4}{27} \int \frac{dx}{\left[\frac{4}{\sqrt{27}}\left(x + \frac{3}{4}\right)\right]^2 + 1}$$

Step 3:

Let $u = \frac{4}{\sqrt{27}}\left(x + \frac{3}{4}\right)$.

Then $\frac{du}{dx} = \frac{4}{\sqrt{27}}(1)$, and

$$dx = \frac{\sqrt{27}}{4} du.$$

Substituting, we obtain

$$= \frac{4}{27} \int \frac{\frac{\sqrt{27}}{4} du}{u^2 + 1}$$

$$= \frac{\sqrt{3}}{9} \int \frac{du}{u^2 + 1}$$

(antiderivative of $\frac{1}{u^2 + 1}$ is $\arctan u$)

$$= \frac{\sqrt{3}}{9}[\arctan u] + C$$

$$= \frac{\sqrt{3}}{9}\left[\arctan \frac{4(x + \frac{3}{4})}{\sqrt{27}}\right] + C$$

$$= \frac{\sqrt{3}}{9}\arctan \frac{4\sqrt{3}(x + \frac{3}{4})}{9} + C$$

$$= \frac{\sqrt{3}}{9}\arctan \frac{4\sqrt{3}}{9}\left(\frac{4x + 3}{4}\right) + C$$

$$= \frac{\sqrt{3}}{9}\arctan\left[\frac{\sqrt{3}}{9}(4x + 3)\right] + C.$$

Example 8b: Find $\int \dfrac{x}{4x^2 + 6x + 9}\,dx$ (A)

Step 1: Complete the square on the denominator.

$$\int \dfrac{x}{4x^2 + 6x + 9}\,dx = \int \dfrac{x}{4[x^2 + \frac{3}{2}x + \frac{9}{4}]}\,dx$$

$$\int \dfrac{x\,dx}{4[x^2 + \frac{3}{2}x + \left(\frac{3}{4}\right)^2 - \left(\frac{3}{4}\right)^2 + \frac{9}{4}]}$$

$$= \dfrac{1}{4}\int \dfrac{x\,dx}{\left(x + \frac{3}{4}\right)^2 + \frac{27}{16}}$$

Step 2: Write denominator in the form $a^2 + 1$

$$= \dfrac{1}{4}\int \dfrac{x\left(\frac{16}{27}\right)}{\left[\left(x + \frac{3}{4}\right)^2 + \frac{27}{16}\right]\frac{16}{27}}\,dx$$

$$= \dfrac{1}{4} \cdot \dfrac{16}{27}\int \dfrac{x\,dx}{\left(x + \frac{3}{4}\right)^2 \cdot \frac{16}{27} + 1}$$

$$= \dfrac{4}{27}\int \dfrac{x\,dx}{\left[\frac{4}{\sqrt{27}}\left(x + \frac{3}{4}\right)\right]^2 + 1}$$ (B)

Step 3: Let $u = \dfrac{4}{\sqrt{27}}\left(x + \frac{3}{4}\right)$. Then

Step 4 $\dfrac{du}{dx} = \dfrac{4}{\sqrt{27}}(1)$, and $dx = \dfrac{\sqrt{27}}{4}\,du$

$x = \dfrac{3(u\sqrt{3} - 1)}{4}$ (from $u = \dfrac{4}{\sqrt{27}}\left(x + \frac{3}{4}\right)$

Substitute for x, and dx in (B), Then

$$= \dfrac{4}{27}\int \dfrac{\frac{3(u\sqrt{3} - 1)}{4} \cdot \frac{\sqrt{27}}{4}\,du}{u^2 + 1}$$

$$= \dfrac{4}{27} \cdot \dfrac{9\sqrt{3}}{4(4)}\int \dfrac{(u\sqrt{3} - 1)\,du}{u^2 + 1}$$ (C)

$$= \dfrac{\sqrt{3}\sqrt{3}}{12}\int \dfrac{u\,dt}{t \cdot 2u} - \dfrac{\sqrt{3}}{12}\int \dfrac{1\,du}{u^2 + 1}$$

Let $t = u^2 + 1$ in the first integral only.
Then $dt = 2u\,du$; and (C) becomes

$$= \dfrac{1}{4 \cdot 2}\int \dfrac{dt}{t} - \dfrac{\sqrt{3}}{12}\int \dfrac{1\,du}{u^2 + 1}$$

$$= \dfrac{1}{8}\int \dfrac{dt}{t} - \dfrac{\sqrt{3}}{12}\int \dfrac{1\,du}{u^2 + 1}$$

$$= \dfrac{1}{8}\ln t - \dfrac{\sqrt{3}}{12}\arctan u + C$$

Step 5 $= \dfrac{1}{8}\ln\left|\left[\frac{4}{\sqrt{27}}\left(x + \frac{3}{4}\right)\right]^2 + 1\right| - \dfrac{\sqrt{3}}{12}\arctan u + C$

$$= \dfrac{1}{8}\ln\left|\left[\frac{4}{\sqrt{27}}\left(x + \frac{3}{4}\right)\right]^2 + 1\right| - \dfrac{\sqrt{3}}{12}\arctan \frac{4}{\sqrt{27}}\left(x + \frac{3}{4}\right) + C$$

$$= \dfrac{1}{8}\ln\left|\frac{16}{27}\left(\frac{16x^2 + 24x + 9}{16}\right) + 1\right| - \dfrac{\sqrt{3}}{12}\arctan \frac{4\sqrt{3}}{9}\left(\frac{4x + 3}{4}\right) + C$$

$$= \dfrac{1}{8}\ln\left|\left(\frac{16x^2 + 24x + 36}{27}\right)\right| - \dfrac{\sqrt{3}}{12}\arctan \frac{\sqrt{3}}{9}(4x + 3) + C$$

$$= \dfrac{1}{8}\ln\left|\left(\frac{4(4x^2 + 6x + 9)}{27}\right)\right| - \dfrac{\sqrt{3}}{12}\arctan \frac{\sqrt{3}}{9}(4x + 3) + C$$

$$= \dfrac{1}{8}\left[\ln|4x^2 + 6x + 9| + \ln\frac{4}{27}\right] - \dfrac{\sqrt{3}}{12}\arctan \frac{\sqrt{3}}{9}(4x + 3) + C$$

$$= \dfrac{1}{8}\left[\ln|4x^2 + 6x + 9|\right] - \dfrac{\sqrt{3}}{12}\arctan\left[\frac{\sqrt{3}}{9}(4x + 3)\right] + C \quad \text{(Note: } \ln\frac{4}{27} \text{ is a constant}$$

and can be absorbed into C))

Case 4: Integrating a Rational Function in which the Denominator is the Difference between two Squares

Using $x = a\sin\theta$

Some integrations can be done by trigonometric substitution as well as by partial fraction decomposition. In Method 1, p.281, we used partial fraction decomposition. In Method 2, below, we use trigonometric substitution.

Example 9 Find $\int \dfrac{dx}{a^2 - x^2}$ (A)

Method 2: Using Trigonometric substitution

Step 1: Let $x = a\sin\theta$ $-\dfrac{\pi}{2} \le \theta \le \dfrac{\pi}{2}$ (B)

Then $\sin\theta = \dfrac{x}{a}$. Draw the corresponding right triangle using $\sin\theta = \dfrac{x}{a}$ to obtain

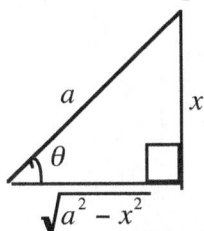

$\text{sine} = \dfrac{\text{opposite side}}{\text{hypotenuse}}$

Place x and a accordingly, and calculate the third side.

Step 2: From (B) $\dfrac{dx}{d\theta} = a\cos\theta$ and from which

$dx = a\cos\theta \, d\theta$.   (D)

Also $x^2 = a^2 \sin^2\theta$ (E)

Step 3: Substitute for dx and for x from (D) and respectively, in (A) to obtain

$$\int \frac{a\cos\theta d\theta}{a^2 - a^2\sin^2\theta} = \int \frac{a\cos\theta d\theta}{a^2(1 - \sin^2\theta)}$$

$$= \int \frac{a\cos\theta d\theta}{a^2\cos^2\theta}$$

$$= \int \frac{d\theta}{a\cos\theta}$$

$$= \frac{1}{a}\int \sec\theta \, d\theta$$

Scrapwork

$\cos^2\theta = 1 - \sin^2\theta$

$$\int \frac{dx}{a^2 - x^2} = \frac{1}{a}\ln|\sec\theta + \tan\theta| + C. \qquad \left(\int \sec\theta \, d\theta = \ln|\sec\theta + \tan\theta| + C\right)$$

Step 4: Now, we convert back from θ to x using the right triangle in Step 1

$$\sec\theta = \frac{a}{\sqrt{a^2 - x^2}}, \quad \tan\theta = \frac{x}{\sqrt{a^2 - x^2}}$$

$$\int \frac{dx}{a^2 - x^2} = \frac{1}{a}\ln\left|\frac{a}{\sqrt{a^2 - x^2}} + \frac{x}{\sqrt{a^2 - x^2}}\right| + C$$

$$= \frac{1}{a}\ln\left|\frac{a + x}{\sqrt{a^2 - x^2}}\right| + C. \text{ which is equivalent to } \frac{1}{2a}\ln\left|\frac{a + x}{a - x}\right| + C \text{ on p.281.}$$

When $a = 1$, we obtain $\int \dfrac{dx}{1 - x^2} = \ln\left|\dfrac{1 + x}{\sqrt{1 - x^2}}\right| + C = \dfrac{1}{2}\ln\left|\dfrac{1 + x}{1 - x}\right| + C.$

Extra: Below, we show that. $\dfrac{1}{a}\ln\left|\dfrac{a+x}{\sqrt{a^2-x^2}}\right| = \dfrac{1}{2a}\ln\left|\dfrac{a+x}{a-x}\right|$,

Approach 1 (see also p.302 for math similarity)

Change $\dfrac{1}{2a}\ln\left|\dfrac{a+x}{a-x}\right|$ to $\dfrac{1}{a}\ln\left|\dfrac{a+x}{\sqrt{a^2-x^2}}\right|$

Step 1:

$$\dfrac{1}{2a}\ln\left|\dfrac{a+x}{a-x}\right| = \dfrac{1}{a}\left|\ln\left(\dfrac{a+x}{a-x}\right)^{\frac{1}{2}}\right|$$

Step 2: $= \dfrac{1}{a}\ln\left|\dfrac{\sqrt{a+x}}{\sqrt{a-x}}\right|$

$= \dfrac{1}{a}\ln\left|\dfrac{(\sqrt{a+x})(\sqrt{a+x})}{(\sqrt{a-x})(\sqrt{a+x})}\right|$

$= \dfrac{1}{a}\ln\left|\dfrac{a+x}{\sqrt{a^2-x^2}}\right|$

Approach 2

Step 1: Rationalize the denominator

$$\dfrac{1}{a}\ln\left|\dfrac{a+x}{\sqrt{a^2-x^2}}\right| + C$$

$= \dfrac{1}{a}\ln\left|\dfrac{(a+x)\left(\sqrt{a^2-x^2}\right)}{\left(\sqrt{a^2-x^2}\right)\left(\sqrt{a^2-x^2}\right)}\right| + C$

$= \dfrac{1}{a}\ln\left|\dfrac{(a+x)\left(\sqrt{a^2-x^2}\right)}{a^2-x^2}\right| + C$

$= \dfrac{1}{a}\ln\left|\dfrac{(a+x)\left(\sqrt{a^2-x^2}\right)}{(a+x)(a-x)}\right| + C$

Step 2: $= \dfrac{1}{a}\ln\left|\dfrac{\left(a^2-x^2\right)^{\frac{1}{2}}}{(a-x)}\right| + C$

$= \dfrac{1}{a}\left[\ln\left(a^2-x^2\right)^{\frac{1}{2}} - \ln(a-x)\right] + C$

$= \dfrac{1}{a}\left[\dfrac{1}{2}\ln\left(a^2-x^2\right) - \ln(a-x)\right] + C$

$= \dfrac{1}{a}\left[\dfrac{1}{2}\left(\ln(a+x)(a-x)\right) - \ln(a-x)\right] + C$

$= \dfrac{1}{a}\left[\dfrac{1}{2}\left(\ln(a+x) + \ln(a-x)\right) - \ln(a-x)\right] + C$

$= \dfrac{1}{a}\left[\dfrac{1}{2}\ln(a+x) - \dfrac{1}{2}\ln(a-x)\right] + C$

$= \dfrac{1}{2a}\ln\left|\dfrac{a+x}{a-x}\right| + C.$

Approach 3

: We change $\dfrac{1}{a}\ln\left|\dfrac{a+x}{\sqrt{a^2-x^2}}\right|$ to $\dfrac{1}{2a}\ln\left|\dfrac{a+x}{a-x}\right|$

Step 1: $\dfrac{1}{a}\ln\left|\dfrac{a+x}{\sqrt{a^2-x^2}}\right| + C$

$= \dfrac{1}{a}\ln\left|\dfrac{\left(\sqrt{a+x}\right)^2}{\left(\sqrt{a^2-x^2}\right)}\right| + C$

$= \dfrac{1}{a}\ln\left|\dfrac{\left(\sqrt{a+x}\right)}{\left(\sqrt{a^2-x^2}\right)} \cdot \dfrac{\left(\sqrt{a+x}\right)}{1}\right| + C$

$= \dfrac{1}{a}\ln\left|\sqrt{\dfrac{a+x}{a^2-x^2}} \cdot \dfrac{\left(\sqrt{a+x}\right)}{1}\right| + C$

Step 2:

$= \dfrac{1}{a}\ln\left|\sqrt{\dfrac{(a+x)}{(a+x)(a-x)}} \cdot \dfrac{\left(\sqrt{a+x}\right)}{1}\right| + C$

$= \dfrac{1}{a}\ln\left|\sqrt{\dfrac{a+x}{a-x}}\right| + C$

$= \dfrac{1}{a}\ln\left|\left(\dfrac{a+x}{a-x}\right)^{\frac{1}{2}}\right| + C$

$= \dfrac{1}{2a}\ln\left|\dfrac{a+x}{a-x}\right| + C$

Lesson 41 Exercises

1. Find $\int \dfrac{dx}{1+x^2}$

2. Find $\int \dfrac{1}{1+4x^2}\,dx$

3. Find $\int \dfrac{dx}{a^2+x^2}$

4. Find $\int \dfrac{dx}{(1+x^2)^2}$

5. Find $\int \dfrac{x}{x^2-6x+13}\,dx$

6. Find $\int \dfrac{x}{4x^2+6x+9}\,dx$

7. Find $\int \dfrac{dx}{a^2-x^2}$

8. Find $\int \dfrac{3x^3+5x^2+9x+12}{x^2+4}\,dx$

9. Find $\int \dfrac{\sqrt{x}}{\sqrt[3]{x}+1}\,dx$

10. $\int \dfrac{1}{4x^2+6x+9}\,dx$

Answers: 1. $\text{Tan}^{-1}x+C$; **2,** $\dfrac{1}{2}\text{Tan}^{-1}2x+C$; **3.** $\dfrac{1}{a}\text{Tan}^{-1}\dfrac{x}{a}+C$

4. $\dfrac{x}{2(1+x^2)}+\dfrac{1}{2}\arctan x+C$; **5.** $\dfrac{1}{2}\ln\left|(x-3)^2+4\right|+\dfrac{3}{2}\tan^{-1}\left(\dfrac{x-3}{2}\right)+C$

6. $=\dfrac{1}{8}\left[\ln\left|(4x^2+6x+9)\right|\right]-\dfrac{\sqrt{3}}{12}\arctan\dfrac{\sqrt{3}}{9}(4x+3)+C$;

7. $\dfrac{1}{a}\ln\left|\dfrac{a+x}{\sqrt{a^2-x^2}}\right|+C$; or $\dfrac{1}{2a}\ln\left|\dfrac{a+x}{a-x}\right|+C$

8. $\dfrac{3}{2}x^2+5x-\dfrac{3}{2}\ln\left|x^2+4\right|-4\text{Tan}^{-1}\left(\dfrac{x}{2}\right)+C$

9. $\dfrac{6}{7}x^{\frac{7}{6}}-\dfrac{6}{5}x^{\frac{5}{6}}+2x^{\frac{1}{2}}-6x^{\frac{1}{6}}+6\text{Tan}^{-1}(x^{\frac{1}{6}}))+C$;

10. $=\dfrac{\sqrt{3}}{9}\arctan\left[\dfrac{\sqrt{3}}{9}(4x+3)\right]+C$.

Lesson 42

Integrals of Rational Functions of Sines and Cosines
(Here, the word rational pertains to the word ratio)

An integrand which is a rational function of $\sin x$ and $\cos x$ can be changed to a rational function of u by the following substitution.

1. $u = \tan \frac{1}{2} x$, and from which using trigonometric identities, we obtain also the following:

2. $\sin x = \dfrac{2u}{1+u^2}$, **3.** $\cos x = \dfrac{1-u^2}{1+u^2}$, 4. $dx = \dfrac{2du}{1+u^2}$ or $dx = \dfrac{2}{1+u^2} \, du$

To **memorize** these substitutions, note that the denominator, $1 + u^2$ is the same for all three substitutions.

Example 1 Find $\displaystyle\int \frac{1}{5 - 4\cos x} \, dx$. Note $u = \tan \frac{1}{2} x$, $\cos x = \dfrac{1-u^2}{1+u^2}$

Step 1: Let $\cos x = \dfrac{1-u^2}{1+u^2}$,

$\qquad$ Let $dx = \dfrac{2}{1+u^2} \, du$ in

$$\int \frac{1}{5 - 4\cos x} \, dx \,.$$

Then $\displaystyle\int \frac{1}{5 - 4\left(\frac{1-u^2}{1+u^2}\right)} \cdot \frac{2}{1+u^2} \, du$

$$\int \frac{1}{5 - \frac{4-4u^2}{1+u^2}} \cdot \frac{2}{1+u^2} \, du$$

Step 2: Simplify

$$= \int \frac{1}{\frac{5+5u^2-4+4u^2}{1+u^2}} \cdot \frac{2}{1+u^2} \, du$$

$$= \int \frac{1}{\frac{1+9u^2}{1+u^2}} \cdot \frac{2}{1+u^2} \, du$$

$$= \int \frac{1+u^2}{1 + 9u^2} \cdot \frac{2}{1+u^2} \, du$$

$$= \int \frac{2}{1 + 9u^2} \, du$$

$$= 2\int \frac{1}{1 + (3u)^2} \, du$$

$$= 2\int \frac{1}{1 + (3u)^2} \, du \quad \text{(A)}$$

Step 3: Let $t = 3u$. Then $\dfrac{dt}{du} = 3$

$\qquad$ and $du = \dfrac{dt}{3}$

Step 4: Substitute t for $3u$;

$\qquad$ $\dfrac{dt}{3}$ for du in (A) to obtain

$$2\int \frac{1}{1 + t^2} \cdot \frac{dt}{3}$$

$$= \frac{2}{3}\int \frac{1}{1 + t^2} \, dt$$

Step 5: * Integrate using trig. substitution to obtain

$$= \frac{2}{3}\operatorname{Tan}^{-1}(t) + C$$

$$= \frac{2}{3}\operatorname{Tan}^{-1}(3u) \qquad (t = 3u)$$

$$\int \frac{1}{5 - 4\cos x} \, dx = \frac{2}{3}\operatorname{Tan}^{-1}\left(3\tan\frac{x}{2}\right) + C$$

$$(u = \tan\tfrac{1}{2} x)$$

Example 2; We can also obtain another form of the integral

$\int \sec x\, dx = \ln|\sec x + \tan x| + C$ using the following:

$u = \tan \frac{1}{2} x$, $\cos x = \frac{1-u^2}{1+u^2}$ and

$dx = \frac{2}{1+u^2}\, du$

$\int \sec x\, dx = \int \frac{1}{\cos x}\, dx$

$= \int \frac{1}{\frac{1-u^2}{1+u^2}} \cdot \frac{2}{1+u^2}\, du$

$= \int \frac{1+u^2}{1-u^2} \cdot \frac{2}{1+u^2}\, du$

$= 2\int \frac{1}{1-u^2}\, du$

$= 2\int \frac{1}{(1+u)(1-u)}\, du$

$= 2\int \frac{1}{2} \frac{1}{(1+u)}\, du + \int \frac{1}{2} \frac{1}{(1-u)}\, du$

$= 2\left(\frac{1}{2}\ln|1+u| - \frac{1}{2}\ln|1-u|\right) + C$

$= \ln|1+u| - \ln|1-u| + C$

$= \ln\left|\frac{1+u}{1-u}\right| + C$

$\int \sec x\, dx = \ln\left|\frac{1+\tan\frac{x}{2}}{1-\tan\frac{x}{2}}\right| + C.$

Derivation of $\sin x = \frac{2u}{1+u^2}$,

using $u = \tan \frac{1}{2} x$.

From $\sin 2x = 2\sin x \cos x$

$\sin x = 2\sin\frac{x}{2}\cos\frac{x}{2}$

$= 2\frac{\sin\frac{x}{2}}{\cos\frac{x}{2}}\cos\frac{x}{2} \cdot \cos\frac{x}{2}$

$= 2\tan\frac{x}{2}\cos^2\frac{x}{2}$

$= 2\tan\frac{x}{2}\frac{1}{\sec^2\frac{x}{2}}$

$= 2\tan\frac{x}{2}\frac{1}{1 + \tan^2\frac{x}{2}}$

$= 2u\frac{1}{1+u^2}$

$= \frac{2u}{1+u^2}$

$\sin x = \frac{2u}{1+u^2}$

For the derivations of $\cos x = \frac{1-u^2}{1+u^2}$,

and $dx = \frac{2}{1+u^2}\, du$, see the Appendix.

Extra: Show that $\int \sec x\, dx = \ln\left|\frac{1+\tan\frac{x}{2}}{1-\tan\frac{x}{2}}\right| + C = \ln\left|\tan\left(\frac{\pi}{4} + \frac{x}{2}\right)\right| + C.$

Solution $\ln\left|\frac{1+\tan\frac{x}{2}}{1-\tan\frac{x}{2}}\right| = \ln\left|\frac{\tan\frac{\pi}{4}+\tan\frac{x}{2}}{1-\tan\frac{x}{2}\tan\frac{\pi}{4}}\right|$ (A)

$\ln\left|\tan\left(\frac{\pi}{4} + \frac{x}{2}\right)\right| = \ln\left|\frac{\tan\frac{\pi}{4}+\tan\frac{x}{2}}{1-\tan\frac{\pi}{4}\tan\frac{x}{2}}\right|$ (B)

Since the right sides of (A) and (B) are identical, the left sides are equal, and

$\ln\left|\frac{1+\tan\frac{x}{2}}{1-\tan\frac{x}{2}}\right| = \ln\left|\tan\left(\frac{\pi}{4} + \frac{x}{2}\right)\right|$

$\therefore \int \sec x\, dx = \ln\left|\frac{1+\tan\frac{x}{2}}{1-\tan\frac{x}{2}}\right| + C = \ln\left|\tan\left(\frac{\pi}{4} + \frac{x}{2}\right)\right| + C.$

(Applying $1 = \tan\frac{\pi}{4}$ to $\ln\left|\frac{1+\tan\frac{x}{2}}{1-\tan\frac{x}{2}}\right|$)

(Similarly,, by applying $\tan(A + B) = \frac{\tan A + \tan B}{1 - \tan A \tan B}$ to $\ln\left|\tan\left(\frac{\pi}{4} + \frac{x}{2}\right)\right|$

Example 3 Also, below, we show that

$$\int \sec x\, dx = \ln\sqrt{\frac{1+\sin x}{1-\sin x}} + C$$

Step 1: $\int \sec x\, dx = \int \frac{1}{\cos x}\, dx$

$$= \int \frac{1}{\cos x} \cdot \frac{\cos x}{\cos x}\, dx$$

$$= \int \frac{\cos x}{\cos^2 x}\, dx$$

$$= \int \frac{\cos x}{1-\sin^2 x}\, dx$$

$(u = \sin x)$ $\quad = \int \frac{\cos x}{1-u^2} \cdot \frac{du}{\cos x}$

$(dx = \frac{du}{\cos x})$ $\quad = \int \frac{1}{1-u^2}\, du$

$$= \int \frac{1}{(1+u)(1-u)}\, du$$

$$= \frac{1}{2}\int \frac{1}{1+u}\, du + \frac{1}{2}\int \frac{1}{1-u}\, du$$

Step 2: $= \frac{1}{2}\ln\left|\frac{1+u}{1-u}\right| + C$

$$= \ln\left|\frac{1+u}{1-u}\right|^{\frac{1}{2}} + C$$

$$= \ln\left|\frac{1+\sin x}{1-\sin x}\right|^{\frac{1}{2}} + C$$

$$= \ln\left(\frac{1+\sin x}{1-\sin x}\right)^{\frac{1}{2}} + C$$

$$\int \sec x\, dx = \ln\sqrt{\frac{1+\sin x}{1-\sin x}} + C$$

(using partial fraction decomposition)

Example 4 Show also that $\int \sec x\, dx = \ln\sqrt{\frac{1+\sin x}{1-\sin x}} + C = \ln|\sec x + \tan x| + C$

$$\int \sec x\, dx = \ln\sqrt{\frac{(1+\sin x)}{(1-\sin x)} \cdot \frac{(1+\sin x)}{(1+\sin x)}} + C$$

$$= \ln\sqrt{\frac{(1+\sin x)^2}{1-\sin^2 x}} + C$$

$$= \ln\sqrt{\frac{(1+\sin x)^2}{\cos^2 x}} + C$$

$$= \ln\left|\frac{1+\sin x}{\cos x}\right| + C$$

$$= \ln\left|\frac{1}{\cos x} + \frac{\sin x}{\cos x}\right| + C$$

$$\int \sec x\, dx = \ln|\sec x + \tan x| + C = \ln\sqrt{\frac{1+\sin x}{1-\sin x}} + C$$

$$\left(\tfrac{1}{\cos x} = \sec x\, ;\ \tfrac{\sin x}{\cos x} = \tan x\right)$$

$\therefore \int \sec x\, dx$

$$= \ln|\sec x + \tan x| + C = \ln\sqrt{\frac{1+\sin x}{1-\sin x}} + C = \ln\left|\frac{1+\tan\frac{x}{2}}{1-\tan\frac{x}{2}}\right| + C = \ln\left|\tan\left(\tfrac{\pi}{4} + \tfrac{x}{2}\right)\right| + C.$$

(Also, see previous page and p.238)

Extra show

Example 5 Show that $\int \csc x\, dx = \ln\sqrt{\frac{1-\cos x}{1+\cos x}} + C$ (See p.298 for math similarity)

Step 1: $\int \csc x\, dx = \int \frac{1}{\sin x}\, dx$

$\qquad = \int \frac{1}{\sin x} \cdot \frac{\sin x}{\sin x}\, dx$

$\qquad = \int \frac{\sin x}{\sin^2 x}\, dx$

$\qquad = \int \frac{\sin x}{1-\cos^2 x}\, dx$

$(\, u = \cos x\,) \qquad = \int \frac{\sin x}{1-u^2} \cdot \left(-\frac{du}{\sin x}\right)$

$(\, dx = -\frac{du}{\sin x}\,) \qquad = -\int \frac{1}{1-u^2}\, du$

$\qquad = -\int \frac{1}{(1+u)(1-u)}\, du$

$\qquad = -\left(\frac{1}{2}\int \frac{1}{1+u}\, du + \frac{1}{2}\int \frac{1}{1-u}\, du \right)$

(using partial fraction decomposition)

Step 2: $\quad = -\frac{1}{2}\ln\left|\frac{1+u}{1-u}\right| + C$

$\qquad = \ln\left|\frac{1+u}{1-u}\right|^{-\frac{1}{2}} + C$

$\qquad = \ln\left|\frac{1+\cos x}{1-\cos x}\right|^{-\frac{1}{2}} + C$

$\qquad = \ln\left(\frac{1-\cos x}{1+\cos x}\right)^{\frac{1}{2}} + C$

$\int \csc x\, dx = \ln\sqrt{\frac{1-\cos x}{1+\cos x}} + C.$

Example 6 Show also that $\int \csc x\, dx = \ln|\csc x - \cot x| + C$

$\int \csc x\, dx = \ln\sqrt{\frac{(1-\cos x)}{(1+\cos x)} \cdot \frac{(1-\cos x)}{(1-\cos x)}} + C \quad$ (also see p. 238)

$\qquad = \ln\sqrt{\frac{(1-\cos x)^2}{1-\cos^2 x}} + C$

$\qquad = \ln\sqrt{\frac{(1-\cos x)^2}{\sin^2 x}} + C$

$\qquad = \ln\left|\frac{1-\cos x}{\sin x}\right| + C$

$\qquad = \ln\left|\frac{1}{\sin x} - \frac{\cos x}{\sin x}\right| + C$

$\int \csc x\, dx = \ln|\csc x - \cot x| + C \qquad \left(\frac{1}{\sin x} = \csc x;\ \frac{\cos x}{\sin x} = \cot x \right)$

$\therefore\ \int \csc x\, dx = \ln|\csc x - \cot x| + C = \ln\sqrt{\frac{1-\cos x}{1+\cos x}} + C$

Lesson 42 Exercises

1. Find $\int \frac{dx}{\cos x - \sin x + 1}\, dx$; **2.** $\int \frac{1}{5-4\cos x}\, dx$; **3.** $\int \frac{1}{\sin x + 1}\, dx$

4. Show that $\int \csc x\, dx = \ln\sqrt{\frac{1-\cos x}{1+\cos x}} + C$.

Answers **1.** $-\ln\left|\tan(\frac{x}{2}) - 1\right|$; **2.** $= \frac{2}{3}\tan^{-1}\left(3\tan\frac{x}{2}\right) + C$; **3.** $-\frac{2}{\tan\frac{x}{2}+1} + C$.

Lesson 43
Integration of Radical Functions II
(by Trigonometric Substitution; Review Lesson 41)

In Lesson 32, we integrated some radical functions using simple u-substitution. In this lesson, we integrate radical functions using trigonometric substitution.

Case 1 : Using $x = a\tan\theta$

Example 1 Find $\int \sqrt{x^2 + 4}\, dx$ (A)

Solution

Step 1: The integrand is of the form $\sqrt{x^2 + a^2}$. We accordingly use the substitution $x = a\tan\theta$

Step 2: By comparison of $\sqrt{x^2 + 4}$ with
$$\sqrt{x^2 + a^2}\ ,\ a^2 = 4,\text{ and } a = 2$$

Step 3: Let $x = 2\tan\theta$.   (B)

Step 4: Draw a right triangle using $\tan\theta = \frac{x}{2}$ (from $x = 2\tan\theta$)

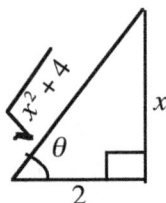

By the Pythagorean theorem,
the hypotenuse $= \sqrt{x^2 + 4}$
$\sec\theta = \frac{\sqrt{x^2+4}}{2}$ (we use this in step 7)
$\tan\theta = \frac{x}{2}$ (we use this in step 7

Step 5: Find dx from $x = 2\tan\theta$.
$$\frac{dx}{d\theta} = 2\sec^2\theta \text{ and}$$
$$dx = 2\sec^2 d\theta$$

Step 6: In $\int \sqrt{x^2 + 4}\, dx$, substitute for $x = 2\tan\theta$; and for $dx = 2\sec^2 d\theta$.

Then $\int \sqrt{4\tan^2\theta + 4}\ \bullet 2\sec^2\theta d\theta$

$= \int 2\sec\theta \bullet 2\sec^2\theta d\theta$

$= 4\int \sec^3\theta\, d\theta$.

Step 7: Integrate by parts to obtain (See p.260, Example 2.)
$$4\int \sec^3\theta d\theta = 4\bullet\tfrac{1}{2}\ln|\sec\theta + \tan\theta| + 4\bullet\tfrac{1}{2}\tan\theta\sec\theta + C$$

$$= \tfrac{1}{2}(4)\ln\left|\frac{\sqrt{x^2 + 4}}{2} + \frac{x}{2}\right| + \tfrac{1}{2}(4)\bullet\frac{x}{2}\bullet\frac{\sqrt{x^2 + 4}}{2} + C$$

$$\int \sqrt{x^2 + 4}\, dx = 2\ln\left|\frac{\sqrt{x^2 + 4}}{2} + \frac{x}{2}\right| + \tfrac{1}{2}x\sqrt{x^2 + 4} + C$$

Example 2 Find $\int \dfrac{1}{\sqrt{a^2 + x^2}}\, dx$ (1)

Solution

Step 1: Let $x = a\tan\theta$

     Then $x^2 = a^2\tan^2\theta$ (2)

     $\dfrac{dx}{d\theta} = a\sec^2\theta$ and

     $dx = a\sec^2\theta\, d\theta$ (3)

Step 2: Substitute for x^2, and dx from (2) and (3) respectively in (1)

$$\int \frac{a\sec^2\theta\, d\theta}{\sqrt{a^2 + a^2\tan^2\theta}}$$

$$\int \frac{a\sec^2\theta\, d\theta}{\sqrt{a^2(1 + \tan^2\theta)}}$$

$$= \int \frac{a\sec^2\theta\, d\theta}{\sqrt{a^2\sec^2\theta}} \qquad \left(\int \frac{a\sec^2\theta\, d\theta}{a\sec\theta} \right.$$

$$= \int \sec\theta\, d\theta$$

$$= \ln\left|\sec\theta + \tan\theta\right| + C \qquad\qquad (4)$$

Step 3: Covert back from θ to x.

From $x = a\tan\theta$, $\tan\theta = \dfrac{x}{a}$; and we draw the corresponding right triangle.

From the figure, $\sec\theta = \dfrac{\sqrt{a^2 + x^2}}{a}$; $\tan\theta = \dfrac{x}{a}$

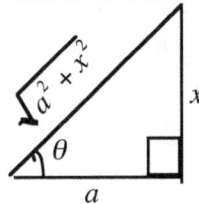

Step 4: Substitute for $\sec\theta$. and $\tan\theta$ from step 3 in (4)

Then $\displaystyle\int \frac{dx}{\sqrt{a^2 + x^2}} = \ln\left|\frac{\sqrt{a^2 + x^2}}{a} + \frac{x}{a}\right| + C$.

$$\int \frac{dx}{\sqrt{a^2 + x^2}} = \ln\left|\frac{x + \sqrt{a^2 + x^2}}{a}\right| + C$$

$$\int \frac{dx}{\sqrt{a^2 + x^2}} = \ln\left|x + \sqrt{a^2 + x^2}\right| - \ln a + C$$

$$\int \frac{dx}{\sqrt{a^2 + x^2}} = \ln\left|x + \sqrt{a^2 + x^2}\right| + C \qquad (\ln a \text{ is absorbed into C})$$

Extra: When $a = 1$, the above becomes

$$\int \frac{dx}{\sqrt{1 + x^2}} = \ln\left|x + \sqrt{1 + x^2}\right| + C$$

See also p.414, 415, where we will call the expression on the right-hand side $\sinh^{-1}x$. So when you get there, come back to reconnect...

Example 3 Find $\int \sqrt{a^2 + x^2}\, dx$ $\qquad$ (1)

Step 1: $\qquad$ Let $x = a\tan\theta$

$\qquad\qquad$ Then $x^2 = a^2 \tan^2\theta$ $\qquad\qquad$ (2)

$\qquad\qquad \dfrac{dx}{d\theta} = a\sec^2\theta$ and

$\qquad\qquad dx = a\sec^2\theta\, d\theta$ $\qquad\qquad$ (3)

Step 2: Substitute for x^2, and dx from (2) and (3) respectively in (1)

Then, $\int \sqrt{a^2 + a^2\tan^2\theta}\; a\sec^2\theta d\theta$

$\qquad \int \sqrt{a^2(1 + \tan^2\theta)}\; a\sec^2\theta d\theta \qquad\qquad (1 + \tan^2\theta = \sec^2\theta)$

$\qquad = \int \sqrt{a^2\sec^2\theta}\; a\sec^2\theta d\theta$

$\qquad = \int a\sec\theta \bullet a\sec^2\, d\theta$

$\qquad = \int a^2\sec^3\theta\, d\theta \qquad\qquad\qquad (4)$

Step 3: We now integrate $a^2 \int \sec^3\theta\, d\theta$ by parts (see also p.260, Example 2)

$a^2 \int \sec^2\theta \bullet \sec\theta\, d\theta = a^2[\sec\theta \bullet \tan\theta - \int (\sec\theta \tan\theta)\tan\, d\theta]$

$\quad (u = \sec\theta) \qquad\qquad = a^2[\sec\theta \bullet \tan\theta - \int \sec\theta \tan^2\theta\, d\theta]$

$\qquad\qquad\qquad\qquad = a^2[\sec\theta \bullet \tan\theta - \int \sec\theta\,(\sec^2\theta - 1)\, d\theta]$

$\qquad\qquad\qquad\qquad = a^2[\sec\theta \bullet \tan\theta - \int \sec^3\theta\, d\theta + \int \sec\theta\, d\theta]$

$a^2 \int \sec^3\theta\, d\theta = a^2 \sec\theta \bullet \tan\theta - a^2 \int \sec^3\theta\, d\theta + a^2 \int \sec\theta\, d\theta]$

$a^2 \int \sec^3\theta\, d\theta + a^2 \int \sec^3\theta\, d\theta = a^2 \sec\theta \bullet \tan\theta + a^2 \int \sec\theta\, d\theta]$

$2a^2 \int \sec^3\theta\, d\theta = a^2 \sec\theta \bullet \tan\theta + a^2 \int \sec\theta\, d\theta] \quad$ (Type **3** integration-by-parts)

$a^2 \int \sec^3\theta\, d\theta = \tfrac{1}{2}a^2 \sec\theta \bullet \tan\theta + \tfrac{1}{2}a^2 \int \sec\theta\, d\theta]$

$\qquad\qquad\qquad = \tfrac{1}{2}a^2 \sec\theta \bullet \tan\theta + \tfrac{1}{2}a^2 \ln|\sec\theta + \tan\theta| + C$

Step 4: Convert back from θ to x.

$\qquad$ From $x = a\tan\theta$, $\tan\theta = \dfrac{x}{a}$; and we draw the corresponding right triangle

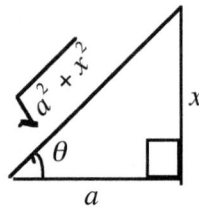

$\qquad$ From the figure, $\sec\theta = \dfrac{\sqrt{a^2 + x^2}}{a}$; $\tan\theta = \dfrac{x}{a}$.

Step 5: Substitute for $\sec\theta$. and $\tan\theta$ from step 3 in (4)

$$a^2 \int \sec^3\theta \, d\theta = \tfrac{1}{2}a^2 \sec\theta \bullet \tan\theta + \tfrac{1}{2}a^2 \ln|\sec\theta + \tan\theta| + C$$

Then $a^2 \int \sec^3\theta \, d\theta = \tfrac{1}{2}a^2 \dfrac{\sqrt{a^2+x^2}}{a} \bullet \dfrac{x}{a} + \tfrac{1}{2}a^2 \ln\left|\dfrac{\sqrt{a^2+x^2}}{a} + \dfrac{x}{a}\right| + C$

Therefore,

$$\int \sqrt{a^2+x^2} \, dx = \tfrac{1}{2}a^2 \dfrac{\sqrt{a^2+x^2}}{a} \bullet \dfrac{x}{a} + \tfrac{1}{2}a^2 \ln\left|\dfrac{\sqrt{a^2+x^2}}{a} + \dfrac{x}{a}\right| + C$$

$$= \tfrac{1}{2}x\sqrt{a^2+x^2} + \tfrac{1}{2}a^2 \ln\left|\dfrac{\sqrt{a^2+x^2}}{a} + \dfrac{x}{a}\right| + C$$

$$= \tfrac{1}{2}x\sqrt{a^2+x^2} + \tfrac{1}{2}a^2 \ln\left|\dfrac{x + \sqrt{a^2+x^2}}{a}\right| + C_1$$

$$= \tfrac{1}{2}x\sqrt{a^2+x^2} + \tfrac{1}{2}a^2 \ln\left|x + \sqrt{a^2+x^2}\right| - \tfrac{1}{2}a^2 \ln a + C_1$$

$$= \tfrac{1}{2}x\sqrt{a^2+x^2} + \tfrac{1}{2}a^2 \ln\left|x + \sqrt{a^2+x^2}\right| + C$$

$$(-\tfrac{1}{2}a^2 \ln a + C_1 \text{ is absorbed into } C)$$

Application: If $a = 2$, $a^2 = 4$, and the above becomes

$$\int \sqrt{4+x^2} \, dx = \tfrac{1}{2}x\dfrac{\sqrt{4+x^2}}{2} + \tfrac{1}{2}(4)\ln\left|\dfrac{\sqrt{4+x^2}}{2} + \dfrac{x}{2}\right| + C$$

$$\int \sqrt{4+x^2} \, dx = \tfrac{1}{2}x\sqrt{4+x^2} + 2\ln\left|\dfrac{\sqrt{4+x^2}}{2} + \dfrac{x}{2}\right| + C \quad \text{(see Example 1, p. 304)}$$

Note the following distinctions:

So now, **1.** $\displaystyle\int \sqrt{a^2+x^2} \, dx = \tfrac{1}{2}x\sqrt{a^2+x^2} + \tfrac{1}{2}a^2 \ln\left|x + \sqrt{a^2+x^2}\right| + C$

2. $\displaystyle\int \dfrac{dx}{\sqrt{a^2+x^2}} = \ln\left|x + \sqrt{a^2+x^2}\right| + C$ (Example 2, p.305)

3. $\displaystyle\int \dfrac{1}{a^2+x^2} \, dx = \tfrac{1}{a}\text{Tan}^{-1}\left(\tfrac{x}{a}\right) + C$ (Example 2, p.287)

$\vdots$

Case 2 (Using $x = \sin\theta$) See also **Lessons 23 and 35**

Example 1a Find $\int \dfrac{1}{\sqrt{a^2 - x^2}}\, dx$ (A)

Step 1. Let $x = a\sin\theta$ $-\dfrac{\pi}{2} \le \theta \le \dfrac{\pi}{2}$ (B)

Then $\sin\theta = \dfrac{x}{a}$.

Draw the corresponding right triangle using $\sin\theta = \dfrac{x}{a}$

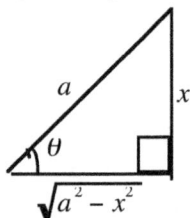

Also, $\theta = \operatorname{Sin}^{-1}\dfrac{x}{a}$ (C)

(We will use (C) this in the last step)

Step 2: From (B) $\dfrac{dx}{d\theta} = a\cos\theta$ and from which

$dx = a\cos\theta\, d\theta$.   (D)

Also, from Step 1, $x^2 = a^2 \sin^2\theta$ (E)

(squaring $x = a\sin\theta$)

Step 3: Substitute for dx and x from (D) and (E), respectively, in

(A) to obtain $\int \dfrac{a\cos\theta\, d\theta}{\sqrt{a^2 - a^2\sin^2\theta}}$

$= \int \dfrac{a\cos\theta\, d\theta}{\sqrt{a^2(1 - \sin^2\theta)}}$

$= \int \dfrac{a\cos\theta\, d\theta}{\sqrt{a^2\cos^2\theta})}$

$(1 - \sin^2\theta = \cos^2\theta\,)$

$= \int \dfrac{a\cos\theta\, d\theta}{a\cos\theta}$ (sq. root of the denom.)

$= \int d\theta$

$= \theta + C$

$= \operatorname{Sin}^{-1}\dfrac{x}{a} + C$

(From (C), $\theta = \operatorname{Sin}^{-1}\dfrac{x}{a}$)

$\int \dfrac{1}{\sqrt{a^2 - x^2}}\, dx = \operatorname{Sin}^{-1}\left(\dfrac{x}{a}\right) + C$

If $a = 1$, $\int \dfrac{1}{\sqrt{1 - x^2}}\, dx = \operatorname{Sin}^{-1}x + C$

Example **1a** was from the general case to the specific. We reverse this order in Example **1b** ,

Example 1b Given : $\int \dfrac{1}{\sqrt{1 - x^2}}\, dx = \operatorname{Sin}^{-1}x + C$ (A); find $\int \dfrac{1}{\sqrt{a^2 - x^2}}\, dx$

1. $\int \dfrac{1}{\sqrt{a^2 - x^2}}\, dx = \int \dfrac{dx}{\sqrt{a^2\left(\frac{a^2}{a^2} - \frac{x^2}{a^2}\right)}}$

$= \int \dfrac{dx}{\sqrt{a^2\left(1 - \left(\frac{x}{a}\right)^2\right)}}$

$= \int \dfrac{dx}{a\sqrt{1 - \left(\frac{x}{a}\right)^2}}$

Step 2: Let $u = \dfrac{x}{a}$. Then $x = au$,

and $\dfrac{dx}{du} = a$, and $dx = adu$

$= \int \dfrac{adu}{a\sqrt{1 - u^2}}$ ($dx = adu$)

Applying (A) , $\int \dfrac{du}{\sqrt{1 - u^2}} = \operatorname{Sin}^{-1}u + C$

Step 3: We change back to x.

(In both sides of the equation)

From Step 2, $u = \dfrac{x}{a}$, $du = \dfrac{dx}{a}$

$\int \dfrac{du}{\sqrt{1 - \frac{x^2}{a^2}}} = \operatorname{Sin}^{-1}\dfrac{x}{a} + C$

$\int \dfrac{dx}{a\sqrt{\frac{a^2 - x^2}{a^2}}} = \operatorname{Sin}^{-1}\dfrac{x}{a} + C$

$\int \dfrac{dx}{\sqrt{a^2 - x^2}} = \operatorname{Sin}^{-1}\dfrac{x}{a} + C$,

In Example **1a**, we used $x = a\sin\theta$. in the next Example, we use $x = a\cos\theta$.

Example 1c (Example **1a** redone): Find $\int \dfrac{1}{\sqrt{a^2 - x^2}}\, dx$ using $x = a\cos\theta$.

Solution (a) $\int \dfrac{1}{\sqrt{a^2 - x^2}}\, dx$ $\qquad\qquad\qquad$ (A)

Step 1: Let $x = a\cos\theta$ $\qquad\quad 0 \le \theta \le \pi$ $\qquad\qquad$ (B)

Then $\cos\theta = \frac{x}{a}$. Draw the corresponding right triangle using $\cos\theta = \frac{x}{a}$ to obtain the figure below.

Then $\theta = \text{Cos}^{-1}\dfrac{x}{a}$ $\qquad$ (C)

(We will use (C) this in the last step)

Step 2: From (B) $\dfrac{dx}{d\theta} = -a\sin\theta$ and

from which $dx = -a\sin\theta\, d\theta$.(D)

Also $x^2 = a^2\cos^2\theta$ $\qquad\qquad$ (E)

(squaring $x = a\cos\theta$ from Step 1)

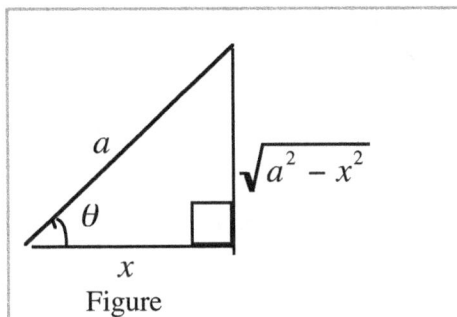

Figure

Step 3: We substitute for dx and for x from (D) and (E), respectively in (A).

$$\int \dfrac{-a\sin\theta\, d\theta}{\sqrt{a^2 - a^2\cos^2\theta}} = -\int \dfrac{a\sin\theta\, d\theta}{\sqrt{a^2(1 - \cos^2\theta)}}$$

$$= -\int \dfrac{a\cos\theta\, d\theta}{\sqrt{a^2\sin^2\theta}} \quad (\, 1 - \cos^2\theta = \sin^2\theta\,)$$

$$= -\int \dfrac{a\cos\theta\, d\theta}{a\cos\theta} \quad \text{(Finding the square root of the denominator)}$$

$$= -\int 1 d\theta$$

$$= -\theta + C$$

$$= -\text{Cos}^{-1}\dfrac{x}{a} + C \qquad\qquad (\text{From (C)}, \theta = \text{Cos}^{-1}\dfrac{x}{a})$$

We also know that $\dfrac{d}{dx}\text{Cos}^{-1}\left(\frac{x}{a}\right) + C = -\dfrac{1}{\sqrt{a^2 - x^2}}$ and

$$\dfrac{d}{dx}\text{Sin}^{-1}\left(\tfrac{x}{a}\right) + C = +\dfrac{1}{\sqrt{a^2 - x^2}}$$

$$-\dfrac{d}{dx}\text{Cos}^{-1}\left(\tfrac{x}{a}\right) + C = -\left(-\dfrac{1}{\sqrt{a^2 - x^2}}\right) = +\dfrac{1}{\sqrt{a^2 - x^2}} \text{and therefore,}$$

$$\int \dfrac{1}{\sqrt{a^2 - x^2}}\, dx = \text{Sin}^{-1}\left(\tfrac{x}{a}\right) + C. \quad \text{(see also Lessons 23 and 35}$$

When $a = 1$, we obtain $\int \dfrac{1}{\sqrt{1 - x^2}}\, dx = \text{Sin}^{-1}(x) + C$

From 1a, 1b, and 1c, we can say that it is not critical whether we use the sine or the cosine in the above problem, but we must note the sign.

Example 2 Find $\int \sqrt{a^2 - x^2}\ dx$ (A)

Step 1. Let $x = a\sin\theta$ $-\frac{\pi}{2} \le \theta \le \frac{\pi}{2}$ (B)

Then $\sin\theta = \frac{x}{a}$.

Draw the corresponding right triangle using $\sin\theta = \frac{x}{a}$

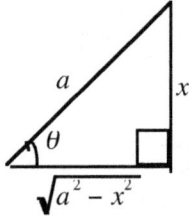

Also, $\theta = \text{Sin}^{-1}\frac{x}{a}$ (C)

(We will use (C) this in the last step)

Step 2: From (B) $\frac{dx}{d\theta} = a\cos\theta$ and from which

$dx = a\cos\theta\ d\theta$.   (D)

From Step 1, $x^2 = a^2 \sin^2\theta$ (E)

        (squaring $x = a\sin\theta$)

Step 3: Substitute for dx and x from (D) and (E), respectively, in (A) to

obtain $\int \sqrt{a^2 - a^2 \sin^2\theta}(a\cos\theta d\theta)$

$= \int \sqrt{a^2(1 - \sin^2\theta)}(a\cos\theta d\theta)$

$= \int \sqrt{a^2 \cos^2\theta}(a\cos\theta d\theta)$

            $(1 - \sin^2\theta = \cos^2\theta)$

$= \int a\cos\theta(a\cos\theta d\theta)$

$= \int a^2 \cos^2\theta\ d\theta$

$= a^2 \int \cos^2\theta\ d\theta$

Step 4:

Applying $\cos^2\theta = \frac{1}{2}\cos 2\theta + \frac{1}{2}$

$a^2 \int \cos^2\theta\ d\theta = a^2 \int(\frac{1}{2}\cos 2\theta + \frac{1}{2})d\theta$

$= a^2\left(\frac{1}{2}(\frac{1}{2})\sin 2\theta + \frac{1}{2}\theta\right) + C$

$= a^2\left(\frac{1}{4}\sin 2\theta + \frac{1}{2}\theta\right) + C$

$= a^2\left(\frac{1}{4}\cdot 2\sin\theta\cos\theta + \frac{1}{2}\theta\right) + C$

$= a^2\left(\frac{1}{2}\sin\theta\cos\theta + \frac{1}{2}\theta\right) + C$

    $(\sin 2\theta = 2\sin\theta\cos\theta)$

$= a^2\left(\frac{1}{2}\cdot \frac{x}{a}\frac{\sqrt{a^2-x^2}}{a} + \frac{1}{2}\text{Sin}^{-1}\frac{x}{a}\right) + C$

$= \frac{1}{2}x\sqrt{a^2 - x^2} + \frac{1}{2}a^2\text{Sin}^{-1}\frac{x}{a} + C$

        (From (C), $\theta = \text{Sin}^{-1}\frac{x}{a}$)

    $(\sin\theta = \frac{x}{a};\ \ \cos\theta = \frac{\sqrt{a^2 - x^2}}{a})$

Therefore, $\int \sqrt{a^2 - x^2}\ dx$

$= \frac{1}{2}x\sqrt{a^2 - x^2} + \frac{1}{2}a^2\text{Sin}^{-1}\frac{x}{a} + C$

(see also Example 1a)

If $a = 1$, we obtain

$\int \sqrt{1 - x^2}\ dx =$

        $\frac{1}{2}x\sqrt{1 - x^2} + \frac{1}{2}\text{Sin}^{-1}x + C$

Example 3 Find $\int \dfrac{x^3}{\sqrt{16 - x^2}}\,dx$ (A)

Step 1: Let $x = 4\sin\theta$ $-\frac{\pi}{2} \le \theta \le \frac{\pi}{2}$ (B)

Then $\sin\theta = \frac{x}{4}$.

Draw the corresponding right triangle using $\sin\theta = \frac{x}{4}$

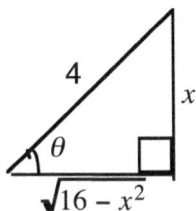

Step 2: From (B) $\frac{dx}{d\theta} = 4\cos\theta$ and from which

$dx = 4\cos\theta\,d\theta$. (C)

Also, from Step 1, $x^2 = 16\sin^2\theta$ (D)

(squaring $x = 4\sin\theta$)

Also $x^3 = 64\sin^3\theta$ (E)

Step 3: Substitute for dx and x^2, x^3 from (C) and (D),, and (E) respectively,

in (A) to obtain $\int \dfrac{64\sin^3\theta \bullet 4\cos\theta\,d\theta}{\sqrt{16 - 16\sin^2\theta}}$

Step 4: Simplify

$\int \dfrac{64\sin^3\theta \bullet 4\cos\theta\,d\theta}{\sqrt{16(1 - \sin^2\theta)}}$

$\int \dfrac{64\sin^3\theta \bullet 4\cos\theta\,d\theta}{\sqrt{16\cos^2\theta}}$

$(1 - \sin^2\theta = \cos^2\theta)$

$\int \dfrac{64\sin^3\theta \bullet 4\cos\theta\,d\theta}{4\cos\theta}$

(sq. root of the denom.)

$= 64\int \sin^3\theta\,d\theta$

Step 5: Integrate $64\int \sin^3\theta\,d\theta$

(see also p. 257-258)

$64\int \sin^3\theta\,d\theta = 64\int \sin^2\theta \bullet \sin\theta\,d\theta$

(Apply $\sin^2\theta = 1 - \cos^2\theta$)

$= 64\int (1 - \cos^2\theta) \bullet \sin\theta\,d\theta$

$= 64\left[\int \sin\theta\,d\theta - \int \cos^2\theta \sin\theta\,d\theta \right]$

$= 64\left[-\cos\theta - (-\int u^2\,du) \right]$

(Let $u = \cos\theta$ for the second integral)

$= 64\left[-\cos\theta + \frac{u^3}{3} \right] + C$

$= 64\left[-\cos\theta + \frac{\cos^3\theta}{3} \right] + C$

$= 64\left[-\frac{\sqrt{16 - x^2}}{4} + \frac{1}{3}\left(\frac{\sqrt{16 - x^2}}{4} \right)^3 \right] + C$

(using the above right triangle)
See also Lesson 37, Case 3, (c)

$\therefore \int \dfrac{x^3}{\sqrt{16 - x^2}}\,dx = 64\left[-\frac{\sqrt{16 - x^2}}{4} + \frac{1}{3}\left(\frac{\sqrt{16 - x^2}}{4} \right)^3 \right] + C\,.$

Case 3 ((Using $x = a\sec\theta$)

Example 1 Show that $\displaystyle\int \frac{1}{x\sqrt{x^2 - a^2}}\, dx \overset{?}{=} \frac{1}{a}\,\text{Sec}^{-1}\!\left(\frac{x}{a}\right) + C$

Solution (See also Lessons 23 and 35)

$$\int \frac{dx}{x\sqrt{x^2 - a^2}} \qquad\qquad\text{(A)}$$

Step 1: Let $x = a\sec\theta$ $\qquad 0 \le \theta \le \frac{\pi}{2}$ or $\pi \le \theta \le \frac{3\pi}{2}$ $\qquad$ (B)

$\qquad$ Then $\theta = \text{Sec}^{-1}\dfrac{x}{a}$ <--------we will use this in the last step (C)

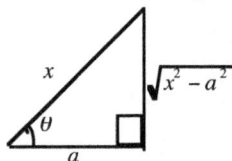

$\qquad$ From (B) $\dfrac{dx}{d\theta} = a\sec\theta\tan\theta$ and from which

$\qquad\qquad dx = a\sec\theta\tan\theta\, d\theta.$ $\qquad\qquad$ D)

$\qquad$ Also $x^2 = a^2\sec^2\theta$ $\qquad\qquad\qquad$ (E)

Step 2: Now, we substitute for dx and for x from (D) and (E) in (A) to obtain

$$\int \frac{a\sec\theta\tan\theta\, d\theta}{a\sec\theta\sqrt{a^2\sec^2\theta - a^2}}$$

$$= \int \frac{\tan\theta\, d\theta}{\sqrt{a^2(\sec^2\theta - 1)}}$$

$$= \int \frac{\tan\theta\, d\theta}{\sqrt{a^2\tan^2\theta}} \qquad (\sec^2\theta - 1 = \tan^2\theta)$$

$$= \int \frac{\tan\theta\, d\theta}{a\tan\theta}$$

$$= \int \frac{d\theta}{a}$$

$$= \frac{1}{a}\int d\theta$$

$$= \frac{1}{a}\theta + C$$

$$= \frac{1}{a}\text{Sec}^{-1}\frac{x}{a} + C \qquad\qquad (\text{From (C)}, \theta = \text{Sec}^{-1}\frac{x}{a})$$

$$\int \frac{1}{x\sqrt{x^2 - a^2}}\, dx = \frac{1}{a}\text{Sec}^{-1}\!\left(\frac{x}{a}\right) + C$$

When $a = 1$, we obtain $\displaystyle\int \frac{1}{x\sqrt{x^2 - 1}}\, dx = \text{Sec}^{-1}x + C$

Example 2 Find $\int \frac{\sqrt{x^2-16}}{x}\,dx$

Method 1 (using $x = 4\sec\theta$)

Solution

$\int \frac{\sqrt{x^2-16}}{x}\,dx$ (A)

Step 1

Let $x = 4\sec\theta$ $(0 \le \theta \le \frac{\pi}{2})$ (B)

Then $\theta = \text{Sec}^{-1}\frac{x}{4}$ (C)

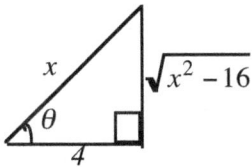

From (B), $\frac{dx}{d\theta} = a\sec\theta\tan\theta$ and from
which $dx = a\sec\theta\tan\theta\,d\theta$. (D)

Also $x^2 = 16\sec^2\theta$ (E)

Step 2: Now, we substitute for dx
  and for x from (D) and (E) in
  (A) to obtain

$= \int \frac{\sqrt{16\sec^2\theta - 16} \cdot 4\sec\theta\tan\theta\,d\theta}{4\sec\theta}$

$= \int \frac{\sqrt{16(\sec^2\theta - 1)} \cdot 4\sec\theta\tan\theta\,d\theta}{4\sec\theta}$

$= \int \frac{\sqrt{16(\tan\theta^2)} \cdot 4\sec\theta\tan\theta\,d\theta}{4\sec\theta}$

$= \int \frac{4\tan\theta \cdot 4\sec\theta\tan\theta\,d\theta}{4\sec\theta}$

$= 4\int \tan^2\theta\,d\theta$

$= 4\int (\sec^2\theta - 1)d\theta$

$= 4\int \sec^2\theta\,d\theta - 4\int d\theta$

$= 4\tan\theta - 4\theta + C$

$= \frac{4\sqrt{x^2-16}}{4} - 4Sec^{-1}\frac{x}{4} + C$

$= \sqrt{x^2-16} - 4Sec^{-1}\frac{x}{4} + C$

Method 2
(Two-step u-subst, plus trig. subst.)

$\int \frac{\sqrt{x^2-16}}{x}\,dx$ (A)

Step 1: Let $u = \sqrt{x^2-16}$ (B)
Then $u^2 = x^2 - 16$

$\frac{2u\,du}{dx} = 2x$; $dx = \frac{u\,du}{x}$

$\int \frac{\sqrt{x^2-16}}{x}\,dx$

$= \int \frac{u}{x} \cdot \frac{u\,du}{x}$

$= \int \frac{u^2}{x^2}\,du$

From (B) $x^2 = u^2 + 16$

$= \int \frac{u^2}{u^2+16}\,du$

(change to proper rational function)

$= \int \frac{u^2+16-16}{u^2+16}\,du$

$= \int \frac{u^2+16}{u^2+16}\,du - \int \frac{16}{u^2+16}\,du$

$= \int 1\,du - 16\int \frac{1}{u^2+16}\,du$

Let $u = 4\tan\theta$ and $\tan\theta = \frac{u}{4}$

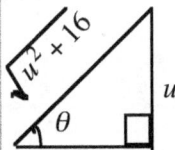

$= u - 16\left[\frac{1}{4}\text{Tan}^{-1}\frac{u}{4}\right] + C$

$= \sqrt{x^2-16} - 16\left[\frac{1}{4}\text{Tan}^{-1}\frac{\sqrt{x^2-16}}{4}\right] + C$

$= \sqrt{x^2-16} - 4\text{Tan}^{-1}\left(\frac{\sqrt{x^2-16}}{4}\right) + C$

As an exercise, we next show that the results of Method **1** and **2** are equivalent.

EXTRA 1: Show that $\operatorname{Sec}^{-1}\frac{x}{4} = \operatorname{Tan}^{-1}\left(\frac{\sqrt{x^2-16}}{4}\right)$

Step 1: Let $\operatorname{Sec}^{-1}\frac{x}{4} = y_1$,   (1)

    Then $\sec y_1 = \frac{x}{4}$

Draw a right triangle using $\sec y_1 = \frac{x}{4}$

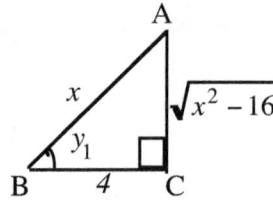

Step 2: Let $\operatorname{Tan}^{-1}\left(\frac{\sqrt{x^2-16}}{4}\right) = y_2$ (2)

    Then $\tan y_2 = \frac{\sqrt{x^2-16}}{4}$

Draw a right triangle using $\tan y_2 = \frac{\sqrt{x^2-16}}{4}$

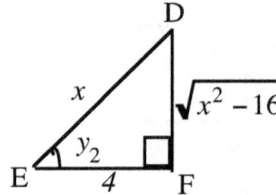

Step 3: Since $\triangle ABC$ and $\triangle DEF$ are congruent by SSS (Side-Side-Side)

    $y_1 = y_2$

Therefore, $\operatorname{Sec}^{-1}\frac{x}{4} = \operatorname{Tan}^{-1}\left(\frac{\sqrt{x^2-16}}{4}\right)$ (From (1) and (2))

EXTRA 2: Exercise in differentiation of inverse trigonometric functions
Let us differentiate each result to see if we obtain the original integrand.

For Method 1

We find $\frac{d}{dx}\left(\sqrt{x^2-16} - 4\operatorname{Sec}^{-1}\frac{x}{4}\right)$

Step 1: $\frac{d}{dx}\left(\sqrt{x^2-16}\right) = \frac{d}{dx}\left(x^2-16\right)^{\frac{1}{2}}$

    $= \frac{1}{2}(2x)\left(x^2-16\right)^{-\frac{1}{2}} = \frac{x}{\sqrt{x^2-16}}$

Step 2: Find $\frac{d}{dx}\left(-4\operatorname{Sec}^{-1}\frac{x}{4}\right)$

  Let $\operatorname{Sec}^{-1}\frac{x}{4} = y$,   (1)

    Then $\sec y = \frac{x}{4}$.

Step 3: $\sec y \tan y \frac{dy}{dx} = \frac{1}{4}$
(implicit differentiation)

$\frac{dy}{dx} = \frac{1}{4\sec y \tan y}$

$\frac{dy}{dx} = \frac{1}{4}\cdot\frac{4}{x}\cdot\frac{4}{\sqrt{x^2-16}}$

    $= \frac{4}{x\sqrt{x^2-16}}$

$\frac{d}{dx}\left(-4\operatorname{Sec}^{-1}\frac{x}{4}\right) = \frac{-(4)4}{x\sqrt{x^2-16}}$

    $= -\frac{16}{x\sqrt{x^2-16}}$

Step 4:

$\frac{d}{dx}\left(\sqrt{x^2-16} - 4\operatorname{Sec}^{-1}\frac{x}{4}\right) = \frac{x}{\sqrt{x^2-16}} - \frac{16}{x\sqrt{x^2-16}}$

$= \frac{x\bullet x}{x\sqrt{x^2-16}} - \frac{16}{x\sqrt{x^2-16}} = \frac{x^2}{x\sqrt{x^2-16}} - \frac{16}{x\sqrt{x^2-16}}$

$= \frac{x^2-16}{x\sqrt{x^2-16}} = \frac{(x^2-16)\bullet\sqrt{x^2-16}}{x\sqrt{x^2-16}\bullet\sqrt{x^2-16}} = \frac{(x^2-16)\bullet\sqrt{x^2-16}}{x(x^2-16)}\boxed{= \frac{\sqrt{x^2-16}}{x}}$

As an exercise, check the differentiation for Method 2.

Example 3 Find $\int \dfrac{1}{\sqrt{x^2-1}}dx$

Step 1: Let $x = \sec\theta$; (trig. substitution)
$$dx = \sec\theta\tan\theta\, d\theta$$

$(\dfrac{dx}{d\theta} = \sec\theta\tan\theta)$

Substitute in $\int \dfrac{1}{\sqrt{x^2-1}}dx$ to obtain

$$\int \frac{1}{\sqrt{\sec^2\theta-1}}\sec\theta\tan\theta d\theta \qquad (x^2=\sec^2\theta)$$

$$=\int \frac{1}{\sqrt{\tan^2\theta}}\sec\theta\tan\theta d\theta$$

$$=\int \frac{1}{\tan\theta}\sec\theta\tan\theta d\theta$$

Step 2: $\quad =\int \sec\theta d\theta \qquad$ (Integrate now)
$$= \ln|\sec\theta + \tan\theta| + C$$
$$\sec\theta = x ;\ \tan\theta = \sqrt{x^2-1}$$
$$\int \frac{1}{\sqrt{x^2-1}}dx = \ln\left|x+\sqrt{x^2-1}\right| + C.$$

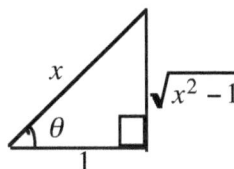

$\sec\theta = x$
$\tan\theta = \sqrt{x^2-1}$

See also p.416, p.429, where we will call the expression on the right-hand side, the

hyperbolic $\cosh^{-1}x$. So when you get there, come back to reconnect.

Example 4 Find $\int \sqrt{x^2-1}dx$.

Step !: Let $x = \sec\theta$; (trig substitution)
Then $\quad \dfrac{dx}{d\theta} = \sec\theta\tan\theta$, and
$$dx = \sec\theta\tan\theta\, d\theta,\ x^2 = \sec^2\theta$$

Substitute in $\int \sqrt{x^2-1}dx$ to obtain
$$\int \sqrt{\sec^2\theta-1}\,\sec\theta\tan\theta d\theta$$
$$=\int \sqrt{\tan^2\theta}\,\sec\theta\tan\theta d\theta$$
$$=\int \tan\theta\sec\theta\tan\theta d\theta$$
$$=\int \tan^2\theta\sec\theta d\theta$$
$$=\int (\sec^2\theta-1)\sec\theta d\theta$$
$$\qquad (\tan^2\theta = \sec^2\theta-1)$$
$$=\int \sec^3\theta d\theta - \int \sec\theta d\theta$$
(For integration of $\int \sec^3\theta d\theta$, see p.260 Example 2.)

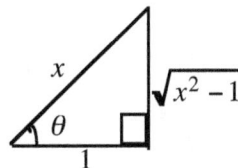

Note $\tan\theta = \sqrt{x^2-1},\ \ \sec\theta = x$

Step 2::Intergrate as on p.260, Ex. 2
$$= \frac{1}{2}\ln|\sec\theta + \tan\theta| + \frac{1}{2}\sec\theta\tan\theta$$
$$\qquad - \ln|\sec\theta + \tan\theta| + C$$
$$= -\frac{1}{2}\ln|\sec\theta + \tan\theta| + \frac{1}{2}\sec\theta\tan\theta$$
$$= \frac{1}{2}\sec\theta\tan\theta - \frac{1}{2}\ln|\sec\theta + \tan\theta|$$
$$= \frac{1}{2}x\sqrt{x^2-1} - \frac{1}{2}\ln\left|x+\sqrt{x^2-1}\right| + C$$
$$\int \sqrt{x^2-1}dx =$$
$$\frac{1}{2}x\sqrt{x^2-1} - \frac{1}{2}\ln\left|x+\sqrt{x^2-1}\right| + C$$

Integrals Involving Quadratic Trinomials: $\sqrt{ax^2 + bx + c}$

Here, we apply simple u-substitution followed by trigonometric substitution

Example 1: Find: $\displaystyle\int \frac{dx}{\sqrt{3 - x^2 - 2x}}$

Step 1: Complete the square

$$\int \frac{dx}{\sqrt{3 - (x^2 + 2x)}}$$

$$= \int \frac{dx}{\sqrt{3 - (x^2 + 2x + 1 - 1)}}$$

$$= \int \frac{dx}{\sqrt{3 - [(x + 1)^2 - 1]}}$$

$$= \int \frac{dx}{\sqrt{3 - (x + 1)^2 + 1}}$$

$$= \int \frac{dx}{\sqrt{4 - (x + 1)^2}}$$

Step 2: Let $u = x + 1$; then $du = dx$

$$\int \frac{dx}{\sqrt{4 - (x + 1)^2}} = \int \frac{dx}{\sqrt{4 - u^2}}$$

Step 3: Let $u = 2\sin\theta$; then $\sin\theta = \frac{u}{2}$

$$\int \frac{dx}{\sqrt{4 - u^2}} = \int \frac{2\cos\theta d\theta}{\sqrt{4 - 4\sin^2\theta}}$$

$$\int \frac{dx}{\sqrt{4 - u^2}} = \int \frac{2\cos\theta\, d\theta}{\sqrt{4(1 - \sin^2\theta)}}$$

$$= \int \frac{2\cos\theta\, d\theta}{\sqrt{4\cos^2\theta}} \quad (1 - \sin^2\theta = \cos^2\theta)$$

$$= \int \frac{2\cos\theta\, d\theta}{2\cos\theta}$$

$$= \int\! d\theta$$

$$= \theta + C$$

$$= \sin^{-1}\!\left(\frac{u}{2}\right) + C$$

$$= \sin^{-1}\!\left(\frac{x + 1}{2}\right) + C$$

$$\int \frac{dx}{\sqrt{3 - x^2 - 2x}} = \sin^{-1}\!\left(\frac{x + 1}{2}\right) + C$$

$\theta = \sin^{-1}\!\left(\dfrac{u}{2}\right)$

$(u^2 = 4\sin^2\theta)$

$\left(\dfrac{du}{d\theta} = 2\cos\theta\right)$

$(du = 2\cos\theta\, d\theta)$

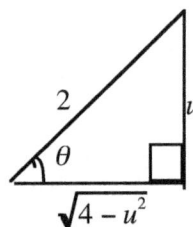

$u = x + 1$

Integration of More Radicals

Example 1 Find $\int \sqrt{\dfrac{1-x}{1+x}}\, dx$

Solution: We begin with simple u-substitution followed by trig. substitution.

$\int \sqrt{\dfrac{1-x}{1+x}}\, dx \qquad (1)$

Step 1 (Simple u-substitution)

Let $u = \sqrt{\dfrac{1-x}{1+x}}$

Then $u^2 = \dfrac{1-x}{1+x}$

$u^2 + u^2 x = 1 - x$

$u^2 x + x = 1 - u^2$

$x(u^2 + 1) = 1 - u^2$

$x = \dfrac{1-u^2}{1+u^2}$

$\dfrac{dx}{du} = \dfrac{-2u(1+u^2) - 2u(1-u^2)}{(1+u^2)^2}$

$\dfrac{dx}{du} = \dfrac{-4u}{(1+u^2)^2}$

$dx = \dfrac{-4u}{(1+u^2)^2}\, du$

Step 2 (Subs for dx)

$\int \sqrt{\dfrac{1-x}{1+x}}\, dx$

$= \int u \cdot \dfrac{-4u}{(1+u^2)^2}\, du$

Step 3: Apply trig substitution

$= -4 \int \dfrac{u^2}{(1+u^2)^2}\, du$

Let $u = \tan\theta$

Then $u^2 = \tan^2\theta$

$\dfrac{du}{d\theta} = \sec^2\theta\,;$

and $du = \sec^2\theta\, d\theta$.

$\left(\tan\theta = \dfrac{u}{1}\right)$

Step 4

$-4 \int \dfrac{u^2}{(1+u^2)^2}\, du =$

$= -4 \int \dfrac{\tan^2\theta \cdot \sec^2\theta\, d\theta}{(\sec^2\theta)^2}$

$= -4 \int \dfrac{\tan^2\theta\, d\theta}{\sec^2\theta}$

$\left[\dfrac{\tan^2\theta}{\sec^2\theta} = \left(\dfrac{\sin^2\theta}{\cos^2\theta}\right)\dfrac{\cos^2\theta}{1}\right]$

$= -4 \int \sin^2\theta\, d\theta$

$= -4 \int \left[\dfrac{1}{2} - \dfrac{1}{2}\cos 2\theta\right] d\theta$

Note: $\left(\sin^2\theta = \dfrac{1}{2} - \dfrac{1}{2}\cos 2\theta\right)$

$= -4\left[\dfrac{\theta}{2} - \dfrac{1}{2}\cdot\dfrac{1}{2}\sin 2\theta\right] + C$

$= -4\left[\dfrac{\theta}{2} - \dfrac{1}{4}\sin 2\theta\right] + c$

(antiderivative of $\cos\theta$ is $\sin\theta$)

$= -2\theta + \sin 2\theta + c$

$= -2\theta + 2\sin\theta\cos\theta + c$

$= -2\,\mathrm{Tan}^{-1}(u) + 2\dfrac{u}{\sqrt{1+u^2}} \cdot \dfrac{1}{\sqrt{1+u^2}} + c$

$= -2\,\mathrm{Tan}^{-1}(u) + \dfrac{2u}{1+u^2} + c$

$= -2\,\mathrm{Tan}^{-1}\left(\sqrt{\dfrac{1-x}{1+x}}\right) + (1+x)\sqrt{\dfrac{1-x}{1+x}} + c$

$= -2\,\mathrm{Tan}^{-1}\left(\sqrt{\dfrac{1-x}{1+x}}\right) + \sqrt{1-x^2} + c$

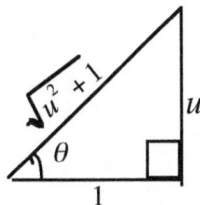

Example 2 Find $\int \sqrt{1 + e^x}\, dx$ **(A)**

Solution We use simple u-substitution with two approaches

Approach 1: Substitute for e^x **Step 1**: Rationalize the given integral	**Approach 2** Take logs and solve for x **Step 1**: Rationalize the given integral

Approach 1: Substitute for e^x
Step 1: Rationalize the given integral

$$\text{Let } u = \sqrt{1 + e^x} \quad \text{(B)}$$
$$\text{Then } u^2 = 1 + e^x, \quad \text{(C)}$$

Step 2: Solve for e^x to obtain

$$e^x = u^2 - 1$$

From (C) $2u\dfrac{du}{dx} = e^x$ and

$$dx = 2u\frac{du}{e^x}$$

Step 3: Subst. $2u\dfrac{du}{e^x}$ for dx,

$$u^2 - 1 \text{ for } e^x \text{ in}$$
$$\int \sqrt{1 + e^x}\, dx \text{ to obtain}$$

$$\int u \bullet \frac{2u\,du}{u^2 - 1}$$

$$= \int \frac{2u^2 du}{u^2 - 1}$$

$$= 2\int \frac{u^2 - 1 + 1\,du}{u^2 - 1}$$

$$= 2\int \frac{u^2 - 1}{u^2 - 1}\,du + \frac{1}{u^2 - 1}\,du$$

$$= 2\left(\int 1\,du + \int \frac{1}{u^2 - 1}\,du \right)$$

Step 4: Integrate second integral by partial fraction decomposition

$$= 2\left(\int 1\,du - \frac{1}{2}\int \frac{1}{u + 1}\,du + \frac{1}{2}\int \frac{1}{u - 1}\,du \right)$$

$$= 2\left(u - \frac{1}{2}\ln(u + 1) + \frac{1}{2}\ln(u - 1) \right) + c$$

$$= 2u - \ln(u + 1) + \ln(u - 1) + c$$

$$= 2u + \ln\left| \frac{u - 1}{u + 1} \right| + c$$

$$= 2\sqrt{1 + e^x} + \ln\left| \frac{\sqrt{1 + e^x} - 1}{\sqrt{1 + e^x} + 1} \right| + c$$

Approach 2 Take logs and solve for x
Step 1: Rationalize the given integral

$$\text{Let } u = \sqrt{1 + e^x} \quad \text{(B)}$$
$$\text{Then } u^2 = 1 + e^x \quad \text{(C)},$$

Step 2: Solve for the exponential equation for x by first taking logs

$$e^x = u^2 - 1$$
$$\ln e^x = \ln(u^2 - 1)$$
$$x \ln e = \ln(u^2 - 1)$$
$$x(1) = \ln(u^2 - 1)$$
$$x = \ln(u^2 - 1)$$

Step 3: $\dfrac{dx}{du} = \dfrac{2u}{(u^2 - 1)}$

$$dx = \frac{2u}{u^2 - 1}\,du$$

Step 4: Substitute u for $\sqrt{1 + e^x}$,

$$\frac{2u}{u^2 - 1}\,du \text{ for } dx \text{ in}$$

$$\int \sqrt{1 + e^x}\, dx \text{ to obtain}$$

$$\int \frac{u \bullet 2u\,du}{u^2 - 1}$$

$$= 2\int \frac{u^2 du}{u^2 - 1}$$

$$= 2\left(\int 1\,du + \int \frac{1}{u^2 - 1}\,du \right)$$

Step 5: Integrate second integrand by partial fraction decomposition

$$= 2\left(\int 1\,du - \frac{1}{2}\int \frac{1}{u + 1}\,du + \frac{1}{2}\int \frac{1}{u - 1}\,du \right)$$

$$= 2\left(u - \frac{1}{2}\ln(u + 1) + \frac{1}{2}\ln(u - 1) \right) + c$$

$$= 2u - \ln(u + 1) + \ln(u - 1) + c$$

$$= 2u + \ln\left| \frac{u - 1}{u + 1} \right| + c$$

$$= 2\sqrt{1 + e^x} + \ln\left| \frac{\sqrt{1 + e^x} - 1}{\sqrt{1 + e^x} + 1} \right| + c.$$

Example 3

Find $\int \frac{\sqrt{x^2+1}}{x} dx$ (A)

Step 1:

Let $u = \sqrt{x^2+1}$ (B)

Then $u^2 = x^2 + 1$

$2u\frac{du}{dx} = 2x$

$u\frac{du}{x} = dx$ (C)

$\int \frac{u \cdot u\, du}{x \cdot x} = \int \frac{u^2 du}{x^2}$

(Subst. (B) & (C) in (A))

From $u^2 = x^2 + 1$

$x^2 = u^2 - 1$

Step 2: $\int \frac{u^2 du}{x \cdot x} = \int \frac{u^2}{u^2-1} du$

$= \int \frac{u^2-1+1}{u^2-1} du$

$= \int \left(\frac{u^2-1}{u^2-1} + \frac{1}{u^2-1}\right) du$

$= \int 1\, du + \int \frac{1}{u^2-1} du$

$= \int 1\, du - \int \frac{1}{2(u+1)} du + \int \frac{1}{2(u-1)} du$

$= u - \frac{1}{2}\ln|u+1| + \frac{1}{2}\ln|u-1| + C$

$\sqrt{x^2+1} + \frac{1}{2}\ln\left|\frac{\sqrt{x^2+1}-1}{\sqrt{x^2+1}+1}\right| + C \Leftarrow$

Note:

In step 1, we use a two step u-substitution.
In step 2, we use Partial fraction decomposition

$\sqrt{x^2+1}$

$-\frac{1}{2}\ln\left|\sqrt{x^2+1}+1\right|$

$+\frac{1}{2}\ln\left|\sqrt{x^2+1}-1\right|$

Lesson 43 Exercises

1.. Is it possible that for some problems, even though we can apply trigonometric substitution, partial fractions decomposition will also work?

2. Write down the two Pythagorean identities which are the basis of the trigonometric substitution techniques..

3. Find $\int \frac{1}{\sqrt{a^2-x^2}} dx$

4. In Problem 3, is it possible to obtain the same results if we let $x = a\cos\theta$, instead of $x = a\sin\theta$

5. Find $\int \frac{1}{x\sqrt{x^2-a^2}} dx$;

6. Find $\int \sqrt{x^2+4}\, dx$

7. Find $\int \frac{1}{\sqrt{a^2+x^2}} dx$

8. Find $\int \sqrt{a^2+x^2}\, dx$

9. Find: $\int \frac{dx}{\sqrt{3-x^2-2x}}$;

10. $\int \sqrt{x^2+5}\, dx$;

11. $\int \frac{1}{\sqrt{x^2-16}} dx$

12. Find $\int \sqrt{\frac{1-x}{1+x}}\, dx$;

13. Find $\int \sqrt{1+e^x}\, dx$

Answers: 3. $\operatorname{Sin}^{-1}\left(\frac{x}{a}\right) + C$; **4.** Yes; **5.** $\frac{1}{a}\operatorname{Sec}^{-1}\left(\frac{x}{a}\right) + C$;

6. $\frac{1}{2}x\sqrt{x^2+4} + 2\ln\left|\frac{\sqrt{x^2+4}}{2} + \frac{x}{2}\right| + C$; **7.** $\ln\left|\frac{\sqrt{a^2+x^2}}{a} + \frac{x}{a}\right| + C$;

8. $\frac{1}{2}x\sqrt{a^2+x^2} + \frac{1}{2}a^2\ln\left|\frac{\sqrt{a^2+x^2}}{a} + \frac{x}{a}\right| + C$; **9.** $\operatorname{Sin}^{-1}\left(\frac{x+1}{2}\right) + C$;

10. $\frac{1}{2}x\sqrt{x^2+5} + \frac{5}{2}\ln\left|\sqrt{\frac{x^2+5}{5}} + \frac{x\sqrt{5}}{5}\right| + C$; **11.** $\ln\left|x + \sqrt{x^2-16}\right| + C$;

12. $-2\operatorname{Tan}^{-1}\left(\sqrt{\frac{1-x}{1+x}}\right) + \sqrt{1-x^2} + C$; **13.** $2\sqrt{1+e^x} + \ln\left|\frac{\sqrt{1+e^x}-1}{\sqrt{1+e^x}+1}\right| + C$

Guidelines: **General Approach to Integration**

So far, we have learned the various techniques, methods or approaches for integrating functions. The problem one may have is how to determine as quickly as possible which technique or approach to use, given an integrand. Below, we assume that the student can handle the integration of the basic functions.

Step 1: Check to see if simple *u*-substitution will work, guided by Lesson 30 and your experience. If it does **not** work, go to Step 2.

Our other options are partial fraction decomposition, trigonometric substitution, integration by parts, and use of trigonometric identities; but note that sometimes you may have to apply a combination of these techniques in order to complete the integration. For a given problem, more than one technique may work; but one of these techniques would be more or most efficient.

Step 2: Check for the possibility of applying **partial fraction decomposition:**

This can handle rational functions such a $\dfrac{1}{a^2 - x^2}$, $\dfrac{1}{1 - x^2}$ in which

the denominator is factorable, but **cannot** handle $\dfrac{1}{a^2 + x^2}$ or $\dfrac{1}{1 + x^2}$

or radical functions directly.

Step 3: If the denominator is not factorable, try **trigonometric substitution**. However, this can also handle rational functions in which the denominator is the difference of two squares. Candidates here are

$\dfrac{1}{a^2 + x^2}$, $\dfrac{1}{1 + x^2}$, $\dfrac{1}{a^2 - x^2}$, $\dfrac{1}{1 - x^2}$ as well as . $\dfrac{1}{\sqrt{a^2 - x^2}}$, $\dfrac{1}{\sqrt{1 - x^2}}$

. $\dfrac{1}{\sqrt{a^2 + x^2}}$, $\dfrac{1}{\sqrt{1 + x^2}}$ and . $\sqrt{a^2 - x^2}$. Trigonometric substitution can

handle more different functions than partial fraction decomposition.

(**a**) For problems involving the **sum of two squares as** denominators, try the **tangent** function for substitution.

Sample candidates are $\dfrac{1}{a^2 + x^2}$, $\dfrac{1}{1 + x^2}$, $\dfrac{1}{\sqrt{a^2 + x^2}}$, $\dfrac{1}{\sqrt{1 + x^2}}$

(b) For problems involving the **square root** of the **difference of two squares**, if the **minus sign** precedes the x^2-term, try the **sine** function . Sample candidates are $\sqrt{a^2 - x^2}$, $\dfrac{1}{\sqrt{1 - x^2}}$, $\sqrt{1 - x^2}$

(C) For the **square root** of the **difference of two squares**, if the **minus sign** precedes the **constant term**, try the **secant** function.

Candidates are: $\sqrt{x^2 - 1}$, $x\sqrt{x^2 - a^2}$, $\dfrac{1}{x\sqrt{x^2 - 1}}$, $\dfrac{1}{\sqrt{x^2 - 1}}$ and $\dfrac{\sqrt{x^2 - 16}}{x}$

Step 4: Consider **integration by parts** if the integrand is the product or quotient of two **different types** of functions; or a single function such as the $\arcsin x$ or $\ln x$ function.

The objective of the above guidelines is to avoid wasting trial-and-error time on methods which may not work when taking examinations. Note that almost all examinations are timed. However, when studying, after successfully completing a problem using a particular method, experiment with the other methods to do the same problem in order to experience what works and what does not work, so that you do not do unnecessary experimentation when taking examinations.

Important: Always keep **Step 1** in mind at any step, since it may show up.

Lesson 44

Applications of Indefinite Integrals
Equations of Curves (and Lines); Equations of Motion

Equations of Curves (and Lines)

In elementary mathematics, given an equation of a line, we can find the slope m of this line. Conversely, given two properties, such as the slope m and the coordinates (x_0, y_0) of one point on the line, we can find an equation of the line satisfying these conditions.

So also, in calculus, given an equation $y = f(x)$ of a curve, we can find the slope m at a point (x_0, y_0) on the curve, by applying differentiation.

Conversely, given the slope $m = \dfrac{dy}{dx}$ at a point (x_0, y_0), we can find a general equation, $y = f(x) + C$ for a family of curves, by integration. For a particular curve, we need more information such as the coordinates of a given point on the curve.

Example 1
(a) Find an equation of the family of lines with slope $= 3$.

(b) Find an equation of the line with slope $= 3$, and containing the point $(2, 5)$.

Solution

(a) From $\dfrac{dy}{dx} = 3$ $\qquad y = 3x + C$ integrating)	a) By elementary math, we can $\quad$ apply $y = mx + b$ to obtain $\qquad y = 3x + b$
(b) Since the line contains the point $(2, 5)$, we can use this information find the constant C. Substituting $x = 2, y = 5,$ in $y = 3x + C$, $\qquad 5 = 3(2) + C$ $\qquad 5 = 6 + C$ $\qquad -1 = C$ $\qquad C = -1,$ and $\qquad y = 3x - 1$	(b) By elementary math, we can $\quad$ apply $y - y_1 = m(x - x_1)$ $\qquad y - 5 = 3(x - 2)$ $\qquad y - 5 = 3x - 6$ $\qquad\quad y = 3x - 6 + 5$ $\qquad\quad y = 3x - 1$ $\quad$ or Find b, with $m = 3$. $\qquad 5 = 3(2) + b$ and $-1 = b$ Applying $y = mx + b$, $y = 3x - 1$

Equations of Motion

If a body moves along a straight line and travels a distance s, the distance s is a function of the time t taken to travel the distance s, and we can write $s = f(t)$, where $t \geq 0$.

Previously (p.198), we learned the following:

The **linear velocity** v at time t is given by $v = \dfrac{ds}{dt}$.

If $v > 0$, the body is moving in the direction of increasing s.

If $v < 0$, the body is moving in the direction of decreasing s.

If $v = 0$, the body is at rest.

The **linear acceleration** a at time t is given by

$$a = \frac{dv}{dt} = \frac{d^2s}{dt^2}$$

If $a > 0$, v is increasing; but if $a < 0$, v is decreasing.
If v and a have the same sign, the speed of the body is increasing.
If v and a have opposite signs, the speed of the body is decreasing.

Finding Equations of Motion

From above, and reversing the steps:
Given or knowing the acceleration, $a(t)$, we can find an equation for the velocity, $v(t)$ by integration:

$$v(t) = \int a(t)dt$$

Similarly, given or knowing the velocity, $v(t)$, we can find an equation for the distance, $s(t)$, by integration:

$$s(t) = \int v(t)dt$$

Example 2 A particle is moving along a straight line according to the
following: $a = 6 - 2t$; when $t = 0$, $v = 3$; when $t = 0$, $s = 0$
where the distance s is a function of the time t taken to travel the
distance s, and a is the acceleration, Find (a) $v(t)$; (b) $s(t)$.

Solution

(a) Velocity v, is given by

Step 1: $v(t) = \int a(t)dt$

$$= \int (6 - 2t)dt$$

$$= 6t - \frac{2t^2}{2} + C_1$$

$$v(t) = 6t - t^2 + C_1$$

Step 2 Find : C_1

When $t = 0$, $v = 3$

$$3 = 6(0) - (0)^2 + C_1$$

$$3 = C_1$$

$$v(t) = 6t - t^2 + 3$$

$$v(t) = -t^2 + 6t + 3.$$

(b) The distance s is *given by*

Step 1: $s(t) = \int v(t)dt$

$$= \int (-t^2 + 6t + 3)dt$$

$$= -\frac{t^3}{3} + \frac{6t^2}{2} + 3t + C_2$$

$$= -\frac{t^3}{3} + 3t^2 + 3t + C_2$$

Step 2 Find : C_2

When $t = 0$, $s = 0$

$$0 = -\frac{(0)^3}{3} + 3(0)^2 + 3(0) + C_2$$

$$0 = C_2$$

$$s(t) = -\frac{t^3}{3} + 3t^2 + 3t + 0$$

$$s(t) = -\frac{t^3}{3} + 3t^2 + 3t$$

Lesson 44 Exercises

1. (a) Find an equation of the family of lines with slope $= 3$.

(b) Find an equation of the line with slope $= 3$, and containing the point $(2, 5)$

2. A particle is moving along a straight line according to the following: $a = 6 - 2t$; when $t = 0$, $v = 3$; when $t = 0$, $s = 0$ where the distance s is a function of the time t taken to travel the distance s, and a is the acceleration, Find (a) $v(t)$; (b) $s(t)$.

Answers:

1. (a) $y = 3x + C$; (b) $y = 3x - 1$; **2,** $v(t) = -t^2 + 6t + 3$;

$s(t) = -\dfrac{t^3}{3} + 3t^2 + 3t$.

CHAPTER 13
Lesson 45
Definite Integral

Definition

If the function f is continuous on the closed interval $[a, b]$, then the definite integral from a to b, symbolized $\int_a^b f(x)\, dx$ (a number) is defined as the net area between the curve $y = f(x)$, the x–axis, and the lines $x = a$ and $x = b$.

Symbolically, $\int_a^b f(x)\, dx = \lim_{n \to \infty} \sum_{k=1}^{n} f(c_k)\Delta x_k$, where $c_k \in [c_{k-1}, c_k]$

$\sum_{k=1}^{n} f(x_k)\Delta x_k$ is called the Riemann sum of f on $[a, b]$.

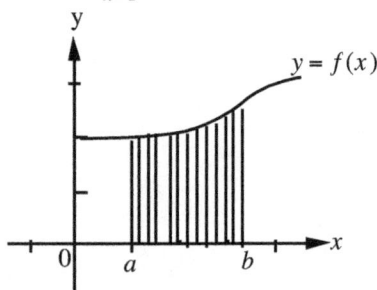

To evaluate $\int_a^b f(x)\, dx$, divide $[a, b]$ into n equal subintervals and sum up the products $f(x_k)\Delta x$ (areas of the n rectangles) as $n \to \infty$, where

$\Delta x = \dfrac{b - a}{n}$, altitude $= f(x_k)$

$a = x_0 < x_1 < x_2 < \ldots < x_n = b$

The numbers a and b are called the limits of integration, where a is the lower limit and b is the upper limit; and $f(x)$ is called the integrand.

Note that the definite integral $\int_a^b f(x)\, dx$ is number but the indefinite integral denoted by $\int f(x)\, dx$ is a family of functions (which are antiderivatives of f)

Thus $\underbrace{\int_a^b f(x)\, dx}_{\substack{\text{definite} \\ \text{integral}}} = $ real number but $\underbrace{\int f(x)\, dx}_{\substack{\text{indefinite} \\ \text{integral}}} = g(x) + C$ where $g'(x) = f(x)$)

Fundamental Theorem of Calculus

If f is continuous on the closed interval $[a,\ b]$, and if $\int f(x)dx = g(x)$, where

g is an antiderivative of f, then $\int_a^b f(x)dx = g(b) - g(a) = \left[\int f(x)dx\right]_a^b$

This equation provides a simple method of evaluating $\int_a^b f(x)dx$, if we can find an antiderivative g for f, and we do not have to use the Riemann sum definition. However, the applications of the definite integral will be motivated by the Riemann definition.

The fundamental theorem of calculus connects the **indefinite integral** (which is purely algebraic) and the **definite integral** (which is purely geometric).

Example 1

Applying the Fundamental Theorem of Calculus

Find $\int_2^5 x^2 dx$

Solution

$\int_2^5 x^2 dx$

$= \left[\dfrac{x^{2+1}}{2+1}\right]_2^5$

$= \left[\dfrac{x^3}{3}\right]_2^5$

$= \left(\dfrac{5^3}{3}\right) - \left(\dfrac{2^3}{3}\right)$

$= \left(\dfrac{125}{3}\right) - \left(\dfrac{8}{3}\right)$

$= 39$

$\int_2^5 x^2 dx = 39.$

Example 2 Find $\int_1^4 (4x^3 + 2x^2 - 8x + 7)dx$

Solution

$\int_1^4 (4x^3 + 2x^2 - 8x + 7)dx$

$= \left[\dfrac{4x^{3+1}}{3+1} + \dfrac{2x^{2+1}}{2+1} - \dfrac{8x^{1+1}}{1+1} + 7x\right]_1^4$

$= \left[\dfrac{4x^4}{4} + \dfrac{2x^3}{3} - \dfrac{8x^2}{2} + 7x\right]_1^4$

$= \left[x^4 + \dfrac{2x^3}{3} - 4x^2 + 7x\right]_1^4$

$= \left(4^4 + \dfrac{2(4)^3}{3} - 4(4)^2 + 7(4)\right) - \left(1^4 + \dfrac{2(1)^3}{3} - 4(1)^2 + 7(1)\right) ($

substitution of the upper limit minus substitution of the lower limit)

$= (262\tfrac{2}{3}) - (4\tfrac{2}{3})$

$= 258.$

$\therefore \int_1^4 (4x^3 + 2x^2 - 8x + 7)dx = 258$

Example 3 Find $\int_{-2}^5 |x - 3| dx$

The absolute value sign compels us to apply the absolute value definition and consider two cases of the integrand.

Case 1: $f(x) = x - 3$ if $x - 3 \geq 0$ or $x \geq 3$ (yielding the limits from 3 to 5)

Case 2: $f(x) = -(x - 3)$ if $x - 3 < 0$ or $x < 3$ (yielding the limits from -2 to 3)

We draw a number line to help us determine which integrand and which limits to use between -2 and 5:

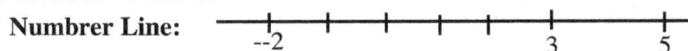

Numbrer Line:

We use the integrand $-(x - 3)$ from -2 to 3. (actually from -2 to 3^-)

We use the integrand $x - 3$ from 3 to 5.

For the integral: $\int_{-2}^3 -(x - 3)\, dx + \int_3^5 (x - 3)\, dx$

$= \int_{-2}^3 (-x + 3)\, dx + \int_3^5 (x - 3)\, dx$

$= \left[-\dfrac{x^2}{2} + 3x\right]_{-2}^3 + \left[\dfrac{x^2}{2} - 3x\right]_3^5$

$= \left[\left(-\dfrac{3^2}{2} + 3(3)\right) - \left(-\dfrac{(-2)^2}{2} + 3(-2)\right)\right] + \left[\left(\dfrac{5^2}{2} - 3(5)\right) - \left(\dfrac{3^2}{2} - 3(3)\right)\right]$

$\int_{-2}^5 |x - 3| dx = 14\tfrac{1}{2}$ or $\dfrac{29}{2}.$

Substitution Method for Definite Integrals

We cover two approaches

Example 4 Find $\displaystyle\int_0^3 \frac{5x}{\left(x^2+1\right)^3}dx$

Approach 1

Step 1: Let $u = x^2 + 1$

Then $\dfrac{du}{dx} = 2x$ and $dx = \dfrac{du}{2x}$

Now, in $\displaystyle\int \frac{5x}{\left(x^2+1\right)^3}dx$ replace

dx by $\dfrac{du}{2x}$; and $x^3 + 1$ by u. to obtain

$$= \int \frac{5x}{u^3} \cdot \frac{du}{2x}$$

$$= \int \frac{5}{u^3} \cdot \frac{du}{2} \text{ (cancel the } x)$$

$$= \frac{5}{2}\int \frac{du}{u^3}$$

$$= \frac{5}{2}\int u^{-3}du$$

$$= \frac{5}{2} \cdot \frac{u^{-3+1}}{-3+1} + C$$

$$= \frac{5}{2} \cdot \frac{u^{-2}}{-2} + C$$

$$= -\frac{5}{4} \cdot u^{-2} + C$$

$$= -\frac{5}{4} \cdot \frac{1}{u^2} + C$$

$$= -\frac{5}{4(x^2+1)^2} + C$$

Step 2:

$$\int_0^3 \frac{5x}{\left(x^2+1\right)^3}dx = \left[-\frac{5}{4(x^2+1)^2} + C \right]_0^3$$

$$= \left(-\frac{5}{4(3^2+1)^2} + C \right) - \left(-\frac{5}{4(0^2+1)^2} + C \right)$$

$$= \left(-\frac{5}{400} + C \right) - \left(-\frac{5}{4} + C \right)$$

$$= -\frac{5}{400} + C + \frac{5}{4} - C$$

$$= -\frac{5}{400} + \frac{5}{4}$$

$$= \frac{99}{80}.$$

Note above that the constant term C drops out and therefore in practice, we will not write it from Step 2. Note also that in Step 1, we used the indefinite integral sign $"\int"$, rather than the definite sign $"\displaystyle\int_0^3"$. In approach 2 below, we take care of the limits by a different approach.

Approach 2 Find $\displaystyle\int_0^3 \frac{5x}{\left(x^2+1\right)^3}dx$

Step 1: Let $u = x^2 + 1$

Then $\dfrac{du}{dx} = 2x$ and $dx = \dfrac{du}{2x}$

Step 2: Express the limits of integration in terms of u.

When $x = 0$, $u = 0^2 + 1 = 1$;

when $x = 3$, $u = 3^2 + 1 = 10$

Therefore, the limits in terms of u are $u = 1$ and $u = 10$

Step 3: In $\displaystyle\int_0^3 \frac{5x}{\left(x^2+1\right)^3}dx$,

replace 0 by 1; 3 by 10; $x^2 + 1$

by u; dx by $\dfrac{du}{2x}$. or $du = 2x\,dx$,

Then $\displaystyle\int_0^3 \frac{5x}{(x^2+1)^3}\,dx = \int_1^{10} \frac{5x}{u^3}\cdot\frac{du}{2x}$:

$\displaystyle= \int_1^{10} \frac{5}{u^3}\cdot\frac{du}{2}$

Step 4:

$\displaystyle= \frac{5}{2}\int_1^{10} \frac{du}{u^3}$

$\displaystyle= \int_1^{10} u^{-3}\,du$

$\displaystyle= \frac{5}{2}\left[\frac{u^{-3+1}}{-3+1}\right]_1^{10}$

$\displaystyle= \frac{5}{2}\left[\frac{u^{-2}}{-2}\right]_1^{10}$

$\displaystyle= -\frac{5}{4}\left[\frac{1}{u^2}\right]_1^{10}$

$\displaystyle= -\frac{5}{4}\left(\frac{1}{10^2} - \frac{1}{1^2}\right)$

$\displaystyle= -\frac{5}{4}\left(\frac{1}{100} - \frac{1}{1}\right)$

$\displaystyle= -\frac{5}{4}\left(\frac{-99}{100}\right) = \frac{99}{80}$

Example 5 Find $\displaystyle\int_0^{\frac{\pi}{4}} \sin^3 2x\cos 2x\,dx$ (Using Approach 2)

Step 1: Let $u = \sin 2x$

Then $\dfrac{du}{dx} = 2\cos 2x$ and

$dx = \dfrac{du}{2\cos 2x}$ or

$du = 2\cos 2x\,dx$

Step 2: Express the limits of integration in terms of u.

When $x = 0$, $u = \sin 0 = 0$;

when $x = \frac{\pi}{4}$,

$u = \sin 2\bullet\frac{\pi}{4} = \sin\frac{\pi}{2} = 1$

Therefore, the limits in terms of u are $u = 0$ and $u = 1$.

Step 3: In $\displaystyle\int_0^{\frac{\pi}{4}} \sin^3 2x\cos 2x\,dx$,

replace $\frac{\pi}{4}$ by 1; $\sin 2x$ by u;

dx by $\dfrac{du}{2\cos 2x}$ .or $du = 2\cos 2x\,dx$

to obtain

Step 4: $\displaystyle\int_0^1 u^3\cos 2x\frac{du}{2\cos 2x}$

$\displaystyle= \frac{1}{2}\int_0^1 u^3\,du$

$\displaystyle= \frac{1}{2}\left[\frac{u^4}{4}\right]_0^1$

$\displaystyle= \frac{1}{2}\left(\frac{1^4}{4} - \frac{0^4}{4}\right)$

$\displaystyle= \frac{1}{2}\left(\frac{1}{4}\right) = \frac{1}{8}$

Example 6 Find $\displaystyle\int_2^5 \frac{x-2}{\sqrt{x-1}}\,dx$ (**Using Approach 2**)

Step 1: Let $u = \sqrt{x-1}$

Then $u^2 = x - 1$

$x = u^2 + 1$

Also, $2\dfrac{du}{dx} = 1$ and $dx = 2u\,du$

Step 2: Express the limits of integration in terms of u.

When $x = 2$, $u = \sqrt{2-1} = 1$;

when $x = 5$, $u = \sqrt{5-1} = 2$

Therefore, the limits in terms of u are $u = 1$ and $u = 2$.

Step 3: Substitute the results from Step 1 and Step 2 in

$\displaystyle\int_2^5 \frac{x-2}{\sqrt{x-1}}\,dx$ to obtain

$\displaystyle\int_1^2 \frac{u^2 + 1 - 2}{u} \cdot 2u\,du$

$= 2\displaystyle\int_1^2 (u^2 - 1)\,du$

Step 4: Integrate now to obtain

$2\left[\dfrac{u^3}{3} - u\right]_1^2$

$= 2\left[\left(\dfrac{2^3}{3} - 2\right) - \left(\dfrac{1^3}{3} - 1\right)\right]$

$= 2\left[\left(\dfrac{8}{3} - 2\right) - \left(\dfrac{1}{3} - 1\right)\right]$

$= 2\left[\left(\dfrac{2}{3}\right) - \left(-\dfrac{2}{3}\right)\right]$

$= 2\left[\dfrac{2}{3} + \dfrac{2}{3}\right]$

$= 2\left[\dfrac{4}{3}\right]$

$\displaystyle\int_2^5 \frac{x-2}{\sqrt{x-1}}\,dx = \dfrac{8}{3}$

Using the indefinite integral to evaluate a definite integral

Example 7 Find $\displaystyle\int_0^{2\sqrt{3}} \frac{x^3}{\sqrt{16 - x^2}}\,dx$

Solution: On p.311,, Example 3, we found the indefinite integral

$\displaystyle\int \frac{x^3}{\sqrt{16 - x^2}}\,dx$ to be $= 64\left[-\dfrac{\sqrt{16 - x^2}}{4} + \dfrac{1}{3}\left(\dfrac{\sqrt{16 - x^2}}{4}\right)^3\right] + C$

We therefore apply the limits of integration to obtain:

$$\int_0^{2\sqrt{3}} \frac{x^3}{\sqrt{16 - x^2}}\,dx = 64\left[-\frac{\sqrt{16 - x^2}}{4} + \frac{1}{3}\left(\frac{\sqrt{16 - x^2}}{4}\right)^3\right]_0^{2\sqrt{3}}$$

$$= 64\left[\left\{-\frac{\sqrt{16 - (2\sqrt{3})^2}}{4} + \frac{1}{3}\left(\frac{\sqrt{16 - (2\sqrt{3})^2}}{4}\right)^3\right\} - \left\{-\frac{\sqrt{16 - 0}}{4} + \frac{1}{3}\left(\frac{\sqrt{16 - 0}}{4}\right)^3\right\}\right]$$

$$= 64\left[\left\{-\frac{\sqrt{4}}{4} + \frac{1}{3}\left(\frac{\sqrt{4}}{4}\right)^3\right\} - \left\{-\frac{\sqrt{16}}{4} + \frac{1}{3}\left(\frac{\sqrt{16}}{4}\right)^3\right\}\right] = 64\left[\left\{-\frac{11}{24}\right\} - \left\{-\frac{2}{3}\right\}\right] = \frac{40}{3}.$$

Mean Value Theorem for Definite Integrals

If $f(x)$ is continuous on the closed interval $a \leq x \leq b$, then there exists at least

number c in the interval $a \leq x \leq b$ such that $\int_a^b f(x)dx = f(c)(b-a)$

The value of $f(c)$ is called the mean value of the function $f(x)$ on the closed

interval $a \leq x \leq b$.

Definition: The average value of f on the closed interval $a \leq x \leq b$ is

$$\frac{1}{b-a}\int_a^b f(x)dx \quad \text{or} \quad \frac{\int_a^b f(x)dx}{b-a}$$

Example 1 Find the average value of $\int_{-1}^{3}(2x^2-1)dx$ on $[-1,3]$.

Solution The average value of f on the closed interval $a \leq x \leq b$ is given by

$$\frac{1}{b-a}\int_a^b f(x)dx$$

Step 1	**Step 2**
Average value of $\int_{-1}^{3}(2x^2-1)dx$	$= \frac{1}{4}\left[(18-3)-\left(-\frac{2}{3}+1\right)\right]$
$= \frac{1}{3-(-1)}\int_{-1}^{3}(2x^2-1)dx$	$= \frac{1}{4}\left[(15)-\left(\frac{1}{3}\right)\right]$
$= \frac{1}{4}\left[\frac{2x^3}{3}-x\right]_{-1}^{3}$	$= \frac{1}{4}\left[\left(14\frac{2}{3}\right)\right]$
$= \frac{1}{4}\left[\left(\frac{2(3)^3}{3}-(3)\right)-\left(\frac{2(-1)^3}{3}-(-1)\right)\right]$	$= \frac{11}{3}$.

More Properties of Definite Integrals

1. $\frac{d}{dx}\int_a^x f(u)du = f(x)$, where u is a dummy variable, We could use t or x.

   (Also, $\frac{d}{dx}\int_a^x f(x)dx = f(x)$)

2. $\int_a^b kf(x)dx = k\int_b^a f(x)dx$.

3. $\int_a^b f(x)dx = -\int_b^a f(x)dx$. (interchange the limits and introduce a minus sign)

4. $\int_a^b f(x)dx = \int_a^c f(x)dx + \int_c^b f(x)dx$, where $a < c < b$ on a closed interval.

5. $\int_a^b [f(x) \pm g(x)]dx = \int_b^a f(x)dx \pm \int_b^a f(x)dx$ <--- sum or difference.

6. If $f(x) \geq 0$ on $[a, b]$, then $\int_a^b f(x)dx \geq 0$.

7. If $f(x) \leq 0$ on $[a, b]$, then $\int_a^b f(x)dx \leq 0$.

Theorem If the function f is continuous on the closed interval $[a,b]$, then

   f is integrable on $[a,b]$.

Lesson 45 Exercises

1. Find $\int_2^5 x^2 dx$

2.. Find $\int_1^4 (4x^3 + 2x^2 - 8x + 7)dx$

3. Find $\int_0^3 \frac{5x}{\left(x^2 + 1\right)^3} dx$

4. Find $\int_0^{\frac{\pi}{4}} \sin^3 2x \cos 2x \, dx$

5. Find $\int_0^2 2x^2 \sqrt{x^3 + 1} \, dx$

6. Find $\int_2^5 \frac{x - 2}{\sqrt{x - 1}} dx$

7. $\int_{-2}^5 |t - 3| dt$

8. $\int_5^8 r\sqrt{r - 4} \, dr$

9, $\int_1^9 \sqrt{r} \, dr$

10. $\int_{-\frac{\pi}{4}}^{\frac{\pi}{4}} \cos x \, dx$

Answers: **1.** 39; **2.** 258; **3.** $\frac{99}{80}$; **4.** $\frac{1}{8}$; **5,** $\frac{104}{9}$; **6.** $\frac{8}{3}$. **7.** $14\frac{1}{2}$ or $\frac{29}{2}$;

8. $\frac{466}{15}$; **9.** $\frac{52}{3}$; **10.** $\sqrt{2}$

CHAPTER 14

Applications of Integration

Lesson 46
Introduction to Areas

Let us begin from what we are familiar with: the area of a rectangle.

Example 1 Find the area bounded above by the line $y = 5$, bounded below by the line $y = 2$, and bounded on the left by the line $x = 1$, and on the right by $x = 6$.

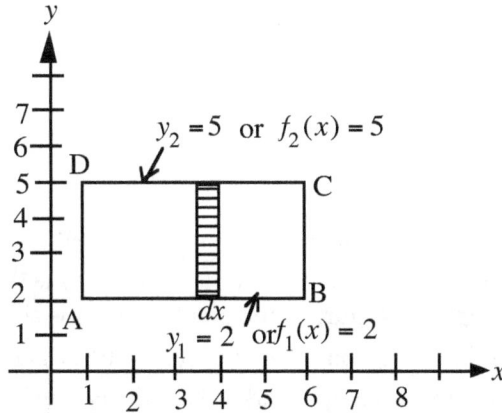

Method 1: Using geometry

Length L of rectangle $ABCD$ is $L = 6 - 1 = 5$
Width W of $ABCD$ is $W = 5 - 2 = 3$
Area of rectangle ABCD = $5(3) = 15$ sq, units

Method 2: Using calculus (of course, this is an overkill since Method 1 does it easily)
To find the area of the region of the specified region, we can inscribe rectangles in the region and sum up the areas of the rectangles as the number of rectangles increases indefinitely. For a short-cut, instead of inscribing an infinite number of rectangles, we inscribe only a representative rectangle with its dimensions, and then apply definite integration. We can inscribe either a vertical or a horizontal rectangle. Each type of rectangle inscribed has its relative merits according to the region enclosed. In this example, we inscribe a vertical rectangle of length

$y_2 - y_1 = f_2(x) - f_1(x) = 5 - 2$ and width dx, and then integrate

$A = \int_a^b [f_2(x) - f_1(x)] dx$ (Note that $f_1(x)$ and $f_2(x)$ are constant functions)

$= \int_1^6 (5 - 2) dx = \int_1^6 3 dx = [3x]_1^6 = [3(6)] - [3(1)] = 15$ \qquad $(18 - 3)$

The area is again 15 sq. units.

Before we proceed further, we cover some phrases and concepts that will be useful in sketching graphs as well as drawing representative rectangles.

Positive and Negative Functions

When we say that a function is **positive** on a certain interval, say, the interval from $x = a$ to $x = b$, we mean that, algebraically, all the y-coordinate values on this interval from a to b are **positive**, and graphically, the entire curve lies a**bove** the x-axis on this interval. In Figure 1, $f(x) > 0$ on the interval from a to b Similarly, when we say that a function is **negative** on the interval from a to b, we mean that all the y-coordinate values are negative (have minus signs) on this interval from a to b, and that, graphically, the entire curve lies **below** the x-axis, In Figure 3, we can write symbolically that $f(x) < 0$ on the interval from a to b.

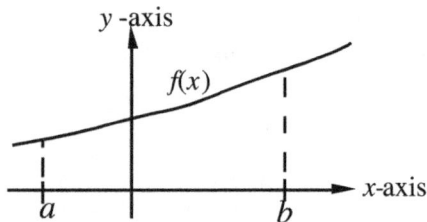

Figure 1

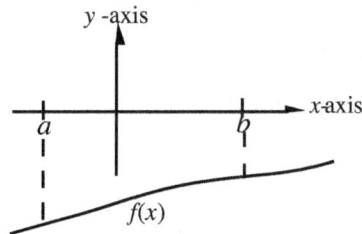

Figure 2

The terms "above" and "below".

In Figure 3,, the graph of $f(x)$ is above the graph of $g(x)$ which implies that the graph of $g(x)$ is below the graph of $f(x)$. We can also say that $f(x)$ is the top graph, and $g(x)$ is the bottom graph. In Figure 4, the graph of $g(x)$ is above the graph of $h(x)$ which implies that the graph of $h(x)$ is below the graph of $f(x)$. We can also say that $g(x)$ is the top graph, and $h(x)$ is the bottom graph. Symbolically, we can say that, in Figure 3, $f(x) \geq g(x)$ for all x.
Similarly, in Figure 4, $g(x) \geq h(x)$ for all x.

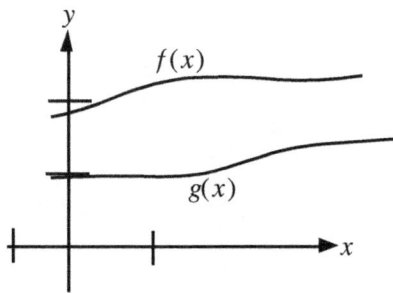

Figure 3

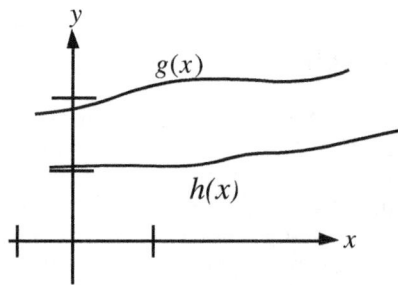

Figure 4

Sometimes, instead of the word "above" we will use appropriately the adjectives "top", or "upper" interchangeably to mean "above". Similarly, instead of the word "below", we will use the words "bottom," or "lower".

We extend the above discussion to Figure 5 below. On the interval between $x = a$ and $x = b$, $f(x)$ is above $g(x)$ (that is, $f(x) \geq g(x)$ for all x on this interval). On the interval between $x = b$ and $x = c$, $g(x)$ is above $f(x)$ (that is, $g(x) \geq h(x)$ for all x. on this interval).

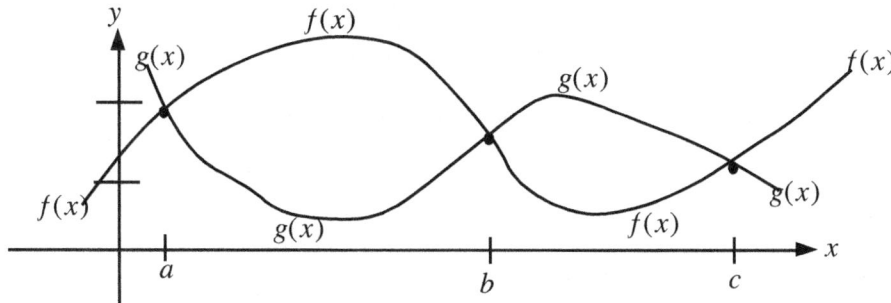

Figure 5

Drawing Representative Rectangles

In problems involving plane areas by integration, the main difficulty a student may have is not the application of integration but rather it is the proper determination of the representative rectangles and their dimensions.

Question: May we use a horizontal rectangle, or a vertical rectangle?

Answer: This depends on the shape of the region whose area we want to find. Sometimes, a single representative rectangle would be sufficient, and sometimes two or more representative rectangles may be needed. For a given region, we may use a single rectangle or two or more rectangles depending on whether we use a horizontal or a vertical rectangle. Note that generally, the number of representative rectangles used equals the number of integrals involved in the problem.

Case 1 Using a Vertical Representative Rectangle (Lengthwise vertical)

The determination of the number of representative rectangles between any two curves would be guided as follows: We agree that whatever is said for a curve also applies to a straight line. **Guidelines:** For a specific set of limits, the lower ends (bottoms) of all possible representative rectangles must be on the same curve C_1, and the other ends, the upper ends, must all be on the same curve C_2, where C_2 could be the same C_1. If this condition is satisfied, then a single representative rectangle is sufficient, otherwise two or more representative rectangles would be needed.

Step 1: Sketch the graph of the region bounded by the two curves (by sketching the graphs of the individual curves; and determine which curve is the upper boundary and which is the lower boundary of the region.

Step 2: From the lower curve to the upper curve draw a representative rectangle of width dx, and length $= y_2 - y_1 = f_2(x) - f_1(x)$.

The length of the representative rectangle is given by the difference between the y-coordinates of points on the upper curve and lower curve.

In using a representative **vertical rectangle**, the area, A, of the region bounded above (upper boundary) by the graph of $y_2 = f_2(x)$ and bounded below (lower boundary) on the bottom by $y_1 = f_1(x)$ is given by

$$A = \int_a^b \left(y_2 - y_1\right) dx = \int_a^b [f_2(x) - f_1(x)] dx$$

Vertical rectangles are summed (added) from the left to the right.

Case 2 Using a Horizontal Representative Rectangle (Lengthwise horizontal)

The determination of the number of representative rectangles between any two curves would be guided as follows: We agree that whatever is said for a curve also applies to a straight line. **Guidelines:** For a specific set of limits, the left ends of all possible representative rectangles must be on the same curve, C_1, and the other ends, the right ends, must all be on the same curve, C_1 or C_2. If this condition is satisfied, then a single representative rectangle is sufficient, otherwise two or more representative rectangles would be needed.

Step 1: Sketch the graph of the region bounded by the two curves (by sketching the graphs of the individual curves; and determine which curve is the right boundary and which curve is the left boundary of the region.

Step 2: From the left curve to the right curve, draw a representative rectangle of width dy, and length $= x_2 - x_1 = g_2(y) - g_1(y)$

The length of the representative rectangle is given by the difference between the x-coordinates of points on the right curve and left curve.

In using a representative **horizontal rectangle**, the area, A, of the region bounded on the right (right boundary) by the graph of $x_2 = g_2(y)$ and bounded on the left (left boundary) by $x_1 = g_1(y)$ is given by

$$A = \int_c^d \left(x_2 - x_1\right)dy = \int_c^d [g_2(y) - g_1(y)]dy$$

Special case 1: Using Vertical Rectangles

To find the area of the region between a curve, the x-axis and the lines $x = a$ and $x = b$, we can consider the x-axis as the line $y = 0$. Then

$$A = \int_a^b \left(y_2 - y_1\right)dx = \int_a^b \left(y_2 - 0\right)dx = \int_a^b y\,dx$$

The length of the representative rectangle is given by the y-coordinate of a point on the curve, and the width of the representative rectangle is dx. Vertical rectangles are summed from the left to the right.

Special case 2: Using Horizontal Rectangles

To find the area of the region between a curve, the y-axis and the lines $y = c$ and $y = d$, we can consider the y-axis as the line $x = 0$. Then

$$A = \int_c^d \left(x_2 - x_1\right)dy = \int_c^d \left(x_2 - 0\right)dy = \int_c^d x\,dy$$

The length of the representative rectangle is given by the x-coordinate of a point on the curve, and the width of the representative rectangle is dy. Horizontal rectangles are summed from the bottom to the top.

Note: $\int_a^b f(x)\,dx$ assumes that $f(x) \geq 0$ for all x in the closed interval $[a, b]$.

($f(x) > 0$ means the curve lies above the x-axis.). 2. f is continuous on the closed interval $[a, b]$

$\int_a^b f(x)\,dx =$ the number of square units bounded by the curve $y = f(x)$, the x-axis, and the lines $x = a$ and $x = b$. However, if $f(x) < 0$ (i.e., the curve lies below the x-axis.); then the area bounded by the curve $y = f(x)$, the x-axis, and the lines $x = a$ and $x = b$ is given by $-\int_a^b f(x)\,dx$ (the y-coordinate of the top of the representative vertical rectangle is given by $y = 0$).

Example 1 Using integrals, we represent the area between $f(x)$ and $g(x)$ from

$x = a$ to $x = c$, as follows: $\int_a^b [f(x) - g(x)]dx + \int_b^c [g(x) - f(x)]dx$

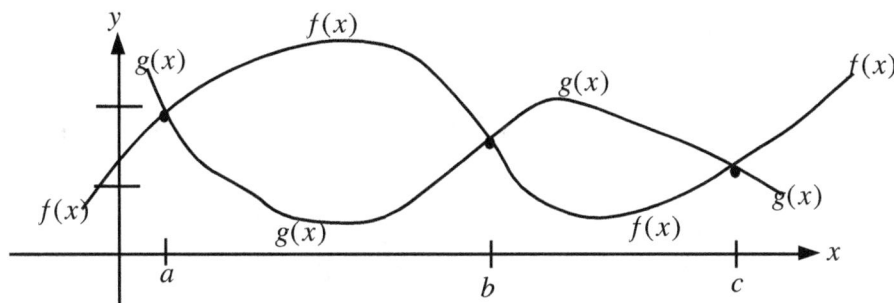

Example 2 Using integrals, represent the area between $f(x)$ and $g(x)$ from $x = a$ to $x = c$.

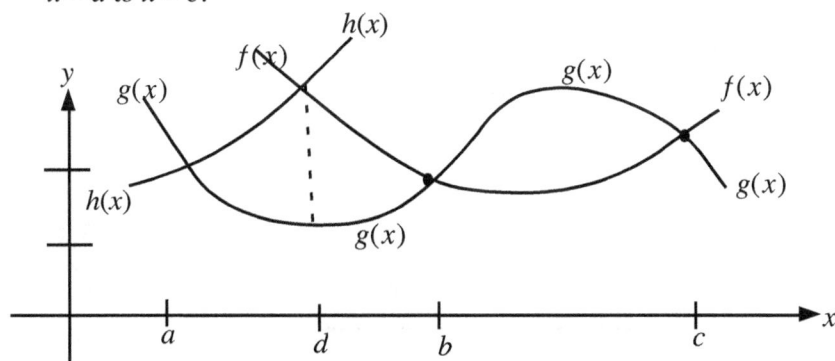

Solution: $\int_a^d [h(x) - g(x)]dx + \int_d^b [f(x) - g(x)]dx + \int_b^c [g(x) - f(x)]dx$

Example 3 Using integrals, represent the area from a to c.

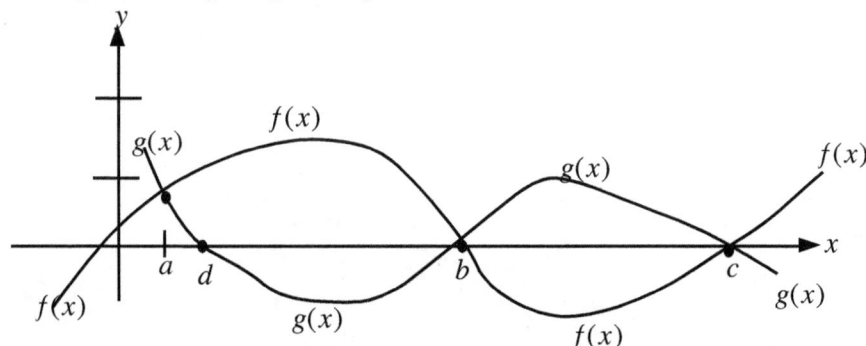

Solution   Note that (Below the x-axis $f(x) < 0$, $g(x) < 0$)

$$\int_a^d [f(x) - g(x)]dx + \int_d^b [f(x)]dx + \int_d^b [0 - g(x)]dx + \int_b^c [g(x)]dx + \int_b^c [0 - f(x)]dx$$

$$= \int_a^d [f(x) - g(x)]dx + \int_d^b [f(x) - g(x)]dx + \int_b^c [g(x) - f(x)]dx$$

$$= \int_a^b [f(x) - g(x)]dx + \int_b^c [g(x) - f(x)]dx \, .$$

Lesson 46 Exercises

1. Using integrals, represent the area between $f(x)$ and $g(x)$ from $x = a$ to $x = c$.

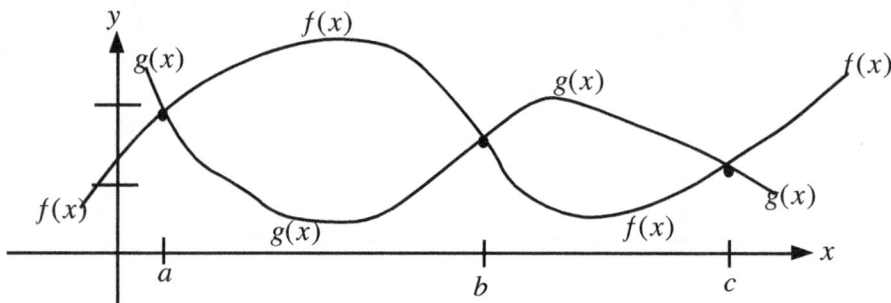

2. Using integrals, represent the area between $f(x)$ and $g(x)$ from $x = a$ to $x = c$.

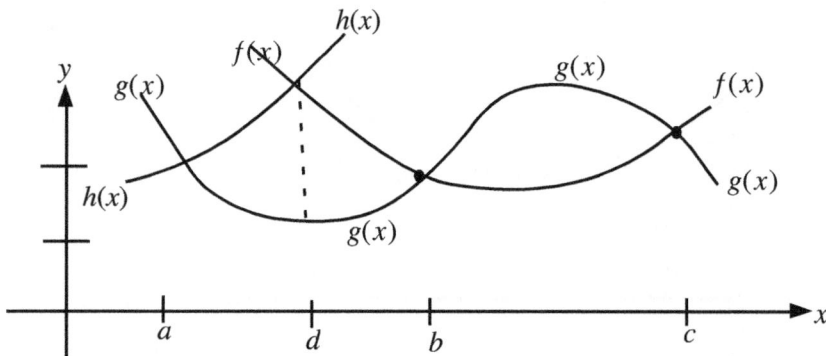

3.

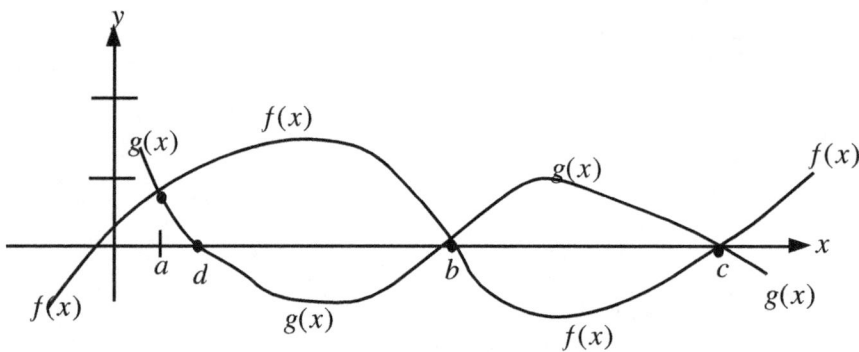

Answers:

1. $\int_a^b [f(x) - g(x)]dx + \int_b^c [g(x) - f(x)]dx$;

2. $\int_a^d [h(x) - g(x)]dx + \int_d^b [f(x) - g(x)]dx + \int_b^c [g(x) - f(x)]dx$

3. $\int_a^b [f(x) - g(x)]dx + \int_b^c [g(x) - f(x)]dx$

Lesson 47
Area of a Region Bounded by Two Lines:
Curved and Straight Lines
(Generalized Approach)

In the examples that follow, we do not have any simple formulas for calculating the areas and so we will resort to calculus.

Example 1 Find the area bounded by the curve $f(x) = -x^2 + 9$ and the x-axis..

Method 1: Using Vertical slicing

Step 1: Sketch the graph of the required region

Let $y = -x^2 + 9$

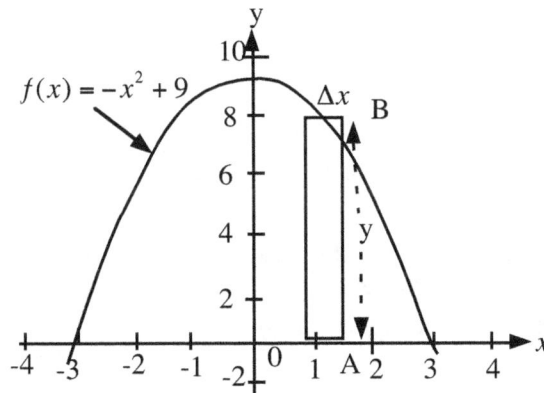

Step 2: Find the x-intercepts

(the points of intersection of $y = -x^2 + 9$ with the x-axis)

Let y = 0 in $y = -x^2 + 9$. Then

$-x^2 + 9 = 0$
$x^2 - 9 = 0$
$(x - 3)(x + 3) = 0$
$x - 3 = 0$, or $x + 3 = 0$
$x = 3$, or $x = -3$

The x-intercepts are -3 and 3. These numbers will be used as the limits of the subsequent integration.

Step 3: Draw a vertical representative rectangle from a point on $y = -x^2 + 9$ to a point on the x-axis (the line $y = 0$).

The y-coordinate, say y_{top} of the top of the representative rectangle is given by $y_{top} = -x^2 + 9$; and the y-coordinate, say y_{bot} of the bottom of representative rectangle is given by $y_{bot} = 0$ (y-value on the x-axis) .

The length of the representative rectangle is

$y_{top} - y_{bot} = (-x^2 + 9) - (0) = -x^2 + 9$; and the width is dx

Area of the representative rectangle = length × width
To find the area of the required region, we use two approaches:

1. We can find the area using Length $= -x^2 + 9$, width $= dx$, together with the limits $x = 0$ and $x = 3$, followed by doubling (for symmetry) the area found; or

2. We can find the area using Length $= -x^2 + 9$, width $= dx$, and the limits $x = 3$ and $x = -3$. Required area A of the region is

$$A = 2\int_0^3 (-x^2 + 9)dx \quad \text{or} \quad A = \int_{-3}^3 (-x^2 + 9)dx$$

Step 4: $= 2\left[-\dfrac{x^3}{3} + 9x\right]_0^3 \quad$ or $\quad A = \left[-\dfrac{x^3}{3} + 9x\right]_{-3}^3$

$= 2\left[\left(-\dfrac{3^3}{3} + 9(3)\right) - (0)\right] \quad$ or $\quad = \left(-\dfrac{3^3}{3} + 9(3)\right) - \left(-\dfrac{(-3)^3}{3} + 9(-3)\right)$

$= 2\left[\left(-\dfrac{27}{3} + 27\right) - (0)\right] \quad$ or $\quad = \left(-\dfrac{27}{3} + 27\right) - \left(-\dfrac{-27}{3} - 27\right)$

$= 2[-9 + 27] \quad$ or $\quad = (-9 + 27) - (9 - 27)$

$= 36 \qquad\qquad\qquad\qquad = 18 + 18 = 36$

The area is s 36 sq. units.

Example 2: Find the area bounded by the curve $f(x) = -x^2 + 9$ and the x-axis.

Method 1: Using Horizontal slicing

Step 1: Sketch the graph of the required region

Let $y = -x^2 + 9$

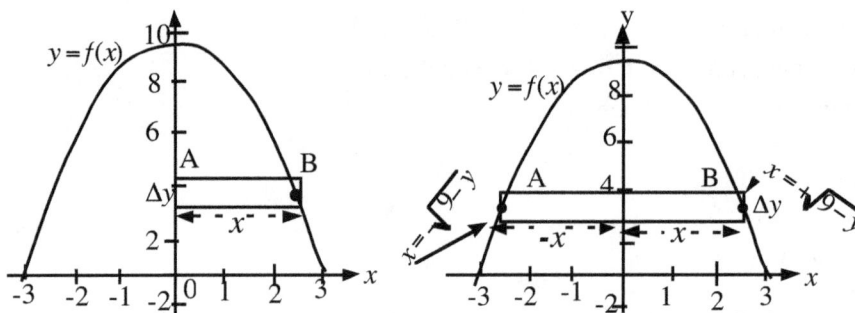

Step 2: Since we are using horizontal slicing the length of the representative rectangle will be horizontal. The length will be in terms of y.

$x^2 = 9 - y$

$x = \pm\sqrt{9 - y}$. We use $x = +\sqrt{9 - y}$ at B.

Step 3: Draw a horizontal rectangle from a point on the y-axis (the line $x = 0$) to a point on $y = -x^2 + 9$. The x-coordinate, say , x_{right} of the right end of the representative rectangle is given by $x_{right} = \sqrt{9 - y}$; and the x-coordinate of the left end of the rectangle, say x_{left} is given by $x_{left} = 0$ ($x = 0$ on the y-axis) The length of the representative rectangle is

$$x_{right} - x_{left} = \sqrt{9 - y} - (0) = \sqrt{9 - y} ; \text{ and the width is } dy.$$

Area of the representative rectangle = length × width

Step 4: We find the limits of integration.

The x-coordinate of the vertex of the parabola is 0, and the corresponding y-coordinate is 9 (from $y = -(0)^2 + 9$)

The limits of the integration are $y = 0$ (on the x-axis, $y = 0$) and $y = 9$.(vertex). To find the area of the required region, we have two approaches:

1. We can find the area using Length = $\sqrt{9-y}$, width = dy, together with the limits $y = 0$ and $y = 9$, followed by doubling (for symmetry) the area found; or

2. We can find the area using Length = $(+\sqrt{9-y}) - (-\sqrt{9-y}) = 2\sqrt{9-y}$, width = dy, and the limits $y = 0$ and $y = 9$.

By either approach, the area A of the required region is

$$A = 2\int_0^9 \sqrt{9-y}\, dy \quad \text{or}$$

Step 5: $\quad A = 2\int_0^9 (9-y)^{\frac{1}{2}}\, dy$

$$= 2\left[-\frac{2}{3}(9-y)^{\frac{3}{2}}\right]_0^9$$

$$= -\frac{4}{3}\left[(9-y)^{\frac{3}{2}}\right]_0^9$$

$$= -\frac{4}{3}\left[(9-9)^{\frac{3}{2}} - (9-0)^{\frac{3}{2}}\right]$$

$$= -\frac{4}{3}\left[0 - (9)^{\frac{3}{2}}\right]$$

$$= -\frac{4}{3}\left[-\left(\sqrt{9}\right)^3\right]$$

$$= -\frac{4}{3}[-27]$$

$$= 36.$$

The area is 36 sq. units.

Scrapwork

Integrate by u-substitution with

$u = 9 - y$

$du = -dy$ and $dy = -du$

$$A = -2\int u^{\frac{1}{2}}\, du$$

$$= \frac{-2u^{\frac{3}{2}}}{\frac{3}{2}}$$

$$= -2 \cdot \frac{2}{3}(9-y)^{\frac{3}{2}}$$

$$= -\frac{4}{3}(9-y)^{\frac{3}{2}}$$

Example 2 Find the area bounded by the curve $f(x) = -x^2 + 6x$ and the   340
x-axis. This graph is the graph in Example 1 with each point shifted three units
to the right.
Solution

Step 1: Sketch the graph of the required region

Let $y = -x^2 + 6x$

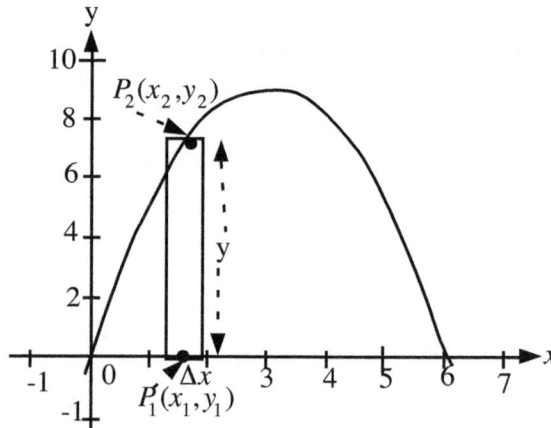

Step 2: Find the x-intercepts

(the points of intersection of $y = -x^2 + 6x$ with the x-axis)

Let y = 0 in $y = -x^2 + 6x$. Then

$x^2 - 6x = 0$
$x(x - 6) = 0$
$x = 0 \quad x - 6 = 0$
$x = 0, \ x = 6$

The x-intercepts are 0 and 6. These numbers will be used as the limits
of the subsequent integration.

Step 3: Draw a vertical representative rectangle from a point on $y = -x^2 + 6x$
to a point on the x-axis (the line $y = 0$)
The y-coordinate, say y_{top} of the top of the representative rectangle is
given by $y_{top} = -x^2 + 6x$; and the y-coordinate, say y_{bot} of the bottom
of representative rectangle is given by $y_{bot} = 0$. The length of the
representative rectangle is

$y_{top} - y_{bot} = (-x^2 + 6x) - (0) = -x^2 + 6x$; and the width is dx

Area of the representative rectangle = length × width
To find the area of the required region, we have two approaches:

1. We can find the area using Length = $-x^2 + 6x$, width = dx, and the limits
$x = 0$ and $x = 6$.

2. We can find the area using Length = $-x^2 + 6x$, width = dx, together with the
limits $x = 0$ and $x = 3$, followed by doubling the area found.

Step 4: Required area A of the region is

$$A = \int_0^6 (-x^2 + 6x)\,dx \qquad \text{or} \qquad A = 2\int_0^3 (-x^2 + 6x)\,dx$$

$$= \left[-\frac{x^3}{3} + 3x^2\right]_0^6 \qquad\qquad = 2\left[-\frac{x^3}{3} + \frac{6x^2}{2}\right]_0^3$$

$$= \left[\left(-\frac{(6)^3}{3} + 3(6)^2\right) - (0)\right] \qquad = 2\left[\left(-\frac{(3)^3}{3} + 3(3)^2\right) - (0)\right]$$

$$= -72 + 108 \qquad\qquad\qquad = 2\left[\left(-\frac{27}{3} + 27\right) - (0)\right]$$

$$= 36 \qquad\qquad\qquad\qquad = 2[-9 + 27]$$

The area is s 36 sq. units. $\qquad\qquad = 2[18]$

$$\qquad\qquad\qquad\qquad\qquad = 36$$

Example 3

Find the area bounded by the curve $f(x) = -x^2 + 6x$ the x-axis and the lines $x = 0$, $x = 5$.

Solution

Method 1: Using Vertical slicing (Method is exactly like that in Example 2, except for the upper limit.

Step 1: Sketch the graph of the required region

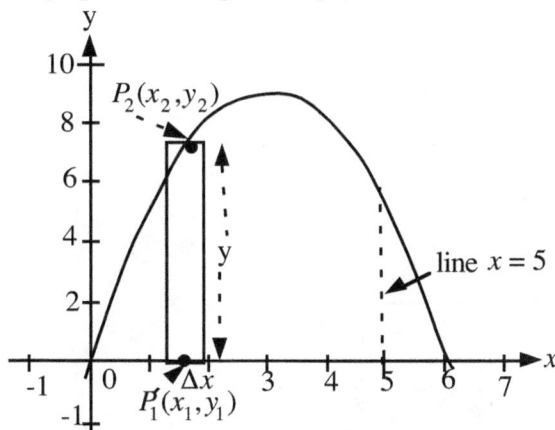

Step 2: Draw a vertical representative rectangle from a point on $y = -x^2 + 6x$ to a point on the x-axis (the line $y = 0$)

The y-coordinate, say y_{top} of the top of the representative rectangle is given by $y_{top} = -x^2 + 6x$; and the y-coordinate, say y_{bot} of the bottom of representative rectangle is given by $y_{bot} = 0$. The length of the representative rectangle is

$y_{top} - y_{bot} = (-x^2 + 6x) - (0) = -x^2 + 6x$; and the width is dx

Area of the representative rectangle = length × width

We find the area using Length = $-x^2 + 6x$, width = dx, together with the limits $x = 0$ and $x = 5$.

Step 3: Required area A of the region is

$$A = \int_0^5 y \, dx \qquad (\sum_{i=1}^{N} y_i \Delta x_1)$$

Step 4: $A = \int_0^5 (-x^2 + 6x) \, dx$

$$= \left[\frac{-x^3}{3} + 3x^2 \right]_0^5$$

Step 5: $= \left[\frac{-(5^3)}{3} + 3(5)^2 \right] - \left[\frac{-(0)^3}{3} + 3(0)^2 \right]$ (substitute the upper limit first)

$$= \left[\frac{-125}{3} + 75 \right] - [0]$$

$$= 33\frac{1}{3}$$

$$= 33\frac{1}{3} \text{ square units}$$

The area is $33\frac{1}{3}$ sq. units.

Method 2: Using Horizontal slicing

Step 1 Here, we use two different representative two rectangles and two corresponding integrals, As in the Figure, we draw a representative rectangle for the region DCBVAD (the inverted-U above the line $y = 5$, and another representative rectangle for the region EFCDE. (region between bounded on the left by the parabola, bounded on the right by the line $x = 5$, and bounded below by the x-axis)

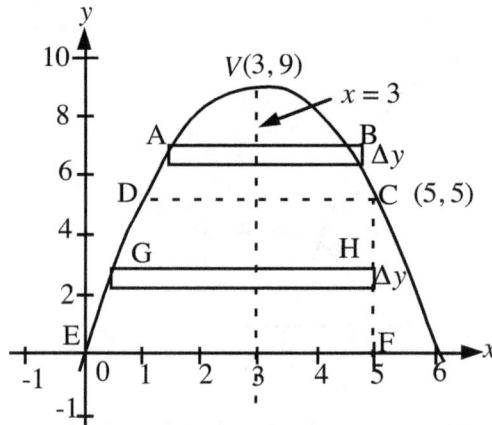

Step 2: **On the top part, from the line $y = 5$ to the line $y = 9$**
Since we are using horizontal slicing the length of the representative rectangle will be horizontal. The length will be in terms of y, and we

solve $y = -x^2 + 6x$ for x.

$$y = -x^2 + 6x$$

$$x^2 - 6x + y = 0$$

$$x = 3 \pm \sqrt{9 - y}$$

$$x = \frac{-(-6) \pm \sqrt{36 - 4(1)y}}{2}$$

We can find the length from A to B by symmetry or by difference.

By symmetry:
The length from A to the axis of symmetry (the line $x = 3$) is
$$x = 3 - (3 - \sqrt{9-y}) = \sqrt{9-y}$$
The length from A to B $= 2\sqrt{9-y}$

By difference: The x-coordinate, say, x_{right} of the right end (B) of the representative rectangle is given by $x_{right} = 3 + \sqrt{9-y}$; and the x-coordinate of the left end (A) of the rectangle, say x_{left} is given by $x_{left} = 3 - \sqrt{9-y}$. The length of the representative rectangle (from A to B) is
$$x_{right} - x_{left} = 3 + \sqrt{9-y} - (3 - \sqrt{9-y}) = 2\sqrt{9-y}; \text{ and the width is } dy.$$
Area of the representative rectangle = length × width

Step 3: We find the limits of integration
The x-coordinate of the vertex of the parabola is 3, and the corresponding y-coordinate is 9 (from $y = -3^2 + 6(3) = 9$)
For the bottom region bounded by the line $y = 5$, the curve, and the x-axis, the length of representative rectangle is $= 5 - (3 - \sqrt{9-y}) = 2 + \sqrt{9-y}$ width $= dy$, and the limits $y = 0$ and $y = 5$ (See the above $x_{right} - x_{left}$ equation.).
Total area of required region is

$$A = 2\int_5^9 \sqrt{9-y}\, dy + \int_0^5 2 + \sqrt{9-y}\, dy$$

$$= 2\int_5^9 (9-y)^{\frac{1}{2}}\, dy + \int_0^5 2 + (9-y)^{\frac{1}{2}}\, dy$$

$$= 2\left[-\frac{2}{3}(9-y)^{\frac{3}{2}}\right]_5^9 + \left[2y - \frac{2}{3}(9-y)^{\frac{3}{2}}\right]_0^5$$

$$= -\frac{4}{3}\left[(9-y)^{\frac{3}{2}}\right]_5^9 + \left[2y - \frac{2}{3}(9-y)^{\frac{3}{2}}\right]_0^5$$

$$= -\frac{4}{3}\left[(9-9)^{\frac{3}{2}} - (9-5)^{\frac{3}{2}}\right] + \left[2(5) - \frac{2}{3}(9-5)^{\frac{3}{2}}\right] - \left[2(0) - \frac{2}{3}(9-0)^{\frac{3}{2}}\right]$$

$$= -\frac{4}{3}\left[0 - (4)^{\frac{3}{2}}\right] + \left[10 - \frac{2}{3}(4)^{\frac{3}{2}}\right] - \left[0 - \frac{2}{3}(9)^{\frac{3}{2}}\right]$$

$$= -\frac{4}{3}\left[\left(-\sqrt{4}\right)^3\right] + \left[10 - \frac{2}{3}\left(\sqrt{4}\right)^3\right] - \left[-\frac{2}{3}\left(\sqrt{9}\right)^3\right]$$

$$= -\frac{4}{3}[-8] + \left[10 - \frac{16}{3}\right] - \left[-\frac{2}{3}(27)\right]$$

$$= \frac{32}{3} + 10 - \frac{16}{3} + 18$$

$$= 10\frac{2}{3} + 10 - 5\frac{1}{3} + 18$$

$$= 33\frac{1}{3}.$$

The area is $= 33\frac{1}{3}$ sq. units (More work in Method 2, **Method 1 is** the better approach)

Example 4

Find the area bounded by the curve $f(x) = \frac{1}{4}x^2$, the x-axis and
the lines $x = 2$, $x = 5$

Method 1: Using vertical slicing

Step 1: Sketch the graph of the required region.

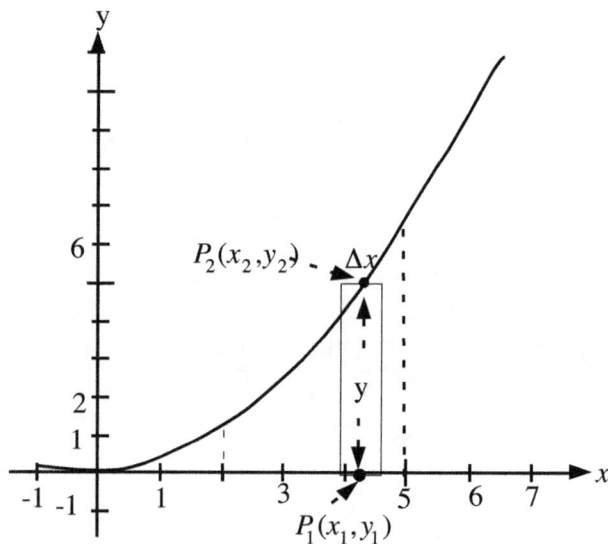

$A = \int_2^5 y\, dx$

$(y = y_2 - y_1 = \frac{1}{4}x^2 - 0; \quad y_2 = \frac{1}{4}x^2, \ y_1 = 0, \text{ the } x\text{-axis})$

Step 2 $A = \int_2^5 \frac{1}{4}x^2\, dx$

$\qquad = \frac{1}{4}\int_2^5 x^2\, dx$

$\qquad = \frac{1}{4}\left[\frac{x^3}{3}\right]_2^5$

Step 3: $= \left[\frac{5^3}{12}\right] - \left[\frac{2^3}{12}\right]$

$\qquad = \frac{125}{12} - \frac{8}{12}$

$\qquad = 9\frac{3}{4} \text{ square units}$

The area $= 9\frac{3}{4}$ square units.

Method 2: Using Horizontal Slicing

Find the area bounded by the curve $f(x) = \frac{1}{4}x^2$, the x-axis and

the lines $x = 2$, $x = 5$

Step 1: Sketch the graph of the required region. We need two representative rectangles and two corresponding integrals

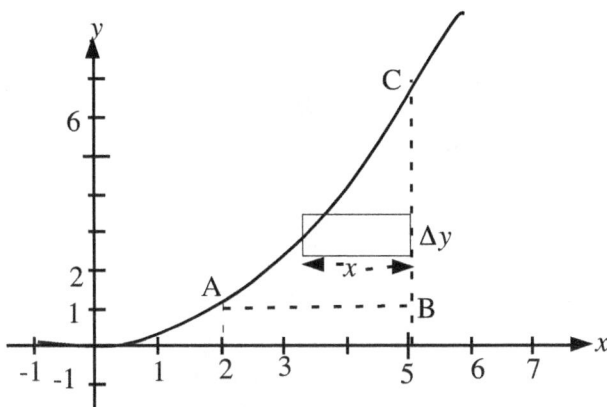

Step 2:   Find coordinates of the points of intersection at A and C

when $x = 2$, $y = \frac{1}{4}x^2 = \frac{1}{4}(2)^2 = 1$: A is at $(2, 1)$

when $x = 5$, $y = \frac{1}{4}x^2 = \frac{1}{4}(5)^2 = \frac{25}{4}$: C is at $(5, \frac{25}{4})$

Step 3: Total area = area enclosed by AB, the lines $x = 2$, $x = 5$ and the x-axis plus area of region ABC. .Since we are using horizontal slicing the length of the representative rectangle will be horizontal. The length will be in terms of y, and we solve $y = \frac{1}{4}x^2$ for x to obtain $x = 2\sqrt{y}$.

Total area = $A = \int_0^1 (5-2)\,dy + \int_1^{\frac{25}{4}} (5 - 2\sqrt{y})\,dy$

$$(\text{Length} = x_{\text{right}} - x_{\text{left}} = 5 - 2y^{\frac{1}{2}})$$

$$= \int_0^1 3\,dy + \int_1^{\frac{25}{4}} \left(5 - 2y^{\frac{1}{2}}\right)dy$$

$$= \left[3y\right]_0^1 + \left[5y - 2 \bullet \frac{2}{3}y^{\frac{3}{2}}\right]_1^{\frac{25}{4}}$$

$$= \left[3y\right]_0^1 + \left[5y - \frac{4}{3}y^{\frac{3}{2}}\right]_1^{\frac{25}{4}}$$

$$= 3 + \left[5(\tfrac{25}{4}) - \frac{4}{3}\left(\frac{25}{4}\right)^{\frac{3}{2}}\right] - \left[5(1) - \frac{4}{3}(1)^{\frac{3}{2}}\right]$$

$$= 3 + 31\tfrac{1}{4} - 20\tfrac{5}{6} - 3\tfrac{2}{3} = 9\tfrac{3}{4} \qquad \left(\frac{4}{3}\left(\frac{25}{4}\right)^{\frac{3}{2}} = \frac{4}{3}\left(\frac{5}{2}\right)^3 = 20\tfrac{5}{6}\right)$$

$$= 9\tfrac{3}{4}$$

The area is $9\frac{3}{4}$ square units.

Example 5

Find the area of the region enclosed by the curve $y = \sqrt{x}$ and the line $y = \frac{1}{3}x$.

Method 1: Using Vertical Slicing

Step 1: Sketch the graphs of $y = \sqrt{x}$ and $y = \frac{1}{3}x$ on the same set of axes. Also draw a vertical representative rectangle.

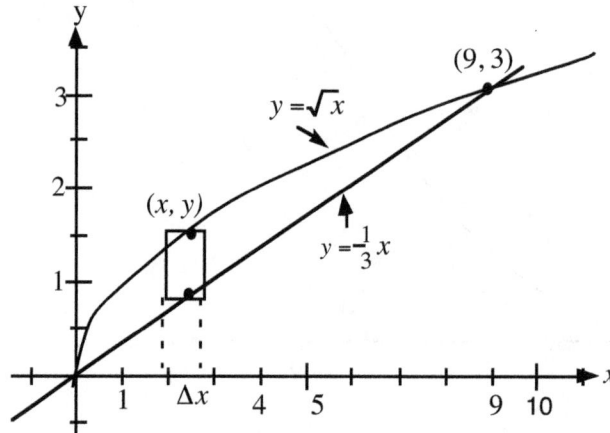

Figure. Graphs of $y = \frac{1}{3}x$, $y = \sqrt{x}$

Step 2: Find the coordinates of the points of intersection of $y = \frac{1}{3}x$ and

$y = \sqrt{x}$ by solving these two equations simultaneously.

$y = \sqrt{x} \qquad (1)$

$y = \frac{x}{3} \qquad (2)$

$\sqrt{x} = \frac{x}{3} \qquad$ (Equating right sides of (1) and (2))

$x = \frac{x^2}{9} \qquad$ (Squaring both sides)

$x^2 - 9x = 0$

$x(x - 9) = 0$

$x = 0, \ x = 9 \quad$ (By checking, 0, and 9 satisfy $\sqrt{x} = \frac{x}{3}$)

When $x = 0$, $y = 0$; when $x = 9$, $y = \sqrt{9} = 3$

The points of intersection are $(0,0)$ and $(9,3)$.

Step 3: Find the length of the representative rectangle.

The length of the representative rectangle is the y-coordinate of the top end of the rectangle minus the y-coordinate of the bottom end of the rectangle. On the top end $\boxed{y = \sqrt{x}}$, and on the bottom end $\boxed{y = \frac{1}{3}x}$.

The length, L, of representative rectangle $= \sqrt{x} - \frac{1}{3}x$.

Step 4: Find the limits of integration. The limits are found by summing up the elements from the left ($x = 0$) to the right ($x = 9$).

The limits of integration are 0 and 9.

Area of the representative rectangle $=$ length $\times$ width.

We find the area using the length, $\sqrt{x} - \frac{1}{3}x$, width $= dx$, together with the limits $x = 0$ and $x = 9$.

Area, A, is given by

$$A = \int_0^9 \left(x^{\frac{1}{2}} - \frac{1}{3}x\right)dx$$

$$= \left[\frac{x^{\frac{1}{2}+1}}{\frac{1}{2}+1} - \frac{1}{3}\cdot\frac{x^2}{2}\right]_0^9$$

$$= \left[\frac{2x^{\frac{3}{2}}}{3} - \frac{x^2}{6}\right]_0^9$$

$$= \left[\frac{2(9)^{\frac{3}{2}}}{3} - \frac{(9)^2}{6}\right] - \left[\frac{2(0)^{\frac{3}{2}}}{3} - \frac{(0)^2}{6}\right]$$

$$= \frac{2(\sqrt{9})^3}{3} - \frac{81}{6} - 0$$

$$= 18 - 13\frac{1}{2}$$

$$= 4\frac{1}{2}.$$

The area is $4\frac{1}{2}$ sq. units.

Method 2 Using Horizontal Slicing

Find the area of the region enclosed by the curve $y = \sqrt{x}$ and the line $y = \frac{1}{3}x$.

Solution

Step 1: Sketch the graph with a horizontal representative rectangle.

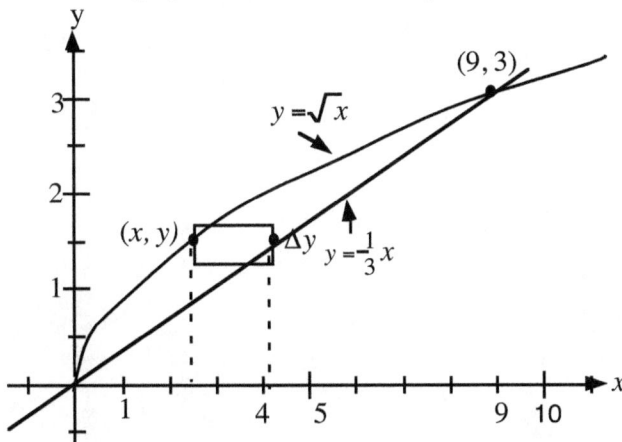

Step 2: Find the length of the representative rectangle.

The length of the representative rectangle is the x-coordinate of the right end of the rectangle minus the x-coordinate of the left end of the rectangle. On the left end $y = \sqrt{x}$ or $\boxed{x = y^2}$, and on the right end $y = \frac{1}{3}x$ or $\boxed{x = 3y}$. Length, L, of representative rectangle $= 3y - y^2$.

Step 3: Find the limits of integration. The limits are found by summing up the elements from the bottom ($y = 0$) to the top ($y = 3$).

The limits of integration are 0 and 3.

Area of the representative rectangle = length × width

We find the area using the length, $3y - y^2$, width = dy, together with the limits $y = 0$ and $y = 3$.

Area, A, is given by

$$A = \int_0^3 (3y - y^2)\,dy$$

$$= \left[\frac{3y^2}{2} - \frac{y^3}{3}\right]_0^3$$

$$= \left[\frac{3(3)^2}{2} - \frac{(3)^3}{3}\right] - [0]$$

$$= \frac{3(3)^2}{2} - \frac{(3)^3}{3}$$

$$= 13\tfrac{1}{2} - 9$$

$$= 4\tfrac{1}{2}$$

Again, the area is $4\tfrac{1}{2}$ sq. units.

Example 6 Find the area bounded by the curve $y^2 = 2x$, the line $y = x - 4$
Solution
Method 1: **Using Horizontal Slicing**
Here, a single horizontal representative rectangle and a corresponding single integral would be sufficient, since the left ends of all possible horizontal rectangles would lie on the same curve $y^2 = 2x$, and all the right ends of possible rectangles would also lie on the same line $y = x - 4$. (see p.334, Case 2)

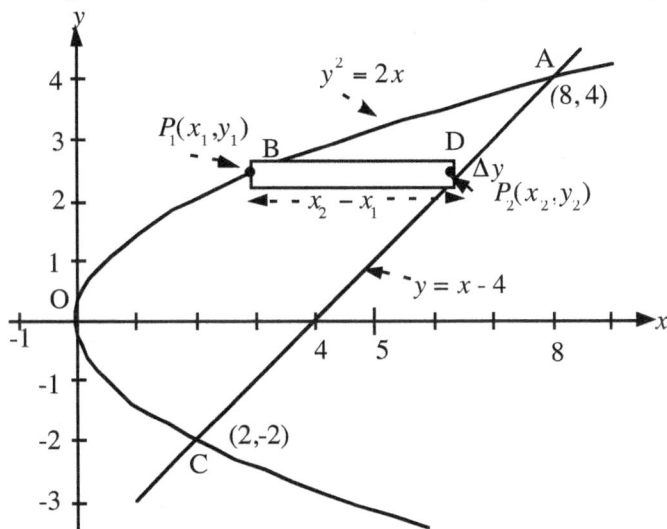

Step 1: Find the points of intersections of $y^2 = 2x$, and $y = x - 4$ at A and C, by solving the two equations simultaneously.

$y^2 = 2x$ (1)

$y = x - 4$ (2)

Substitute right side of (2) for the left side of (1)

Then $(x - 4)^2 = 2x$

$x^2 - 8x + 16 = 2x$

$x^2 - 10x + 16 = 0$

$(x - 2)(x - 8) = 0$

$x = 2, x = 8$

When $x = 2$, $y = 2 - 4 = -2$, The point C is the point $(2, -2)$

When $x = 8$. $y = 8 - 4 = 4$. The point A is the point $(8, 4)$.

The limits of integration are $y = -2$ and $y = 4$ (Sum-up from the bottom to the top)

Step 2: Since we are using horizontal slicing the length of the representative rectangle will be horizontal. The length will be in terms of y, and we solve $y^2 = 2x$ for x; and also solve $y = x - 4$ for x.

From $y^2 = 2x$, $x = \dfrac{y^2}{2}$; and from $y = x - 4$, $x = y + 4$

With reference to Figure, $x_1 = \dfrac{y^2}{2}$, $x_2 = y + 4$.

The length of the representative rectangle $= x_2 - x_1 = (y+4) - \dfrac{y^2}{2}$.

The width of rectangle $= dy$

Area of the representative rectangle $=$ length $\times$ width

We find the area using length, $(y+4) - \dfrac{y^2}{2}$, width $= dy$, together with the limits $y = -2$ and $y = 4$.

Step 3: Required area , A, of the region is given by

$A = \displaystyle\int_{-2}^{4} (y + 4 - \dfrac{y^2}{2})\,dy$

$A = \displaystyle\int_{-2}^{4} (-\dfrac{y^2}{2} + y + 4)\,dy$ (rewriting)

$= \left[-\dfrac{y^3}{2(3)} + \dfrac{y^2}{2} + 4y \right]_{-2}^{4}$

$= \left[-\dfrac{y^3}{6} + \dfrac{y^2}{2} + 4y \right]_{-2}^{4}$

$= \left[-\dfrac{4^3}{6} + \dfrac{4^2}{2} + 4(4) \right] - \left[-\dfrac{(-2)^3}{6} + \dfrac{(-2)^2}{2} + 4(-2) \right]$

$= \left[-10\tfrac{2}{3} + 24 \right] - \left[1\tfrac{1}{3} - 6 \right]$

$= \left[13\tfrac{1}{3} \right] - \left[-4\tfrac{2}{3} \right]$

$= \left[13\tfrac{1}{3} \right] - \left[-4\tfrac{2}{3} \right]$

$= 13\tfrac{1}{3} + 4\tfrac{2}{3}$.

$= 18$

The area is 18 square units.

Method 2: Using Vertical Slicing

Here, a single vertical representative rectangle would **not** be sufficient, since even though the top ends of all possible vertical rectangles would lie on the same curve $y^2 = 2x$, some of the bottom ends would **not** lie on the same curve or line; some on $y = x - 4$, and some on $y^2 = 2x$. Therefore, we need **two** vertical representative rectangles and two integrals. One rectangle is for the region AEBCA, and another rectangle for the region BOCB. See guidelines on p.333.

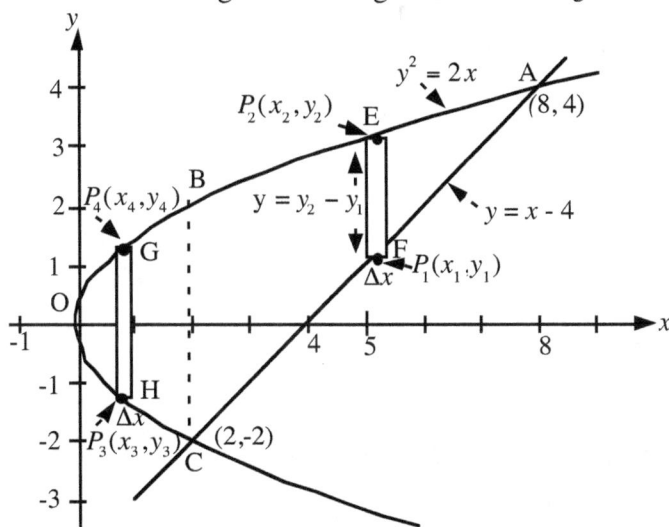

Step 2: Since we are using vertical slicing, the length of the representative rectangle will be vertical. The length will be in terms of x, and we solve $y^2 = 2x$ for y.

From $y^2 = 2x$, $y = \pm\sqrt{2}x^{\frac{1}{2}}$. (Equation of the upper parabola is $\sqrt{2}x^{\frac{1}{2}}$)

With reference to Figure, $y_1 = x - 4$, $y_2 = +\sqrt{2}x^{\frac{1}{2}}$.
For the region AEBCA:

Length of representative rectangle $= y_2 - y_1 = \sqrt{2}x^{\frac{1}{2}} - (x - 4)$.

Width of rectangle $= dx$

Area of the representative rectangle $=$ length $\times$ width

We find the area using length $= \sqrt{2}x^{\frac{1}{2}} - (x - 4)$, width $= dx$, together with the limits $x = 2$ and $x = 8$. Area A_1 of the region AEBCA is given by

$$A_1 = \int_{2}^{8} (\sqrt{2}x^{\frac{1}{2}} - (x - 4))dx$$

Step 3: For the region BOCB.

We can find the length of the representative rectangle by either doubling, $y_4 = +\sqrt{2}x^{\frac{1}{2}}$ to obtain $2\sqrt{2}x^{\frac{1}{2}}$ or by

$y_4 - y_3 = +\sqrt{2}x^{\frac{1}{2}} - (-\sqrt{2}x^{\frac{1}{2}})$ to obtain $2\sqrt{2}x^{\frac{1}{2}}$, where $y_3 = -\sqrt{2}x^{\frac{1}{2}}$.

(Equation of the lower parabola is $-\sqrt{2}x^{\frac{1}{2}}$)

We can find the area using $\text{Length} = 2\sqrt{2}x^{\frac{1}{2}}$, width $= dx$, and the limits $x = 0$ and $x = 2$, Area, A_2, of the region BOCB is given by

$$A_2 = \int_0^2 2\sqrt{2}x^{\frac{1}{2}}dx$$

Total area, $A_1 + A_2 = \int_2^8 \sqrt{2}x^{\frac{1}{2}} - (x - 4)dx + \int_0^2 2\sqrt{2}x^{\frac{1}{2}}dx$.

$$= \sqrt{2}\int_2^8 x^{\frac{1}{2}} - \int_2^8 x\,dx + \int_2^8 4\,dx + 2\sqrt{2}\int_0^2 x^{\frac{1}{2}}dx$$

$$= \left[\sqrt{2} \cdot \frac{2}{3}x^{\frac{3}{2}} - \frac{x^2}{2} + 4x\right]_2^8 + 2\sqrt{2} \cdot \left[\frac{2}{3}x^{\frac{3}{2}}\right]_0^2$$

$$= \left[\frac{2\sqrt{2}}{3}(8)^{\frac{3}{2}} - \frac{8^2}{2} + 4(8)\right] - \left[\frac{2\sqrt{2}}{3}(2)^{\frac{3}{2}} - \frac{2^2}{2} + 4(2)\right] + 2\sqrt{2}\left[\frac{2}{3}(2)^{\frac{3}{2}}\right] - [0]$$

$$= \left[\frac{2\sqrt{2}}{3}(16\sqrt{2}) - \frac{64}{2} + 32\right] - \left[\frac{2\sqrt{2}}{3}(2\sqrt{2}) - \frac{4}{2} + 8\right] + 2\sqrt{2}\left[\frac{2}{3}(2\sqrt{2})\right]$$

$$= \left[\frac{64}{3} - 32 + 32\right] - \left[\frac{8}{3} - 2 + 8\right] + \left[\frac{16}{3}\right]$$

$$= 21\frac{1}{3} - 8\frac{2}{3} + 5\frac{1}{3} = 18. \text{ The area is 18 square units.}$$

Lesson 47 Exercises A

1. Find the area bounded by the curve $f(x) = -x^2 + 9$ and the x-axis..
2. Find the area bounded by the curve $f(x) = -x^2 + 6x$ and the x-axis.
3. Find the area bounded by the curve $f(x) = -x^2 + 6x$ the x-axis and ,the lines $x = 0$, $x = 5$..
4. Find the area bounded by the curve $f(x) = \frac{1}{4}x^2$, the x-axis and the lines $x = 2$, $x = 5$
5. Find the area bounded by the curve $y^2 = 2x$, the line $y = x - 4$

Answers: 1. 36 sq. units; **2.** 36 sq. units; **3.** $33\frac{1}{3}$ sq. units; **4.** $9\frac{3}{4}$ sq. units
5. 18 square units; .

Lesson 47 Exercises B

1. Find the area bounded by the curve $f(x) = -x^2 + 4$ and the x-axis..
2. Find the area bounded by the curve $f(x) = -x^2 + 4x$ and the x-axis.
3. Find the area bounded by the curve $f(x) = -x^2 + 4x$ the x-axis and ,the lines $x = 0$, $x = 4$..
4. Find the area bounded by the curve $f(x) = \frac{1}{4}x^2$, the x-axis and the lines $x = 1$, $x = 5$
5. Find the area bounded by the curve $y^2 = 2x - 2$, and the line $y = x - 5$

Answers: **1.** $10\frac{2}{3}$ sq. units; **2.** $10\frac{2}{3}$ sq. units; **3.** $10\frac{2}{3}$ sq. units; **4.** $10\frac{1}{3}$ sq. units. **5.** 18. square units; .

1 Find the area bounded by the curve $f(x) = x^2$, the x-axis and ,the lines $x = 3$, $x = 5$.

2. Find the area bounded by the curve $f(x) = \dfrac{x-2}{\sqrt{x}}$ the x-axis and ,the lines $x = 1$, $x = 4$.

3 Find the area bounded by the curve $f(x) = \cos x$ the x-axis and ,the lines $-\dfrac{\pi}{4}$, $\dfrac{\pi}{4}$.

4. Find the area bounded by the graph of $f(x) = |x - 3|$ the x-axis and ,the lines $x = -2$, $x = 5$..

5. Find the area enclosed by the curve $f(x) = \sqrt{x}$ the x-axis and ,the lines $x = 1$, $x = 9$.

6. Find the area of the region enclosed by the curves $y = \sqrt{x}$
$y = \dfrac{1}{10}x^2$, over the interval $[0, 4]$.

7. Find the area of the region enclosed by the curve $y = \sqrt{x}$ and the line $y = \tfrac{1}{3}x$ over the interval $[4, 9]$

Answers in sq. units

: **1.** $\dfrac{98}{3}$; **2.** $\dfrac{2}{3}$ **3.** $\sqrt{2}$; **4.** $14\tfrac{1}{2}$; **5.** $\dfrac{52}{3}$; **6.** $\dfrac{16}{5}$. **7.** $\dfrac{11}{6}$

Lesson 48
Applications of Integration to Volumes of Revolution

Preliminaries:

If an area is moved continuously in a certain direction, a **volume** is generated, in the same way that if a **point** is moved continuously in a certain direction, a **line** is produced. Producing a volume this way may be viewed as piling up very thin sheets of millions of papers on one another.

We generate a **volume of revolution** by revolving a plane area about a line in the plane. The line about which we revolve the area is called the axis of revolution or rotation. We note that in producing a volume, an area is moved continuously in a certain direction. Volume has **three** dimensions.

The rectangular volume, V is given by V = length × width × height

There are two main methods for **calculating** the volume of a solid of revolution, (with each method having its relative merits) namely,

1. The **disk method** in which the representative rectangle is **perpendicular** (lengthwise) to the axis of rotation, and there are two cases:

    Case 1 The axis of revolution is part of the boundary of the plane area.
    Case 2: The axis of revolution is **not** part of the boundary of the plane area.
            and a washer is formed; and

2. The **cylindrical shell method** in which the representative rectangle is **parallel** (lengthwise) to the axis of revolution. When two concentric right-circular cylinders enclose a solid, a cylindrical shell is formed.

Basic Formulas

Even though there are a number of formulas for various cases of volumes of solids of revolution (in most textbooks), two basic formulas need to be memorized, and the other formulas could be derived once a student has mastered the principles involved in these problems. By viewing each problem as a physics problem, the two basic formulas presented would be sufficient.

Disk Method: (Cross sections of Solids obtained are circular in shape)
For a right circular cylinder generated by moving a plane region of area A through a distance h, the volume V of the cylinder is given by $V = A \times h$.

General disk formula to remember

If r is the radius of revolution, dx = width of vertical representative rectangle, dy = width of horizontal representative rectangle, then the volume V generated is

$$V = \int_m^n \pi r^2 \bullet dx \text{ or } dy. \qquad (\text{ Area of a circle} \times dx \text{ or } dy)$$

(where m and n are respectively the lower and upper limits to be determined)

General cylindrical shell formula:

If r is the radius of revolution, dx = width of vertical representative rectangle, dy = width of horizontal representative rectangle, then the volume V generated

is $V = \int_m^n 2\pi rL \bullet dx \text{ or } dy.$ (Perimeter × L × dx or dy)

(In most textbooks, $L = h$. Using L for length (instead of h for height) makes for easy recall when the representative rectangle is horizontal).

Determining the radius of revolution without having to memorize formulas for the different cases (Assuming a physics problem)

For a vertical representative rectangle
Imagine the x-axis is a horizontal pin passing through a horizontal hole at the base of the rectangle, and imagine puling the top of the rectangle towards you and visualize anticlockwise circular motion of the rectangle about the x-axis (revolution about the x-axis). You would observe that the distance from the top (also a point on the curve) to the axis is the radius of revolution or rotation.

For a horizontal representative rectangle
Imagine the y-axis is a vertical pin passing through a vertical hole at the base of the rectangle, and imagine pushing the right end of the rectangle forwards into the page and visualize the anticlockwise circular motion of the rectangle about the y-axis (revolution about the y-axis). You would observe that the distance from the right end of this rectangle to the y-axis is the radius of revolution.

Also
If the representative rectangle is vertical, the distance (y-coordinate) from the top end (also a point on the curve) of this rectangle to the axis of revolution is the radius of revolution; but if the representative rectangle is horizontal, the distance (x-coordinate) from right end of this rectangle to the axis of revolution is the radius of revolution. (See p.359-368 for examples)

Disk Method

1. Revolution about the *x-axis*

(a) Axis of revolution is part of the boundary of the plane area
If the region R between a curve, $y = f(x)$, the x-axis and the lines $x = a$ and $x = b$, is revolved about the x-axis, then the volume V of the solid generated is

given by $V = \int_a^b \pi r^2 dx$ (area of a circle $= \pi r^2$)

$$V = \pi \int_a^b y^2 dx$$

$$V = \pi \int_a^b [f(x)]^2 dx \quad \text{(radius of cross-section } r = y = f(x))$$

(b)
Axis of revolution is not part of the boundary of the plane area (Washer)
The shape formed has an inner radius and outer radius.
The volume V of the solid generated when the region between the graphs of $f(x)$ and $g(x)$ and the lines $x = a$ and $x = b$, is revolved about the x-axis is

$$V = \pi \int_a^b [g(x)]^2 - [f(x)]^2) dx \quad \text{(outer radius } = g(x), \text{ inner radius } f(x))$$

(The element of volume is the difference between the volume generated by the graphs of $g(x)$ and $f(x)$. We may recall that $f(x)$ and $g(x)$ are the radii of revolution. Note also that in the formula for V, $f(x)$ and $g(x)$ are **squared before subtracting**.)

2. Revolution about the *y-axis* (Length of representative rectangle ⊥ the y-axis)
(a) If the cross sections are perpendicular to the y-axis:

$$V = \pi \int_c^d x^2 \, dy$$

$$V = \pi \int_c^d \left[f(y) \right]^2 dy \quad \text{(radius of cross-section } r = x = f(y))$$

(b) If the axis of revolution is **not** part of the boundary of the plane area and the cross-sections are perpendicular to the y-axis, the volume V is given by

$$V = \pi \int_c^d \left[g(y) \right]^2 - \left[f(y) \right]^2 dy$$

where the region between the graphs of $g(y)$ and $f(y)$ and the lines $y = c$ and $y = d$, is revolved about the y-axis or a line parallel to the y-axis.

The following diagrams are to help visualize both the disk and shell methods.
Representative rectangle perpendicular or parallel to an axis or a line

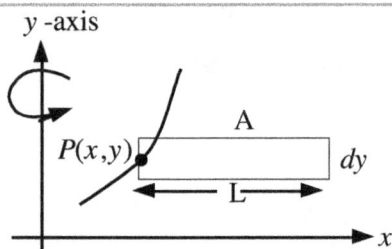

Fig, 1
For **Disk Method.**
Rectangle **perpendicular** (lengthwise) to the y-axis, the axis of revolution, but width-wise **parallel** to the y-axis.
(As is, you cannot use the shell method for revolution about the y-axis but can use the **shell method** for revolution about the x-axis. Why?)

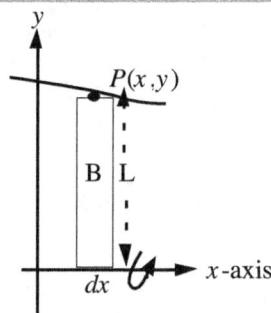

Fig. 2
For **Disk Method.**
Rectangle **perpendicular** (lengthwise) to the x-axis, the axis of revolution
(As is, you cannot use the shell method for revolution about the x-axis but can use the **shell method** for revolution about the y-axis. Why?)

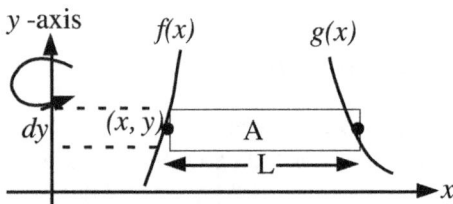

Fig, 3
For **Disk Method** (Washer)
Rectangle is perpendicular (lengthwise) to the y-axis, the axis of revolution.
As is, you cannot use the shell method for revolution about the y-axis but can use the **shell method** for revolution about the x-axis. Why?

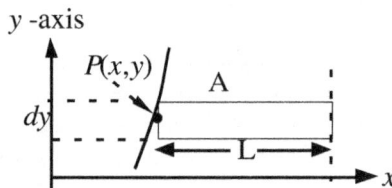

Fig, 4
For **Disk Method** (Washer)
Rectangle is **perpendicular** (lengthwise) to the y-axis, the axis of revolution
As is, you cannot use the shell method for revolution about the y-axis but can use the shell method for revolution about the x-axis. Why?

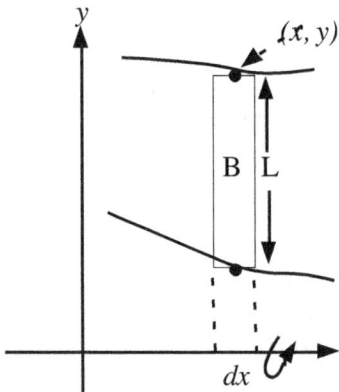

Fig, 5
For **Disk Method** (Washer)
Rectangle perpendicular (lengthwise) to the x-axis, the axis of revolution. As is, you cannot use the shell method for revolution about the x-axis but can use the **shell method** for revolution about the y-axis. Why?

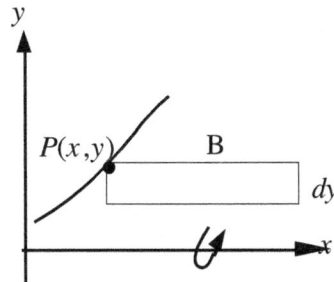

Fig, 6
Cylindrical shell method
Rectangle parallel (lengthwise) to the x-axis, the axis of revolution. As is, you cannot use the Disk method for revolution about the x-axis but can use the Disk method for revolution about the y-axis. Why?

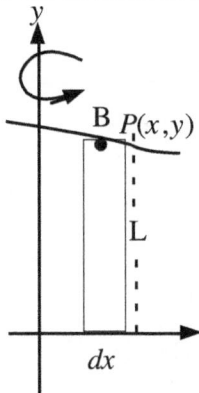

Fig, 7
Cylindrical shell method
Rectangle parallel (lengthwise) to the y-axis, the axis of revolution
As is, you cannot use the Disk method for revolution about the y-axis but can use the Disk method for revolution about the x-axis. Why?

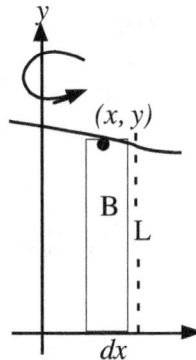

Fig, 8
Cylindrical shell method
Rectangle parallel (lengthwise) to the y-axis, the axis of revolution. As is, you cannot use the Disk method for revolution about the y-axis but can use the Disk method for revolution about the x-axis. Why?

Cylindrical Shell Method

In the **cylindrical shell** method, the representative rectangle is parallel (lengthwise) to the axis of revolution.

If the region R between a curve, $y = f(x)$, the x-axis and the lines $x = a$ and $x = b$, is revolved about the y-axis, then the volume V of the solid generated is

given by $V = \int_a^b 2\pi xy\,dx$

$$V = \int_a^b 2\pi x f(x)\,dx$$

General form to remember:

If r is the radius of revolution and dx is width for vertical representative rectangle and dy for horizontal representative rectangle, or L = length of representative rectangle, then the volume, V, generated is

$$\boxed{V = \int_m^n 2\pi rL \bullet dx \text{ or } dy}.(m \text{ and } n \text{ are, respectively, the lower and upper limits})$$

In most textbooks $L = h$, where h is the height. Using L makes for easy recall when the representative rectangle is horizontal.

Another view:

$2\pi r$ = the circumference or perimeter of the circular part

L = the length or height of the cylinder

$2\pi rL$ = surface area of the lateral side of the cylinder.

We convert whatever volume generated into cylindrical form and then determine the volume.

Memory Device
Perpendicular or parallel orientation to the axis of revolution

To remember the difference between the **disk method** and the **shell method**, associate "shell" with "parallel" since both words have double " ll " (double el) in their spellings. Therefore, for the **shell method,** the representative rectangle is parallel (lengthwise) to the axis of revolution. Then of course, the perpendicular orientation is for the disk method. Therefore, for the **disk method**, the representative rectangle is perpendicular (lengthwise) to the axis of revolution.

For the basic formulas:

Disk Method: $V = \int_m^n \pi r^2 \bullet dx \text{ or } dy$ (Area of a circle $\times dx$ or dy)

Shell Method: $V = \int_m^n 2\pi rL \bullet dx \text{ or } dy$ (Perimeter $\times L \times dx$ or dy)

(Note the "L" in the cylindrical shell method)   (Lateral surface area $\times dx$ or dy)

Example 1 (a) Use the **disk method**; (b) Use the **cylindrical shell method**
Find the volume of the solid obtained by revolving the region enclosed
by the curves $y = x^2$, the x-axis, the line $x = 2$ about the x-axis.

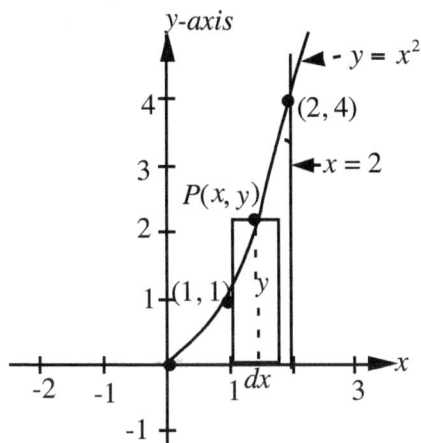

Solution (a) **Using disk method**

Step 1: Sketch the graph of the curve of $y = x^2$, and draw a vertical
representative rectangle, since for the disk method, the representative
rectangle must be perpendicular (lengthwise) to the x-axis (the axis of
revolution, in this case). The length of the rectangle is y (It happens that
in this example, y is also the **radius** of revolution)

Step 2: Find the radius r of revolution .

$r = y$ Imagine the x-axis is a horizontal pin passing through a horizontal
hole at the base of the rectangle, and imagine pulling the top of the
rectangle towards you and visualize anticlockwise circular motion of
the rectangle about the x-axis (revolution about the x-axis). You
would observe that the distance from the top (also a point on the
curve) to the axis is the radius of revolution or rotation.

Step 3: Find the limits of integration

First, find the point of intersection of $y = x^2$ and the line $x = 2$ by
substituting for $x = 2$ in $y = x^2$ to obtain $y = 4$. For the limits of
integration, the lower limit is $x = 0$, and the upper limit is $x = 2$ The
limits of integration are 0 and 2.

Step 4: Write an equation for the volume V of the solid obtained by revolving
the region enclosed by the curves $y = x^2$, the x-axis, and the line $x = 2$,
about the x-axis.

$V = \int_a^b \pi r^2 \bullet dx$ (Area of a circle is πr^2). "dx" means with respect to x.

$V = \int_0^2 \pi y^2 dx$ ($r = y$).

$V = \int_0^2 \pi \left(x^2\right)^2 dx$ ($y = x^2$. All variables must be in terms of x. because of dx)

$= \pi \int_0^2 x^4 dx$

$= \pi \left[\dfrac{x^5}{5}\right]_0^2$

$$= \pi\left[\left(\frac{2^5}{5}\right) - \left(\frac{0^5}{5}\right)\right]$$

$$= \pi\left[\frac{32}{5} - 0\right]$$

$$= \frac{32\pi}{5} \quad \text{cubic units.}$$

(b) Using the cylindrical shell method

Step 1: In using the cylindrical shell method, the representative rectangle must be parallel (lengthwise) to the x-axis (the axis of revolution in this case)   Sketch the graph with a horizontal representative rectangle.

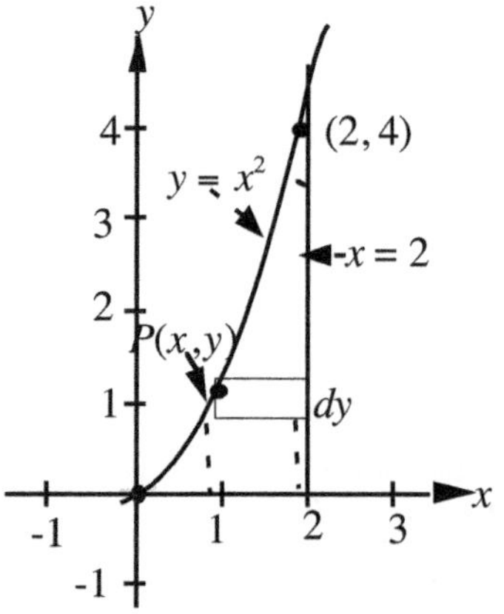

Step 2: Find the radius r of revolution. The radius r of revolution is $r = y$

$r = y$

Imagine the x-axis is a horizontal pin passing through two horizontal holes at the ends of two light rods attached to the ends of the representative rectangle, and imagine pulling the rectangle towards you and visualize anticlockwise circular motion of the rectangle about the x-axis (revolution about the x-axis). You would observe that the distance from the left side (also a point on the curve) to the axis is the radius of revolution or rotation.

Step 3: Find the limits of integration The point of intersection of $y = x^2$ and the line $x = 2$ is found by substituting for $x = 2$ in $y = x^2$ to obtain $y = 4$. The limits are found by summing up the elements from the bottom ($y = 0$) to the top ($y = 4$). The limits of integration are 0 and 4. (Figure below)

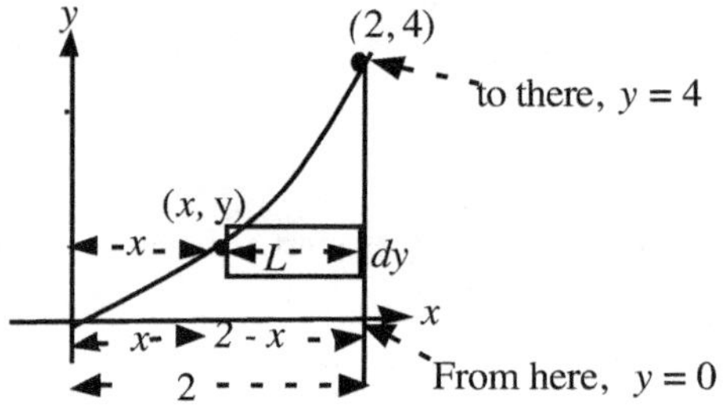

Step 4: Find the length of the representative rectangle. The length is $2 - x$

$L = 2 - x$
or
$h = 2 - x$

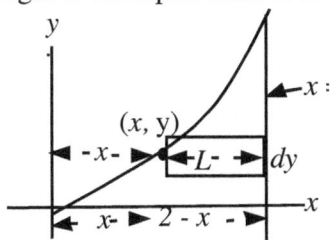

Step 5: Write an equation for the volume V of the solid obtained by revolving the region enclosed by the curves $y = x^2$, the x-axis, and the line $x = 2$, about the x-axis. The total volume V is given by

$$V = \int_a^b 2\pi r L\, dy \qquad (r = \text{radius}, \ L = \text{length of rectangle})$$

$$V = \int_0^4 2\pi y(2 - x)dy \quad (\text{Note: } r = y = f(x); \ L = 2 - x)$$

$$V = 2\pi \int_0^4 y(2 - x)dy$$

$$= 2\pi \int_0^4 y(2 - \sqrt{y})dy \quad (\text{Express } x \text{ in terms of } y \text{ and integrate with respect to y.})$$

$$= 2\pi \int_0^4 (2y - y^{\frac{3}{2}})dy$$

$$= 2\pi \left[\frac{2y^2}{2} - \frac{2}{5} y^{\frac{5}{2}} \right]_0^4$$

$$= 2\pi \left[y^2 - \frac{2}{5} y^{\frac{5}{2}} \right]_0^4$$

$$= 2\pi[(4^2 - \frac{2}{5} 4^{\frac{5}{2}}) - (0)]$$

$$= 2\pi \left[16 - \frac{64}{5} \right]$$

$$= 2\pi \left[\frac{16}{5} \right]$$

$$= \frac{32\pi}{5} \text{ cubic units}$$

The volume is $\frac{32\pi}{5}$ cubic units.

Example 2

Find the volume of the solid obtained by revolving the region enclosed by the curves $y = x^2$, the x-axis, the lines $x = 2$ about the line $x = 2$.

Method 1 Using the disk method

Step 1: Sketch the graph of the curve of $y = x^2$, and the line $x = 2$.

Draw a horizontal representative rectangle, since for the disk method, the representative rectangle must be perpendicular (lengthwise) to the line $x = 2$ (the axis of revolution in this case).

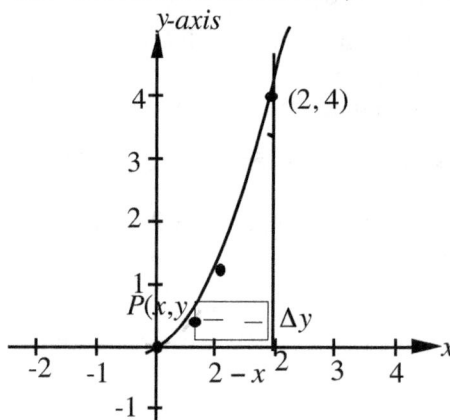

Step 2: Find the radius r of revolution. The radius of revolution is $r = 2 - x$

$r = 2 - x$ Imagine the line $x = 2$ is a vertical pin passing through a vertical hole at the right end of the rectangle, and imagine pulling the left end of the rectangle towards you and visualize anticlockwise circular motion of the rectangle about the line $x = 2$. You would observe that the distance from the left end (also a point on the curve) to the line $x = 2$ is the radius of revolution or rotation.

Step 3: Find the limits of integration

First, find the points of intersection of $y = x^2$ and the line $x = 2$.

For the limits, when $x = 0$, $y = 0$; and when $x = 2$, $y = 2^2 = 4$

We will integrate with respect to y (because of Δy or dy) from the bottom ($y = 0$) to the top ($y = 4$). The limits of integration are 0 and 4.

Step 4: Write the equation for the volume, V, of the solid obtained by revolving the region of radius $2 - x$, about the line $x = 2$, from $y = 0$ to $y = 4$.

(Basic formula: $V = \int_c^d \pi r^2 dy$)

$$V = \int_0^4 \pi (2 - x)^2 dy \qquad\qquad (V = \int_0^4 \pi [f(y)]^2 dy) \qquad\qquad (1)$$

$$V = \pi \int_0^4 (x^2 - 4x + 4) dy \qquad\qquad\qquad\qquad (2)$$

We express x in terms of y (i.e., find $f(y)$) by solving $y = x^2$ for x.

From $y = x^2$, $x = \sqrt{y}$.   (We are in the first quadrant)

In (2) replace x^2 by y and x by $\sqrt{y}$ to obtain

$$V = \pi \int_0^4 (y - 4\sqrt{y} + 4) dy.$$

$$V = \pi \int_0^4 (y - 4y^{\frac{1}{2}} + 4)\,dy$$

$$= \pi \left[\frac{y^2}{2} - 4 \cdot \frac{2}{3} y^{\frac{3}{2}} + 4y \right]_0^4$$

$$= \pi \left[\frac{y^2}{2} - \frac{8}{3} y^{\frac{3}{2}} + 4y \right]_0^4$$

$$= \pi \left[\left(\frac{16}{2} - \frac{8}{3} (\sqrt{4})^3 + 4(4) \right) - (0) \right]$$

$$= \pi \left[8 - \frac{64}{3} + 16 \right]$$

$$= \pi \left[\frac{8}{3} \right]$$

$$= \frac{8\pi}{3} \text{ cubic units.}$$

Method 2 Using the cylindrical shell method

Step 1: In using the cylindrical shell method, the representative rectangle must be parallel (lengthwise) to the *line* $x = 2$ (the axis of revolution in this case) Sketch the graph with a vertical representative rectangle.

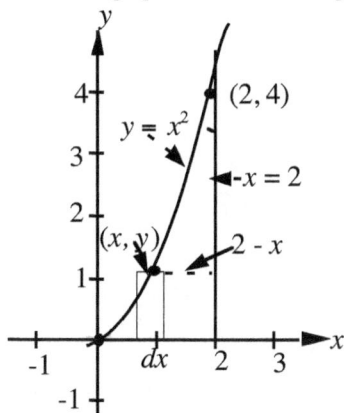

Step 2: Find the radius r of revolution. The radius of revolution is $r = 2 - x$

Imagine the line $x = 2$ is a vertical pin passing through two vertical holes at the ends of two light rods attached to the ends of the representative rectangle, and imagine pulling the rectangle (rotating the rectangle) about the line $x = 2$. You would observe that the distance from the left side (also a point on the curve) to the line $x = 2$, $2 - x$ is the radius of revolution or rotation.

Step 3: Find the length of the representative rectangle. The length is y.

Step 4: Find the limits of integration The limits are found by summing up the elements from the left ($x = 0$) to the right ($x = 2$).

The limits of integration are 0 and 2.

Step 5: Write an equation for the volume V of the solid obtained by revolving the region with representative rectangle of radius $2 - x$ and length y about the line $x = 2$.

The total volume, V, is given by

$$V = \int_a^b 2\pi r L\, dx \quad (r = \text{radius}, \ L = \text{length of rectangle})$$

$$V = 2\pi \int_0^2 (2 - x) y\, dx \quad (\text{Note: } r = 2 - x; \ L = y \ )$$

$$V = 2\pi \int_0^2 (2 - x) x^2\, dx \quad (y = x^2. \text{All variables in terms of } x, \text{ because of } dx)$$

$$= 2\pi \int_0^2 (2x^2 - x^3)\, dx$$

$$= 2\pi \left[2\frac{x^3}{3} - \frac{x^4}{4} \right]_0^2$$

$$= 2\pi \left[\left(2\frac{(2)^3}{3} - \frac{(2)^4}{4} \right) - (0) \right]$$

$$= 2\pi \left[2 \cdot \frac{8}{3} - \frac{16}{4} \right]$$

$$= 2\pi \left[\frac{16}{3} - 4 \right]$$

$$= 2\pi \left[\frac{4}{3} \right]$$

$$= \frac{8\pi}{3} \text{ cubic units.}$$

Example 3
Find the volume of the solid obtained by revolving the region enclosed by the curves $y = \sqrt{x}$ and $y = \frac{1}{3}x$, about the x-axis.

Solution

Method 1: Using the disk Method

Step 1: Sketch the graphs of $y = \sqrt{x}$ and $y = \frac{1}{3}x$ on the same set of axes, Also draw a vertical representative rectangle since for the disk method, this rectangle must be perpendicular (lengthwise) to the axis of revolution, the x-axis.

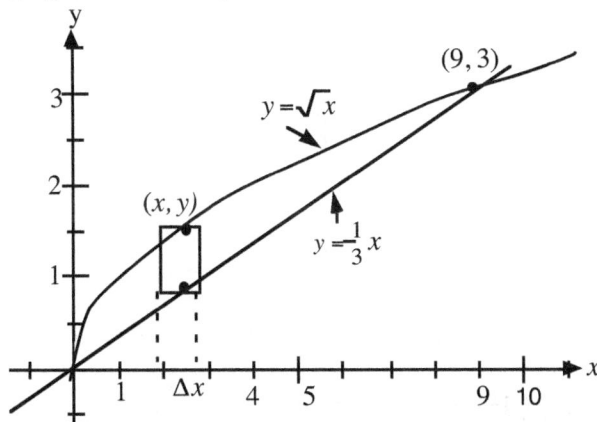

Figure. Graphs of $y = \frac{1}{3}x$, $y = \sqrt{x}$

Step 2: Find the coordinates of the points of intersection of $y = \frac{1}{3}x$ and $y = \sqrt{x}$ by solving these two equations simultaneously.

$y = \sqrt{x}$ (1)

$y = \frac{x}{3}$ (2)

$\sqrt{x} = \frac{x}{3}$ (Equating right sides of (1) and (2))

$x = \frac{x^2}{9}$ (Squaring both sides)

$x^2 - 9x = 0$

$x(x - 9) = 0$

$x = 0, \ x = 9$

When $x = 0$, $y = 0$; when $x = 9$, $y = \sqrt{9} = 3$

Step 3: Find the radii of revolution (for the washer)

For the curve $y = \sqrt{x}$: $r_1 = \sqrt{x}$; For the line $y = \frac{x}{3}$: $r_2 = \frac{x}{3}$

Step 4: Find the limits of integration.
The limits are 0 and 9 (From Step 2)

Step 5: Write an equation for the volume V of the solid obtained by revolving the region enclosed by the curves $y = \sqrt{x}$ ($r_1 = \sqrt{x}$); $y = \frac{1}{3}x$ ($r_2 = \frac{x}{3}$)

about the x-axis   Basic formula $V = \int_a^b \pi r^2 \cdot dx$

For two different radii (**washer**),

$$V = \int_a^b (\pi r_1^2 - \pi r_2^2)dx \quad \text{(square before subtracting)}$$

$$= \pi \int_a^b (r_1^2 - r_2^2)dx$$

$$= \pi \int_0^9 \left[(\sqrt{x})^2 - (\tfrac{x}{3})^2 \right]dx$$

$$= \pi \int_0^9 \left[x - \tfrac{x^2}{9} \right]dx$$

$$= \pi \left[\frac{x^2}{2} - \frac{x^3}{(3)(9)} \right]_0^9$$

$$= \pi \left[\left(\frac{9^2}{2} - \frac{(9)(9)(9)}{(3)(9)} \right) - \left(\frac{0^2}{2} - \frac{(0)^3}{(3)(9)} \right) \right]$$

$$= \pi \left[\left(\frac{81}{2} - 27 \right) - (0) \right]$$

$$= \pi \left[\frac{81}{2} - \frac{54}{2} \right] \qquad \left(\pi \left[\frac{81}{2} - \frac{27(2)}{2} \right] \right)$$

$$= \frac{27\pi}{2} \text{ cubic units.}$$

Method 2 Using the cylindrical shell method

Find the volume of the solid obtained by revolving the region enclosed by the curves $y = \sqrt{x}$ $y = \tfrac{1}{3}x$, about the x-axis

Step 1: In using the cylindrical shell method, the representative rectangle must be parallel (lengthwise) to the *x-axis* (the axis of revolution in this case) Sketch the graph with a horizontal representative rectangle.

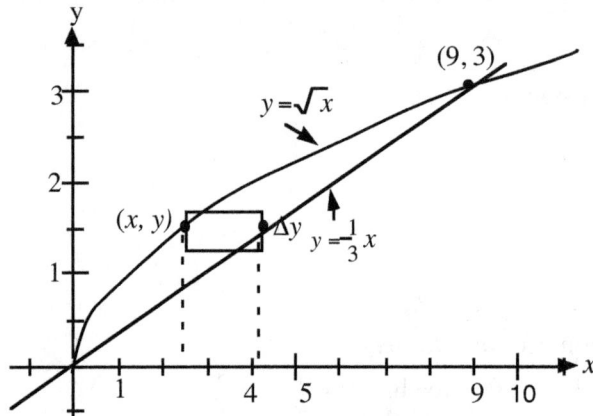

Step 2: Find the radius r of revolution. The radius of revolution is $r = y$

Imagine the x-axis is a horizontal pin passing through two horizontal holes at the ends of two light rods attached to the ends of the representative rectangle, and imagine rotating the rectangle about the x-axis. You would observe that the vertical distance from the left side (also a point on $y = \sqrt{x}$) or from the right side (also a point on $y = \tfrac{1}{3}x$) to the x-axis, y, is the radius of revolution or rotation.

Step 3: Find the length of the representative rectangle.
The length of the representative rectangle is the x-coordinate of the right end of the rectangle minus the x-coordinate of the left end of the rectangle. On the left end $y = \sqrt{x}$ or $y^2 = x$, and on the right end $y = \frac{1}{3}x$ or $3y = x$. Length, L, of representative rectangle $= 3y - y^2$

Step 4: Find the limits of integration   The limits are found by summing up the elements from the bottom ($y = 0$) to the top ($y = 3$).
The limits of integration are 0 and 3.

Step 5: Write an equation for the volume V of the solid obtained by revolving about the x-axis, the region with the representative rectangle of length $3y - y^2$ and radius of revolution y and length $L = 3y - y^2$,
The total volume V is given by

$$V = \int_a^b 2\pi\, rL\, dx \quad (\, L = \text{length of rectangle})$$

$$V = 2\pi\int_0^3 y(3y - y^2)dy \quad (r = y;\ L = 3y - y^2)$$

$$= 2\pi\int_0^3 (3y^2 - y^3)dy \quad \text{(All variables in terms of } y, \text{ because of } dy)$$

$$= 2\pi\int_0^3 (3y^2 - y^3)dy$$

$$= 2\pi\left[\frac{3y^3}{3} - \frac{y^4}{4}\right]_0^3$$

$$= 2\pi\left[y^3 - \frac{y^4}{4}\right]_0^3$$

$$= 2\pi\left[3^3 - \frac{3^4}{4}\right] - 2\pi[0 - 0]$$

$$= 2\pi\left[27 - \frac{81}{4}\right] - [0]$$

$$= \frac{27\pi}{2} \text{ cubic units}$$

Note above: $V = 2\pi\int_0^3 [y\{g(y) - f(y)\}]dy$.

Example 3b

Find the volume of the solid obtained by revolving the region enclosed by the curves $y = \sqrt{x}$ $y = \frac{1}{3}x$, over the closed interval $[1, 6]$, about the x-axis

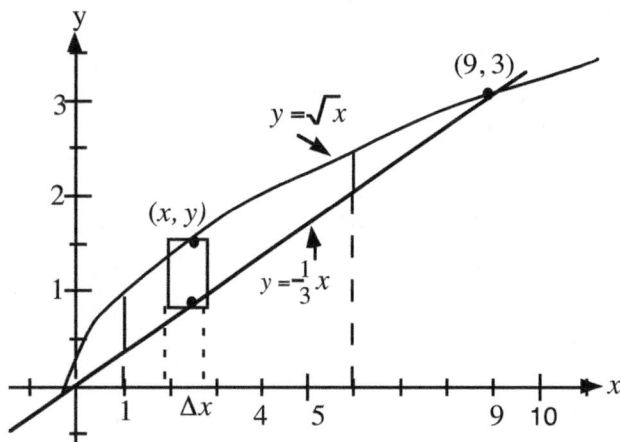

Solution   Using the Disk Method

All the steps are the same as in Example 3, Method 1, except for the limits of integration. Using some of the results from Example 3, Method 1. the volume V of the solid obtained by revolving the region enclosed by the curves $y = \sqrt{x}$ $y = \frac{1}{3}x$, over the closed interval $[1, 6]$ about the x-axis is given by

$$V = \pi\left[\frac{x^2}{2} - \frac{x^3}{(3)(9)}\right]_1^6$$

$$= \pi\left[\left(\frac{6^2}{2} - \frac{6^3}{(3)(9)}\right) - \left(\frac{1^2}{2} - \frac{1^3}{(3)(9)}\right)\right]$$

$$= \pi\left[(18 - 8) - \left(\frac{1}{2} - \frac{1}{27}\right)\right]$$

$$= \pi\left[10 - \frac{25}{54}\right]$$

$$= \frac{515\pi}{54}.$$

What about using the Shell Method?

To use the shell method, the representative rectangle should be parallel to the x-axis. Observe that in Example 3b, if we were to draw vertical lines from the top boundary to the bottom boundary, all the top ends would be on the same boundary and all the bottom ends would be on the same boundary. However, if we draw horizontal lines on the interval $[1, 6]$, **not** all the left ends would be on the same left boundary and **not** all the right ends would be on the same right boundary. In fact to use the shell method, we would need several representative rectangles This approach is more involved and we will not use it here. See also p.333

Lesson 48 Exercises

1. Distinguish between the disk method and the cylindrical shell method in finding solids of revolution

2. For a given area do we get the same answer if we use either method?

3. Write down the two basic formulas we must remember for the two methods if we do not wish to memorize formulas for various cases.

4 (a) Use the disk method; (b) Use the cylindrical shell method
Find the volume of the solid obtained by revolving the region enclosed by the curves $y = x^2$, the x-axis, the line $x = 2$ about the x-axis.

5. Find the volume of the solid obtained by revolving the region enclosed by the curves $y = x^2$, the x-axis, the lines $x = 2$ about the line $x = 2$.

6. Find the volume of the solid obtained by revolving the region enclosed by the curves $y = \sqrt{x}$ $y = \frac{1}{10}x^2$, over the interval $[0, 4]$ about the x-axis

7. Find the volume of the solid obtained by revolving the region enclosed by the curves $y = \sqrt{x}$ $y = \frac{1}{3}x$, about the x-axis

7b. Find the volume of the solid obtained by revolving the region enclosed by the curves $y = \sqrt{x}$ $y = \frac{1}{3}x$, over the interval $[1, 6]$ about the x-axis

---

Answer: 3. $V = \int_a^b \pi r^2 \cdot dx$ or dy ; shell formula: $V = \int_a^b 2\pi rL \cdot dx$ or dy;

4. $\frac{32\pi}{5}$ cubic units; **5.** $\frac{8\pi}{3}$ cubic units; **6.** $\frac{744\pi}{125}$ cubic units

7. $\frac{27\pi}{2}$ cubic units; **7b.** $\frac{515\pi}{54}$ cubic units.

Discuss the relative merits of the methods for finding volumes of revolution.

Read also the guidelines on page 333, regarding whether to use vertical representative rectangles or horizontal representative rectangles.

Lesson 49
Applications of Integration to Volumes of
Solids of Known Cross Sections $A(x)$

There are practical applications of this lesson in problems such as the following:

1. Two cuts are made on a circular cylinder of radius r units. One cut is along a plane perpendicular to the axis of the cylinder, and the other plane intersects the first plane at an inclined angle of $\theta°$, the two cuts meeting on a line through the center of the cylinder. Find the volume of the cut.

2 If the axes of two circular cylinders of equal radii intersect at right angles, find the common volume of intersection.

3 A hole of radius r is drilled through the center of a sphere of radius R. Find an expression of the volume of the remaining sphere, assuming that the axis of the hole is a diameter of the sphere.

In finding the mathematical models in the problems of known cross-sections, we will be guided by the following conditions.

Conditions:
1. The area of a **plane section is perpendicular to a fixed line,** and
2. The area is expressed in terms of the **perpendicular distance** of the plane section **from a fixed point**.

Case 1: **Cross-sections Perpendicular to the** x–axis

Let a solid be bounded by two parallel planes perpendicular to the x–axis at $x = a$ and $x = b$; and let $A(x)$ be the cross-sectional area of the solid perpendicular to the x–axis for each x in the closed interval $a \leq x \leq b$. Then if A is continuous on $a \leq x \leq b$, then the volume V of the solid is given by

$$V = \int_a^b A(x)\,dx$$

Case 2: **Cross-sections Perpendicular to the** y–axis

Let a solid be bounded by two parallel planes perpendicular to the y–axis at $y = c$ and $y = d$; and let $A(y)$ be the cross-sectional area of the solid perpendicular to the y–axis for each y in the closed interval $c \leq y \leq d$. Then if A is continuous on $c \leq y \leq d$, then volume V of the solid is given by

$$V = \int_c^d A(y)\,dy$$

Wording of the problem: Consider two usual given possibilities:
1. Given the solid: You select the similar cross-sectional representation.
2. Given the base: Use the base to construct the specified types of cross-sections such as square cross-sections, triangular cross-sections, or semicircular cross-sections.

Observe also: One part of the problem wording provides information for determining the perpendicular cross-sectional area, and the other part of the wording guides one to obtain a relationship or relationships between the variables involved, and from the relationship(s), one expresses the cross-sectional area in terms of a single variable for the integration process. The variable of integration is that of the axis perpendicular to the cross sections.

General Procedure
Step 1: Draw a representative cross-section.
Step 2: Find an expression for the area of the cross-section.
(a) Write down the general formula for the cross-sectional area
(b) Write down the expression for the cross-sectional area.
Step 3: Find an equation for the boundary of the base.
Step 4: Specify the x–y coordinate system of axes.

Let one of the axes (x–axis or y –axis) correspond to the fixed line in condition #1. (previous page)

Try to express $A(x)$ (or $A(y)$) as a function of the distance x (or y) from the origin.

Example: The equation of the boundary of a circle is given by $x^2 + y^2 = r^2$
Step 5: Write down the integral for the volume, and integrate.

Some Statements and Possible Implications
1. A solid with a **circular base** of radius r units. Use the base to find a representative cross sectional area or the boundary equation.

(We may apply the equation of the circle: $x^2 + y^2 = r^2$)
Boundary of circular base. Use the base to find a representative cross sectional area or the boundary equation.
2. All plane sections are perpendicular to a fixed diameter (condition 1 is satisfied)
(Let either the x–axis or y–axis coincide with the fixed diameter of the circle).
3. All plane sections are **equilateral triangles.**
(Use this to find the cross sectional area $A(x)$

(Area A of an **equilateral triangle** of side s, is given by $A = \frac{s^2\sqrt{3}}{4}$)

4. All plane sections are **isosceles right triangles** $A = \frac{s^2}{2}$,

where s = the length of one of the non-hypotenuse side (leg)
Use this to find the cross sectional area $A(x)$

All plane sections are **isosceles triangles** $A = \frac{1}{2}bh$

5. Each plane section is perpendicular to the y–axis.(**Condition 1 satisfied**)

Find $V = \int_{c}^{d} A(y)\,dy$

6. Each cross-section is perpendicular to the major axis **(fixed line condition)**
Let either the x-axis or the y-axis coincide with the major axis and find

accordingly, either $V = \int_{a}^{b} A(x)\,dx$ or $V = \int_{c}^{d} A(y)\,dy$
All cross-sections perpendicular to x-axis are **squares. (Condition 1 satisfied)**

$V = \int_{a}^{b} A(x)\,dx$

7. Sections of the solid are right triangles with bases along the **base** of solid
Use the base to build cross-sections which are right triangles.
8. Frustum of a cone: Given this solid, you select the cross-sectional representation.
9. The hypotenuse is in the plane of the **base.** Use the hypotenuse and the Pythagorean theorem to find the cross-sectional area or the boundary equation.

10. Isosceles triangles have **one leg** in the plane of the **base** (Cross-sections are **isosceles triangles**.)

11. Tetrahedron has three mutually perpendicular faces.

12. A chord of a circle as a **base**
(Build the cross-sections using this base)

13. Cut wedge is in the shape of a **right circular cone**
Consider a cone and its properties.
On problems involving the cone, consider using similarity of triangles to eliminate the introduced unknowns.

14. Base of a solid is region bounded by the parabola $8y = x^2$ and the line $y = 4$
Use this information to find an equation of the boundary

15. The base of a solid is in the form of an ellipse, with major axis a and minor axis b. Each section is an isosceles triangle of altitude h.
Use the equation of the ellipse to find the base of this triangle and then find

$A(x)$ using the area of the triangle formula, area $A = \frac{1}{2}bh$.

Ellipse equation: $\dfrac{x^2}{a^2} + \dfrac{y^2}{b^2} = r^2$

Pyramid

A **pyramid** is a solid bounded by plane (flat) surfaces, and whose base is a polygon, and whose surfaces meet at a point called its vertex. The base may have any number of sides, while all other surfaces are triangles.

Examples: 1. A pyramid has a **square** base. **2.** A pyramid has **rectangular** base.

On problems involving the pyramid, consider using similarity of triangles to eliminate the introduced unknowns.

A **regular pyramid** is a pyramid whose base is a polygon with equal sides, and whose altitude connects the vertex and the center of the base.

Right pyramid (Look for right triangles and similarity of triangles)
In a right pyramid, the vertex is directly above the centroid of the base.

Example 1 Two cuts are made on a circular cylinder of radius r units, One cut is along a plane perpendicular to the axis of the cylinder, and the other plane intersects the first plane at an inclined angle of $\theta°$, the two cuts meeting on a line through the center of the cylinder. Find the volume of the cut.

Solution: Draw an x–y–z coordinate system of axes and let the origin be at the center of the cylinder and let the intersection of the two cuts be along the y–axis. Also, let the positive side of the x–axis face the first cut. Consider a cut made perpendicular to the y–axis. Such a cut section is a right triangle with an angle of measure $\theta°$, with the adjacent side of length x (the base). The other side of the right triangle is of length $x \tan \theta°$ (the height) .

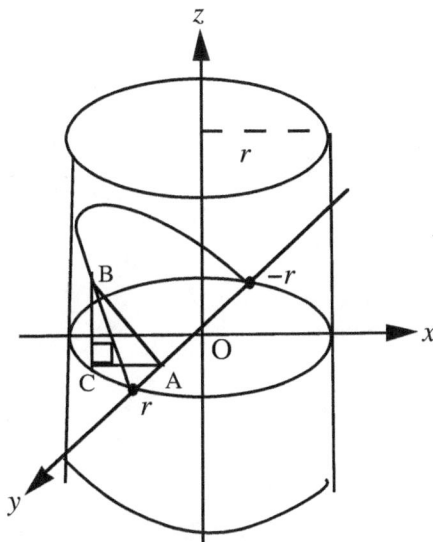

In right $\triangle ABC$, adjacent side $= \overline{CA}$, $m\angle A = \theta°$, $CA = x$, $BC = x\tan\theta°$

The area, A, of a cut section using $A = \frac{1}{2}bh$ is

$A = \frac{1}{2}(x)(x\tan\theta°)$ (base $= CA$, height $= BC$

$= \frac{1}{2}x^2 \tan\theta°$

$= \frac{1}{2}(r^2 - y^2)\tan\theta°$ (From $r^2 = x^2 + y^2$ for the circular part, $x^2 = r^2 - y^2$)

Volume V of the cut is given by

$V = \frac{1}{2}\tan\theta° \int_{-r}^{r} (r^2 - y^2)dy$

$= \frac{1}{2}\tan\theta° \int_{-r}^{r} (r^2 - y^2)dy$

$= \frac{1}{2}\tan\theta° \left[r^2 y - \frac{y^3}{3} \right]_{-r}^{r}$ (r^2 is a constant)

$$= \frac{1}{2}\tan\theta^\circ\left[\left(r^2(r) - \frac{r^3}{3}\right) - \left(r^2(-r) - \frac{(-r)^3}{3}\right)\right]$$

$$= \frac{1}{2}\tan\theta^\circ\left[\left(r^3 - \frac{r^3}{3}\right) - \left(-r^3 + \frac{r^3}{3}\right)\right]$$

$$= \frac{1}{2}\tan\theta^\circ\left[\left(\frac{3r^3 - r^3}{3}\right) - \left(\frac{-3r^3 + r^3}{3}\right)\right]$$

$$= \frac{1}{2}\tan\theta^\circ\left[\left(\frac{2r^3}{3}\right) - \left(\frac{-2r^3}{3}\right)\right]$$

$$= \frac{1}{2}\tan\theta^\circ\left[\left(\frac{2r^3}{3} + \frac{2r^3}{3}\right)\right]$$

$$= \frac{1}{2}\tan\theta^\circ\left[\frac{4r^3}{3}\right]$$

$$= \frac{2}{3}r^3\tan\theta^\circ.$$

Example 2 If the axes of two circular cylinders of equal radii intersect at right
angles, find the common volume of intersection.

Solution
Step !;

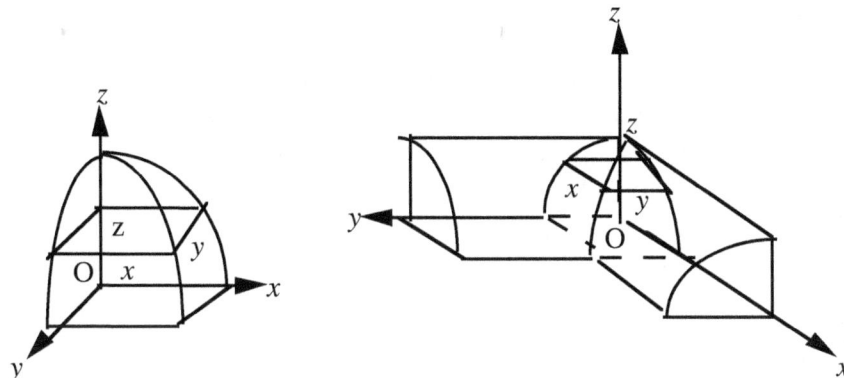

Step 2: The common cross-section
perpendicular to the z –axis is
a square of side $2y$ or $2x$
$(2x = 2y)$.
An equation of the cylinder
involving y is $y^2 + z^2 = r^2$ and
from which $y = \sqrt{r^2 - z^2}$

Step 3: Cross-section area =
$$(2y)^2 = (2\sqrt{r^2 - z^2})^2$$
$$= 4(r^2 - z^2)$$
Common volume V, is given by
$$V = \int_{-r}^{r} 4(r^2 - z^2)\,dz$$
$$= 4\int_{-r}^{r} (r^2 - z^2)\,dz$$
$$= 4\left[r^2 z - \frac{z^3}{3} \right]_{-r}^{r}$$

Step 4::

$$= 4\left[\left(r^2 (r) - \frac{r^3}{3} \right) - \left(r^2 (-r) - \frac{(-r)^3}{3} \right) \right]$$

$$= 4\left[\left(r^3 - \frac{r^3}{3} \right) - \left(-r^3 + \frac{r^3}{3} \right) \right]$$

$$= 4\left[\left(\frac{3r^3 - r^3}{3} \right) - \left(\frac{-3r^3 + r^3}{3} \right) \right]$$

$$= 4\left[\left(\frac{2r^3}{3} \right) - \left(\frac{-2r^3}{3} \right) \right]$$

$$= 4\left[\left(\frac{2r^3}{3} + \frac{2r^3}{3} \right) \right]$$

$$= 4\left[\frac{4r^3}{3} \right]$$

$$= \frac{16r^3}{3}$$

The common volume is

$$\frac{16r^3}{3}$$ cubic units.

Example 3 The base of a right cone is in the form of an ellipse, with major axis $2a$ and minor axis $2b$. If the height of the cone is h, find the volume of this cone.

Step 1: Draw an x–y–z coordinate system of axes and let the origin be at the center of the cone.

Consider a plane section perpendicular to the height (that is, parallel to the base). Such a section is an ellipse of major axis $2x$ and minor axis $2y$. The area of this section is πxy. We use similar triangles to obtain relationships between a, x, z and h and between b, y, z and h. (Recall how we used similar triangles in related rates problems.)

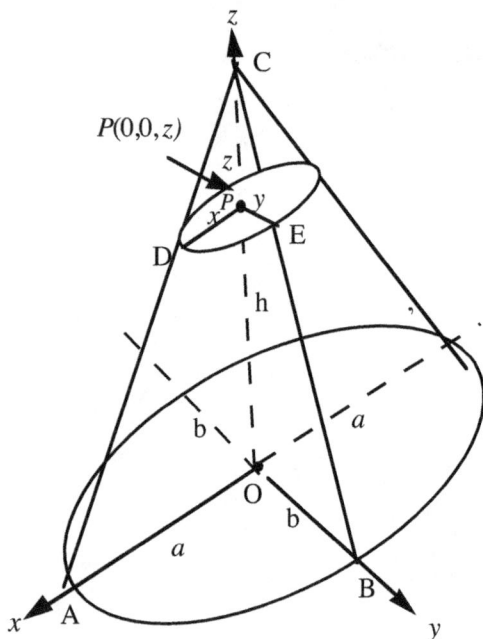
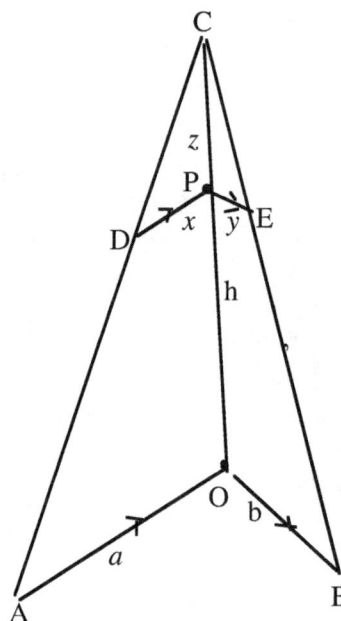

| Fig. 1 | Fig. 2 |

Step 2: In Fig. 1, $\triangle DPC$ and $\triangle AOC$ are similar ($\overline{DP}$ is parallel to $\overline{AO}$)

$$\therefore \quad \frac{x}{a} = \frac{h-z}{h} \quad \text{and from which} \quad x = \frac{a(h-z)}{h} \, a$$

Similarly, In Fig. 2, $\triangle ECP$ and $\triangle BCO$ are similar ($\overline{PE}$ is parallel to $\overline{OB}$

$$\therefore \quad \frac{y}{b} = \frac{h-z}{h} \quad \text{and from which} \quad y = \frac{b(h-z)}{h}$$

Step 3: Substitute for x and y in the area formula $A = \pi xy$, to obtain the area of a section as

$$A = \pi \frac{a(h-z)}{h} \cdot \frac{b(h-z)}{h}$$
$$= \frac{\pi ab(h-z)^2}{h^2}$$

Assume that we are translating the variable cross-sectional area up along the z-axis from the bottom $z = 0$ to the $z = h$,

Volume, V, of the cone is $V = \dfrac{\pi ab}{h^2} \displaystyle\int_0^h (h-z)^2 \, dz$ (and we will integrate by two methods).

Method #1	**Method #2**
Using change of limits	**Using indefinite integration**

Method #1

Using change of limits

Step 5 Let $u = h - z$

Then $\dfrac{du}{dz} = -1$, and

$dz = -du$

Step 6: Express the limits of integration in terms of u

Lower limit: When $z = 0$,

$u = h - 0 = h$ (lower)

Upper limit: When $z = h$,

$u = h - h = 0$ (upper)

Step 7: $= \dfrac{\pi ab}{h^2} \displaystyle\int_h^0 u^2(-du)$

$= -\dfrac{\pi ab}{h^2} \displaystyle\int_h^0 u^2\, du$

$= -\dfrac{\pi ab}{h^2} \left[\dfrac{u^3}{3} \right]_h^0$

$= -\dfrac{\pi ab}{h^2} \left[\left(\dfrac{0^3}{3} \right) - \left(\dfrac{h^3}{3} \right) \right]$

$= -\dfrac{\pi ab}{h^2} \left[(0) - \left(\dfrac{h^3}{3} \right) \right]$

$= -\dfrac{\pi ab}{h^2} \left[-\dfrac{h^3}{3} \right]$

$V = \dfrac{1}{3}\pi abh.$

Method #2

Using indefinite integration

Step 4: Let $u = h - z$

Then $\dfrac{du}{dz} = -1$, and $dz = -du$

Step 5: Substitute u for $h - z$; $-du$ for dz

in $V = \dfrac{\pi ab}{h^2} \displaystyle\int (h - z)^2\, dz$ to obtain

$V = \dfrac{\pi ab}{h^2} \left[-\displaystyle\int u^2\, du \right]$

$= -\dfrac{\pi ab}{h^2} \left[\dfrac{(h-z)^3}{3} + C \right]$

Step 5: Introduce the original limits. Then

$V = -\dfrac{\pi ab}{h^2} \left[\dfrac{(h-z)^3}{3} + C \right]_0^h$

$= -\dfrac{\pi ab}{h^2} \left[\left(\dfrac{(h-h)^3}{3} + C \right) - \left(\dfrac{(h-0)^3}{3} + C \right) \right]$

$= -\dfrac{\pi ab}{h^2} \left[(0 + C) - \left(\dfrac{h^3}{3} + C \right) \right]$

$= -\dfrac{\pi ab}{h^2} \left[C - \dfrac{h^3}{3} - C \right]$

$= -\dfrac{\pi ab}{h^2} \left[-\dfrac{h^3}{3} \right]$

$V = \dfrac{1}{3}\pi abh.$

Example 4.

The base of a solid is a circle of radius r. All plane sections perpendicular to a fixed diameter (fixed line) of the base are squares. Find the volume of the solid

Solution

Following the procedure suggested on p. 371:

Step 1: Draw a circle in an x– y – z coordinate system of axes, and let the center of the circle be at the origin. Also let the fixed diameter be along the x–axis.

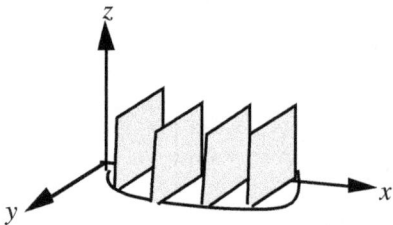

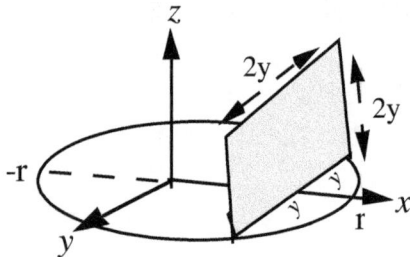

Step 2:

Since plane sections are squares,

area of a plane section $= (2y)^2 = 4y^2$

Step 3: Find the equation of the circular boundary base.

The equation is that of a circle

$x^2 + y^2 = r^2$. We solve for y

$y^2 = r^2 - x^2$, We want $A(x)$

Step 4: Area of a plane section $4y^2$

becomes $4(r^2 - x^2)$, and

Step 5: The volume, V of the solid

is $V = 4\int_{-r}^{r} (r^2 - x^2)\,dx$.

Step 6:

$$V = 4\left[r^2 x - \frac{x^3}{3}\right]_{-r}^{r}$$

Note: $(\int(r^2 - x^2)dx = \int r^2 dx - \int x^2 dx)$

$$= 4\left[\left(r^2(r) - \frac{r^3}{3}\right) - \left(r^2(-r) - \frac{(-r)^3}{3}\right)\right]$$

$$= 4\left[\left(r^3 - \frac{r^3}{3}\right) - \left(-r^3 + \frac{r^3}{3}\right)\right]$$

$$= 4\left[r^3 - \frac{r^3}{3} + r^3 - \frac{r^3}{3}\right]$$

$$= 4\left[2r^3 - \frac{2r^3}{3}\right]$$

$$= 4\left[\frac{6r^3}{3} - \frac{2r^3}{3}\right]$$

$$= 4\left[\frac{4r^3}{3}\right] = \frac{16r^3}{3}.$$

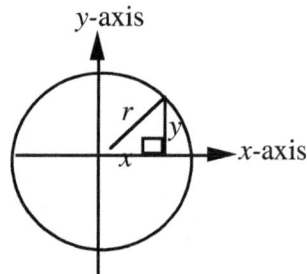

Example 5 A hole of radius r is drilled through the center of a sphere of radius R. Find an expression of the volume of the remaining sphere, assuming that the axis of the hole is a diameter of the sphere..

Method 1 Following the procedure on p.370-372:

Step 1: The plane section is horizontal and the fixed line is the y-axis. This cross-section is perpendicular to the y-axis. The variable of integration is y. We apply the Pythagorean theorem to find the cross sectional area as well as to determine the limits of integration.

Find an expression for the area cross-section, A, of the remaining volume.

$$R^2 = y^2 + x^2; \text{ and } x^2 = R^2 - y^2 \quad (A)$$

$A = \pi x^2 - \pi r^2$ (using horizontal radii)

(cross-section of sphere - cross-section of cylinder)

$$= \pi(R^2 - y^2) - \pi r^2) \quad (B) \qquad \longleftarrow$$

$\left(\text{Replacing } x^2 \text{ by } R^2 - y^2 \text{ to express in terms of } y \right)$

Step 2: Find the limits of integration; (-d to d)

$$R^2 = d^2 + r^2 \qquad (2d \text{ is the height of the hole; } r = \text{radius of the hole})$$

$$d = \sqrt{R^2 - r^2} = (R^2 - r^2)^{\frac{1}{2}}$$

Step 3: Find an expression $\int A(y)dy$ for the remaining volume.

$$\int A(y)dy = \pi \int (R^2 - y^2 - r^2)dy \quad \text{(Applying equation (B))}$$

Step 4 Take the limits from 0 to d, and multiply the integral by 2, (by symmetry). See also method 2.

$$= 2\pi \int_0^d (R^2 - y^2 - r^2)dy$$

$$= 2\pi \left[R^2 y - \frac{y^3}{3} - r^2 y \right]_0^d$$

horizontal cross-section :

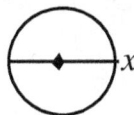

Fig 2

Step 5: Substitute the limits of integration (Step 2)

$$V = 2\pi [R^2(R^2 - r^2)^{\frac{1}{2}} - \frac{(R^2 - r^2)^{\frac{3}{2}}}{3} - r^2(R^2 - r^2)^{\frac{1}{2}}]$$

$$= 2\pi(R^2 - r^2)^{\frac{1}{2}} [R^2 - \frac{(R^2 - r^2)}{3} - r^2]$$

$$= 2\pi(R^2 - r^2)^{\frac{1}{2}} [R^2 - \frac{R^2}{3} + \frac{r^2}{3} - r^2]$$

$$= 2\pi(R^2 - r^2)^{\frac{1}{2}} \left[\frac{2R^2}{3} - \frac{2r^2}{3} \right]$$

$$= \frac{2(2)\pi(R^2 - r^2)^{\frac{1}{2}}}{3} [R^2 - r^2]$$

$$\boxed{V = \frac{4\pi}{3}(R^2 - r^2)^{\frac{3}{2}}}. \quad \longleftarrow \text{in terms of } R \text{ and } r$$

To express V in terms of the length, L, of the cylindrical hole through the sphere, go to **Step 6**.

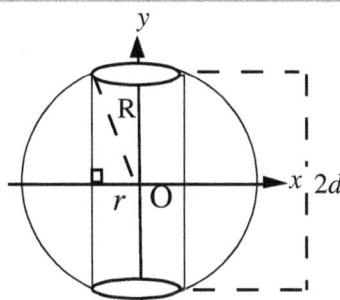

Step 6: From Step 2,

$$R^2 = d^2 + r^2$$

Let $2d = L$. Then $d = \frac{L}{2}$

$$R^2 = \left(\frac{L}{2}\right)^2 + r^2$$

Replace R^2 by $\left(\frac{L}{2}\right)^2 + r^2$

in $V = \frac{4\pi}{3}(R^2 - r^2)^{\frac{3}{2}}$,

$$V = \frac{4\pi}{3}(\left(\frac{L}{2}\right)^2 + r^2 - r^2)^{\frac{3}{2}}$$

$$V = \frac{4\pi}{3}\left[\left(\frac{L}{2}\right)^2\right]^{\frac{3}{2}}$$

$$\boxed{V = \frac{\pi L^3}{6}}(L = \text{Hole length})$$

Method 2   Steps 1 to 3 are the same as in Method 1

Step 4: Find an expression for the remaining volume.

$$V = \pi \int_{-d}^{d} (R^2 - y^2 - r^2)\,dy$$

Step 5: $V = \pi \left[R^2 y - \dfrac{y^3}{3} - r^2 y \right]_{-d}^{d}$

Step 6: Substitute the limits of integration

$$V = \pi \left[\left(R^2 d - \frac{d^3}{3} - r^2 d \right) - \left(R^2(-d) - \frac{(-d)^3}{3} - r^2(-d) \right) \right]$$

$$V = \pi \left[\left(R^2 d - \frac{d^3}{3} - r^2 d \right) - \left(-R^2 d + \frac{d^3}{3} + r^2 d \right) \right]$$

$$V = \pi \left[R^2 d - \frac{d^3}{3} - r^2 d + R^2 d - \frac{d^3}{3} - r^2 d \right]$$

$$V = \pi \left[2dR^2 - \frac{2d^3}{3} - 2dr^2 \right]$$

$$= 2\pi \left[\left(\sqrt{R^2 - r^2} \right) R^2 - \frac{\left(\sqrt{R^2 - r^2} \right)^3}{3} - \left(\sqrt{R^2 - r^2} \right) r^2 \right]. \ (d = \sqrt{R^2 - r^2})$$

$$= 2\pi \left[\left(R^2 - r^2 \right)^{\frac{1}{2}} R^2 - \frac{(R^2 - r^2)^{\frac{3}{2}}}{3} - \left(R^2 - r^2 \right)^{\frac{1}{2}} r^2 \right]$$

$$= 2\pi \left(R^2 - r^2 \right)^{\frac{1}{2}} \left[\frac{3R^2}{3} - \frac{(R^2 - r^2)}{3} - \frac{3r^2}{3} \right]$$

$$= 2\pi \left(R^2 - r^2 \right)^{\frac{1}{2}} \left[\frac{3R^2 - R^2 + r^2 - 3r^2}{3} \right]$$

Step 7: $= 2\pi (R^2 - r^2)^{\frac{1}{2}} \left[\dfrac{2R^2}{3} - \dfrac{2r^2}{3} \right]$

Step 8: $= \dfrac{2(2)\pi (R^2 - r^2)^{\frac{1}{2}}}{3} \left[R^2 - r^2 \right]$

$$\boxed{V = \frac{4\pi}{3}(R^2 - r^2)^{\frac{3}{2}}}$$

Another Approach for Example 5: We can also find the remaining volume of the sphere by the cylindrical shell method (p.358) by revolution about the axis of the cylinder.

$$y = \sqrt{R^2 - x^2}, \ r < x < R; \qquad L = 2y \quad V = \int_{m}^{n} 2\pi\, xL \bullet dx \,.\ (\text{rad, of rev..} = x)$$

Step 1: $V = 2 \left[2\pi \int_{r}^{R} x \left(R^2 - x^2 \right)^{\frac{1}{2}} dx \right]$

$$= 4\pi \int_{r}^{R} x \left(R^2 - x^2 \right)^{\frac{1}{2}} dx$$

$$= 4\pi \int_{r}^{R} x u^{\frac{1}{2}} \frac{du}{-2x} \quad (u = R^2 - x^2)$$

$$= -2\pi \left[\frac{2}{3} u^{\frac{3}{2}} \right] + C$$

Step 2:

$$= -\frac{4\pi}{3} \left[(R^2 - x^2)^{\frac{3}{2}} \right]_{r}^{R}$$

$$= -\frac{4\pi}{3} \left[\left(R^2 - R^2 \right)^{\frac{3}{2}} - \left(R^2 - r^2 \right)^{\frac{3}{2}} \right]$$

$$= -\frac{4\pi}{3} \left[0 - \left(R^2 - r^2 \right)^{\frac{3}{2}} \right]$$

$$\boxed{V = \frac{4\pi}{3} \left(R^2 - r^2 \right)^{\frac{3}{2}}}$$

Extra 1: Volume removed = Volume of sphere - volume remaining

$$= \frac{4\pi}{3} R^3 - \frac{4\pi}{3} \left(R^2 - r^2 \right)^{\frac{3}{2}} \boxed{= \frac{4\pi}{3} [R^3 - \left(R^2 - r^2 \right)^{\frac{3}{2}}]}$$

Extra 2: In terms of d, $V = \dfrac{4\pi}{3}(R^2 - [R^2 - d^2])^{\frac{3}{2}} = \dfrac{4\pi}{3}(R^2 - R^2 + d^2)^{\frac{3}{2}} \boxed{= \dfrac{4\pi}{3} d^3}$

Example 6

A right pyramid has a square base of length b units. If the altitude of this pyramid is h units, find a formula for the volume of the pyramid.

Let us be guided by the following conditions suggested previously:

1. The area of a **plane section is perpendicular to a fixed line,** and

2. The area is expressed in terms of the **perpendicular distance** of the plane section **from a fixed point.**

Step 1: We want to satisfy **condition #1** of the guidelines.

The **basis** of the cross-sectional area is the square base of the pyramid. Let a side of the square base $= s$

Then the cross-sectional area $= s^2$

Also, let this plane section of area s^2 be perpendicular to the y-axis (the fixed line condition).

Step 2: We want to satisfy **condition #2** of the guidelines.

Let the perpendicular distance from the fixed point $A(0,h)$ at any point between O and A $= y$

Also, express s^2 in terms of y. (to eliminate s)

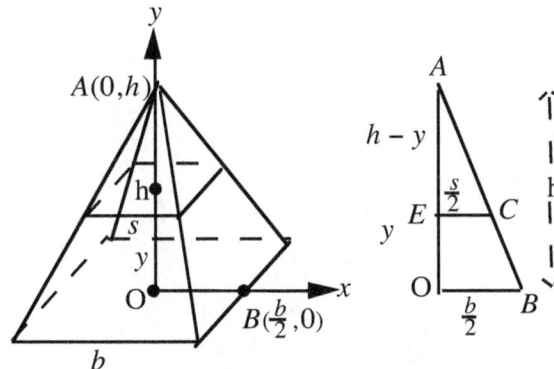

We use similarity of triangles to eliminate s.

Since $\triangle AOB$ and $\triangle AEC$ are similar ($\overline{OB}$ is parallel $\overline{EC}$)

$$\frac{h}{h-y} = \frac{\frac{b}{2}}{\frac{s}{2}} \text{ and from which } s = \frac{b(h-y)}{h} \text{ or } s = \frac{b}{h}(h-y)$$

Step 3: Write down the integral for the volume and integrate.

Assume that we are translating the variable cross-sectional area up along the y-axis from the bottom , $y = 0$ to the top , $y = h$.,

Then, we integrate with respect to y from $y = 0$ to $y = h$

Step 4: $V = \int_c^d A(y)\,dy$ (where c, and d are the limits of integration)

$$= \int_0^h \frac{b^2}{h^2}(h-y)^2\,dy$$

$$= \frac{b^2}{h^2}\int_0^h (h-y)^2\,dy \ .$$

We use simple u-substitution to integrate, and we use two approaches.

Approach #1	**Approach #2**

Approach #1

Step 5 Let $u = h - y$

Then $\dfrac{du}{dy} = -1$, and

$dy = -du$

Step 6: Express the limits of integration in terms of u

Lower limit: When $y = 0$,

$u = h - 0 = h$ (lower)

Upper limit: When $y = h$,

$u = h - h = 0$ (upper)

Step 7: $= \dfrac{b^2}{h^2} \displaystyle\int_h^0 u^2 (-du)$

$= -\dfrac{b^2}{h^2} \displaystyle\int_h^0 u^2 \, du$

$= -\dfrac{b^2}{h^2} \left[\dfrac{u^3}{3} \right]_h^0$

$= -\dfrac{b^2}{h^2} \left([0] - \left[\dfrac{h^3}{3} \right] \right)$

$= -\dfrac{b^2}{h^2} \left(-\dfrac{h^3}{3} \right)$

$= \dfrac{1}{3} b^2 h .$

Approach #2

Step 5 Let $u = h - y$

Then $\dfrac{du}{dy} = -1$, and $dy = -du$

Step 6 $= \dfrac{b^2}{h^2} \displaystyle\int u^2 (-du)$

$= -\dfrac{b^2}{h^2} \displaystyle\int u^2 \, du$

$= -\dfrac{b^2}{h^2} \cdot \dfrac{u^3}{3} + C$

$= -\dfrac{b^2}{3h^2} u^3 + C$

$= -\dfrac{b^2}{3h^2} (h - y)^3 + C$

Step 7 $= \left[-\dfrac{b^2}{3h^2} (h - y)^3 + C \right]_0^h$

$= \left[-\dfrac{b^2}{3h^2} (h - h)^3 + C \right] - \left[-\dfrac{b^2}{3h^2} (h - 0)^3 + C \right]$

$= \left[-\dfrac{b^2}{3h^2} (0)^3 + C \right] - \left[-\dfrac{b^2}{3h^2} (h)^3 + C \right]$

$= [0 + C] - \left[-\dfrac{b^2 h^3}{3h^2} + C \right]$

$= C - \left[-\dfrac{b^2 h^3}{3h^2} + C \right]$

$= C + \dfrac{b^2 h^3}{3h^2} - C$

$= \dfrac{1}{3} b^2 h .$

Lesson 49 Exercises

1. Two cuts are made on a circular cylinder of radius r units One cut is along plane perpendicular to the axis of the cylinder., and the other plane intersects the first plane at an inclined angle of $\theta°$, the two cuts meeting on a line through the center of the cylinder. Find the volume of the cut.

2. If the axes of two circular cylinders of equal radii intersect at right angles, find the common volume of intersection.

3. The base of a right cone is in the form of an ellipse, with major axis $2a$ and minor axis $2b$. If the height of the cone is h, find the volume of this cone..

4. The base of a solid is a circle of radius r. All plane sections perpendicular to a fixed diameter (fixed line) of the base are squares. Find the volume of the solid

5. A hole of radius r is drilled through the center of a sphere of radius R. Find an expression of the volume of the remaining sphere, assuming that the axis of the hole is a diameter of the sphere..

6. A right pyramid has a square base of length b units.. If the altitude of this pyramid is h units,, find a formula for the volume of the pyramid,

---

Answer: 1. $V = \frac{2}{3}r^3 \tan\theta°$; **2.** $V = \frac{16r^3}{3}$ **3.** $V = \frac{1}{3}\pi abh$ **4.** $V = \frac{16r^3}{3}$

5. $V = \frac{4\pi}{3}(R^2 - r^2)^{\frac{3}{2}}$ **6.** $V = \frac{1}{3}b^2 h$

Lesson 50
Finding the length of a Plane Curve
(Application of Integration)

Arc Length

In Lesson 27, we found the derivative of arc length. Here, we find the arc length.

Definition
The arc length of a curve is the limit of the sum of the lengths of consecutive chords connecting points on the arc when the number of points is increased indefinitely, such that the length of each chord approaches zero.

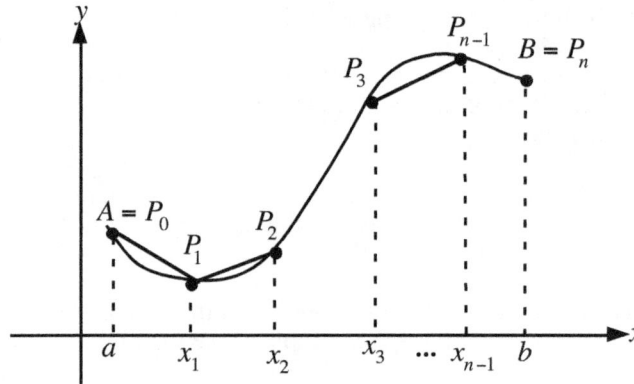

Given: $y = f(x)$, the arc length, L, is given by

$$L = \int_a^b \sqrt{1 + \left(\frac{dy}{dx}\right)^2}\ dx \quad \text{(Note that in Lesson 27, we denoted the arc length by } S\text{)}$$

Example 1 Find the distance between the points $(-2, 3)$ and $(4, -5)$

Method 1   Apply the distance formula:

$$d = \sqrt{(x_2 - x_1)^2 + (y_2 - y_1)^2} \tag{1}$$

where d is the distance between the points $P_1(x_1, y_1)$ and $P_2(x_2, y_2)$.

Substituting $x_1 = -2,\ y_1 = 3, x_2 = 4,\ y_2 = -5$ in equation (1) above,

$$d = \sqrt{(4 - (-2))^2 + (-5 - 3)^2}$$

$$d = \sqrt{(4 + 2)^2 + (-8)^2}$$

$$= \sqrt{(6)^2 + (-8)^2}$$

$$= \sqrt{36 + 64}$$

$$= \sqrt{100}$$

$$= 10$$

∴ the distance between the given points is 10 units.

Method 2: Using $L = \int_a^b \sqrt{1 + \left(\dfrac{dy}{dx}\right)^2}\, dx$

Step 1. Find an equation of the line connecting the given points

$$m = \frac{y_2 - y_1}{x_2 - x_2} = \frac{-5 - 3}{4 - (-2)} = \frac{-8}{6} = -\frac{4}{3}$$

$$y - 3 = -\frac{4}{3}(x + 2) \quad \text{(applying } y - y_1 = m(x - x_1)\text{)}$$

$$y = -\frac{4}{3}x + \frac{1}{3}$$

Step 2: Apply. $\quad L = \int_a^b \sqrt{1 + \left(\dfrac{dy}{dx}\right)^2}\, dx \qquad$ (arc length formula)

From $y = -\dfrac{4}{3}x + \dfrac{1}{3}$

$$\frac{dy}{dx} = -\frac{4}{3}$$

$$L = \int_{-2}^4 \sqrt{1 + \left(-\frac{4}{3}\right)^2}\, dx$$

$$= \int_{-2}^4 \sqrt{1 + \frac{16}{9}}\, dx$$

$$= \int_{-2}^4 \sqrt{\frac{25}{9}}\, dx$$

Step 3:

$$L = \int_{-2}^4 \frac{5}{3}\, dx$$

$$= \frac{5}{3}\int_{-2}^4 dx$$

$$= \frac{5}{3}[x]_{-2}^4$$

$$= \frac{5}{3}[4 - (-2)]$$

$$= \frac{5}{3}[6]$$

$$= 10 \text{ units}$$

What about applying $x = g(y)$

Solution From $y = -\dfrac{4}{3}x + \dfrac{1}{3}$

$$x = -\frac{3}{4}y + \frac{1}{4}, \text{ and } \frac{dx}{dy} = -\frac{3}{4}:$$

Now, apply $L = \int_c^d \sqrt{1 + \left(\dfrac{dx}{dy}\right)^2}\, dy \qquad$ (arc length formula)

Step 1: $L = \int_{-5}^3 \sqrt{1 + \left(-\dfrac{3}{4}\right)^2}\, dy$

$$= \int_{-5}^3 \sqrt{1 + \frac{9}{16}}\, dy$$

$$= \int_{-5}^3 \sqrt{\frac{25}{16}}\, dy$$

$$= \int_{-5}^3 \frac{5}{4}\, dy$$

Step 2

$$L = \frac{5}{4}\int_{-5}^3 dy$$

$$= \frac{5}{4}[y]_{-5}^3$$

$$= \frac{5}{4}[3 - (-5)]$$

$$= \frac{5}{4}[8]$$

$$= 10 \text{ units.}$$

Yes, we obtain the same result.

Example 2 Find the length of the graph of $y = x^2$ between $x = 1$ and $x = 3$.

The length, L, is given by $L = \int_a^b \sqrt{1 + \left(\dfrac{dy}{dx}\right)^2}\, dx$ (A)

Step 1: $y = x^2$ and $\dfrac{dy}{dx} = 2x$; $\left(\dfrac{dy}{dx}\right)^2 = (2x)^2 = 4x^2$

$L = \int_1^3 \sqrt{1 + (2x)^2}\, dx$ (B)   (substituting accordingly)

Let $u = 2x$. Then $\dfrac{du}{dx} = 2$, and $dx = \dfrac{du}{2}$

For the limits: When $x = 1$, $u = 2(1) = 2$

When $x = 3$, $u = 2(3) = 6$

Substitute the above in (B).

Then $L = \dfrac{1}{2}\int_2^6 \sqrt{1 + u^2}\, du$ (C)

Let $u = \tan\theta$. $u^2 = \tan^2\theta$

$\dfrac{du}{d\theta} = \sec^2\theta$ and $du = \sec^2\theta\, d\theta$

For the limits: $u = \tan\theta$.

When $u = 2$, $\theta = \tan^{-1}(2)$; When $u = 6$, $\theta = \tan^{-1}(6)$

Step 2: Substituting the above in (A) accordingly,

$L = \dfrac{1}{2}\int_{\tan^{-1}(2)}^{\tan^{-1}(6)} \sqrt{1 + \tan^2\theta}\,\sec^2\theta\, d\theta$

$= \dfrac{1}{2}\int_{\tan^{-1}(2)}^{\tan^{-1}(6)} \sqrt{\sec^2\theta}\,\sec^2\theta\, d\theta$

$= \dfrac{1}{2}\int_{\tan^{-1}(2)}^{\tan^{-1}(6)} \sec^3\theta\, d\theta$

(To integrate $\sec^3\theta$, see page 260, Example 2.)

Scrapwork
(Pre-calculus)

$\theta = \tan^{-1}(2)$

$\tan\theta = \dfrac{2}{1}$

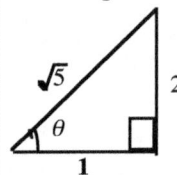

$\sec(\tan^{-1}(2))$

$= \sec\theta$

$= \sqrt{5}$

$\theta = \tan^{-1}(6)$

$\tan\theta = \dfrac{6}{1}$

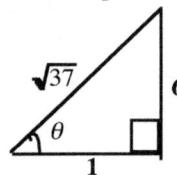

$\sec(\tan^{-1}(6))$

$= \sec\theta = \sqrt{37}$

$(1 + \tan^2\theta = \sec^2\theta)$

Step 3: $= \dfrac{1}{2}\left[\dfrac{1}{2}\ln|\sec\theta + \tan\theta| + \dfrac{1}{2}\tan\theta\sec\theta\right]_{\tan^{-1}(2)}^{\tan^{-1}(6)}$

$= \dfrac{1}{4}\left[\ln\left|\sec(\tan^{-1}(6)) + \tan(\tan^{-1}(6))\right| + \tan(\tan^{-1}(6))\sec(\tan^{-1}(6))\right] -$

$\left[\ln\left|\sec(\tan^{-1}(2)) + \tan(\tan^{-1}(2))\right| + \tan(\tan^{-1}(2))\sec(\tan^{-1}(2))\right]$

$= \dfrac{1}{4}\left\{\left[\ln\left|\sqrt{37}) + 6\right| + 6\sqrt{37}\right] - \left[\ln\left|\sqrt{5}) + 2\right| + 2\sqrt{5}\right]\right\}$

$= \dfrac{1}{4}\left\{\left[\ln|6.082 + 6| + 6(6.082)\right] - \left[\ln|2.24 + 2| + 2(2.24)\right]\right\}$

$= \dfrac{1}{4}\left\{\left[\ln|12.082| + 36.492\right] - \left[\ln|4.24| + 4.48\right]\right\}$

$= \dfrac{1}{4}\left\{38.984 - [5.924]\right\}$

$= \dfrac{1}{4}\left\{33.06\right\}$ $= 8.27$ units

The length of the graph of $y = x^2$ between $x = 1$ and $x = 3$ is 8.27 units.

Area of Surface of Revolution

Given: $y = f(x)$. The area , S, of the surface of revolution about the x–axis is

$$\text{given by } S = 2\pi \int_a^b y \sqrt{1 + \left(\frac{dy}{dx}\right)^2}\, dx \qquad (1)$$

Example 1: Find the area of the surface of revolution by revolving about the x–axis, the curve given by $y = \frac{1}{3}x^3$ from $x = 0$ to $x = 3$.

Step 1: Find $\left(\frac{dy}{dx}\right)^2$

$$y = \frac{1}{3}x^3 \qquad (0 \le x \le 3)$$

$$\frac{dy}{dx} = \frac{1}{3}(3)x^2 = x^2; \qquad \left(\frac{dy}{dx}\right)^2 = x^4$$

Step 2 Substitute in (1)

$$S = 2\pi \int_0^3 y \sqrt{1 + \left(x^2\right)^2}\, dx$$

Step 3: $S = 2\pi \int_0^3 \frac{1}{3}x^3 \sqrt{1 + x^4}\, dx$ <-

$$(y = \frac{1}{3}x^3)$$

$$= \frac{2\pi}{3} \int_0^3 x^3 \sqrt{1 + x^4}\, dx$$

Step 4: Integrate.

Let $u = 1 + x^4$; Then $\frac{du}{dx} = 4x^3$.

$$dx = \frac{du}{4x^3}dx \text{ or } du = 4x^3 dx$$

Step 5: Method 1

$$S = \frac{2\pi}{3} \int x^3 u^{\frac{1}{2}} \frac{du}{4x^3}$$

$$= \frac{\pi}{6} \int u^{\frac{1}{2}}\, du$$

$$= \frac{\pi}{6}\left[\frac{2}{3}u^{\frac{3}{2}}\right] + C$$

$$= \frac{2\pi}{18}\left[u^{\frac{3}{2}}\right] + C = \frac{\pi}{9}\left[\left(\sqrt{1+x^4}\right)^3\right]_0^3$$

$$= \frac{\pi}{9}\left[\left(\sqrt{1+3^4}\right)^3 - \left(\sqrt{1+0^4}\right)^3\right]$$

$$= \frac{\pi}{9}\left[\left(\sqrt{82}\right)^3 - \left(\sqrt{1}\right)\right] \quad \text{(C drops out)}$$

$$= \pi\left[\frac{(82\sqrt{82})-1}{9}\right] \approx 258.9 \text{ sq. units.}$$

Step 5: Method 2

From $u = 1 + x^4$. when $x = 0$, $u = 1$
when $x = 3$, $u = 1 + 3^4 = 1 + 81 = 82$

$$S = \frac{\pi}{6}\int_1^{82} u^{\frac{1}{2}}\, du$$

$$= \frac{2\pi}{3 \cdot 6}\left[u^{\frac{3}{2}}\right]_1^{81}$$

$$= \frac{\pi}{9}\left(82^{\frac{3}{2}} - 1^{\frac{3}{2}}\right)$$

$$= \pi\left[\frac{82^{\frac{3}{2}} - 1}{9}\right] \text{ square units.}$$

$$= 258.9 \text{ sq. units.}$$

Revolution about the y–axis (For more clarification, see Appendix C)

Example 2: Find the area of the surface of revolution by revolving about the y–axis, the curve given by $y = \frac{1}{3}x^3$ from $x = 0$ to $x = 3$.

$$\text{Given: } x = g(y), \qquad S = 2\pi \int_c^d x \sqrt{1 + \left(\frac{dx}{dy}\right)^2}\, dy. \qquad (A)$$

Step 1: From $y = \frac{1}{3}x^3$

$$\boxed{x = (3y)^{\frac{1}{3}}} \qquad (B)$$

For the limits: From $y = 0$ to $y = 9$.

Step 2: $\frac{dx}{dy} = \frac{1}{\frac{dy}{dx}} = \boxed{\frac{1}{x^2}}$ $\qquad (\frac{dy}{dx} = x^2)$

$$\left(\frac{dx}{dy}\right)^2 = \boxed{\frac{1}{x^4}}$$

Step 3: Substitute the above in equation (A) to obtain

$$= 2\pi \int_0^9 (3y)^{\frac{1}{3}} \sqrt{1 + \frac{1}{(3y)^{\frac{4}{3}}}}\, dy \qquad \text{(C}$$

Step 4: Let $u = (3y)^{\frac{1}{3}}$. Then

$$\frac{du}{dy} = 3^{\frac{1}{3}}(\tfrac{1}{3})y^{-\frac{2}{3}}; \quad dy == 3^{\frac{2}{3}}y^{\frac{2}{3}}du$$

Step 5: $u = (3y)^{\frac{1}{3}} = 3^{\frac{1}{3}}y^{\frac{1}{3}}; \quad \dfrac{u^2}{3^{\frac{1}{3}}} = y^{\frac{2}{3}};$

$$\boxed{dy = u^2 du}$$

For the limits: when $y = 0$, $u = 0$; when

$y = 9$, $u = (3(9))^{\frac{1}{3}} = 3$

Step 6: $2\pi \int_0^3 u \sqrt{1 + \dfrac{1}{u^4}}\, u^2 du$

$$2\pi \int_0^3 u \sqrt{u^4 + 1}\, du \qquad \text{((D)}$$

Step 7: Integrate by trig. substitution.

Let $u^2 = \tan\theta$. Then $2u\dfrac{du}{d\theta} = \sec^2\theta$

$du = \dfrac{\sec^2 d\theta}{2u}$.

Using $u^2 = \tan\theta$, we draw the corresponding right triangle

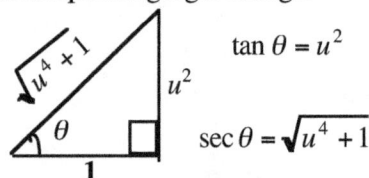

$\tan\theta = u^2$

$\sec\theta = \sqrt{u^4 + 1}$

Step 8: We use indefinite integration now and express the integration results in terms of u later; and use the limits in (D)

$$2\pi \int u \sqrt{\tan^2\theta + 1}\, \frac{\sec^2\theta d\theta}{2u}$$

$$= \pi \int \sqrt{\tan^2\theta + 1} \bullet \sec^2\theta d\theta$$

$$= \pi \int \sqrt{\sec^2\theta} \bullet \sec^2\theta d\theta$$

$$= \pi \int \sec\theta \bullet \sec^2\theta d\theta$$

$$= \pi \int \sec^3\theta d\theta \qquad \text{(E)}$$

Step 9: $\int \sec^3\theta\, d\theta = \frac{1}{2}\ln|\sec\theta + \tan\theta| + \frac{1}{2}\tan\theta\sec\theta + C$ (See p.260, Case 6,)

$$\pi \int \sec^3\theta d\theta = \frac{\pi}{2}\ln|\sec\theta + \tan\theta| + \frac{\pi}{2}\tan\theta\sec\theta + C$$

Express the right side in terms of u. ($\sec\theta = \sqrt{u^4 + 1}$; $\tan\theta = u^2$)

$$2\pi \int_0^3 u\sqrt{u^4 + 1}\, du = \left[\frac{\pi}{2}\left(\ln\left|\sqrt{u^4 + 1} + u^2\right| + u^2\sqrt{u^4 + 1} \right) \right]_0^3$$

$$= \frac{\pi}{2}\left\{ \left(\ln\left|\sqrt{3^4 + 1} + 3^2\right| + 3^2\sqrt{3^4 + 1} \right) - \left(\ln\left|\sqrt{0^4 + 1} + 0^2\right| + 0^2\sqrt{0^4 + 1} \right) \right\}$$

$$= \frac{\pi}{2}\left(\ln\left|\sqrt{82} + 9\right| + 9\sqrt{82} \right) - (0) = \frac{\pi}{2}\left(\ln\left|\sqrt{82} + 9\right| + 9\sqrt{82} \right) \approx 132.6 \text{ square units}$$

Lesson 50 Exercises

1. Write down from memory the arc length formula.

2. Use the arc length formula to find the distance between the points $(-2, 3)$ and $(4, -5)$

3. Write down from memory the area of surface of revolution formula .

4. Find the area of the surface of revolution by revolving about the x–axis, the curve given by $y = \frac{1}{3}x^3$ for $0 \le x \le 3$

5.: Find the area of the surface of revolution by revolving about the y–axis, the curve given by $y = \frac{1}{3}x^3$ for $0 \le x \le 3$.

Answers: 2. 10 units; **4.** ≈ 258.9 sq. units.. 5. ≈ 132.6 sq. units

CHAPTER 15

Lesson 51: L'Hôpital's Rule: Limits of Indeterminate Forms
(Differentiate and find the limit)
Lesson 52: Improper Integrals: Infinite limits; discontinuities
(Integrate and find the limit)

Lesson 51
L'Hôpital's Rule
Limits of Indeterminate Forms

$$\frac{0}{0},\ \frac{\infty}{\infty},\ 0 \bullet \infty,\ \infty - \infty,\ 0^0,\ \infty^0,\ 1^\infty$$

If $\lim_{x \to a} f(x) = 0$, and $\lim_{x \to a} g(x) = 0$, then we say that the function $\frac{f(x)}{g(x)}$ has the indeterminate form $\frac{0}{0}$. Similarly, if $\lim_{x \to a} f(x) = +\infty$, and $\lim_{x \to a} g(x) = +\infty$, then the function $\frac{f(x)}{g(x)}$ has the indeterminate form $\frac{\infty}{\infty}$.

L'Hôpital's Rule is a rule for evaluating the limit of an **indeterminate quotient** of functions of form $\frac{0}{0}$ or $\frac{\infty}{\infty}$. In addition, L'Hôpital's rule can be applied to forms such as $0 \bullet \infty$, $\infty - \infty$, 0^0, ∞^0, and 1^∞, but each of these forms must be changed to one of the quotient forms, $\frac{0}{0}$ or $\frac{\infty}{\infty}$ before applying L'Hôpital's rule. With the exponential forms 0^0, ∞^0, and 1^∞, we first take natural logarithms of the functions, change to product form and then to quotient form. In this case of taking the natural logarithm before finding the limit, if the limit of the logarithm function is L, then the limit of the original function is e^L.

L'Hôpital's rule is a generalized method for handling indeterminate forms.

L'Hôpital's Rule

Let lim denote any of the following symbols:

$$\lim_{x \to a},\ \lim_{x \to a^+},\ \lim_{x \to a^-},\ \lim_{x \to +\infty},\ \lim_{x \to -\infty}$$

If $f(x)$ and $g(x)$ are two functions such that
Case 1: $\lim f(x) = 0$, $\lim g(x) = 0$, and $g'(x) \neq 0$ **or**
Case 2: $\lim f(x) = \infty$, $\lim g(x) = \infty$, and $g'(x) \neq 0$,

then $\lim \frac{f(x)}{g(x)} = \lim \frac{f'(x)}{g'(x)}$, if $\lim \frac{f'(x)}{g'(x)}$ exists (That is, a real number), or it is infinite ($+\infty$ or $-\infty$).

In words, in attempting to find the limit of a function, if we obtain any of the above indeterminate forms, we can **separately differentiate** the numerator, and **separately differentiate** the denominator and then find the limit. If we again obtain an indeterminate form, we can differentiate again and then evaluate; and we may repeat the process several times.

Note: Do **not** use the differentiation quotient rule for the above differentiation.

Warning: Misapplication of L'Hôpital's **rule**

Before applying L'Hôpital's rule, first determine if it is applicable by checking the conditions in Case 1 or Case 2 above. If L'Hôpital's rule is not applicable and it is applied, we may obtain a limit which is not the limit of the given function. For example, suppose we are asked to find $\lim\limits_{x \to 0} \dfrac{\cos x}{x}$. Let us check to see if the rule is applicable. Let $f(x) = \cos x$, and let $g(x) = x$

Then $\lim\limits_{x \to 0} \cos x = 1$; $\lim\limits_{x \to 0} x = 0$; and therefore $\lim\limits_{x \to 0} \dfrac{\cos x}{x}$ is **not** of indeterminate form and L'Hôpital's rule is **not** applicable.

In fact , $\lim\limits_{x \to 0} \dfrac{\cos x}{x}$ is undefined . ($\lim\limits_{x \to 0^+} \dfrac{\cos x}{x} = \infty$; $\lim\limits_{x \to 0^-} \dfrac{\cos x}{x} = -\infty$)

Suppose, we apply L'Hôpital's rule to $\lim\limits_{x \to 0} \dfrac{\cos x}{x}$, then we obtain

$$\lim\limits_{x \to 0} \dfrac{\cos x}{x} = \lim\limits_{x \to 0} \dfrac{\dfrac{d}{dx}(\cos x)}{\dfrac{d}{dx}(x)} = \lim\limits_{x \to 0} \dfrac{-\sin x}{1} = 0 \text{, which is } \textbf{not} \text{ the limit because}$$

$\lim\limits_{x \to 0} \dfrac{\cos x}{x}$ is undefined . ($\lim\limits_{x \to 0^+} \dfrac{\cos x}{x} = \infty$; $\lim\limits_{x \to 0^-} \dfrac{\cos x}{x} = -\infty$)

Procedure for applying L'Hôpital's **rule**

Step 1: Determine if $\lim \dfrac{f(x)}{g(x)}$ is of indeterminate form and $g'(x) \neq 0$. If these conditions are satisfied, go to Step 2; otherwise L'Hôpital's rule is **not** applicable, and therefore, try other techniques.

Step 2: Differentiate $f(x)$

Step 3: Differentiate $g(x)$

Step 4 Find $\lim \dfrac{f'(x)}{g'(x)}$, If this limit, exists (i.e., it is a real number); or it is infinity ($+\infty$, or $-\infty$), then

Step 5: $\lim \dfrac{f(x)}{g(x)} = \lim \dfrac{f'(x)}{g'(x)}$. (as in Step 4).

If we obtain an indeterminate form, we will differentiate again and then evaluate.

Now, let us apply the above steps to do some examples.

Examples on L'Hôpital's Rule

Indeterminate form $\frac{0}{0}$

Example 1 Find $\lim_{x \to 0} \frac{\sin x}{x}$ (In Lesson 3, we found this using the Pinching Theorem)

Step 1: Let $f(x) = \sin x$, and let $g(x) = x$. $\lim_{x \to 0} \sin x = 0$; $\lim_{x \to 0} x = 0$,

$\quad g'(x) = 1 \neq 0$

Therefore, $\lim_{x \to 0} \frac{\sin x}{x}$ is of indeterminate form $\frac{0}{0}$, and $g'(x) = 1 \neq 0$ and

L'Hôpital's rule is applicable.

Step 2: $f'(x) = \frac{d}{dx} \sin x = \cos x$; $g'(x) = \frac{d}{dx}(x) = 1$; and

$$\lim_{x \to 0} \frac{f'(x)}{g'(x)} = \lim_{x \to 0} \frac{\cos x}{1} = \frac{1}{1} = 1$$

Therefore , $\lim_{x \to 0} \frac{\sin x}{x} = 1$.

Example 2 Find $\lim_{x \to 0} \frac{1 - \cos x}{x}$

Step 1: Let $f(x) = 1 - \cos x$, and let $g(x) = x$.

$\quad \lim_{x \to 0} 1 - \cos x = 0$; $\lim_{x \to 0} x = 0$; $g'(x) = 1 \neq 0$.

Therefore, $\lim_{x \to 0} \frac{1 - \cos x}{x}$ is of indeterminate form $\frac{0}{0}$, and $g'(x) = 1 \neq 0$ and

L'Hôpital's rule is applicable.

Step 2: $f'(x) = \frac{d}{dx}(1 - \cos x) = \sin x$; $g'(x) = \frac{d}{dx}(x) = 1$.

$$\lim_{x \to 0} \frac{f'(x)}{g'(x)} = \lim_{x \to 0} \frac{\sin x}{1} = \frac{0}{1} = 0$$

Therefore , $\lim_{x \to 0} \frac{1 - \cos x}{x} = 0$

Indeterminate form $\frac{\infty}{\infty}$

Example 3 Find $\lim_{x \to +\infty} \frac{x^3 + 4}{3x^3 + x}$

Solution We will use two methods: First the usual method; followed by
L'Hôpital's Rule.

Method 1: Divide each term in the numerator and denominator by the highest power of x in the **denominator** and then consider large values of x. The highest power in the denominator is x^3

$$\lim_{x \to +\infty} \frac{\frac{x^3}{x^3} + \frac{4}{x^3}}{3\frac{x^3}{x^3} + \frac{x}{x^3}} = \lim_{x \to +\infty} \frac{1 + \frac{4}{x^3}}{3 + \frac{1}{x^2}} = \frac{1}{3} \quad \text{(For large values of } x, \frac{1}{x^2} \approx 0, \frac{4}{x^3} \approx 0)$$

$$\lim_{x \to +\infty} \frac{x^3 + 4}{3x^3 + x} = \frac{1}{3}.$$

Method 2 Using L'Hôpital's Rule

Step 1: Let $f(x) = x^3 + 4$, and let $g(x) = 3x^3 + x$.

$$\lim_{x \to +\infty} x^3 + 4 = +\infty \ , \ \lim_{x \to +\infty} 3x^3 + x = +\infty \ , \text{and } g'(x) = 9x^2 + 1 \neq 0$$

Therefore, $\lim\limits_{x \to +\infty} \dfrac{x^3 + 4}{3x^3 + x}$ is of indeterminate form $\dfrac{\infty}{\infty}$, and $g'(x) = 9x^2 + 1 \neq 0$

and L'Hôpital's rule is therefore applicable.

Step 2: $f'(x) = \dfrac{d}{dx}(x^3 + 4) = 3x^2$; $g'(x) = \dfrac{d}{dx}(3x^3 + x) = 9x^2 + 1$.

$$\lim_{x \to +\infty} \frac{f'(x)}{g'(x)} = \lim_{x \to +\infty} \frac{3x^2}{9x^2 + 1} = \frac{\infty}{\infty}, \text{which is still of indeterminate form } \frac{\infty}{\infty}$$

Step 3: We apply L'Hôpital's rule again

$$\lim_{x \to +\infty} \frac{f''(x)}{g''(x)} = \lim_{x \to +\infty} \frac{6x}{18x} = \lim_{x \to +\infty} \frac{6}{18} = \frac{1}{3}.$$

Therefore , $\lim\limits_{x \to +\infty} \dfrac{x^3 + 4}{3x^3 + x} = \dfrac{1}{3}$ (same as before).

Other Indeterminate Forms (Five forms)

L'Hôpital's rule can be applied to forms such as $0 \bullet \infty$, $\infty - \infty$, 0^0, ∞^0, and 1^∞, but each of these forms must be changed to one of the quotient forms, $\frac{0}{0}$ or $\frac{\infty}{\infty}$ before applying L'Hôpital's rule.

Indeterminate form $\infty - \infty$ (change to form $\frac{0}{0}$ or $\frac{\infty}{\infty}$)

Example 4 Find $\lim\limits_{x \to 0} \left(\frac{1}{x} - \frac{1}{\sin x} \right)$

Step 1: Check for the applicability of L'Hôpital's **rule**

As $x \to 0$, $\lim \left(\frac{1}{x} - \frac{1}{\sin x} \right)$ is of the form $\infty - \infty$.

Therefore, L'Hôpital's rule is applicable.

Step 2: Change to form $\frac{0}{0}$ by combining the two fractions:

$$\lim\limits_{x \to 0} \left(\frac{1}{x} - \frac{1}{\sin x} \right)$$

$$= \lim\limits_{x \to 0} \left(\frac{\sin x - x}{x \sin x} \right) \quad (\text{form } \frac{0}{0} \text{ <-- say }, x = 0, \sin 0 = 0)$$

Step 3: Let $f(x) = \sin x - x$, and let $g(x) = x \sin x$..

Then $f'(x) = \cos x - 1$; $g'(x) = x \cos x + \sin x$

$$\lim\limits_{x \to 0} \frac{f'(x)}{g'(x)} = \lim\limits_{x \to 0} \frac{\cos x - 1}{x \cos x + \sin x} = \frac{0}{0} \ (= \frac{1-1}{0(1)+0})$$

We apply L'Hôpital's rule again.

Step 4: Now, $f(x)$ for this step is $f'(x) = \cos x - 1$, and

$g(x)$ for this step is $g'(x) = x \cos x + \sin x$; and we differentiate again

Then $f''(x) = -\sin x$; $g''(x) = -x \sin x + \cos x + \cos x$

$$g''(x) = -x \sin x + 2 \cos x$$

Step 5: $\lim\limits_{x \to 0} \frac{f''(x)}{g''(x)} = \lim\limits_{x \to 0} \frac{-\sin x}{-x \sin x + 2 \cos x} = 0$ (Note: $\frac{0}{0+2} = \frac{0}{2} = 0$)

Step 6: $\lim\limits_{x \to 0} \left(\frac{1}{x} - \frac{1}{\sin x} \right) = 0$.

Indeterminate form $(0 \bullet \infty)$ (change to form $\frac{\infty}{\infty}$ or $\frac{0}{0}$)

Example 5 Find $\lim\limits_{x \to 0^+} x \ln x$

Step 1: Check for the applicability of L'Hôpital's rule.

As $x \to 0^+$, $\lim x \ln x$ is of form $0 \bullet \infty$. (see the graph of $\ln x$)

Step 2: Change to form $\frac{\infty}{\infty}$ as follows:

$$\lim\limits_{x \to 0^+} x \ln x = \lim\limits_{x \to 0^+} \frac{\ln x}{\frac{1}{x}} \ (\text{which is now of form } \frac{\infty}{\infty} \text{ as } x \to 0^+).$$

Step 3: Now, $f(x) = \ln x$, $g(x) = \dfrac{1}{x}$

$$f'(x) = \frac{1}{x}, \ g'(x) = -\frac{1}{x^2}, \ \frac{f'(x)}{g'(x)} = -x$$

Step 4: $\displaystyle\lim_{x \to 0^+} x \ln x = \lim_{x \to 0^+} \frac{f'(x)}{g'(x)} = \lim_{x \to 0^+} (-x) = 0.$

Indeterminate form ∞^0

Example 6 Find $\displaystyle\lim_{x \to +\infty} x^{e^{-x}}$

Step 1: Check for the applicability of L'Hôpital's **rule.**

As $x \to +\infty$, $\lim x^{e^{-x}} = \lim x^{\frac{1}{e^x}} = \infty^0$ (see Appendix: graph of e^{-x})

(since when x is very large, $\dfrac{1}{e^x} = 0$)

Therefore, $\displaystyle\lim_{x \to +\infty} x^{e^{-x}}$ is of type ∞^0, and L'Hôpital's rule is applicable. However,

Step 2: Change the limit form ∞^0 to form $\dfrac{0}{0}$ or $\dfrac{\infty}{\infty}$

Let $y = \displaystyle\lim_{x \to +\infty} x^{e^{-x}}$, and take natural logs of both sides of this equation

Then $\ln y = \ln \displaystyle\lim_{x \to +\infty} x^{e^{-x}}$

$\qquad = \displaystyle\lim_{x \to +\infty} \ln x^{e^{-x}}$ (since the logarithmic and exponential

$\qquad\qquad$ functions are continuous, $f[\lim(g(x)] = \lim f[g(x)]$

$\qquad = \displaystyle\lim_{x \to +\infty} e^{-x} \ln x$ $\qquad (\log M^p = p \log M)$

$\ln y = \displaystyle\lim_{x \to +\infty} \frac{\ln x}{e^x}$ (which is of indeterminate form $\dfrac{\infty}{\infty}$)

Step 3: Let $f(x) = \ln x$, and let $g(x) = e^x$. Then differentiating.:

$$f'(x) = \frac{1}{x}, g'(x) = e^x \text{ and } \frac{f'(x)}{g'(x)} = \frac{\frac{1}{x}}{e^x}$$

Step 4: $\displaystyle\lim_{x \to +\infty} x^{e^{-x}} = \lim_{x \to +\infty} \frac{f'(x)}{g'(x)} = \frac{\frac{1}{x}}{e^x} = 0$ $\quad (\frac{\frac{1}{x}}{e^x} = \frac{1}{xe^x} = \frac{1}{\infty} = 0)$

Step 5: $\ln y = 0$ (same as $\log_e y = 0$)

Take antilogs or evoke the equivalent exponential definition to obtain $y = e^0 = 1$ (Note that we took logarithms in Step 2 and we undo it now)

Step 6: $\displaystyle\lim_{x \to +\infty} x^{e^{-x}} = 1.$ $\qquad$ (From Step 2, $y = \displaystyle\lim_{x \to +\infty} x^{e^{-x}}$).

Example 7 Find $\lim\limits_{x \to +\infty} x^{\frac{1}{x}}$

Step 1: Check for the applicability of L'Hôpital's rule.

As $x \to +\infty$, $\lim x^{\frac{1}{x}} \to \infty^0$

$(x \to +\infty$ and $\frac{1}{x} \to 0)$

Therefore, $\lim\limits_{x \to +\infty} x^{\frac{1}{x}}$ is of form ∞^0, and L'Hôpital's rule is applicable. However,

Step 2: Change this limit form ∞^0 to form $\frac{0}{0}$ or $\frac{\infty}{\infty}$.

Let $y = \lim\limits_{x \to +\infty} x^{\frac{1}{x}}$, and take natural logs of both sides of this equation. Then

$\ln y = \ln\left(\lim\limits_{x \to +\infty} x^{\frac{1}{x}} \right)$

$\ln y = \lim\limits_{x \to +\infty} \left(\ln x^{\frac{1}{x}} \right)$ $(f[\lim(g(x)] = \lim f[g(x)])$

$\ln y = \lim\limits_{x \to +\infty} \left(\frac{1}{x} \ln x \right)$ (Writing the exponent as a factor)

$\ln y = \lim\limits_{x \to +\infty} \left(\frac{\ln x}{x} \right)$ <---of form $\frac{\infty}{\infty}$)

Step 3: Let $f(x) = \ln x$; $g(x) = x$

     $f'(x) = \frac{1}{x}$; $g(x) = 1$

Step 4: $\dfrac{f'(x)}{g'(x)} = \dfrac{\frac{1}{x}}{1} = \dfrac{1}{x}$

    $\ln y = \lim\limits_{x \to +\infty} \frac{1}{x} = 0$ $(\ln y = \lim\limits_{x \to +\infty} \frac{f'(x)}{g'(x)} = \frac{1}{x})$

    $\ln y = 0$ (same as $\log_e y = 0$)

    $y = e^0 = 1$ (equivalent exponential definition; If $\log_e y = 0$, then $e^0 = y$)

Step 5: $\lim\limits_{x \to +\infty} x^{\frac{1}{x}} = 1.$ (From Step 3, $y = \lim\limits_{x \to +\infty} x^{\frac{1}{x}}$).

Indeterminate form 1^{∞}

Example 8 Find $\lim\limits_{x \to 0} (ax + 1)^{\frac{1}{x}}$

Step 1: Check for the applicability of L'Hôpital's rule.

As $x \to 0$, $\lim (ax + 1)^{\frac{1}{x}} \to 1^{\infty}$

$(ax + 1 \to 1$ and $\frac{1}{x} \to \infty)$

Therefore, $\lim\limits_{x \to 0} (ax + 1)^{\frac{1}{x}}$ is of type 1^{∞}, and L'Hôpital's rule is applicable. However,

Step 2: Change this limit form 1^{∞} to form $\frac{0}{0}$ or $\frac{\infty}{\infty}$.

Let $y = \lim\limits_{x \to 0} (ax + 1)^{\frac{1}{x}}$, and take natural logs of both sides of this equation. Then

$\ln y = \ln \left(\lim\limits_{x \to 0} (ax + 1)^{\frac{1}{x}} \right)$

$\ln y = \lim\limits_{x \to 0} \left(\ln (ax + 1)^{\frac{1}{x}} \right)$ $\quad$ $(f[\lim(g(x)] = \lim f[g(x)])$

$\ln y = \lim\limits_{x \to 0} \left(\frac{1}{x} \ln (ax + 1) \right)$

$\ln y = \lim\limits_{x \to 0} \frac{\ln (ax + 1)}{x}$

Step 3: Let $f(x) = \ln (ax + 1);$ $\quad g(x) = x$

$f'(x) = \frac{a}{ax + 1};$ $\quad g(x) = 1$

Step 4: $\frac{f'(x)}{g'(x)} = \frac{\frac{a}{ax + 1}}{1} = \frac{a}{ax + 1}$

$\ln y = \lim\limits_{x \to 0} \frac{\ln (ax + 1)}{x}$

$= \lim\limits_{x \to 0} \frac{a}{ax + 1}$ $\quad\quad$ $(\lim\limits_{x \to 0} \frac{f'(x)}{g'(x)})$

$\ln y = a$ $\quad\quad$ (same as $\log_e y = a$)

$y = e^a$ $\quad$ (equivalent exponential definition)

Step 5: $\lim\limits_{x \to 0} (ax + 1)^{\frac{1}{x}} = e^a$. $\quad$ (From Step 3, $y = \lim\limits_{x \to 0} (ax + 1)^{\frac{1}{x}}$)

Indeterminate form 0^0

Example 9 Find $\lim\limits_{x \to 0^+} x^{\sin x}$

Step 1: Check for the applicability of L'Hôpital's rule

As $x \to 0^+$, $\lim x^{\sin x} \to 0^0$

$(x \to 0$ and $\sin x \to 0)$

Therefore, $\lim\limits_{x \to 0^+} x^{\sin x}$ is of indeterminate form 0^0, and L'Hôpital's rule is applicable. However,

Step 2: Change this limit form 0^0 to form $\frac{0}{0}$ or $\frac{\infty}{\infty}$.

Let $y = \lim\limits_{x \to 0^+} x^{\sin x}$, and take natural logs of both sides of this equation. Then

$\ln y = \ln\left(\lim\limits_{x \to 0^+} x^{\sin x} \right)$

$\ln y = \lim\limits_{x \to 0^+} \left(\ln x^{\sin x} \right)$ (Writing the exponent as a factor)

$\quad = \lim\limits_{x \to 0^+} (\sin x \ln x)$

$\quad = \lim\limits_{x \to 0^+} \left(\dfrac{\ln x}{\left(\frac{1}{\sin x}\right)} \right)$

$\quad = \lim\limits_{x \to 0^+} \left(\dfrac{\ln x}{\csc x} \right)$

Step 3: Let $f(x) = \ln x$, and let $g(x) = \csc x$. Then

$f'(x) = \dfrac{1}{x}$; $g'(x) = -\csc x \cot x$

$\ln y = \lim\limits_{x \to 0^+} \dfrac{\frac{1}{x}}{-\csc x \cot x}$ $\left(\lim\limits_{x \to 0^+} \dfrac{f'(x)}{g'(x)} \right)$

$\ln y = \lim\limits_{x \to 0^+} \dfrac{\sin^2 x}{-x \cos x} \to \dfrac{0}{0}$ $\left(\dfrac{\frac{1}{x}}{-\frac{1}{\sin x} \frac{\cos x}{\sin x}} = -\dfrac{\sin^2 x}{x \cos x} \right)$

Step 4 We apply L'Hôpital's rule again

$\dfrac{d}{dx}\sin^2 x = \boxed{2\sin x \cos x}$, (Let $u = \sin x$; $\dfrac{du}{dx} = \cos x$; $y = u^2$; $\dfrac{dy}{du} = 2u = 2\sin x$)

$\dfrac{d}{dx}(-x\cos x) = -(-x\sin x + \cos x) = \boxed{x\sin x - \cos x}$

Step 5 $\ln y = \lim\limits_{x \to 0^+} \dfrac{2\sin x \cos x}{x\sin x - \cos x} = 0$

$\quad (x \to 0^+,\ 2\sin x \cos x \to (0)(1) = 0)$; $\ (x \to 0^+,\ x\sin x - \cos x \to (0)(0) - 1 = -1)$

$\quad \ln y = 0$ (same as $\log_e y = 0$)

$\quad y = e^0 = 1$ (equivalent exponential definition)

Step 6: Therefore $\lim\limits_{x \to 0^+} x^{\sin x} = 1$. (From Step 2, $y = \lim\limits_{x \to 0^+} x^{\sin x}$).

Lesson 51 Exercises

1. When do we use L'Hôpital's rule?

2. Is possible that we may apply L'Hôpital's rule when it is not applicable? Give an example.

3. What are the two main cases for the application of L'Hôpital's rule?

4. Find $\lim\limits_{x \to 0} \dfrac{\sin x}{x}$

5. Find $\lim\limits_{x \to 0} \dfrac{1 - \cos x}{x}$

6. Find $\lim\limits_{x \to +\infty} \dfrac{x^3 + 4}{3x^3 + x}$

7. Find $\lim\limits_{x \to 0} \dfrac{8^x - 2^x}{4x}$

8. To handle an indeterminate form $\infty - \infty$, we change to which form first?

9. Find $\lim\limits_{x \to 0} \left(\dfrac{1}{x} - \dfrac{1}{\sin x} \right)$

10. Find $\lim\limits_{x \to 0^+} x \ln x$

11. Find $\lim\limits_{x \to +\infty} x^{e^{-x}}$

12. Find $\lim\limits_{x \to +\infty} x^{\frac{1}{x}}$

13. Find $\lim\limits_{x \to 0} (ax + 1)^{\frac{1}{x}}$

14. Find $\lim\limits_{x \to 0^+} x^{\sin x}$

15. Show that $\lim\limits_{x \to 0} \dfrac{8^x - 2^x}{4x} = \dfrac{1}{2} \ln 2$

16. Find $\lim\limits_{x \to 0} \dfrac{3^x - 5^x}{2x}$

Answers: 4. 1 ; **5.** 0 ; **6.** $\dfrac{1}{3}$; **7.** $\dfrac{1}{2} \ln 2$ **9.** 0; **10.** 0; **11.** 1; **12.** 1; **13.** e^a; **14.** 1;

16. $\dfrac{1}{2} \ln \dfrac{3}{5}$.

Lesson 52
Improper Integrals

(Integrate and find the limit)

Proper Integral: The definite integral $\int_a^b f(x)dx$ is called a proper integral, where it is assumed that f is continuous on the closed interval $[a, b]$. Here, we assume that the interval of integration $[a, b]$ is finite and the integrand $f(x)$ is finite over this interval. We can extend the concept of the definite integral to three other types of integrals, namely **improper integrals**.

Type 1

Infinite Limits of Integration

Here, at least, one of the limits of integration is infinite and we cover three cases. We know how to integrate when the limits are finite, say, a and b. We also know how to handle limits at infinity (Lesson 4). Here, we use finite numbers a and b to temporarily hold places for $-\infty$ and $+\infty$, so that we can integrate as usual, followed by taking the limits at infinity.

In each of the types and cases covered below, the integral on the left is said to be convergent if the limit on the right exists. If the limit does not exist, the integral is said to be divergent. Note below that when we write a or b is approaching $\pm\infty$, we are using a and b as variables.

Case 1:
$$\int_a^{+\infty} f(x)dx = \lim_{b \to +\infty} \int_a^b f(x)dx$$

Memory Device:

The lower limit is a, and the upper limit is a number b approaching $+\infty$.

Case 2: $\int_{-\infty}^b f(x)dx = \lim_{a \to -\infty} \int_a^b f(x)dx$

Memory Device:

The lower limit is a number a approaching $-\infty$, and the upper limit is b.

Case 3 We have two integrals here

$$\int_{-\infty}^{+\infty} f(x)dx = \lim_{a \to -\infty} \int_a^0 f(x)dx + \lim_{b \to +\infty} \int_0^b f(x)dx$$
(From $-\infty$ to 0, and from 0 to $+\infty$)

Memory Device:

For the first integral: The lower limit is a number a approaching $-\infty$;. and the upper limit is 0.

For the second integral: The lower limit is 0 and the upper limit is a number b approaching $+\infty$.

Type 2

Discontinuous Integrand

Here, the integrand $f(x)$ has one or more points of discontinuity on the interval , $[a, b]$, of integration.

(Discontinuity at either end point or at an interior point. See also p.79)

Case 1: $f(x)$ is continuous on the interval half open on the left $(a, b]$

(that is, continuous on the interval from a to b, excluding a, but including b) **and** $\lim_{x \to a^+} f(x) = \pm\infty$. (That is, infinitely discontinuous at a, the lower end) Here, we define:

$$\int_a^b f(x)dx = \lim_{t \to a+} \int_t^b f(x)dx \qquad \text{(Review p.8-9)}$$

Memory Device:

The lower limit is a number t approaching a from the right, and the upper limit is b.

Case 2: $f(x)$ is continuous on the interval half open on the right $[a, b)$

(that is, continuous on the interval from a to b, including a but excluding b and $\lim_{x \to b^-} f(x) = \pm\infty$ (Infinitely discontinuous at b)

Here , we define: $\int_a^b f(x)dx = \lim_{t \to b^-} \int_a^t f(x)dx$

Memory Device:

The lower limit is a, and the upper limit is a number t approaching b from the left.

Case 3: $f(x)$ is continuous on the interval $[a, b]$ (except at c, where $a < c < b$,

and $\lim_{x \to c} |f(x)| = \pm\infty$. (that is unbounded at interior point c; same as infinitely discontinuous at c)

Here, we have two integrals.. Both integrals must converge for convergence.

$$\int_a^b f(x)dx = \lim_{t \to c-} \int_a^t f(x)dx + \lim_{t \to c+} \int_t^b f(x)dx, \text{ provided both limits exist.}$$

Memory Device:

(This figure is the author's symbol for infinite discontinuity at c.)

For the first integral: The lower limit is a, and the upper limit is a number t approaching c from the left.

For the second integral: The lower limit is a number t approaching c from the right, and the upper limit is b.

Type 3

Integrals of Type 1 and Type 2 Simultaneously

(Integrals with infinite limits and discontinuities in the integrand)

Here, we combine the methods of Type 1 and Type 2.

Examples

Type 1: Infinite Limits of Integration

Case 1:

Example 1: Find $\int_{1}^{+\infty} \frac{1}{x^2} dx$

$\int_{a}^{+\infty} f(x)dx = \lim_{b \to +\infty} \int_{a}^{b} f(x)dx$

$\int_{1}^{+\infty} \frac{1}{x^2} dx = \lim_{b \to +\infty} \int_{1}^{b} \frac{1}{x^2} dx$

$= \lim_{b \to +\infty} \left[-\frac{x^{-1}}{1} \right]_{1}^{b}$

$= \lim_{b \to +\infty} \left[\left(-\frac{1}{b}\right) - \left(-\frac{1}{1}\right) \right]$

$= \lim_{b \to +\infty} \left[-\frac{1}{b} + 1 \right]$

$= [0 + 1]$

$= 1$

Case 2:

Example 2: Find $\int_{-\infty}^{0} e^x dx$

$\int_{-\infty}^{b} f(x)dx = \lim_{a \to -\infty} \int_{a}^{b} f(x)dx$

$\int_{-\infty}^{0} e^x dx = \lim_{a \to -\infty} \int_{a}^{0} e^x dx$

$= \lim_{a \to -\infty} \left[e^x \right]_{a}^{0}$

$= \lim_{a \to -\infty} \left[\left(e^0\right) - \left(e^a\right) \right]$

$= \lim_{a \to -\infty} \left[1 - e^a \right]$

$= 1 - 0$

$= 1$

Note: See and memorize the graphs of e^x and e^{-x} and apply them.

Case 3: $\int_{-\infty}^{+\infty} f(x)dx = \int_{-\infty}^{0} f(x)dx + \int_{0}^{+\infty} f(x)dx$ ($-\infty$ to 0, 0 to $+\infty$)

Example 3: Find $\int_{-\infty}^{+\infty} \frac{1}{1+4x^2} dx = \int_{-\infty}^{0} \frac{1}{1+4x^2} dx + \int_{0}^{+\infty} \frac{1}{1+4x^2} dx$

$\int \frac{1}{1+4x^2} dx = \frac{1}{2} \text{Tan}^{-1} 2x + C$ (Using trigonometric substitution. See p.288)

$\int_{-\infty}^{+\infty} \frac{1}{1+4x^2} dx = \lim_{a \to -\infty} \int_{a}^{0} \frac{1}{1+4x^2} dx + \lim_{b \to +\infty} \int_{0}^{b} \frac{1}{1+4x^2} dx$

$= \frac{1}{2} \lim_{a \to -\infty} \left[\text{Tan}^{-1}(2x) \right]_{a}^{0} + \frac{1}{2} \lim_{b \to +\infty} \left[\text{Tan}^{-1}(2x) \right]_{0}^{b}$

$= \frac{1}{2} \lim_{a \to -\infty} \left[\left(\text{Tan}^{-1}(0)\right) - \left(\text{Tan}^{-1}(2a)\right) \right] + \frac{1}{2} \lim_{b \to +\infty} \left[\left(\text{Tan}^{-1}(2b)\right) - \left(\text{Tan}^{-1}(0)\right) \right]$

$= \frac{1}{2} \left[0 - (-\frac{\pi}{2}) \right] + \frac{1}{2} \left[\frac{\pi}{2} - 0 \right]$ (see graph)

$= \frac{1}{2} \left[\frac{\pi}{2} \right] + \frac{1}{2} \left[\frac{\pi}{2} \right]$ (memorize this graph)

$= \frac{\pi}{4} + \frac{\pi}{4} = \frac{\pi}{2}$

Therefore, $\int_{-\infty}^{+\infty} \frac{1}{1+4x^2} dx = \frac{\pi}{2}$

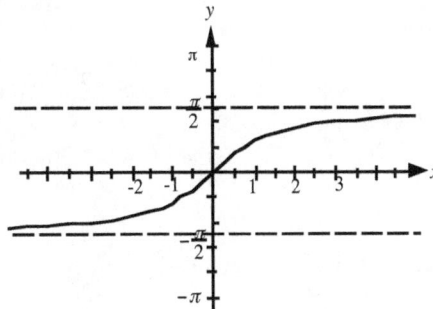

Figure: $y = \text{Tan}^{-1} x$

Examples **Type 2**

Case 1

Example 4: Find $\int_0^3 \frac{1}{\sqrt{x}}\,dx$

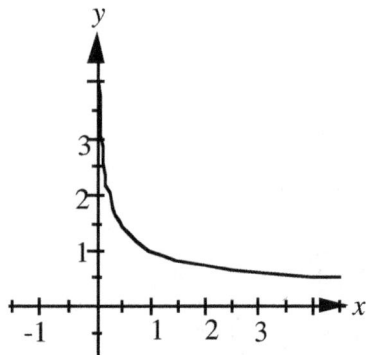

Figure: $f(x) = \frac{1}{\sqrt{x}}$

Step 1:

Check # 1: The integrand $\frac{1}{\sqrt{x}}$ is continuous on $(0, 3]$, that is, it is continuous on the interval from 0 to 3, excluding 0 but including 3 and

Check # 2 $\lim_{x \to 0^+} \frac{1}{\sqrt{x}} = +\infty$

Therefore, conditions for Case 1 have been satisfied, and we proceed.

Step 2: $\int_0^3 \frac{1}{\sqrt{x}}\,dx = \lim_{t \to 0^+} \int_t^3 \frac{1}{\sqrt{x}}\,dx$

$= \lim_{t \to 0^+} \int_t^3 x^{-\frac{1}{2}}\,dx$

$= \lim_{t \to 0^+} \left[\frac{x^{-\frac{1}{2}+1}}{-\frac{1}{2}+1} \right]_t^3$

$= 2 \lim_{t \to 0^+} \left[x^{\frac{1}{2}} \right]_t^3$

$= 2 \lim_{t \to 0^+} \left[\left(3^{\frac{1}{2}}\right) - \left(t^{\frac{1}{2}}\right) \right]$

$= 2\left[\left(\sqrt{3}\right) - (0) \right]$

$= 2\sqrt{3}$

Note above: Square root of a very very small positive number is nearly zero.

Case 2:

Example 5: Find $\int_0^2 \frac{1}{\sqrt{2-x}}\,dx$

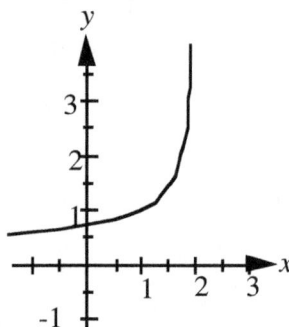

Step 1:

Check # 1: The integrand $\frac{1}{\sqrt{2-x}}$ is continuous on $[0, 2)$. that is, it is continuous on the interval from 0 to 2, including 0 but excluding 2.

Check # 2 $\lim_{x \to 2^-} \frac{1}{\sqrt{2-x}} = +\infty$

Therefore, conditions for Case 2 have been satisfied, and we proceed. Here , we define

Step 2:

$\int_0^2 \frac{1}{\sqrt{2-x}}\,dx = \lim_{t \to 2^-} \int_0^2 \frac{1}{\sqrt{2-x}}\,dx$

$= \lim_{t \to 2^-} \int_0^2 \frac{1}{\sqrt{2-x}}\,dx$ (Integ. by u-sub.)

$= \lim_{t \to 2^-} \left[-2(2-x)^{\frac{1}{2}} \right]_0^t$ $(u = 2-x)$

$= -2 \lim_{t \to 2^-} \left[(2-x)^{\frac{1}{2}} \right]_0^t$ $(dx = -du)$

$= -2 \lim_{t \to 2^-} \left[(2-x)^{\frac{1}{2}} \right]_0^t$

$= -2 \lim_{t \to 2^-} \left[(2-t)^{\frac{1}{2}} - (2-0)^{\frac{1}{2}} \right]$

$= -2 \left[0 - (2)^{\frac{1}{2}} \right]$

$= -2 \left[-\sqrt{2} \right]$

$= 2\sqrt{2}$

Case 3:

Example 6: Find $\int_0^3 \frac{1}{x-2}\, dx$

Step 1: **Check #1** $\frac{1}{x-2}$ is continuous on $[0, 3]$, but discontinuous at 2,
where $0 < 2 < 3$, and .

Check # 2 $\lim\limits_{x \to 2} |f(x)| = \pm\infty$, that is unbounded at an interior point 2. (same

as infinitely discontinuous at 2)

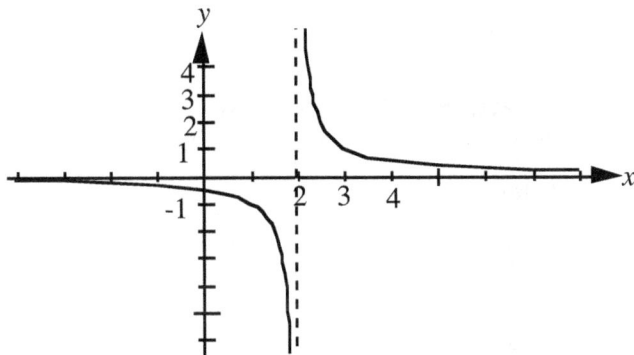

Step 2: $\int_a^b f(x)dx = \int_a^c f(x)dx + \int_c^b f(x)dx$

$$= \lim_{t \to c-} \int_a^t f(x)dx + \lim_{t \to c+} \int_t^b f(x)dx$$

$\int_0^3 \frac{1}{x-2}\, dx = \lim\limits_{t \to 2-} \int_0^t \frac{1}{x-2}\, dx + \lim\limits_{t \to 2+} \int_t^3 \frac{1}{x-2}\, dx$, (if both limits exist.)

$\int_0^3 \frac{1}{x-2}\, dx = \lim\limits_{t \to 2-} \left[\ln|x - 2| \right]_0^t + \lim\limits_{t \to 2+} \left[\ln|x - 2| \right]_t^3$

$$= \lim_{t \to 2-} \left[\left(\ln|t - 2| \right) - \left(\ln|0 - 2| \right) \right] + \lim_{t \to 2+} \left[\left(\ln|3 - 2| \right) - \left(\ln|t - 2| \right) \right]$$

For the first limit , $\lim\limits_{t \to 2-} \left[\left(\ln|t - 2| \right) - \left(\ln|0 - 2| \right) \right] = -\infty - \ln 2 = -\infty$ and the limit

does not exist and therefore, $\lim\limits_{t \to 2-} \int_0^t \frac{1}{x-2}\, dx$ diverges; and it is not necessary to
evaluate the second integral, since for convergence both integrals on the right
side of the equation in Step 2 must exist.

Therefore, the $\int_0^3 \frac{1}{x-2}\, dx$ does not exist.

Example
Type 3

Integrals with infinite limits and integrand discontinuities

Here, we separate this mixed type into Type 1 and Type 2 and then apply the methods of Type 1 and Type 2.

Example $\int_{-\infty}^{3} \frac{1}{x-2} dx$, which is undefined

Lesson 52 Exercises

1. What is the difference between a proper integral and an improper integral?

2. Find $\int_{1}^{+\infty} \frac{1}{x^2} dx$

6. Find $\int_{0}^{2} \frac{1}{\sqrt{2-x}} dx$

3. $\int_{-\infty}^{0} e^x dx$;

7. $\int_{0}^{3} \frac{1}{x-2} dx$

4. $\int_{-\infty}^{+\infty} \frac{1}{1+4x^2} dx$

8. $\int_{-\infty}^{2} \frac{1}{(4-t)^2} dt$

5. Find $\int_{0}^{3} \frac{1}{\sqrt{x}} dx$

9. $\int_{0}^{+\infty} e^{-t} dt$; **10.** $\int_{0}^{1} \frac{1}{\sqrt{1-r}} dr$

Answers:

2. 1; **3.** 1; **4.** $\frac{\pi}{2}$; **5.** $= 2\sqrt{3}$; **6.** $= 2\sqrt{2}$. **7.** Does not exist; **8.** $\frac{1}{2}$; **9.** 1 ; **10.** 2

CHAPTER 16

Hyperbolic Functions I

Lesson 53: **Introduction to Hyperbolic Functions**
Lesson 54: **Differentiation of Hyperbolic Functions**
Lesson 55: **Differentiation of Inverse Hyperbolic Functions**

Lesson 53

From Trigonometric Functions to Hyperbolic Functions

There are six hyperbolic functions, and they are analogous to the six
trigonometric functions, The hyperbolic functions are related to the hyperbola
in a similar way that the trigonometric functions (circular functions) are
related to the circle. Hyperbolic functions occur in the solutions of some linear
differential equations.

To name a hyperbolic function, we attach the letter " h " to the end of the name
of an analogous trigonometric function. For example, the trigonometric
functional name "sin" becomes the hyperbolic name "sinh". However we
define the basic hyperbolic functions in terms of the exponential functions.
Apply your knowledge of trigonometric functions to make a smooth
transition from trigonometric functions to hyperbolic functions. Note below
for example that in $\sinh x$, x is a number and corresponds to angular measure
in radians.

Basic definitions

1. $\sinh x = \dfrac{e^x - e^{-x}}{2}$

2. $y = \cosh x = \dfrac{e^x + e^{-x}}{2}$

3. $\tanh x = \dfrac{\sinh x}{\cosh x} = \dfrac{e^x - e^{-x}}{e^x + e^{-x}}$

4. $\coth x = \dfrac{1}{\tanh x} = \dfrac{\cosh x}{\sinh x} = \dfrac{e^x + e^{-x}}{e^x - e^{-x}}$

5. $\operatorname{sech} x = \dfrac{1}{\cosh x} = \dfrac{2}{e^x + e^{-x}}$

6. $\operatorname{csch} x = \dfrac{1}{\sinh x} = \dfrac{2}{e^x - e^{-x}}$

7. $\cosh x + \sinh x = e^x$;

8. $\cosh x - \sinh x = e^{-x}$

By applying Osborn's rule which says that a trigonometric identity can be
converted to an analogous hyperbolic identify as follows:
Step 1: Expand the identity in terms of integral powers of sines and cosines.
Step 2: Change the sign of every term that contains a product of two
hyperbolic sines. That is, the sign of any term that contains $\sinh^2 x$ or
$\sinh x \sinh y$ is changed; the signs of the other terms remain unchanged.

Comparison of Trigonometric Identities and Hyperbolic Identities

Trigonometric Identities	Hyperbolic Identities
Reciprocal Identities	
$\sec x = \dfrac{1}{\cos x}$	$\operatorname{sech} x = \dfrac{1}{\cosh x}$ ⎤
$\csc x = \dfrac{1}{\sin x}$	$\operatorname{csch} x = \dfrac{1}{\sinh x}$ ⎬ no sign changes
$\cot x = \dfrac{1}{\tan x}$	$\coth x = \dfrac{1}{\tanh x}$ ⎦
Ratio Identities	
$\tan x = \dfrac{\sin x}{\cos x}$	$\tanh x = \dfrac{\sinh x}{\cosh x}$ ⎤ no sign changes
$\cot x = \dfrac{\cos x}{\sin x}$	$\coth x = \dfrac{\cosh x}{\sinh x}$ ⎦

Trigonometric Identities Pythagorean Identities	**Hyperbolic Identities**
$\sin^2 x + \cos^2 x = 1$	$\cosh^2 x - \sinh^2 x = 1$ (sign change for $\sinh^2 x$)
$1 + \tan^2 x = \sec^2 x$	$1 - \tanh^2 x = \mathrm{sech}^2\, x$ (change for $\tanh^2 x = \frac{\sinh^2 x}{\cosh^2 x}$)
$1 + \cot^2 x = \csc^2 x$	$1 - \coth^2 x = -\mathrm{csch}^2\, x$ (changes for $\coth^2 x$ & $\mathrm{csch}^2\, x$) or $\coth^2 x - \mathrm{csch}^2\, x = 1$ or $\coth^2 x - 1 = \mathrm{csch}^2\, x$ (rearranging)

Trigonometric Identities	**Hyperbolic Identities**
1. $\sin 2x = 2\sin x \cos x$	**1.** $\sinh 2x = 2\sinh x \cosh x$ (no sign change)
2. $\cos 2x = \cos^2 x - \sin^2 x$	**2.** $\cosh 2x = \cosh^2 x + \sinh^2 x$ (change for $\sinh^2 x$)
3. $\cos 2x = 2\cos^2 x - 1$	**3.** $\cosh 2x = 2\cosh^2 x - 1$ (no sign change)
4. $\cos 2x = 1 - 2\sin^2 x$	**4.** $\cosh 2x = 2\sinh^2 x + 1$ (change for $\sinh^2 x$)
5. $\cos(-x) = \cos x$	**5.** $\cosh(-x) = \cosh x$ ⎫ (no sign change)
6. $\sin(-x) = -\sin x$	**6.** $\sinh(-x) = -\sinh x$ ⎬
7. $\tan 2x = \dfrac{2\tan x}{1 - \tan^2 x}$	**7.** $\tanh 2x = \dfrac{2\tanh x}{1 + \tanh^2 x}$ (sign change)

Trigonometric Identities	**Hyperbolic Identities** (change in first two)
$\cos(x - y) = \cos x \cos y + \sin x \sin y$	$\cosh(x - y) = \cosh x \cosh y - \underline{\sinh x \sin y}$
$\cos(x + y) = \cos x \cos y - \sin x \sin y$	$\cosh(x + y) = \cosh x \cosh y + \underline{\sinh x \sin y}$
$\sin(x + y) = \sin x \cos y + \sin y \cos x$	$\sinh(x + y) = \sinh x \cosh y + \sinh y \cosh x$
$\sin(x - y) = \sin x \cos y - \sin y \cos x$	$\sinh(x - y) = \sinh x \cosh y - \sinh y \cosh x$

Trigonometric Identities	**Hyperbolic Identities**
1. $\cos\frac{1}{2}x = \pm\sqrt{\dfrac{1 + \cos x}{2}}$	**1.** $\cosh\frac{1}{2}x = \sqrt{\dfrac{1 + \cosh x}{2}}$ no change
2. $\sin\frac{1}{2}x = \pm\sqrt{\dfrac{1 - \cos x}{2}}$	**2.** $\sinh\frac{1}{2}x = \pm\sqrt{\dfrac{\cosh x - 1}{2}}$ sign change
3. $\tan\frac{1}{2}x = \dfrac{1 - \cos x}{\sin x} = \dfrac{\sin x}{1 + \cos x}$	**3.** $\tanh\frac{1}{2}x = \dfrac{\cosh x - 1}{\sinh x} = \dfrac{\sinh x}{1 + \cosh x}$

$$\sin^2 \tfrac{1}{2}x = \frac{1 - \cos x}{2}$$

Note in **2:** Reason for sign change

$$-\sinh^2 \tfrac{1}{2}x = \frac{1 - \cosh x}{2}$$

$$\sinh^2 \tfrac{1}{2}x = \frac{\cosh x - 1}{2}$$

Drawing the Graphs of

$y = \sinh x$; $y = \cosh x$; and $y = \tanh x$

(A picture is worth a thousand words)

Method: By addition of y-coordinates

To draw the graph of $y = \sinh x$,

sketch the graphs of $y_1 = \frac{1}{2}e^x$ and

$y_2 = -\frac{1}{2}e^{-x}$. Then add the y-coordinates of important points, while keeping the x-coordinates unchanged, to obtain ordered pairs (x, y) which are then plotted and the points connected.

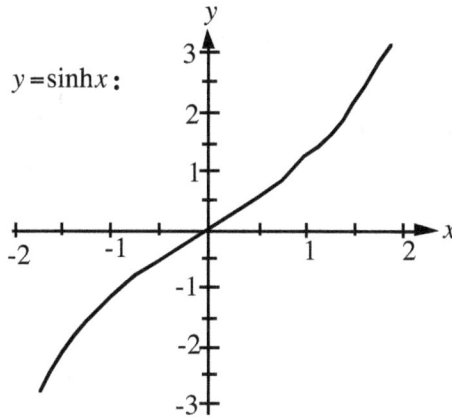

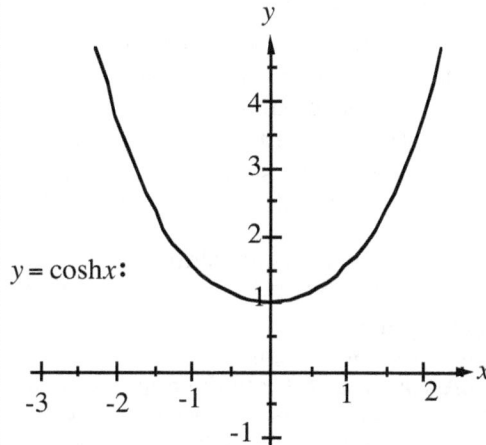

$y = \sinh x$:

Method: By addition of y-coordinates

To draw the graph of $y = \sinh x$,

sketch the graphs of $y_1 = \frac{1}{2}e^x$ and

$y_2 = \frac{1}{2}e^{-x}$. Then add the y-coordinates of important points, while keeping the x-coordinates unchanged, to obtain ordered pairs (x, y) which are then plotted and the points connected.

$y = \cosh x$:

The graph of $y = \tanh x$ can similarly be drawn by applying the definition of $y = \tanh x$

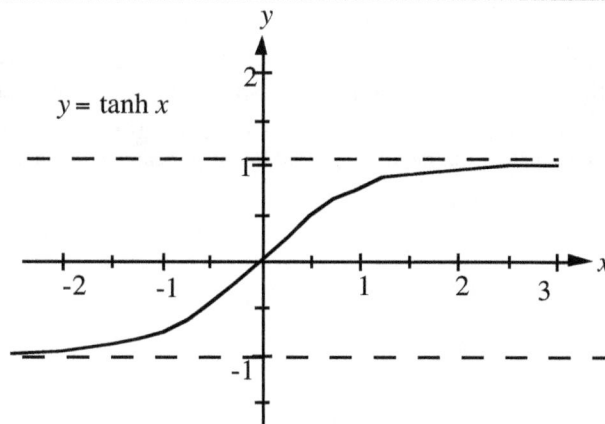

$y = \tanh x$

Lesson 54
Differentiation of Hyperbolic Functions

We present the hyperbolic functions and their corresponding derivatives.
We will derive some of the basic derivatives later. **Memorize** these basic formulas.

Table of hyperbolic **Functions and Corresponding Derivatives**

Function: $y = f(x)$	**Derivative**: $\frac{dy}{dx} = f'(x)$
1. $y = \sinh x$	1. $\cosh x$
2. $y = \cosh x$	2. $\sinh x$
3. $y = \tanh x$	3. $\operatorname{sech}^2 x$
4. $y = \coth x$	4. $-\operatorname{csch}^2 x$
5. $y = \operatorname{sech} x$	5. $-\operatorname{sech} x \tanh x$
6. $y = \operatorname{csch} x$	6. $-\operatorname{csch} x \coth x$

Memory Help: Note the similarities between the hyperbolic functions and
the trigonometric functions. (p.177).

1. Rule of signs: The signs for the derivatives of $\sinh x$, $\cosh x$, $\tanh x$ (on the
right side) are all **plus signs** whereas the signs of the
derivatives of the reciprocal identities, $\coth x$, $\operatorname{sech} x$, $\operatorname{csch} x$
(on the right side) are all **minus signs**. Some of the rules of
signs are different from those of trigonometric functions.

The naming of the derivatives are similar to the device for the corresponding
trigonometric functions (see p.177). Note the "h" in the spellings.

1. **tanh   sech square.**
2. **coth   minus cosech square**
3. **sech   minus sech tanh** (note the minus sign and the repetition of sech)
4 cosech minus **cosech cotanh** (Repetition of cosech is similar to that of sech x in **3**)
Also, for sinh, cosh tanh, sech, coth, and cosech, observe the relationships
between the terms of the following hyperbolic identities. Each derivative is
expressed in terms of the other function in each identity.

1. $\cosh^2 x - \sinh^2 x = 1$; **2.** $1 - \tanh^2 x = \operatorname{sech}^2 x$; **3.** $\coth^2 x - 1 = \operatorname{csch}^2 x$
(sinh goes with cosh; tanh goes with sech; and coth goes with csch.)

Say aloud **1-4** over and over and as you say them, look at the table. Take a sheet of paper and
try to write the above table, from memory, guided by this mnemonic device.

Examples on Differentiation of Hyperbolic Functions

Example 1 Given that $y = \sinh x$,

show that $\dfrac{d}{dx}(\sinh x) = \cosh x$.

Solution

$$y = \sinh x = \frac{e^x - e^{-x}}{2}$$

$$\frac{dy}{dx} = \frac{d}{dx}\left(\frac{e^x - e^{-x}}{2}\right)$$

$$= \frac{d}{dx}\left(\frac{e^x - e^{-x}}{2}\right)$$

$$= \frac{1}{2}\left(\frac{d}{dx}(e^x) - \frac{d}{dx}(e^{-x})\right)$$

$$= \frac{1}{2}\left(e^x + e^{-x}\right) \mathrel{<\!\!-\!-}(-e^{-x})$$

$$\frac{d}{dx}(\sinh x) = \cosh x.$$

(since by definition ,

$$\cosh x = \frac{e^x + e^{-x}}{2})$$

Example 2 Given that $y = \cosh x$,

show that $\dfrac{d}{dx}(\cosh x) = \sinh x$.

Solution $y = \cosh x = \dfrac{e^x + e^{-x}}{2}$

$$\frac{dy}{dx} = \frac{d}{dx}\left(\frac{e^x + e^{-x}}{2}\right)$$

$$= \frac{d}{dx}\left(\frac{e^x + e^{-x}}{2}\right)$$

$$= \frac{1}{2}\left(\frac{d}{dx}(e^x) + \frac{d}{dx}(e^{-x})\right)$$

$$= \frac{1}{2}\left(e^x - e^{-x}\right)$$

$$\frac{d}{dx}(\cosh x) = \sinh x$$

(since by definition ,

$$\sinh x = \frac{e^x - e^{-x}}{2})$$

Example 3 Given that $y = \tanh x$,

show that $\dfrac{d}{dx}(\tanh x) = \operatorname{sech}^2 x$.

Solution

$$y = \frac{\sinh x}{\cosh x} \quad (\tanh x = \frac{\sinh x}{\cosh x})$$

$$\frac{dy}{dx} =$$

$$= \frac{\cosh x \dfrac{d}{dx}[\sinh x] - \sinh x \dfrac{d}{dx}[\cosh x]}{\cosh^2 x}$$

(Using the quotient rule)

$$= \frac{\cosh x \cosh x - \sinh x(\sinh x)}{\cosh^2 x}$$

$$= \frac{\cosh^2 x - \sinh^2 x}{\cosh^2 x}$$

$$= \frac{1}{\cosh^2 x}$$

$$(\cosh^2 x - \sinh^2 x = 1)$$

$$= \operatorname{sech}^2 x \quad (\frac{1}{\cosh x} = \operatorname{sech} x)$$

$$\therefore \frac{d}{dx}(\tanh x) = \operatorname{sech}^2 x.$$

Example 4 Given that $y = \coth x$,

show that $\dfrac{d}{dx}(\coth x) = -\operatorname{csch}^2 x$.

Solution

$$y = \frac{\cosh x}{\sinh x} \quad (\coth x = \frac{\cosh x}{\sinh x})$$

$$\frac{dy}{dx} =$$

$$= \frac{\sinh x \dfrac{d}{dx}[\cosh x] - \cosh x \dfrac{d}{dx}[\sinh x]}{\sinh^2 x}$$

(Using the quotient rule)

$$= \frac{\sinh x(\sinh x) - \cosh x(\cosh x)}{\sinh^2 x}$$

$$= \frac{-[-\sinh^2 x + \cosh^2 x]}{\sinh^2 x}$$

$$= \frac{-[\cosh^2 x - \sinh^2 x]}{\sinh^2 x}$$

$$= -\frac{1}{\sinh^2 x} = -\left(\frac{1}{\sinh x}\right)^2$$

$$= -\operatorname{csch}^2 x \quad (\frac{1}{\sinh} = \operatorname{csch} x)$$

$$\frac{d}{dx}(\coth x) = -\operatorname{csch}^2 x.$$

Example 5 Given that $y = \operatorname{sech} x$,
show that $\dfrac{d}{dx}(\operatorname{sech} x) = -\operatorname{sech} x \tanh x$.
Solution (We apply the chain rule)

$$y = \frac{1}{\cosh x} \qquad (\operatorname{sech} x = \frac{1}{\cosh x})$$

$$y = (\cosh x)^{-1}$$

Let $u = \cosh x$

Then $y = u^{-1}$

$$\frac{dy}{du} = -u^{-2}$$

$$= -\frac{1}{u^2}$$

$$= -\frac{1}{\cosh^2 x}$$

$(u = \cosh x)$

$$\frac{du}{dx} = \sinh x \qquad (u = \cosh x)$$

$$\frac{dy}{dx} = \frac{dy}{du}\frac{du}{dx} \quad \text{(Chain rule)}$$

$$\frac{dy}{dx} = (-\frac{1}{\cosh^2 x})(\sinh x)$$

$$\frac{dy}{dx} = -\frac{\sinh x}{\cosh^2 x}$$

$$\frac{dy}{dx} = -\frac{\sinh x}{\cosh x} \bullet \frac{1}{\cosh x}$$

$$\frac{dy}{dx} = -\tanh x \bullet \operatorname{sech} x$$

$\therefore \dfrac{d}{dx}(\operatorname{sech} x) = -\operatorname{sech} x \tanh x.$

Example 6 Given that $y = \operatorname{csch} x$,
show that $\dfrac{d}{dx}(\operatorname{csch} x) = -\operatorname{csch} x \coth x$.
Solution (We apply the chain rule)

$$y = \frac{1}{\sinh} \qquad (\operatorname{csch} x = \frac{1}{\sinh x})$$

$$y = (\sinh x)^{-1}$$

Let $u = \sinh x$

Then $y = u^{-1}$

$$\frac{dy}{du} = -u^{-2}$$

$$= -\frac{1}{u^2}$$

$$= -\frac{1}{\sinh^2 x} \qquad (u = \sinh x)$$

$$\frac{du}{dx} = \cosh x \qquad (u = \sinh x)$$

$$\frac{dy}{dx} = \frac{dy}{du}\frac{du}{dx} \quad \text{(Chain rule)}$$

$$\frac{dy}{dx} = (-\frac{1}{\sinh^2 x})(\cosh x)$$

$$\frac{dy}{dx} = -\frac{\cosh x}{\sinh^2 x}$$

$$\frac{dy}{dx} = -\frac{\cosh x}{\sinh x} \bullet \frac{1}{\sinh x}$$

$$\frac{dy}{dx} = -\coth x \bullet \operatorname{csch} x$$

$\therefore \dfrac{d}{dx}(\operatorname{csch} x) = -\operatorname{csch} x \coth x.$

Example 7 If $y = -\frac{1}{3}\cosh^3 x$, find $\frac{dy}{dx}$.

Solution We will apply the chain rule. (since the given function is composite)

Let $u = \cosh x$.

Then $y = -\frac{1}{3}u^3$

$\frac{du}{dx} = \sinh x$ \qquad (1)

$\frac{dy}{du} = -\frac{1}{3}\left(3u^{3-1}\right)$

$\qquad = -\frac{1}{3}\left(3u^2\right)$

$\qquad = -u^2$

$\frac{dy}{du} = -\cosh^2 x$ \qquad (2)

Substituting (**2**) and (**1**) in $\frac{dy}{dx} = \frac{dy}{du}\frac{du}{dx}$

$\frac{dy}{dx} = (-\cosh^2 x)(\sinh x)$

$\qquad = -\cosh^2 x \sinh x$. or $= -\sinh x \cosh^2 x$

Example 8 Application of Implicit Differentiation

$$\text{Find } \frac{dy}{dx} \text{ if } \cosh(xy) = x^2 \sinh y \quad (A)$$

Solution

We differentiate both sides of the equation implicitly with respect to x in steps

Step 1: Differentiate left side of the equation implicitly with respect to x

Let $t = \cosh xy$ and

let $u = xy$

Then $t = \cosh u$

$$\frac{du}{dx} = x\frac{dy}{dx} + y(1) \quad (du = xdy + ydx; \frac{du}{dx} = x\frac{dy}{dx} + y\frac{dx}{dx}; \frac{du}{dx} = x\frac{dy}{dx} + y)$$

$$\frac{du}{dx} = x\frac{dy}{dx} + y$$

$$\frac{dt}{du} = \sinh u$$

$$= \sinh xy$$

$$\frac{dt}{dx} = \frac{dt}{du}\frac{du}{dx}$$

$$= \sinh(xy)[x\frac{dy}{dx} + y]$$

$$\therefore \frac{d}{dx}(\cosh xy) = \sinh(xy)[x\frac{dy}{dx} + y] \quad \text{<--differentiating the left side of equation (A)}$$

Step 2: Differentiate the right side of the equation implicitly with respect to x.

(Keep x^2 constant and differentiate $\sin y$ implicitly followed by keeping $\sin y$ constant and differentiating x^2.)

$$\frac{d}{dx}(x^2 \sinh y) = x^2 \cosh y\frac{dy}{dx} + \sinh y(2x)$$

$$= x^2 \cosh y\frac{dy}{dx} + 2x\sinh y \text{(differentiating the right side of equation (A))}$$

Step 3: Equate the results from Step 1 and Step 2 to each other. Then

$$\sinh(xy)[x\frac{dy}{dx} + y] = x^2 \cosh y\frac{dy}{dx} + 2x\sinh y$$

Note: You could skip showing Steps 1 and 2 and work on both sides of (A) simultaneously.

Step 4: Solve for $\frac{dy}{dx}$:

$$x\sinh(xy)\frac{dy}{dx} + y\sinh(xy) = x^2 \cosh y\frac{dy}{dx} + 2x\sinh y$$

$$x\sinh(xy)\frac{dy}{dx} - x^2 \cosh y\frac{dy}{dx} = 2x\sinh y - y\sinh(xy)$$

$$\left(x\sinh(xy) - x^2 \cosh y\right)\frac{dy}{dx} = 2x\sinh y - y\sinh(xy) \quad \text{(Factoring-out } \frac{dy}{dx}\text{)}$$

$$\frac{dy}{dx} = \frac{2x\sinh y - y\sinh(xy)}{x\sinh(xy) - x^2 \cosh y} . \quad \text{(Solving for } \frac{dy}{dx}\text{)}$$

Lesson 54 Exercises

Complete the derivative part from memory

Function: $y = f(x)$ **Derivative:** $\frac{dy}{dx} = f'(x)$

1. $y = \sinh x$ 1.

2. $y = \cosh x$ 2.

3. $y = \tanh x$ 3.

4. $y = \coth x$ 4.

5. $y = \operatorname{sech} x$ 5.

6. $y = \operatorname{csch} x$ 6.

7. If $y = -\frac{1}{3}\cosh^3 x$, find $\frac{dy}{dx}$

8. Find $\frac{dy}{dx}$ if $\cosh(xy) = x^2 \sinh y$

9, From questions 1-6, derive $\frac{dy}{dx} = f'(x)$

Answers: 7. $= -\cosh^2 x \sinh x$; **8.** $\frac{dy}{dx} = \frac{2x\sinh y - y\sinh(xy)}{x\sinh(xy) - x^2\cosh y}$

Lesson 55

Differentiation of Inverse Hyperbolic Functions

Introduction

Given the equation, $y = \sinh x \qquad -\infty < x < \infty$, what is its inverse?

Solution

The inverse is found (by definition) by interchanging the roles of x and y.

Therefore, the inverse of $y = \sinh x$ for $-\infty < x < \infty$ is

$$x = \sinh y \text{ for } -\infty < x < \infty, \qquad -\infty < y < \infty,.$$

If we solve $x = \sinh y$ for y, and use the notation, $\sinh^{-1}x$, we obtain

$y = \sinh^{-1}x$ (Recall previously that the inverse of $f(x)$ was symbolized $f^{-1}(x)$).

Another notation for this inverse is $y = \text{Arcsinh } x$.

We may therefore use either $y = \text{Sinh}^{-1}x$ or $y = \text{Arcsinh } x$ for the inverse of $y = \sinh x$. Similarly, the inverse of $y = \cosh x$ is symbolized $y = \text{Cosh}^{-1}x$ or $y = \text{Arccosh } x$,; for the inverse of $y = \tanh x$, the symbol is $y = \text{Tanh}^{-1}x$; for the inverse of $y = \text{sech } x$, the symbol is $y = \text{Sech}^{-1}x$; for the inverse of $y = \text{csch } x$, the symbol is $y = \text{Csch}^{-1}x$. for $y = \coth x$, we have $y = \text{Coth}^{-1}x$. There are explicit formulas in terms of natural log functions for inverse hyperbolic functions; and they are presented below.

These formulas are the results of the fact that the hyperbolic functions were defined in terms of the exponential functions, and the natural logarithmic function is the inverse of the exponential function,

Inverse Hyperbolic Function	Explicit formula (using logarithms)	Domain
↓↓	↓↓	↓↓
1. $\text{Sinh}^{-1}x = \ln(x + \sqrt{1 + x^2})$		$-\infty < x < \infty$
2. $\text{Cosh}^{-1}x = \ln(x + \sqrt{x^2 - 1})$		$x \geq 1$
3. $\text{Tanh}^{-1}x = \frac{1}{2}\ln\left(\frac{1+x}{1-x}\right)$		$|x| < 1$
4. $\text{Sech}^{-1}x = \ln\left(\frac{1 + \sqrt{1 - x^2}}{x}\right)$		$0 < x \leq 1$
5. $\text{Csch}^{-1}x = \ln\left(\frac{1}{x} + \frac{\sqrt{1 + x^2}}{|x|}\right)$		$x \neq 0$
6. $\text{Coth}^{-1}x = \frac{1}{2}\ln\left(\frac{x+1}{x-1}\right)$		$|x| > 1$

Below, we derive some of the above explicit formulas

(A) If $\sinh^{-1}x = y$, then $x = \sinh y$

By definition, $\sinh y = \dfrac{e^y - e^{-y}}{2}$ and

Step 1: $x = \dfrac{e^y - e^{-y}}{2}$ $\quad (x = \sinh y)$

$2x = e^y - e^{-y}$

$0 = e^y - 2x - e^{-y}$ or

$e^y - 2x - e^{-y} = 0$

$e^y\left(e^y - 2x - e^{-y} = 0\right)$

(multiplying by e^y. See also scrapwork)

$e^{2y} - 2xe^y - 1 = 0$ $\quad$ (A)

Equation (A) is quadratic in form.

Step 2: Let $e^y = t$

Then equation (A) becomes

$t^2 - 2xt - 1 = 0$

$t = \dfrac{-(-2x) \pm \sqrt{(-2x)^2 - 4(-1)}}{2}$

$t = x + \sqrt{x^2 + 1}$ $\quad (e^y = t > 0)$

Now, $e^y = x + \sqrt{x^2 + 1}$. Solve for y.

Step 3: $\log_e e^y = \log_e(x + \sqrt{x^2 + 1})$

$\quad$ (taking logs)

$y\log_e e = \log_e(x + \sqrt{x^2 + 1})$

$y(1) = \log_e(x + \sqrt{x^2 + 1})$

$y = \log_e(x + \sqrt{x^2 + 1})$

$y = \ln(x + \sqrt{x^2 + 1})$ $\quad (\log_e = \ln)$

$\therefore \; \sinh^{-1}x = \ln(x + \sqrt{x^2 + 1})$

See also p.305 to reconnect.

Scrapwork

or $e^y - 2x - \frac{1}{e^y} = 0$

$e^y\left(e^y - 2x - \frac{1}{e^y} = 0\right)$

$e^{2y} - 2xe^y - \frac{e^y}{e^y} = 0$

$e^{2y} - 2xe^y - 1 = 0$

Scrapwork

$t^2 - 2xt - 1 = 0$

$t = \dfrac{-(-2x) \pm \sqrt{(-2x)^2 - 4(-1)}}{2}$

$= \dfrac{2x \pm \sqrt{4x^2 + 4}}{2}$

$= \dfrac{2x \pm \sqrt{4(x^2 + 1)}}{2}$

$= \dfrac{2x \pm 2\sqrt{x^2 + 1}}{2}$

$= \dfrac{2x}{2} \pm \dfrac{2\sqrt{x^2 + 1}}{2}$

$= x + \sqrt{x^2 + 1}$ $\quad ((e^y = t > 0)$

Below, we derive more of the above explicit formulas.

(B) If $\cosh^{-1}x = y$, then $x = \cosh y$

By definition, $\cosh y = \dfrac{e^y + e^{-y}}{2}$ and

Step 1 $x = \dfrac{e^y + e^{-y}}{2}$ $\quad(x = \cosh y)$

$2x = e^y + e^{-y}$

$0 = e^y - 2x + e^{-y}$ or

$e^y - 2x + e^{-y} = 0$

$e^y\left(e^y - 2x + e^{-y} = 0\right)$

(multiplying by e^y. See also srrapwork)

$\quad e^{2y} - 2xe^y + 1 = 0$ $\quad$ (A)

Equation (A) is quadratic in form.

Step 2 Let $e^y = t$. Then equation (A)

becomes $t^2 - 2xt + 1 = 0$

$t = \dfrac{-(-2x) \pm \sqrt{(-2x)^2 - 4(1)}}{2}$

$t = x + \sqrt{x^2 - 1}$

Now, $e^y = x + \sqrt{x^2 - 1}$. Solve for y.

Step 3: $\log_e e^y = \log_e(x + \sqrt{x^2 - 1})$

$\quad\quad$ (taking logs)

$y\log_e e = \log_e(x + \sqrt{x^2 - 1})$

$y(1) = \log_e(x + \sqrt{x^2 - 1})$ $\quad(\log_e e = 1)$

$y = \log_e(x + \sqrt{x^2 - 1})$

$y = \ln(x + \sqrt{x^2 - 1})$ $\quad(\log_e = \ln)$

$\therefore \ \cosh^{-1}x = \ln(x + \sqrt{x^2 - 1})$

See also p.315 to reconnect

Scrapwork

or $e^y - 2x + \dfrac{1}{e^y} = 0$

$e^y\left(e^y - 2x + \dfrac{1}{e^y} = 0\right)$

$e^{2y} - 2xe^y + \dfrac{e^y}{e^y} = 0$

$e^{2y} - 2xe^y + 1 = 0$

Scrapwork

$t^2 - 2xt + 1 = 0$

$t = \dfrac{-(-2x) \pm \sqrt{(-2x)^2 - 4(1)}}{2}$

$= \dfrac{2x \pm \sqrt{4x^2 - 4}}{2}$

$= \dfrac{2x \pm \sqrt{4(x^2 - 1)}}{2}$

$= \dfrac{2x \pm 2\sqrt{x^2 - 1}}{2}$

$= \dfrac{2x}{2} \pm \dfrac{2\sqrt{x^2 - 1}}{2}$

$= x + \sqrt{x^2 - 1}$

(C) If $\text{Tanh}^{-1}x = y$, then $x = \tanh y$

By definition, $\tanh y = \dfrac{e^y - e^{-y}}{e^y + e^{-y}}$ and

$$x = \dfrac{e^y - e^{-y}}{e^y + e^{-y}} \qquad (x = \tanh y)$$

Step 1: $x(e^y + e^{-y}) = e^y - e^{-y}$

$xe^y + xe^{-y} - e^y + e^{-y} = 0$

$e^y\left(xe^y + xe^{-y} - e^y + e^{-y} = 0\right)$

$xe^{2y} + x - e^{2y} + 1 = 0$ (A)

(Multiplying both sides of the equation by x)

(Multiplying by e^y to eliminate the negative exponent)

Step 2: Let $e^{2y} = t$ .Then equation (A) becomes $tx + x - t + 1 = 0$

Solve for t

$tx - t + x + 1 = 0$

$t(x-1) + x + 1 = 0$

$$t = \frac{-x-1}{x-1}$$

$$t = \frac{x+1}{1-x}.$$

(Replace t by e^{2y}.)

Then $e^{2y} = \dfrac{x+1}{1-x}$.

Step 3: Solve for y.

$$\log_e e^{2y} = \log_e \frac{x+1}{1-x})$$

(taking logs)

$$2y\log_e e = \log_e \frac{x+1}{1-x}$$

$$2y(1) = \log_e \frac{x+1}{1-x}$$

$$2y = \log_e \frac{x+1}{1-x}$$

$$y = \frac{1}{2}\log_e \frac{x+1}{1-x}$$

$$y = \frac{1}{2}\ln\frac{x+1}{1-x} \quad (\log_e = \ln)$$

$$\therefore \ \tanh^{-1}x = \frac{1}{2}\ln\left(\frac{x+1}{1-x}\right)$$

$\tanh^{-1}x = y$

Below, we present the inverse hyperbolic functions and their derivatives. We will derive some of the derivatives later. **Memorize** them.

Table of Inverse Hyperbolic Functions and Derivatives

Inverse Hyperbolic Function	Derivative:		Each derivative is $\frac{1}{\text{something}}$
1. $y = \text{Sinh}^{-1}x$;	$\dfrac{dy}{dx} = \dfrac{1}{\sqrt{1+x^2}}$		Note the $\sqrt{1+x^2}$.
2. $y = \text{Cosh}^{-1}x$;	$\dfrac{dy}{dx} = \dfrac{1}{\sqrt{x^2-1}}$	$x > 1$	Note the $x^2 - 1$.
3. $y = \text{Tanh}^{-1}x$;	$\dfrac{dy}{dx} = \dfrac{1}{1-x^2}$	$|x| < 1$	Note the $1 - x^2$.
4. $y = \text{Sech}^{-1}x$;	$\dfrac{dy}{dx} = -\dfrac{1}{x\sqrt{1-x^2}}$	$0 < x < 1$	Note the $x\sqrt{1-x^2}$
5. $y = \text{Csch}^{-1}x$;	$\dfrac{dy}{dx} = -\dfrac{1}{|x|\sqrt{1+x^2}}$	$x \neq 0$	Note the $|x|\sqrt{1+x^2}$
6. $y = \text{Coth}^{-1}x$;	$\dfrac{dy}{dx} = \dfrac{1}{1-x^2}$	$|x| > 1$	Note the $1 - x^2$

Memorize **1, 3** and **4**, and note the sign change between **1 and 2**. Note also that **3** and **6** are the same, except for the domains. The derivatives of the first three functions all have plus signs. **Note** that the derivative of an inverse hyperbolic function is an algebraic function, (the same way that the derivative of an inverse trigonometric function is an algebraic function), and **not** a hyperbolic function.

Note that the derivatives in **3** and **6** are the same; but the corresponding domains are different.

We can also say , in advance, that we can integrate the derivatives of the inverse hyperbolic functions using trigonometric substitution.

Finding the Derivatives of Some Inverse Hyperbolic Functions

Example 1 Show that $\dfrac{d}{dx}(\sinh^{-1}x) = \dfrac{1}{\sqrt{1+x^2}}$ By deriving some basic formulas we get insight into the approaches used in finding derivatives

Method 1a	Method 1b
Step 1: Let $\sinh^{-1}x = y$ Then $x = \sinh y$ (1) $\dfrac{dx}{dy} = \cosh y; \quad \dfrac{dy}{dx} = \dfrac{1}{\frac{dx}{dy}}$ $\dfrac{dy}{dx} = \dfrac{1}{\cosh y}$ (2) **Step 2**: From $\cosh^2 y - \sinh^2 y = 1$ $\cosh y = \sqrt{1+\sinh^2 y}$, and $\dfrac{dy}{dx} = \dfrac{1}{\sqrt{1+\sinh^2 y}}$ (3) **Step 3**: From (1) $x = \sinh y$, and squaring, $x^2 = \sinh^2 y$ $\therefore \dfrac{d}{dx}(\sinh^{-1}x) = \dfrac{1}{\sqrt{1+x^2}}$ (Replacing $\sinh^2 y$ by x^2 in (3)) .	**Step 1:** Let $\sinh^{-1}x = y$ Then $\sinh y = x$ (1) Using implicit differentiation $\cosh y \dfrac{dy}{dx} = 1;$ and $\dfrac{dy}{dx} = \dfrac{1}{\cosh y}$ **Step 2**: From $\cosh^2 y - \sinh^2 y = 1$, $\cosh y = \sqrt{1+\sinh^2 y}$, and $\dfrac{dy}{dx} = \dfrac{1}{\sqrt{1+\sinh^2 y}}$ **Step 3**: From (1) $x = \sinh y$, and squaring, $x^2 = \sinh^2 y$ $\therefore \dfrac{d}{dx}(\sinh^{-1}x) = \dfrac{1}{\sqrt{1+x^2}}$ (Replacing $\sinh^2 y$ by x^2)

Method 2a: If we can remember that $\sinh^{-1}x = \ln(x + \sqrt{x^2+1})$, we can show

that $\dfrac{d}{dx}(\sinh^{-1}x) = \dfrac{1}{\sqrt{1+x^2}}$ by showing that $\dfrac{d}{dx}[\ln(x + \sqrt{x^2+1})] = \dfrac{1}{\sqrt{1+x^2}}$

	Step 2:
Step 1: Let $y = \sinh^{-1}x$ Let $u = x + \sqrt{x^2+1}$ Then $y = \ln u$ $\dfrac{du}{dx} = 1 + \frac{1}{2}\left(1+x^2\right)^{-\frac{1}{2}}(2x)$ $= 1 + \dfrac{x}{\sqrt{1+x^2}}$ $\dfrac{du}{dx} = \dfrac{x + \sqrt{1+x^2}}{\sqrt{1+x^2}}$ $\dfrac{dy}{du} = \dfrac{1}{u} = \dfrac{1}{x + \sqrt{x^2+1}}$	$\dfrac{dy}{dx} = \dfrac{dy}{du}\dfrac{du}{dx} = \dfrac{1}{\left(x+\sqrt{x^2+1}\right)} \bullet \left(\dfrac{x+\sqrt{x^2+1}}{\sqrt{x^2+1}}\right)$ $= \dfrac{1}{\sqrt{x^2+1}}$

Method 2b: Using implicit differentiation

Step 1: ($\log_e = \ln x$)	**Step 2:**
Let $\log_e(x + \sqrt{x^2 + 1}) = y$, Then equivalently, $e^y = x + \sqrt{x^2 + 1}$ Perform implicit differentiation $e^y \dfrac{dy}{dx} = 1 + \frac{1}{2}\left(1 + x^2\right)^{-\frac{1}{2}}(2x)$ $= 1 + \dfrac{x}{\sqrt{1 + x^2}}$ $= \dfrac{x + \sqrt{1 + x^2}}{\sqrt{1 + x^2}}$ $\dfrac{dy}{dx} = \dfrac{1}{e^y} \cdot \dfrac{x + \sqrt{1 + x^2}}{\sqrt{1 + x^2}}$	$\dfrac{dy}{dx} = \dfrac{1}{\left(x + \sqrt{x^2 + 1}\right)} \cdot \left(\dfrac{x + \sqrt{x^2 + 1}}{\sqrt{x^2 + 1}}\right)$ $\qquad\qquad (e^y = x + \sqrt{x^2 + 1})$ $= \dfrac{1}{\sqrt{x^2 + 1}}$

Note: If $y = \cosh^{-1}x$, we can similarly show that $\dfrac{dy}{dx} = \dfrac{1}{\sqrt{x^2 - 1}}$. Try it.

Example 2 Find $\dfrac{dy}{dx}$ if $y = \text{sech}^{-1}x$

Step 1: $y = \text{sech}^{-1}x$ (A)
         is equivalent to
         $\text{sech}\, y = x$ (B)

Step 2: Perform implicit differentiation with respect to x in equation (B)

$$\text{Then} \quad \frac{d}{dx}(\text{sech}\, y) = \frac{d}{dx}(x)$$

$$-\text{sech}\, y \bullet \tanh y \frac{dy}{dx} = 1 \qquad \text{(C)}$$

Step 3: $\dfrac{dy}{dx} = -\dfrac{1}{\text{sech}\, y \bullet \tanh y}$ (Solving equation (C) for $\dfrac{dy}{dx}$)

Step 4: From the hyperbolic identity $1 - \tanh^2 y = \text{sech}^2 y$

        $\tanh y = +\sqrt{1 - \text{sech}^2 y}$,

        $\tanh y = \sqrt{1 - x^2}$ (substituting for $\text{sech}\, y = x$ from (B))

Step 5: $\dfrac{dy}{dx} = -\dfrac{1}{x\sqrt{1 - x^2}}$ (Substitute $\text{sech}\, y = x$; $\tanh y = \sqrt{1 - x^2}$) in Step 3).

Example 3 Find $\dfrac{dy}{dx}$ if $y = \text{Tanh}^{-1}x$

Solution

Step 1: $y = \text{Tanh}^{-1}x$ (A)

        is equivalent to

        $\tanh y = x$ (B)

Step 2: Perform implicit differentiation with respect x in equation (B)

     Then $\text{sech}^2 y \dfrac{dy}{dx} = 1$ (C)

Step 3: $\dfrac{dy}{dx} = \dfrac{1}{\text{sech}^2 y}$ (Solving equation (C) for $\dfrac{dy}{dx}$)

Step 4: Using the hyperbolic identity $\text{sech}^2 y = 1 - \tanh^2 y$

$\dfrac{dy}{dx} = \dfrac{1}{1 - \tanh^2 y}$, and also, substituting x for $\tanh y$ from equation (B)

     $\dfrac{dy}{dx} = \dfrac{1}{1 - x^2}$. $|x| < 1$ ($\tanh y = x$)

Memory device (I use the following concrete example for recall)

My memory device for remembering for example that $y = \text{Sin}^{-1}x$ is equivalent to $\sin y = x$ is that I recall from the trigonometric tables that $\text{Sin}^{-1}(\tfrac{1}{2}) = \tfrac{\pi}{6} = 30°$ is equivalent to $\sin\tfrac{\pi}{6} = \tfrac{1}{2}$.

Example 4 Find $\dfrac{dy}{dx}$ if $y = \text{csch}^{-1}x$

Solution:

Step 1: $y = \text{csch}^{-1}x$ (A)

        is equivalent to

        $\text{csch } y = x$ (B)

Step 2: Perform implicit differentiation with respect to x in equation (B).

     Then $\dfrac{d}{dx}(\text{csch } y) = \dfrac{d}{dx}(x)$

        $-\text{csch } y \bullet \coth y \dfrac{dy}{dx} = 1$ (C)

Step 3: $\dfrac{dy}{dx} = -\dfrac{1}{\text{csch } y \bullet \coth y}$ (D)   (Solving equation (C) for $\dfrac{dy}{dx}$)

Step 4: From the hyperbolic identity $\coth^2 y - 1 = \text{csch}^2 y$

     $\coth y = +\sqrt{\text{csch}^2 y + 1}$,

     $\coth y = \sqrt{x^2 + 1}$ (squaring (B) and substituting x^2 for $\text{csch}^2 y$)

Step 5: $\dfrac{dy}{dx} = -\dfrac{1}{|x|\sqrt{x^2+1}}$ (Substitute x for $\text{csch } y$; $\sqrt{x^2+1}$ for $\coth y$ in (D.)

Lesson 55 Exercises

A Symbolize the inverse of each of the following:

1. $y = \sinh x$, **2.** $y = \cosh x$; **3.** $y = \tanh x$,; **4.** $y = \operatorname{sech} x$.

B Complete the derivative part from memory.

1. $y = \sinh^{-1} x$; $\dfrac{dy}{dx} =$

2. $y = \cosh^{-1} x$; $\dfrac{dy}{dx} =$

3. $y = \tanh^{-1} x$; $\dfrac{dy}{dx} =$

4. $y = \operatorname{sech}^{-1} x$; $\dfrac{dy}{dx} =$

5. $y = \operatorname{csch}^{-1} x$; $\dfrac{dy}{dx} =$

6. $y = \coth^{-1} x$; $\dfrac{dy}{dx} =$

If $y = \sinh^{-1} x$, derive $\dfrac{dy}{dx} = \dfrac{1}{\sqrt{x^2 + 1}}$

8. If $y = \operatorname{sech}^{-1} x$, derive $\dfrac{dy}{dx} = -\dfrac{1}{x\sqrt{1 - x^2}}$;

9. If $y = \operatorname{Sinh}^{-1}(\sqrt{1 - x})$, find $\dfrac{dy}{dx}$

Answers: A: 1. $y = \sinh^{-1} x$; **2.** $y = \cosh^{-1} x$;; **3.** $y = \tanh^{-1} x$; 4. $y = \operatorname{sech}^{-1} x$

9. $-\dfrac{1}{2\sqrt{1 - x}\sqrt{2 - x}} \, . = -\dfrac{1}{2\sqrt{x^2 - 3x + 2}}$

CHAPTER 17

Hyperbolic Functions II

Lesson 56: Integration of Hyperbolic Functions
Lesson 57: Integration of Algebraic Functions whose
    Antiderivatives are Inverse Hyperbolic Functions
Lesson 58: Integration of Inverse Hyperbolic Functions

Lesson 56

Integration of Hyperbolic Functions

We present the hyperbolic functions and their corresponding integrals. Use the hyperbolic derivatives, learned previously, to help you recall some of the entries in the table. Memorize these basic formulas. Later, we will derive some of them.

Hyperbolic Functions and Corresponding Antiderivatives

Function: $f(x)$	**Antiderivative** (Indefinite Integral) $\int f(x)dx$
1. $y = \sinh x$	1. $\cosh x + C$
2. $y = \cosh x$	2. $\sinh x + C$
	$\qquad\qquad A \qquad\qquad\qquad\qquad\qquad B$
3. $y = \tanh x$	3. $-\ln|\operatorname{sech} x| + C$ or $\ln|\cosh x| + C$
4. $y = \coth x$	4. $-\ln|\operatorname{csch} x| + C$ or $\ln|\sinh x| + C$
5. $y = \operatorname{sech} x$	5. $\tan^{-1}(\sinh x) + C$ or $\cot^{-1}(\operatorname{csch} x) + C$
6. $y = \operatorname{csch} x$	6. $\ln\sqrt{\dfrac{\cosh x - 1}{\cosh x + 1}} + C = \ln\left|\tanh\dfrac{x}{2}\right| + C$
7. $y = \operatorname{sech}^2 x$	7. $\tanh x + C$ (from the derivative table of Lesson)
8. $y = \operatorname{csch}^2 x$	8. $-\coth x + C$ (from the derivative table of Lesson)
9. $y = \operatorname{sech} x \tanh x$	9. $-\operatorname{sech} x + C$ (from the derivative table of Lesson)
10. $y = \operatorname{csch} x \coth x$	10. $-\operatorname{csch} x + C$ (from the derivative table of Lesson)

Mnemonic Help:
The memory pattern is similar to that for the trigonometric functions .

 For the last four entries in the table (from $\operatorname{sech}^2 x$ to $\operatorname{csch} x \coth x$), interchange the entries in the table for the derivatives of the hyperbolic functions.

Example A: Show that $\int \sinh x = \cosh x + C$,
          By definition, $\sinh x = \frac{1}{2}(e^x - e^{-x})$, $\cosh x = \frac{1}{2}(e^x + e^{-x})$

$\int \sinh x\,dx = \ = \frac{1}{2}[\int e^x dx - \int e^{-x} dx]$	**Scrapwork** for $\int e^{-x}dx$: Let $u = -x$,
$= \frac{1}{2}[e^x - (-e^{-x})] + C$	Then $du = -dx$ and $dx = -du$
$= \frac{1}{2}[e^x + e^{-x}] + C$	Substituting, $\int e^{-x}dx = \int -e^u du$
	$= -e^u = -e^{-x}$
$\therefore \int \sinh x\,dx = \cosh x + C$ <----	(By definition , $\cosh x = \frac{1}{2}(e^x + e^{-x})$)

Derivation of some basic integration formulas

Example 1 Show that $\int \tanh x\, dx = -\ln|\operatorname{sech} x| + C$ (Apply simple u-substitution)

Step 1: $\int \tanh x\, dx = \int \dfrac{\sinh x}{\cosh x}\, dx$

Let $u = \cosh x$

Then $\dfrac{du}{dx} = \sinh x$, and

$\dfrac{du}{\sinh x} = dx$ or $du = \sinh x\, dx$

Step 2: Substitute u for $\cosh x$,

and $\dfrac{du}{\sinh x}$ for dx or $du = \sinh x\, dx$

Step 3: Then

$\int \dfrac{\sinh x}{u} \bullet \dfrac{du}{\sinh x} = \int \dfrac{du}{u}$

$= \ln|u| + C$

$= \ln|\cosh x| + C \quad (u = \cosh x)$

$= \ln\left|\dfrac{1}{\operatorname{sech} x}\right| + C$

$\int \tanh x\, dx = \ln\left|(\operatorname{sech} x)^{-1}\right| + C$

$= -\ln|\operatorname{sech} x| + C$

$(\operatorname{sech} x)^{-1} = \dfrac{1}{\operatorname{sech} x} = \cosh x$

Example 2 Show that $\int \coth x\, dx = \ln|\sinh x| + C = -\ln|\operatorname{csch} x| + C$.

Solution We apply simple u-substitution

Step 1:

$\int \coth x\, dx = \int \dfrac{\cosh x}{\sinh x}\, dx$

Let $u = \sinh x$

Then $\dfrac{du}{dx} = \cosh x$, and

$\dfrac{du}{\cosh x} = dx$ or

$du = \cosh x\, dx$

Step 2: Substitute u for $\sinh x$, and

$\dfrac{du}{\cosh x}$ for dx (That is, $du = \cosh x\, dx$)

Then $\int \dfrac{\cosh x}{u}\dfrac{du}{\cosh x} = \int \dfrac{du}{u} = \ln|u| + C$

$\therefore \int \coth x\, dx = \ln|\sinh x| + C$

$\left(= \ln\left|\dfrac{1}{\operatorname{csch} x}\right| = \ln\left|(\operatorname{csch} x)^{-1}\right| = -\ln|\operatorname{csch} x|\right)$

Example 3 Find $\int \operatorname{sech} x\, dx$

Step 1: $\int \operatorname{sech} x\, dx = \int \dfrac{1}{\cosh x} \bullet \dfrac{\cosh x}{\cosh x}\, dx$

(multiplying both numerator and

denominator by $\dfrac{\cosh x}{\cosh x}$)

$= \int \dfrac{\cosh x}{\cosh^2 x}\, dx = \int \dfrac{\cosh x}{1 + \sinh^2 x}\, dx$

Step 2: Let $u = \sinh x$.

$\dfrac{du}{dx} = \cosh x$ or $dx = \dfrac{du}{\cosh x}$

Substitute u for $\sinh x$, and

$\dfrac{du}{\cosh x}$ for dx. Then

$\int \operatorname{sech} x\, dx = \int \dfrac{\cosh x}{1 + u^2} \bullet \dfrac{du}{\cosh x}$

$= \int \dfrac{du}{1 + u^2}$

Step 3 Apply trigonometric

substitution. Let $u = \tan\theta$, Then

$u^2 = \tan^2\theta$, $\dfrac{du}{d\theta} = \sec^2\theta$,

$du = \sec^2\theta\, d\theta$

$\int \operatorname{sech} x\, dx = \int \dfrac{\sec^2\theta\, d\theta}{1 + \tan^2\theta}$

$= \int \dfrac{\sec^2\theta\, d\theta}{\sec^2\theta}$

$= \int d\theta$

$= \theta + C$

$= \tan^{-1} u$

$= \tan^{-1}(\sinh x)$

$\therefore \int \operatorname{sech} x\, dx = \tan^{-1}(\sinh x) + C$

$(u = \sinh x)$

Also, $\int \operatorname{sech} x\, dx = \cot^{-1}(\operatorname{csch} x) + C = \csc^{-1}(\coth x) + C = 2\tan^{-1}(e^x) + C$

Example 4 Find $\int \operatorname{csch} x\, dx$

Step 1:

$\int \operatorname{csch} x\, dx = \int \frac{1}{\sinh x} \cdot \frac{\sinh x}{\sinh x}\, dx$

(multiplying both numerator and denominator by $\frac{\sinh x}{\sinh x}$)

$= \int \frac{\sinh x}{\sinh^2 x}\, dx$

$= \int \frac{\sinh x}{\cosh^2 x - 1}\, dx$

Step 2:   Let $u = \cosh x$

$\frac{du}{dx} = \sinh x$ or $dx = \frac{du}{\sinh x}$

Substitute u for $\cosh x$, and

$\frac{du}{\sinh x}$ for dx Then

$\int \operatorname{csch} x\, dx = \int \frac{\sinh x}{u^2 - 1} \cdot \frac{du}{\sinh x}$

$= \int \frac{du}{u^2 - 1}$

Step 3: We now apply partial fraction decomposition

$= \int \frac{du}{u^2 - 1} = \int \frac{du}{(u-1)(u+1)}$ (see p.281)

$= \int -\frac{1}{2} \frac{du}{(u+1)} + \int \frac{1}{2} \frac{du}{(u-1)}$

$= \frac{1}{2} \int \frac{du}{u-1} - \frac{1}{2} \int \frac{du}{u+1}$

$= \frac{1}{2} \big[\ln(u-1) - \ln(u+1) \big] + C$

$= \frac{1}{2} \big[\ln(\cosh x - 1) - \ln(\cosh x + 1) \big] + C$

$= \frac{1}{2} \left[\ln \frac{\cosh x - 1}{\cosh x + 1} \right] = \ln \left[\frac{\cosh x - 1}{\cosh x + 1} \right]^{\frac{1}{2}}$

$\int \operatorname{csch} x\, dx = \ln \sqrt{\frac{\cosh x - 1}{\cosh x + 1}}$

Also, $= \ln \left| \tanh \frac{x}{2} \right| + C$

(by applying $\tan \frac{x}{2} = \frac{\cosh x - 1}{\sinh x}$)

More Examples on Integration of Hyperbolic Functions

Example 5. Find $\int 5 \sinh x\, dx$

Solution

$\int 5 \sinh x\, dx = 5 \int \sinh x\, dx$

$\qquad = 5(\cosh x) + C$

$\qquad = 5 \cosh x + C$

Example 6. Find $\int \frac{\sinh x}{\cosh^2 x}\, dx$

Solution

Method 1: Since the numerator is the **derivative** of $\cosh x$ (in the denominator) we can use the u-substitution method.

Step 1: Let $u = \cosh x$, then $\cosh^2 x = u^2$,

and $\frac{du}{dx} = \sinh x$ and from which $dx = \frac{du}{\sinh x}$

Step 2: Substitute $\frac{du}{\sinh x}$ for dx and u^2 for $\cosh^2 x$ in $\int \frac{\sinh x}{\cosh^2 x}\, dx$ to obtain

$\int \frac{\sinh x}{u^2} \left(\frac{du}{\sinh x} \right)$

$= \int \frac{1}{u^2}\, du$

$$= \int u^{-2} du$$

$$= \frac{u^{-2+1}}{-2+1} + C$$

$$= \frac{u^{-1}}{-1} + C$$

$$= -\frac{1}{u} + C$$

$$\int \frac{\sinh x}{\cosh^2 x} dx = -\frac{1}{\cosh x} + C \quad \text{(replacing } u \text{ by } \cosh x\text{)}$$

$$= -\operatorname{sech} x + C.$$

Method 2

Step 1: $\dfrac{\sinh x}{\cosh^2 x} = \dfrac{1}{\cosh x} \cdot \dfrac{\sinh x}{\cosh x}$

$$= \operatorname{sech} x \tanh x$$

Step 2: Since the derivative of $\operatorname{sech} x$ is $-\operatorname{sech} x \tanh x$, the antiderivative of $\operatorname{sech} x \tanh x$ is $-\operatorname{sech} x + C$.

Therefore, $\displaystyle\int \frac{\sin x}{\cos^2 x} dx = -\operatorname{sech} x + C.$

Example 7. Find $\displaystyle\int \cosh^2 x \sinh x \, dx$

Solution We apply simple u-substitution.

Step 1: Since $\sinh x$ is the derivative of $\cosh x$,

let $u = \cosh x$

Then $\dfrac{du}{dx} = \sinh x$, and from which $dx = \dfrac{du}{\sinh x}$

Step 2: Substitute $\dfrac{du}{\sinh x}$ for dx and u^2 for $\cosh^2 x$ in $\displaystyle\int \cosh^2 x \sinh x \, dx$ to

obtain

$$\int u^2 \sinh x (\frac{du}{\sinh x})$$

$$= \int u^2 du$$

$$= \frac{u^3}{3} + C$$

$$= \frac{1}{3} \cosh^3 x + C$$

Example 8. Find $\int \text{sech}\, 3x \tanh 3x \, dx$

Solution

Step 1: Let $u = \text{sech}\, 3x$.   (since the derivative of $\text{sech}\, x$ is $-\text{sech}\, x \tanh x$)

Then $\dfrac{du}{dx} = -3\,\text{sech}\, 3x \tanh 3x$ and from which $dx = -\dfrac{du}{3\sec 3x \tan 3x}$

Step 2: Substitute $-\dfrac{du}{3\sec 3x \tan 3x}$ for dx in $\int \text{sech}\, 3x \tanh 3x \, dx$ to obtain

$$-\int \frac{\text{sech}\, 3x \tanh 3x}{1} \cdot \frac{du}{3\,\text{sech}\, 3x \tanh 3x}$$

$$= -\int \frac{du}{3} \quad \text{(sech}\, 3x \tanh 3x \text{ in the numerator and denominator cancel out)}$$

$$= -\frac{1}{3} u + C$$

$$= -\frac{1}{3}\,\text{sech}\, 3x + C$$

$$\therefore \int \text{sech}\, 3x \tanh 3x \, dx = -\frac{1}{3}\,\text{sech}\, 3x + C.$$

Example 9 Show that $\int \sinh x \, dx = \cosh x$,

By definition, $\sinh x = \dfrac{e^x - e^{-x}}{2}$, $\cosh x = \dfrac{e^x + e^{-x}}{2}$

$$\int \sinh x \, dx = \int \tfrac{1}{2}(e^x - e^{-x})\,dx$$

$$= \tfrac{1}{2}\!\left(\int e^x dx - e^{-x} dx\right)$$

$$= \tfrac{1}{2}[e^x - (-e^{-x})] + C$$

$$= \tfrac{1}{2}[e^x + e^{-x}] + C$$

$$= \frac{e^x + e^{-x}}{2} + C$$

Since by definition , $\cosh x = \dfrac{e^x + e^{-x}}{2}$

$$\int \sinh x \, dx = \cosh x + C.$$

Scrapwork for $\int e^{-x} dx$
Let $u = -x$,

Then $\dfrac{du}{dx} = -1$ and

$dx = -du$ . .Substituting,

$$\int e^{-x} dx = \int -e^u du$$

$$= -e^u$$

$$= -e^{-x} \quad (u = -x,)$$

Lesson 56 Exercises

In 1-10, complete the antiderivative part from memory

Function: $f(x)$ $\int f(x)dx + C$ (Antiderivative)

1. $y = \sinh x$

    1.

2. $y = \cosh x$

    2.

3. $y = \text{sech}^2 x$

    3.

4. $y = \text{csch}^2 x$

    4.

5. $y = \text{sech}\, x \tanh x$

    5.

6. $y = \text{csch}\, x \coth x$

    6.

7. $y = \tanh x$

    7.

8. $y = \coth x$

    8.

9. $y = \text{sech}\, x$

    9.

10. $y = \text{csch}\, x$

    10.

11.. Find $\int 5\sinh x\, dx$

12. Find $\int \dfrac{\sinh x}{\cosh^2 x}\, dx$

13. Find $\int \cosh^2 x \sinh x\, dx$

14. Find $\int \text{sech}\, 3x \tanh 3x\, dx$

We can use Simple U-substitution for 15-26. Do it

15. Find $\int \dfrac{\sinh x}{\cosh x}\, dx$

16. Find $\int \sinh x \cosh x\, dx$

17, Find $\int \sinh^5 x \cosh x\, dx$

18. Find $\int \cosh^5 x \sinh x\, dx$

19. $\int \sinh^n x \cosh x\, dx$

20. $\int \cosh^n x \sinh x\, dx$

21. $\int \dfrac{\cosh x}{1 + \sinh x}\, dx$

22. $\int \dfrac{\sinh x}{1 + \cosh x}\, dx$

23. $\int \dfrac{\sinh x}{\cosh^2 x}\, dx$

24. $\int \dfrac{\cosh x}{\sinh x}\, dx$

25. $\int \dfrac{\sinh 2x}{1 - \cosh 2x}\, dx$

26. $\int \coth 2x\, dx$

Answers: 11. $5\cosh x + C$; **12.** $-\text{sech}\, x + C$; **13.** $\frac{1}{3}\cosh^3 x + C$;

14. $-\frac{1}{3}\text{sech}\, 3x + C$; **15.** $\ln|\cosh x| + C$; **16.** $\frac{1}{2}\cosh^2 x + C$ or $\frac{1}{2}\sinh^2 x + C$;

17. $\frac{1}{6}\sinh^6 x + C$; **18.** $\frac{1}{6}\cosh^6 x + C$; **19.** $\dfrac{\sinh^{n+1} x}{n+1} + C$; **20.** $\dfrac{\cosh^{n+1} x}{n+1} + C$;

21. $\ln|1 + \sinh x| + C$; **22.** $\ln|1 + \cosh x| + C$; **23.** $-\text{sech}\, x + C$;

24. $\ln|\sinh x| + C$; **25.** $-\frac{1}{2}\ln|1 - \cosh 2x| + C$; or $-\ln|\sinh x| + C$;

26. $\frac{1}{2}\ln|\sinh 2x| + C$.

Lesson 57

Integration of Algebraic Functions whose Antiderivatives are Inverse Hyperbolic Functions

(Antiderivatives of the derivatives of Inverse Hyperbolic Functions.)
The corresponding algebraic functions for trigonometric functions are presented for comparison.

Hyperbolic Functions		**Trigonometric Functions**	
Algebraic Integrand	**Antiderivative**	**Algebraic Integrand**	**Antiderivative**
1. $\dfrac{1}{\sqrt{1+x^2}}$	$\mathrm{Sinh}^{-1}x$	1. $\dfrac{1}{\sqrt{1-x^2}}$	$\mathrm{Sin}^{-1}x + C$
2. $\dfrac{1}{\sqrt{x^2-1}}$	$\mathrm{Cosh}^{-1}x \quad x>1$	2. $-\dfrac{1}{\sqrt{1-x^2}}$	$\mathrm{Cos}^{-1}x + C$
3. $\dfrac{1}{1-x^2}$	$\mathrm{Tanh}^{-1}x \quad |x|<1$	3. $\dfrac{1}{1+x^2}$	$\mathrm{Tan}^{-1}x + C$
4. $-\dfrac{1}{x\sqrt{1-x^2}}$	$\mathrm{Sech}^{-1}x \quad 0<x<1$	4. $\dfrac{1}{x\sqrt{x^2-1}}$	$\mathrm{Sec}^{-1}x + C$
5. $-\dfrac{1}{|x|\sqrt{1+x^2}}$	$\mathrm{Csch}^{-1}x \quad x\neq 0$	5. $-\dfrac{1}{x\sqrt{x^2-1}}$	$\mathrm{Csc}^{-1}x + C$
6. $\dfrac{1}{1-x^2}$	$\mathrm{Coth}^{-1}x \quad |x|>1$	6. $-\dfrac{1}{1+x^2}$	$\mathrm{Cot}^{-1}x + C$

Example 1 Find $\displaystyle\int \frac{dx}{\sqrt{1+x^2}}$ (See also Lesson 43, Example 2, p. 305)

If we can remember that an antiderivative of $\dfrac{1}{\sqrt{1+x^2}}$ is $\sinh^{-1}x$

remember also that $\sinh^{-1}x = \ln(x + \sqrt{x^2+1})$.,p.415, then we can deduce that

$$\int \frac{dx}{\sqrt{1+x^2}} = \ln(x + \sqrt{x^2+1}) + C .$$ However, if we cannot

remember both of the above information, we integrate $\dfrac{1}{\sqrt{1+x^2}}$ using

trigonometric substitution. to obtain $\displaystyle\int \frac{dx}{\sqrt{1+x^2}} = \ln(x + \sqrt{x^2+1}) + C$

See also p.305. Similarly, to integrate $\dfrac{1}{\sqrt{x^2-1}}$, we can say

$\cosh^{-1}x = \ln(x + \sqrt{x^2-1})$, or use trigonometric substitution to obtain

$$\int \frac{dx}{\sqrt{x^2-1}} = \ln(x + \sqrt{x^2-1}) + C \quad \text{(see also p.315, p.416)}$$

Since the hyperbolic functions were defined in terms of the exponential functions, and the natural logarithmic function is the inverse of the exponential function, the inverse hyperbolic functions can be expressed in terms of the natural logarithms This is a repetition from Lesson 55.

Formulas for inverse Hyperbolic Functions in terms of Logarithms

Hyperbolic Functions		Formulas
Algebraic Integrand	**Antiderivative**	**Formulas**
1. $\int \dfrac{dx}{\sqrt{x^2+1}}$	$\text{Sinh}^{-1}x$	1. $= \ln(x + \sqrt{x^2+1})$
2. $\int \dfrac{dx}{\sqrt{x^2-1}}$	$\text{Cosh}^{-1}x \quad x>1$	2. $= \ln(x + \sqrt{x^2-1}) \qquad x>1$
3. $\int \dfrac{dx}{1-x^2}$	$\text{Tanh}^{-1}x \quad |x|<1$	3. $= \frac{1}{2}\ln\left(\dfrac{1+x}{1-x}\right) \qquad |x|<1$
4. $\int \dfrac{dx}{x\sqrt{1-x^2}}$	$\text{Sech}^{-1}x \quad 0<x<1$	4. $= \ln\left(\dfrac{1+\sqrt{1-x^2}}{x}\right) \quad 0<x\le1$
5. $\int \dfrac{dx}{|x|\sqrt{1+x^2}}$	$\text{Csch}^{-1}x \quad x\ne0$	5. $= \ln\left(\dfrac{1}{x} + \dfrac{\sqrt{1+x^2}}{|x|}\right) \quad x\ne0$
6. $\int \dfrac{dx}{1-x^2}$	$\text{Coth}^{-1}x \quad |x|>1$	6. $= \frac{1}{2}\ln\left(\dfrac{x+1}{x-1}\right) \qquad x\ne1$

From above, we can write for example that

$$\int \frac{dx}{\sqrt{x^2+1}} = \text{Sinh}^{-1}x + C = \ln(x + \sqrt{x^2+1}) + C$$

Lesson 57 Exercises

Complete the antiderivative part from memory.

Algebraic Integrand	Indefinite Integral
$f(x)$	$\int f(x)dx + C$
1. $\dfrac{1}{\sqrt{1-x^2}}$	1.
2. $-\dfrac{1}{\sqrt{1-x^2}}$	2.
3. $\dfrac{1}{1+x^2}$	3.
4. $\dfrac{1}{x\sqrt{x^2-1}}$	4.
5. $\dfrac{1}{x\sqrt{x^2-1}}$	5.
6. $-\dfrac{1}{1+x^2}$	6.

7. Show that $\dfrac{d}{dx}(\text{Sin}^{-1}x) = \dfrac{1}{\sqrt{1-x^2}}$

Lesson 58
Integration of Inverse Hyperbolic Functions
(Integrate-by-parts)

In lesson 39, we applied integration-by-parts to integrate inverse trigonometric functions. Similarly, here, we apply integration-by parts to integrate inverse hyperbolic functions. Review Lesson 36 (integration by parts) The integrals are presented first , followed by the derivation of some of the integrals

1. $\int \text{Sinh}^{-1}x\,dx = x\,\text{Sinh}^{-1}x - \sqrt{1+x^2} + C$

2. $\int \text{Cosh}^{-1}x\,dx = x\,\text{Cosh}^{-1}x - \sqrt{x^2-1} + C$

3. $\int \text{Tanh}^{-1}x\,dx = x\,\text{Tanh}^{-1}x + \frac{1}{2}\ln(1-x^2) + C$ **Note:** $\frac{1}{2}\ln(1-x^2) = \ln\sqrt{1-x^2}$

4. $\int \text{Sech}^{-1}x\,dx = x\,\text{Sech}^{-1}x + \text{Sin}^{-1}x + C$ if $\text{Sech}^{-1}x > 0$ or
$\qquad\qquad = x\,\text{Sech}^{-1}x - \text{Sin}^{-1}x + C$ if $\text{Sech}^{-1}x < 0$

5. $\int \text{Csch}^{-1}x\,dx = x\,\text{Csch}^{-1}x + \ln\left|\,x + \sqrt{x^2+1}\,\right| + C$

6. $\int \text{Coth}^{-1}x\,dx = x\,\text{Coth}^{-1}x + \frac{1}{2}\ln\left|1-x^2\right|$ Note: $\frac{1}{2}\ln(1-x^2) = \ln\sqrt{1-x^2}$

Example 1: Find $\int \text{Sinh}^{-1}x\,dx$ (Integrating by parts; process Type 1)

Step 1: Let $u = \sinh^{-1}x$, and let $dv = 1\,dx$ (version 1) or let $v = 1$ (for version 2)

Then $\int \text{Sinh}^{-1}x\,dx = x\text{Sinh}^{-1}x - \int \dfrac{x}{\sqrt{1+x^2}}\,dx$

Integration by parts Formula **(Version 1)**
$\int u\,dv = uv - \int v\,du$
$dv = dx;\ \ v = x$
$du = \dfrac{1}{\sqrt{1+x^2}}\,dx$
Version 2
$\int (uv)dx = u\int v - \int\left(\dfrac{du}{dx}\cdot\int v\right)dx$
$\dfrac{du}{dx} = \dfrac{1}{\sqrt{1+x^2}}$

Step 2: For $\int \dfrac{x}{\sqrt{1+x^2}}\,dx$ (A)

Let $t = 1+x^2$, $\dfrac{dt}{dx} = 2x$, $dx = \dfrac{dt}{2x}$
substituting in (A), we obtain

$\int \dfrac{x}{t^{\frac{1}{2}}}\bullet\left(\dfrac{dt}{2x}\right)$

Now, putting everything together,

Step 3 $\int \text{Sinh}^{-1}x\,dx = x\text{Sinh}^{-1}x - \int \dfrac{x}{t^{\frac{1}{2}}}\bullet\left(\dfrac{dt}{2x}\right)$

$= x\text{Sinh}^{-1}x - \frac{1}{2}\int t^{-\frac{1}{2}}\,dt$

$= x\,\text{Sinh}^{-1}x - \frac{1}{2}\bullet\dfrac{t^{-\frac{1}{2}+1}}{-\frac{1}{2}+1} + C$

Step 4: $= x\,\text{Sinh}^{-1}x - \frac{1}{2}\bullet\dfrac{t^{\frac{1}{2}}}{\frac{1}{2}} + C$

$= x\,\text{Sinh}^{-1}x - t^{\frac{1}{2}} + C$

$= x\,\text{Sinh}^{-1}x - (1+x^2)^{\frac{1}{2}} + C$

$= x\,\text{Sinh}^{-1}x - \sqrt{1+x^2} + C$

$$\therefore \int \text{Sinh}^{-1}x\,dx = x\,\text{Sinh}^{-1}x - \sqrt{1+x^2} + C$$

(**Note** $\frac{d}{dx}(\text{Sinh}^{-1}x) = \dfrac{1}{\sqrt{1+x^2}}$; $\int \dfrac{1}{\sqrt{1+x^2}}\,dx = \text{Sinh}^{-1}x + C$)

Example 2: Find $\int \text{Tanh}^{-1} x \, dx$

Step 1: Let $u = \text{Tanh}^{-1} x$, and let $dv = 1 dx$ (version 1) or let $v = 1$ (for version 2)

Then $\int \text{Tanh}^{-1} x \, dx = x \text{Tanh}^{-1} x - \int \dfrac{x}{1 - x^2} \, dx$

Step 2: For $\int \dfrac{x}{1 - x^2} \, dx$ (integrate by u-subst.)

Let $t = 1 - x^2$. $\dfrac{dt}{dx} = -2x$, $dx = -\dfrac{dt}{2x}$.

Substituting in $\int \dfrac{x}{1 - x^2} \, dx$ we obtain

$$\int \dfrac{x}{t} \bullet \left(-\dfrac{dt}{2x} \right) = -\dfrac{1}{2} \int \dfrac{dt}{t}$$

Now, putting everything together,

$\int \text{Tanh}^{-1} x \, dx = x \text{Tanh}^{-1} x + \dfrac{1}{2} \int \dfrac{dt}{t}$

$\qquad\qquad = x \text{Tanh}^{-1} x + \dfrac{1}{2} \ln t + C$

$\int \text{Tanh}^{-1} x \, dx = x \text{Tanh}^{-1} x + \dfrac{1}{2} \ln(1 - x^2) + C$

> **Integration by parts Formula**
> **(Version 1)**
> $$\int u \, dv = uv - \int v \, du$$
> $dv = dx$
> $v = x$
> $du = \dfrac{1}{1 - x^2} \, dx$
>
> **Version 2**
> $$\int (uv) \, dx = u \int v - \int \left(\dfrac{du}{dx} \bullet \int v \right) dx$$
> $v = 1$
> $\int v \, dx = \int 1 \, dx = x$
> $\dfrac{du}{dx} = \dfrac{1}{1 - x^2}$ (version 2)
>
> Note: $\dfrac{1}{2} \ln(1 - x^2) = \ln \sqrt{1 - x^2}$

Note above that the derivative of $\text{Tanh}^{-1} x$ is $\dfrac{1}{1 - x^2}$; and the integral

$\int \dfrac{1}{1 - x^2} \, dx = \text{Tanh}^{-1} x + C$; but $\int \text{Tanh}^{-1} x \, dx = x \text{Tanh}^{-1} x + \dfrac{1}{2} \ln(1 - x^2) + C$.

Example 3: Find $\int \text{Cosh}^{-1} x \, dx$

Step 1: Let $u = \text{Cosh}^{-1} x$, and let $dv = 1 dx$ (version 1) or let $v = 1$ (for version 2)

$\int \text{Cosh}^{-1} x \, dx = x \text{Cosh}^{-1} x - \int \left(\dfrac{x}{\sqrt{x^2 - 1}} \right) dx$

$\int \text{Cosh}^{-1} x \, dx = x \text{Cosh}^{-1} x - \int \dfrac{x}{\sqrt{x^2 - 1}} \, dx$

Step 2: For $\int \dfrac{x}{\sqrt{x^2 - 1}} \, dx$

Let $t = x^2 - 1$; $\dfrac{dt}{dx} = 2x$, $dx = \dfrac{dt}{2x}$

Substitute in $\int \dfrac{x}{\sqrt{x^2 - 1}} \, dx$ to obtain

$$\int \dfrac{x}{t^{\frac{1}{2}}} \bullet \left(\dfrac{dt}{2x} \right)$$

Now, putting everything together,

$\int \text{Cosh}^{-1} x \, dx = x \text{Cosh}^{-1} x + \int \dfrac{x}{t^{\frac{1}{2}}} \bullet \left(\dfrac{dt}{2x} \right)$

$\left(\dfrac{du}{dx} = \dfrac{1}{\sqrt{x^2 - 1}} \right.$ (version 2))

Step 3:

$= x \text{Cosh}^{-1} x - \dfrac{1}{2} \bullet \dfrac{t^{-\frac{1}{2} + 1}}{-\frac{1}{2} + 1} + C$

$= x \text{Cosh}^{-1} x - \dfrac{1}{2} \int t^{-\frac{1}{2}} \, dt$

$= x \text{Cosh}^{-1} x - \dfrac{1}{2} \bullet \dfrac{t^{\frac{1}{2}}}{\frac{1}{2}} + C$

$= x \text{Cosh}^{-1} x - t^{\frac{1}{2}} + C$

$= x \text{Cosh}^{-1} x - (x^2 - 1)^{\frac{1}{2}} + C$

$= x \text{Cosh}^{-1} x - \sqrt{x^2 - 1} + C$.

Example 4: Find $\int \text{Sech}^{-1}x\,dx$. We integrate by parts.

Step 1: Let $u = \text{Sech}^{-1}x$, and let $dv = 1dx$ (version 1) or let $v = 1$ (for version 2)

Then $\int \text{Sech}^{-1}x\,dx = x\,\text{Sech}^{-1}x - \int -\dfrac{x}{x\sqrt{1-x^2}}\,dx$

$\int \text{Sech}^{-1}x\,dx = x\,\text{Sech}^{-1}x + \int \dfrac{1}{\sqrt{1-x^2}}\,dx$

$(\dfrac{du}{dx} = -\dfrac{1}{x\sqrt{1-x^2}}$

(version 2))

Step 2: For $\int \dfrac{1}{\sqrt{1-x^2}}\,dx$. (A)

Let $x = \sin\theta$; (trigonometric substitution)

(and from which $\theta = \text{Sin}^{-1}x$)

Then $\dfrac{dx}{d\theta} = \cos\theta$, and $dx = \cos\theta\,d\theta$. Substitute in (A)

$\int \dfrac{1}{\sqrt{1-\sin^2\theta}}\cos\theta\,d\theta = \int \dfrac{1}{\sqrt{\cos^2\theta}}\cos\theta\,d\theta$

$= \int \dfrac{1}{\cos\theta}\cos\theta\,d\theta$

$= \int d\theta = \theta + C = \text{Sin}^{-1}x + C$

$\int \dfrac{1}{\sqrt{1-x^2}}\,dx = \text{Sin}^{-1}x + C$

It does not matter that we use trigonometric substitution, even though we are covering hyperbolic substitution since the expression is algebraic and satisfies the criterion for this application.

$\int \text{Sech}^{-1}x\,dx = x\,\text{Sech}^{-1}x + \text{Sin}^{-1}x + C$ if $\text{Sech}^{-1}x > 0$ or
$= x\,\text{Sech}^{-1}x - \text{Sin}^{-1}x + C$ if $\text{Sech}^{-1}x < 0$

Example 5: Find $\int \text{Coth}^{-1}x\,dx$

Step 1: Let $u = \text{Coth}^{-1}x$, and let $dv = 1dx$ (version 1) or let $v = 1$ (for version 2)

Then $\int \text{Coth}^{-1}x\,dx = x\text{Coth}^{-1}x - \int \dfrac{x}{1-x^2}\,dx$

Step 2: For $\int \dfrac{x}{1-x^2}\,dx$ (A) (integrate by u-subst.)

Let $t = 1-x^2$. $\dfrac{dt}{dx} = -2x$, $dx = -\dfrac{dt}{2x}$.

Substituting in (A) we obtain

$\int \dfrac{x}{t}\cdot\left(-\dfrac{dt}{2x}\right) = -\dfrac{1}{2}\int \dfrac{dt}{t}$

Now, putting everything together,

$\int \text{Coth}^{-1}x\,dx = x\text{Coth}^{-1}x + \dfrac{1}{2}\int \dfrac{dt}{t}$

$= x\,\text{Coth}^{-1}x + \dfrac{1}{2}\ln t + C$

$\int \text{Coth}^{-1}x\,dx = x\,\text{Coth}^{-1}x + \dfrac{1}{2}\ln(1-x^2) + C$

Integration by parts Formula
(Version 1)

$\int u\,dv = uv - \int v\,du$

$dv = dx$
$v = x$
$du = \dfrac{1}{1-x^2}\,dx$

Version 2

$\int (uv)dx = u\int v - \int\left(\dfrac{du}{dx}\cdot\int v\right)dx$

$v = 1$
$\int v\,dx = \int 1dx = x$
$\dfrac{du}{dx} = \dfrac{1}{1-x^2}$ (version 2)

Example 6: Find $\int \text{Csch}^{-1}x\,dx$ (Integrating by parts; process Type 1)

Step 1: Let $u = \text{csch}^{-1}x$, and let $dv = 1dx$ (version 1) or let $v = 1$ (for version 2)

Then $\int \text{Csch}^{-1}x\,dx = x\text{Csch}^{-1}x - \int -\dfrac{x}{|x|\sqrt{1+x^2}}\,dx$

$\int \text{Csch}^{-1}x\,dx = x\text{Csch}^{-1}x + \int \dfrac{1}{\sqrt{1+x^2}}\,dx$

Step 2: For $\int \dfrac{1}{\sqrt{1+x^2}}\,dx$.   (A)

 Let $x = \tan\theta$; (trigonometric substitution)

  $x^2 = \tan^2\theta$

 Then $\dfrac{dx}{d\theta} = \sec^2\theta$, and $dx = \sec^2\theta\,d\theta$

 Substitute in $\int \dfrac{1}{\sqrt{1+x^2}}\,dx$, (A) ,to obtain

$\int \dfrac{1}{\sqrt{1+\tan^2\theta}}\sec^2\theta\,d\theta = \int \dfrac{1}{\sqrt{\sec^2\theta}}\sec^2\theta\,d\theta$

$= \int \dfrac{1}{\sec\theta}\sec^2\theta\,d\theta$

$= \int \sec\theta\,d\theta$

$= \ln|\sec\theta + \tan\theta| + C$

$= \ln\left|\sqrt{1+x^2} + x\right| + C$

 Now, putting everything together,

$\int \text{Csch}^{-1}x\,dx = x\text{Csch}^{-1}x + \ln\mid x +\sqrt{x^2+1} \mid + C$

Integration by parts Formula
(Version 1)

$\int u\,dv = uv - \int v\,du$

$dv = dx$
$v = x$

$du = \dfrac{1}{\sqrt{1+x^2}}\,dx$

Version 2

$\int (uv)dx = u\int v - \int \left(\dfrac{du}{dx}\cdot\int v\right)dx$

$\dfrac{du}{dx} = -\dfrac{1}{|x|\sqrt{1+x^2}}$

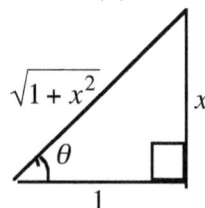

$\dfrac{x}{1} = \tan\theta$ $\left(\dfrac{\text{opposite side}}{\text{adjacent side}}\right)$

$1 + \tan^2\theta = \sec^2\theta$

$\sec\theta = \sqrt{1+x^2}$

Lesson 58 Exercises

Find the following:

1. $\int \text{Sinh}^{-1}x\,dx$;	**4.** $\int \text{Sech}^{-1}x\,dx$	**7.** $\int x\,\text{Tanh}^{-1}x\,dx$
2. $\int \text{Cosh}^{-1}x\,dx$;	**5.** $\int \text{Csch}^{-1}x\,dx$	**8.** $\int x\,\text{Sinh}^{-1}x\,dx$
3. $\int \text{Tanh}^{-1}x\,dx$	**6.** $\int \text{Coth}^{-1}x\,dx$	**9.** $\int \text{Sinh}^{-1}\left(\dfrac{x}{2}\right)dx$

Ans: **7.** $\dfrac{x^2}{2}\text{Tanh}^{-1}x + \dfrac{1}{2}x - \dfrac{1}{2}\text{Tanh}^{-1}x + C$; or $\dfrac{x^2}{2}\text{Tanh}^{-1}x + \dfrac{1}{2}x - \dfrac{1}{4}\ln\left|\dfrac{1+x}{1-x}\right| + C$

8. $\dfrac{x^2}{2}\text{Sinh}^{-1}x - \dfrac{1}{4}x\sqrt{1+x^2} + \dfrac{1}{4}\text{Sinh}^{-1}x + C$;

9. $x\text{Sinh}^{-1}\left(\dfrac{x}{2}\right) - 2\sqrt{1+\dfrac{x^2}{4}} + C$ or $x\text{Sinh}^{-1}\left(\dfrac{x}{2}\right) - \sqrt{4+x^2} + C$

Appendix A

Infinite Series and Series Representations

A series may be convergent or divergent with respect to the sum of its terms. For a convergent series, the sum of n terms remains finite as n increases indefinitely. A finite series is always convergent. Series which are not convergent are said to be divergent,

In integration, when we cannot obtain antiderivatives using elementary functions, we can obtain approximations for the integrals using convergent **infinite series.**

Below, are presented some convergent series with the interval of convergence. It is the author's opinion that because of time constraints, in Calculus 2, students accept the convergence of these series and apply them in integration. In Calculus 3, or perhaps, in Advanced Calculus, students will determine the convergence or divergence of these series and others.

Series Representations

Examples of convergent infinite series

Exponential series:

$$e^x = \sum_{n=0}^{+\infty} \frac{x^n}{n!} = 1 + x + \frac{x^2}{2!} + \frac{x^3}{3!} + \dots + \frac{x^n}{n!} \qquad -\infty < x < \infty$$

Logarithmic Series:

$$\ln(1 + x) = x - \frac{x^2}{2} + \frac{x^3}{3} - \frac{x^4}{4} + \dots \qquad -1 < x \le 1$$

Trigonometric Series:

1. $\sin x = x - \frac{x^3}{3!} + \frac{x^5}{5!} - \frac{x^7}{7!} + \dots \qquad -\infty < x < \infty$

   Memory device: Note that $\sin x$, an odd function, is expressed as a series of odd powers of x.

2. $\cos x = 1 - \frac{x^2}{2!} + \frac{x^4}{4!} - \frac{x^6}{6!} + \dots \qquad -\infty < x < \infty$

   Memory device: Note that $\cos x$, an even function, is expressed as a series of even powers of x.

3. $\tan x = x + \frac{x^3}{3} + \frac{2x^5}{15} + \frac{17x^7}{315} + \dots \qquad |x| < \frac{\pi}{2}$

Hyperbolic Functions:

$$\sinh x = x + \frac{x^3}{3!} + \frac{x^5}{5!} + \frac{x^7}{7!} + \frac{x^9}{9!} + \dots$$

$$\cosh x = 1 + \frac{x^2}{2!} + \frac{x^4}{4!} + \frac{x^6}{6!} + \dots$$

Note: Observe the similarities between $\sin x$ and $\sinh x$; and between $\cos x$ and $\cosh x$

Inverse Trigonometric Series:

$$\tan^{-1} x = x - \frac{x^3}{3} + \frac{x^5}{5} - \frac{x^7}{7} + \dots \qquad |x| \le 1$$

Appendix B

Mathematical Induction

Mathematical induction deals with the natural numbers (or positive integers).

Axiom: An axiom is a statement in mathematics that we willingly accept or assume to be true so as to deduce other statements.

Deductive reasoning: This is a process of drawing a particular conclusion based on a general statement.

Inductive reasoning: This is a process of drawing a general conclusion based on a number of specific cases. Such conclusions are probable. Also, such a general conclusion can be made to become more probable if we consider more and better choice of cases.

In mathematical induction, we follow a logical procedure which allows us to prove the validity of a general conclusion.

The axiom of mathematical induction

Let P be a set of positive integers (natural numbers). Then

(1) If the first positive integer, 1, belongs to the set P and

(2) If the positive integer k belongs to P, then all positive integers belongs to P.

In using the axiom of mathematical induction to construct proofs we must always verify the above two properties of the axiom to complete the proof. We must thus verify that:

(1) The integer 1 satisfies the statement to be proved and

(2) Assuming that the positive integer k belongs to P, the positive integer $(k + 1)$ satisfies the   statement to be proved.

The proofs we shall cover in this chapter will involve only the six basic operations of addition, subtraction, multiplication, division, powers and roots.

Example 1: Prove that the sum of the first n positive even integers is $n(n+1)$ 438
.   That is prove that $2 + 4 + 6 + ... + 2n = n(n+1)$ (1)

Requirement: We must show that

(a) when $n = 1, 2,...$ LHS of (1) = RHS of (1) and

(b) when $n = n + 1$ or $k + 1$, LHS of (1) = RHS. of (1)

(LHS and RHS mean Left-Hand Side and Right-Hand Side of the equation, respectively)

Proof

The general term is $2n$, where n = term-number.

Step 1: Verify that when $n =1, 2$, both LHS and RHS are equal.
          ($n =1$ means considering only the **first** term on the LHS and on the RHS
          replace n by 1) Substituting $n = 1$ in (1); that is, in $2n = n(n+1)$), we obtain
For $n = 1$:

$$2(1) \overset{?}{=} 1(1 + 1)$$

$$2 = 2$$
                    (Note that $n = 2$ means that on the LHS consider the sum of only the first
                    two terms and on the right hand side replace n by 2.)
For $n = 2$

$$\text{Then } 2\underset{n=1}{(1)} + 2\underset{n=2}{(2)} \overset{?}{=} \underset{n=2}{2} (\underset{n=2}{2} +1)$$

$$2 + 4 \overset{?}{=} 3(3)$$

$$6 = 6$$

For curiosity, we try $n =3$, although in practice verification for $n =1$ is sufficient.
$n =3$ means consider the sum of only the first **three** terms on LHS. and substitute
$n = 3$ on RHS.
For $n = 3$:

$$\text{Then } 2\underset{n=1}{(1)} + 2\underset{n=2}{(2)} + 2\underset{n=3}{(3)} \overset{?}{=} \underset{n=3}{3} (\underset{n=3}{3} +1)$$

$$2 + 4 + 6 \overset{?}{=} 3(4)$$

$$12 = 12$$

Step 2:   Prove that when $n = n + 1$, $2 + 4 + 6 + ... + 2n = n(n+1)$ (1)
          Let $n = k$ in (1): $2 + 4 + 6 + ... + 2k = k(k+1)$ (2)

$$\boxed{2 + 4 + 6 + ... + 2k = k^2 + k}$$ (3)

Also, let $n = (k + 1)$ in (1) i.e. substitute $(k + 1)$ for n in (1).
          Then $2 + 4 + 6 + ... + 2(k + 1) = (k + 1)(k + 1 + 1)$ (4)

$$\boxed{2 + 4 + 6 + ... + 2(k + 1) = k^2 + 3k + 2}$$ (5)

Now, we will show that LHS and RHS of (5) are equal.

Equation (3) is accepted as a true statement since we have shown that this is true when $n = k$ We want to manipulate equation (3) by some basic algebraic operations so that both sides of the equation are equal.) In this proof, we add equals to both sides of the equation until the LHS of equation (3) is identical with the LHS of equation (5) and also that the RHS of equation (3) is identical with the RHS of (5).

We write the LHS of equation (5) in another form to indicate explicitly the general term $2k$.

Then $2 + 4 + 6 + ... + 2(k + 1) = 2 + 4 + 6 + ...2k + 2(k + 1)$ \hfill (6)

So now replace LHS. of equation (5) by RHS. of equation (6) and equation (5) becomes

$$2 + 4 + 6 + ... + 2k + 2(k + 1) = k^2 + 3k + 2 \qquad (7)$$

So now we prove equation (7) instead of equation (5).

Observe the LHS of equation (3) and the LHS of equation (7), and notice that if we add $2(k+ 1)$ to LHS of equation (3), that side will be identical with LHS. of equation (7), but we should also add $2(k+ 1)$ to RHS of equation (3), that is adding equals to both sides of an equation.

Adding $2(k+ 1)$ to both sides of equation (3), we obtain:

$2 + 4 + 6 + ... + 2k + 2(k + 1) = k^2 + k + 2(k + 1)$ \hfill (8)

$2 + 4 + 6 + ... + 2k + 2(k + 1) = k^2 + 3k + 2$ \hfill (9)

Now, replacing LHS of equation (9) by LHS of equation (6)   (since LHS of (9) = RHS of (6)

$2 + 4 + 6 + ... + 2(k + 1) = k^2 + 3k + 2$

Hence, we have shown that equation (5) is true.

We have therefore shown that equation (1) holds for $n = 1$ and $n = k + 1$,

$\therefore 2 + 4 + 6 + ... + 2n = n(n + 1)$ and the proof is complete.

Note: Some of the comments and explanations in the above proof are usually not part of the proof, and may therefore be omitted from the proof. They were added to aid the student understand the principles involved in the proof.

Example 2 Prove that $1 \times 2 + 2 \times 3 + ... + n(n+1) = \dfrac{n(n+1)(n+2)}{3}$ (1)

Proof: For $n = 1$,

       LHS of (1) $= 1(1+1) = 1(2) = 2$

RHS of (1) $= \dfrac{1(1+1)(1+2)}{3} = \dfrac{1(2)(3)}{3} = 2$

(Quantities equal to he same quantity are equal to each other)

Therefore, LHS of (1) = RHS of (1).

Let $n = k$ in (1). Then

$1 \times 2 + 2 \times 3 + ... + k(k+1) = \dfrac{k(k+1)(k+2)}{3}$ true   (2)

Also, let $n = k+1$ in (1): $1 \times 2 + 2 \times 3 + ... + (k+1)(k+2) = \dfrac{(k+1)(k+2)(k+3)}{3}$ (3)

Now, we want to prove that the LHS of (3) = RHS of (3).

We rewrite LHS of (3) so that the general term is explicitly indicated.

Then equation (3) becomes

$1 \times 2 + 2 \times 3 + ... + \underbrace{k(K+1)}_{\text{general term}} + (k+1)(k+2) = \dfrac{(k+1)(k+2)(k+3)}{3}$ (4)

Now, instead of proving (3), we prove (4). Expanding equation (4):

$1 \times 2 + 2 \times 3 + ... + 2k^2 + 4k + 2 \overset{?}{=} \dfrac{k^3 + 6k^2 + 11k + 6}{3}$ true   (5)

Similarly, expanding (2), we obtain

$1 \times 2 + 2 \times 3 + ... + k^2 + k = \dfrac{k^3 + 3k^2 + 2k}{3}$ true   (6)

Since, equation (6) is true (why?) we can add equal quantities to both sides of the equation to obtain an equivalent. It is up to us now to obtain equation (5) from equation (6) by algebraic operations. Thus, we add $\dfrac{3k^2 + 9k + 6}{3}$ to both sides of (6). (RHS of (5) minus RHS of (6) yields this addend) Then we obtain

$1 \times 2 + 2 \times 3 + \dfrac{3k^2 + 9k + 6}{3} + ... + k^2 + k = \dfrac{3k^2 + 9k + 6}{3} + \dfrac{k^3 + 3k^2 + 2k}{3}$ (7)

Simplifying (7): $1 \times 2 + 2 \times 3 + ... + 2k^{2i} + 4k + 2 = \dfrac{k^3 + 6k^2 + 11k + 6}{3}$ (8)

Comparing equations (5) and (8), we observe that the LHS's are identical and their RHS's are also identical. We have therefore proved (4) and also the original equation (3). Since we have shown that equation (1) holds for $n = 1$ and $n = k+1$, $1 \times 2 + 2 \times 3 + ... + n(n+1) = \dfrac{n(n+1)(n+2)}{3}$, and the proof is complete.

Note above: In deciding on adding $\dfrac{3k^2 + 9k + 6}{3}$ to both sides of (6), we asked:

"what can be done to the RHS of (6) so that it is the same as the RHS of (5)?".
Practice this technique.

Example 3: Use mathematical induction to prove that $: \dfrac{n+1}{n!} < \dfrac{8}{2^n}$ (1)

Proof

$$\text{For } n = 1: \quad \frac{2}{1} \overset{?}{<} \frac{8}{2}$$

$$2 < 4 \quad \text{True}$$

$$\text{For } n = 2: \quad \frac{3}{2} \overset{?}{<} \frac{8}{4}$$

$$\frac{3}{2} < 2 \quad \text{True}$$

Thus, inequality (1) above holds for $n = 1$, and 2.

Now, assume that inequality (1) holds for $n = k$ (where k is an integer)

If it can be shown that inequality (1) holds for $n = k + 1$, the proof would be complete.

Replacing n by $k + 1$ in inequality (1), we obtain

$$\frac{k+2}{(k+1)!} \overset{?}{<} \frac{8}{2^{k+1}} \tag{2}$$

Rewriting inequality (2) in a different form: $\dfrac{k+2}{(k+1)!} \overset{?}{<} 2^{2-k}$ (3)

Similarly, rewriting inequality (1): $\dfrac{k+1}{k!} \overset{?}{<} 2^{3-k}$ (1)

Assuming that inequality (1) is true, we would show that inequality (3) is true.

Multiplying inequality (1) by 2^{-1}: $\dfrac{k+1}{2k!} \overset{?}{<} 2^{2-k}$ (4)

We will now show that left-hand side of inequality (3) < left-hand side of inequality (4):

$$\frac{k+2}{(k+1)k!} \overset{?}{<} \frac{k+1}{2k!} \quad \textbf{(Note:} \ (k+1)! = (k+1)(K+1-1)! = (k+1)k!\textbf{)}$$

$$2k + 4 \overset{?}{<} k^2 + 2k + 1 \quad \text{(dividing out the } k! \text{ and undoing the denominators)}$$

$$4 < k^2 + 1 \tag{5}$$

Clearly, inequality (5) is true for $k \geq 2$.

$$\text{Now, } \frac{k+2}{(k+1)k!} < \frac{k+1}{2k!} < 2^{2-k}$$

$$\therefore \quad \frac{k+2}{(k+1)k!} < 2^{2-k} \text{ and inequality (3) above holds.}$$

Consequently, inequality (1) holds for $n = k + 1$.

It has also been shown that inequality (1) holds for $n = 1$.

We have therefore shown that inequality (1) holds for $n = 1$ and $n = k + 1$.

$$\therefore \quad \frac{n+1}{n!} < \frac{8}{2^n}$$

$$\text{QED}$$

Appendix B Exercises

Prove each of the following using mathematical induction:

1. $1 + 3 + 5 + \ldots + 2n - 1 = n^2$;

2. $2 + 4 + 6 + \ldots + 2n = n(n + 1)$

3. $1 + 2 + 3 \ldots + n = \dfrac{n(n + 1)}{2}$;

4. $1^3 + 2^3 + 3^3 + \ldots + n^3 = \dfrac{n^2(n + 1)^2}{4}$

5. $4 + 7 + 10 + \ldots + 3n + 1 = \dfrac{n(3n + 5)}{2}$;

6. : $\dfrac{n + 1}{n!} < \dfrac{8}{2^n}$

Appendix C

Graphs of exponential functions

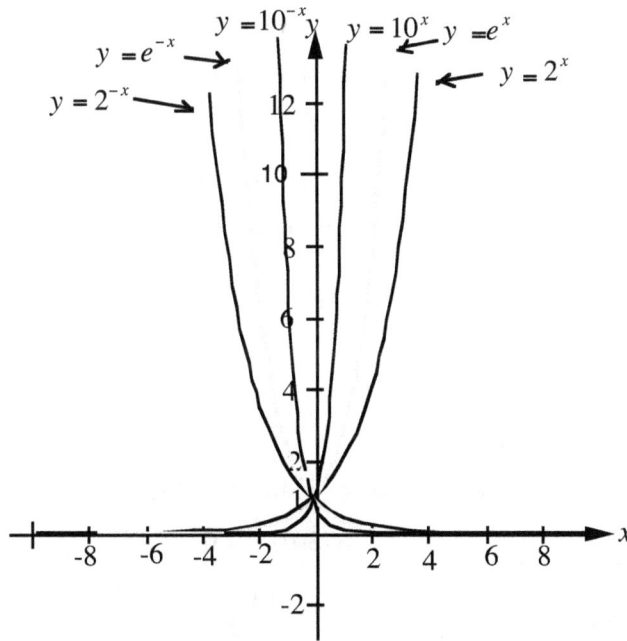

Revolution about the y–axis

Example 2: Find the area of the surface of revolution by revolving about the

y–axis, the curve given by $y = \frac{1}{3}x^3$ from $x = 0$ to $x = 3$

Given: $x = g(y)$, $S = 2\pi \int_c^d x\sqrt{1 + \left(\frac{dx}{dy}\right)^2}\, dy$.   (A)

Step 1: From $y = \frac{1}{3}x^3$

$$\boxed{x = (3y)^{\frac{1}{3}}}$$ (B)

For the limits: When $x = 0$. $y = 0$.

When $x = 3$. $y = \frac{1}{3}(3)^3 = 9$. The

limits are from $y = 0$ to $y = 9$,

Step 2: $\dfrac{dx}{dy} = \dfrac{1}{\frac{dy}{dx}} = \boxed{\dfrac{1}{x^2}}$ $(\frac{dy}{dx} = x^2)$

$$\left(\frac{dx}{dy}\right)^2 = \boxed{\frac{1}{x^4}}$$

Step 3: Substitute for $x = (3y)^{\frac{1}{3}}$,

$\left(\dfrac{dx}{dy}\right)^2 = \boxed{\dfrac{1}{x^4}}$ and the limits $y = 0$,

$y = 9$ in equation (A) to obtain

$$S = 2\pi \int_0^9 (3y)^{\frac{1}{3}} \sqrt{1 + \left(\frac{1}{(3y)^{\frac{1}{3}}}\right)^4}\, dy$$

$$= 2\pi \int_0^9 (3y)^{\frac{1}{3}} \sqrt{1 + \left(\frac{1}{(3y)^{\frac{4}{3}}}\right)}\, dy \quad (C)$$

Step 4 Let $u = (3y)^{\frac{1}{3}}$. Then

$$\frac{du}{dy} = 3^{\frac{1}{3}}(\tfrac{1}{3})y^{-\frac{2}{3}}; \quad dy == 3^{\frac{2}{3}} y^{\frac{2}{3}} du$$

Express y in terms of u.

From $u = (3y)^{\frac{1}{3}} = 3^{\frac{1}{3}} y^{\frac{1}{3}}; \quad \dfrac{u}{3^{\frac{1}{3}}} = y^{\frac{1}{3}}$

$$\left(\frac{u}{3^{\frac{1}{3}}}\right)^2 = \left(y^{\frac{1}{3}}\right)^2; \quad \frac{u^2}{3^{\frac{1}{3}}} = y^{\frac{2}{3}}$$

$$dy = \frac{3^{\frac{2}{3}} u^2 du}{3^{\frac{2}{3}}}; \quad dy = u^2 du$$

For the limits: when $y = 0$,

$u = (3(0))^{\frac{1}{3}} = 0$; when

$y = 9, u = (3(9))^{\frac{1}{3}} = 3$

Step 5: $2\pi \int_0^3 u \sqrt{1 + \frac{1}{u^4}}\, u^2 du$

$$2\pi \int_0^3 u^3 \sqrt{\frac{u^4 + 1}{u^4}}\, du$$

$$2\pi \int_0^3 \frac{u^3}{u^2} \sqrt{u^4 + 1}\, du$$

$$2\pi \int_0^3 u\sqrt{u^4 + 1}\, du \qquad ((D)$$

Step 6: Integrate by trigonometric
substitution

Let $u^2 = \tan\theta$. Then $2u\frac{du}{d\theta} = \sec^2\theta$

$du = \dfrac{\sec^2 d\theta}{2u}$.

Using $u^2 = \tan\theta$, we draw the
corresponding right triangle

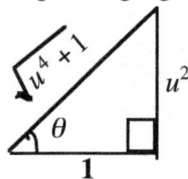

$\sec\theta = \sqrt{u^4 + 1}$; $\tan\theta = u^2$

Step 7: We use indefinite integration now
and express the integration results in terms
of u later; and use the limits in (D)

$$2\pi \int u\sqrt{\tan^2\theta + 1}\, \frac{\sec^2\theta d\theta}{2u}$$

$$= \pi \int \sqrt{\tan^2\theta + 1} \bullet \sec^2\theta d\theta$$

$$= \pi \int \sqrt{\sec^2\theta} \bullet \sec^2\theta d\theta$$

$$= \pi \int \sec\theta \bullet \sec^2\theta d\theta$$

$$= \pi \int \sec^3\theta d\theta \qquad (E)$$

Step 8:
From page 260, Case 6, Example

$$\int \sec^3 \theta \, d\theta = \tfrac{1}{2} \ln|\sec\theta + \tan\theta| + \tfrac{1}{2} \tan\theta \sec\theta + C \quad \text{(F)}$$

$$\pi \int \sec^3 \theta d\theta = \tfrac{\pi}{2} \ln|\sec\theta + \tan\theta| + \tfrac{\pi}{2} \tan\theta \sec\theta + C$$

Express the right side in terms of u. $(\sec\theta = \sqrt{u^4 + 1}, \ \tan\theta = u^2)$

$$2\pi \int_0^3 u\sqrt{u^4 + 1} \, du = \left[\tfrac{\pi}{2} \left(\ln\left|\sqrt{u^4 + 1} + u^2\right| + u^2\sqrt{u^4 + 1} \right) \right]_0^3$$

$$= \tfrac{\pi}{2} \left\{ \left(\ln\left|\sqrt{3^4 + 1} + 3^2\right| + 3^2\sqrt{3^4 + 1} \right) - \left(\ln\left|\sqrt{0^4 + 1} + 0^2\right| + 0^2\sqrt{0^4 + 1} \right) \right\}$$

$$= \tfrac{\pi}{2} \left\{ \left(\ln\left|\sqrt{81 + 1} + 9\right| + 9\sqrt{81 + 1} \right) - \left(\ln\left|\sqrt{1}\right| + 0 \right) \right\}$$

$$= \tfrac{\pi}{2} \left\{ \left(\ln\left|\sqrt{82} + 9\right| + 9\sqrt{82} \right) - (0) \right\}$$

$$= \tfrac{\pi}{2} \left(\ln\left|\sqrt{82} + 9\right| + 9\sqrt{82} \right)$$

About Infinite Limit Calculations

Operations involving infinity do not obey all the properties of the basic algebraic operations. Do not operate on infinity as you operate on real numbers. In operations involving infinite limits, we sometimes meet situations in which we divide by a quantity approaching zero. This situation occurs, sometimes, when using the method of dividing every term by the highest power of the variable in the expression. Note for example that when we write

$\lim\limits_{x \to \pm\infty} \dfrac{1}{x} = 0$, we mean as x approaches $\pm$ infinity (x increases or decreases

without bound,) , $\dfrac{1}{x}$ approaches 0, and it is **not** that $\dfrac{1}{x}$ equals zero, and we

can appropriately let $\dfrac{1}{x} = 0^+$ or 0^- , where 0^+ is a very small positive

number, and 0^- is a very small negative number. In addition to the Infinity Table below, the author has added a new table, the Zero Table which will be useful in handling operations involving "zeros only" in the denominator.

Guidelines for handling zeros

In the evaluation of limits, if we obtain zeros in a substitution process, we can treat the zeros obtained as follows:

1. Zeros resulting in the numerator will remain as zeros, as usual.

2. Zeros resulting in the denominator will be treated as either 0^+ or as 0^- according to whether the approach to zero is from positive or negative values.

Infinity Table For $+\infty$ and $-\infty$		Zero Table (Author's) For 0^+ and 0^-	
Sum : **1.** $+\infty + \infty = +\infty$ **2.** $-\infty - \infty = -\infty$ **3.** $+\infty + a = \infty$ **4.** $-\infty + a = -\infty$ **Product** **5.** $(+\infty)(+\infty) = +\infty$ **6.** $(-\infty)(-\infty) = +\infty$ **7.** $(-\infty)(+\infty) = -\infty$ **8.** $(+\infty)(-\infty) = -\infty$	**9.** $(a)(+\infty) = +\infty$ if $a > 0$ **10.** $(a)(-\infty) = -\infty$ if $a > 0$ **11.** $(-a)(-\infty) = +\infty$ **Division** **12.** $\dfrac{a}{+\infty} = \dfrac{a}{-\infty} = 0$ $a \neq 0$	**Sum** **13.** $0^+ + 0^+ = 0^+$ **14.** $0^- + 0^- = 0^-$ **15.** $0^+ + a = a$ **16.** $0^- + a = a$ **Division** **17.** $\dfrac{a}{0^+} = +\infty$ **18.** $\dfrac{a}{0^-} = -\infty$ $a > 0$	**Product** **19.** $(0^+)(0^+) = 0^+$ **20** $(0^-)(0^-) = 0^+$ **21** $(0^+)(0^-) = 0^-$ **22** $(0^-)(0^+) = 0^-$ **23.** $(0^-)a = 0^-$ **24.** $(0^+)a = 0^+$

Do not use the following, since no meanings are assigned to them.

Try to convert the original expression to a form which on substitution will

result in one of the above; or use other techniques such as L' H$\hat{o}$pital' s **Rule.**

Addition	Multiplication & Division	Powers
25. $(+\infty) + (-\infty)$ **26.** $\infty - \infty$ **27.** $0^+ \pm 0^-$	**28.** $0 \bullet \infty$; **29.** $\dfrac{0}{0}$; **30.** $\dfrac{\infty}{\infty}$; **31.** $\dfrac{0^+}{0^+}$	**32** 0^0; **33.** ∞^0 **34.** 1^∞

$\lim\limits_{n \to \pm\infty} \dfrac{a}{x^n} = 0;$ $\lim\limits_{n \to +\infty} x^n = \infty;$ $\lim\limits_{n \to -\infty} x^n = (-1)^n \infty$

Zero Table Note that the sums with the same sign follow the usual rules of signs. Note also that the product with the same sign and different signs follow the usual rules of signs.

Analogy Involving $+\infty$ and $-\infty$

Let us look at the following from a layperson's point of view.
We first look at some of the permissible operations in the Infinity Table.

1. For $+\infty + \infty = +\infty$: If you have a lot of money and you add a lot of money, the result is that you still have a lot of money.

2. For $-\infty - \infty = -\infty$: If you lose a lot of money and lose a lot of money, you have lost a lot of money.

3. For $+\infty + a = +\infty$:. If you have a lot of money and you add some money, however small or large, you still have a lot of money.

4. $-\infty + a = -\infty$: If you borrow a lot money and you payback some of what you owe, however large or small, you still owe a lot of money.

5. For $(-\infty)(-\infty) = +\infty$: Apply the rules of multiplication of signed numbers.

Now, let us look at a non-permissible operation with meaningless result.
For $+\infty - \infty$. If you were paid lot of money, and you borrowed a lot of money, you cannot tell if you have money or you owe money.

About Division Involving Zero

We distinguish between division by $x = 0$ and by x approaching zero from the right or from the left.

1. When $x = 0$ (exactly), $\dfrac{3}{x} = \dfrac{3}{0}$ is undefined.

2. When x approaches 0 from the right,

$\dfrac{3}{x} = \dfrac{3}{0^+} = +\infty$ (Note that $+\infty$ is not a number)

That is, when x approaches 0 from the right

$\dfrac{3}{x}$ increases without bound.

Formally, $\displaystyle\lim_{x \to 0^+} \dfrac{3}{x} = +\infty$

Illustration: $\dfrac{3}{\frac{1}{1000,000,000}} = 3,000,000,000$

2. When x approaches 0 from the left,

$\dfrac{3}{x} = \dfrac{3}{0^-} = -\infty$ (Note that $-\infty$ is not a number)

That is, when x approaches 0 from the left,

$\dfrac{3}{x}$ decreases without bound.

Formally, $\displaystyle\lim_{x \to 0^-} \dfrac{3}{x} = -\infty$

Illustration: $\dfrac{3}{-\frac{1}{1000,000,000}} = -3,000,000,000$

By $\displaystyle\lim_{x \to +\infty} f(x) = +\infty$, we mean as x increases without bound, $f(x)$ increases without bound. and the limit does not exist.
(x increases without bound implies using larger and larger values of x)

We distinguish between small negative numbers and large negative numbers by examples.
Example
-0.000000001 is a small negative number, but $-1,000,000,000$ is a large negative number.

Comments

($\dfrac{3}{0}$ is undefined because there is no number such that "that number" $\times \, 0 = 3$)

(where 0^+ is a small positive number.
Example: $\frac{1}{1000,000,000}$)

where 0^- is a small negative number)
Example: $-\frac{1}{1000,000,000}$)
Example: For large values of x, say 10^9 or $1,000,000,000$,

$\dfrac{1}{x} = \frac{1}{1000,000,000} \approx 0$ (actually, a very, very small positive number.

Similarly, for large positive values of x

$\dfrac{1}{x^3} \approx 0^+$. For large negative

values of x $\dfrac{1}{x^3} \approx 0^-, \dfrac{1}{x^2} \approx 0^+,$

$\dfrac{1}{x} = 0^-$

$= \dfrac{1}{0^+} = \infty$

Similarly, $\dfrac{3}{0^+} = \infty$

Also, $\dfrac{3}{0^-} = -\infty$

Extended Zero Table

a^+ and a^-	Examples
$a^+ - a = 0^+$	$2^+ - 2 = 0^+$
$a^- - a = 0^-$	$2^- - 2 = 0^-$
$a - a^- = 0^+$	$2 - 2^- = 0^+$
$a - a^+ = 0^-$	$2 - 2^+ = 0^-$

Note (From the Zero Table) that

$$\frac{a}{0^+} = +\infty \qquad a > 0$$

$$\frac{a}{0^-} = -\infty \qquad a > 0$$

Logarithms

$\ln b = \log_e b = \dfrac{\log_b b}{\log_b e} = \dfrac{1}{\log_b e}$. Also, $\boxed{\log_b x = \dfrac{\ln x}{\ln b}}$ <--base b to base e

$\ln b = \dfrac{1}{\log_b e}$, and $\log_b e = \dfrac{1}{\ln b}$. Therefore, $\log_b e$ and $\ln b$ are reciprocals of

each other. To show that $\log_b e$ and $\ln b$ are reciprocals of each other, we

show that their product equals 1. First, we change $\log_b e$ to base e. Then

$\log_b e = \dfrac{\log_e e}{\log_e b}$, and $\log_b e \bullet \log_e b = \dfrac{\log_e e}{\log_e b} \bullet \dfrac{\log_e b}{1} = \dfrac{1}{\ln b} \bullet \dfrac{\ln b}{1} = 1.$

Thus $\log_b e$ and $\log_e b$ are reciprocals of each other.

Method 2 for Case 6, Example 2 (p.260)

Example 2 Find $\int \sec^3 x \, dx$ (**Method 1**)

$\int \sec^3 x \, dx = \int \sec^2 x \sec x \, dx$

  (Integrate by parts: $u = \sec x$. $dv = \sec^2 x \, dx$; $v = \int \sec^2 x \, dx = \tan x$

$\qquad\qquad = \sec x \tan x - \int \sec x \tan x \bullet \tan x \, dx$

$\qquad\qquad = \sec x \tan x - \int \sec x \tan^2 x \, dx$

$\qquad\qquad = \sec x \tan x - \int \sec x (\sec^2 x - 1) \, dx \quad (\tan^2 x = \sec^2 x - 1)$

$\qquad\qquad = \sec x \tan x - \int \sec^3 x \, dx + \int \sec x \, dx$

$2 \int \sec^3 x \, dx = \sec x \tan x + \int \sec x \, dx$ (Note the Type 3 Integration-by-parts)

$\quad \int \sec^3 x \, dx = \tfrac{1}{2} \sec x \tan x + \tfrac{1}{2} \int \sec x \, dx$

$\qquad\qquad = \tfrac{1}{2} \sec x \tan x + \tfrac{1}{2} |\ln \sec x + \tan x| + C$

Derivation of $\cos x = \dfrac{1-u^2}{1+u^2}$, using $u = \tan\frac{1}{2}x$.

From $\cos 2x = 2\cos^2 x - 1$

$\cos x = 2\cos^2 \frac{x}{2} - 1$

$\qquad = 2\dfrac{1}{\sec^2 \frac{x}{2}} - 1$

$\qquad = 2\dfrac{1}{\tan^2 \frac{x}{2}+1} - 1$

$\qquad = 2\dfrac{1}{\tan^2 \frac{x}{2}+1} - 1$

$\qquad = 2\dfrac{1}{u^2+1} - 1$

$\qquad = \dfrac{2}{u^2+1} - \dfrac{u^2+1}{u^2+1}$

$\qquad = \dfrac{2-u^2-1}{u^2+1}$

$\cos x = \dfrac{1-u^2}{1+u^2}$.

Derivation of $dx = \dfrac{2du}{1+u^2}$ using $u = \tan\frac{1}{2}x$.

Let $v = \frac{x}{2}$

Then $u = \tan v$

$\qquad \dfrac{dv}{dx} = \dfrac{1}{2}$

$\qquad \dfrac{du}{dv} = \sec^2 v = \sec^2 \frac{x}{2}$

$\qquad \dfrac{du}{dx} = \dfrac{du}{dv} \bullet \dfrac{dv}{dx}$

$\qquad\qquad = \sec^2 \frac{x}{2} \bullet \frac{1}{2}$

$\qquad\qquad = \frac{1}{2}\sec^2 \frac{x}{2}$

$\qquad\qquad = \frac{1}{2}(1 + \tan^2 \frac{x}{2})$

$\qquad \dfrac{du}{dx} = \dfrac{1+u^2}{2}$

$\qquad dx = \dfrac{2du}{1+u^2}$.

INDEX

A

C

G

H

I

T

V

Trigonometric Identities

Reciprocal Identities

1. $\sec\theta = \dfrac{1}{\cos\theta}$; **2.** $\csc\theta = \dfrac{1}{\sin\theta}$

3. $\cot\theta = \dfrac{1}{\tan\theta}$

Ratio Identities

1. $\tan\theta = \dfrac{\sin\theta}{\cos\theta}$

2. $\cot\theta = \dfrac{\cos\theta}{\sin\theta}$

Pythagorean Identities

1. $\sin^2\theta + \cos^2\theta = 1$

2. $1 + \tan^2\theta = \sec^2\theta$

3. $1 + \cot^2\theta = \csc^2\theta$

Sum and Difference Identities (Addition and Subtraction Formulas)

1. $\cos(A - B) = \cos A\cos B + \sin A\sin B$; **4.** $\sin(A - B) = \sin A\cos B - \sin B\cos A$

2. $\cos(A + B) = \cos A\cos B - \sin A\sin B$; **5.** $\tan(A + B) = \dfrac{\tan A + \tan B}{1 - \tan A\tan B}$

3. $\sin(A + B) = \sin A\cos B + \sin B\cos A$, **6.** $\tan(A - B) = \dfrac{\tan A - \tan B}{1 + \tan A\tan B}$

Double Angle Identities

1. $\sin 2\theta = 2\sin\theta\cos\theta$; **2.** $\cos 2\theta = \cos^2\theta - \sin^2\theta$; **3.** $\cos 2\theta = 2\cos^2\theta - 1$;

4. $\cos 2\theta = 1 - 2\sin^2\theta$; **5.** $\tan 2\theta = \dfrac{2\tan\theta}{1 - \tan^2\theta}$

Half Angle Identities

1. $\cos\dfrac{1}{2}\theta = \pm\sqrt{\dfrac{1 + \cos\theta}{2}}$; **2.** $\sin\dfrac{1}{2}\theta = \pm\sqrt{\dfrac{1 - \cos\theta}{2}}$; **3.** $\tan\dfrac{1}{2}\theta = \dfrac{1 - \cos\theta}{\sin\theta} = \dfrac{\sin\theta}{1 + \cos\theta}$

The $\pm$ sign indicates the sign to use depending on the quadrant location of $\dfrac{\theta}{2}$

Product or Product as a Sum Formulas

1. $\cos(A - B) + \cos(A + B) = 2\cos A\cos B$ $\Big\}$

2. $\cos(A - B) - \cos(A + B) = 2\sin A\sin B$ $\Big\}$ Derived from sum and difference identities $\Big\}$

Sum Identities

1. $\cos A + \cos B = 2\cos\dfrac{A + B}{2}\cos\dfrac{A - B}{2}$ $\Big]$

2. $\cos A - \cos B = -2\sin\dfrac{A + B}{2}\sin\dfrac{A - B}{2}$ $\Big]$

3. $\sin A + \sin B = 2\sin\dfrac{A + B}{2}\cos\dfrac{A - B}{2}$ $\Big]$

4. $\sin A - \sin B = 2\sin\dfrac{A - B}{2}\cos\dfrac{A + B}{2}$ $\Big]$

Cofunction Relationships

$\sin A = \cos(90° - A)$ $\Big\}$
$\cos A = \sin(90° - A)$ $\Big\}$ cofunctions.

$\tan A = \cot(90° - A)$ $\Big\}$
$\cot A = \tan(90° - A)$ $\Big\}$ cofuntions.

$\sec A = \csc(90° - A)$ $\Big\}$
$\csc A = \sec(90° - A)$ $\Big\}$ cofuntions.

Other identities (functional values of negative angles) For any θ,

1. $\cos(-\theta) = \cos\theta$ **4.** $\sec(-\theta) = \sec\theta$

2. $\sin(-\theta) = -\sin\theta$ **5.** $\csc(-\theta) = -\csc\theta$

3. $\tan(-\theta) = -\tan\theta$ **6.** $\cot(-\theta) = -\cot\theta$

Inverse cofunction identities

7. $\text{Sin}^{-1}x + \text{Cos}^{-1}x = \dfrac{\pi}{2}$

8. $\text{Tan}^{-1}x + \text{Cot}^{-1}x = \dfrac{\pi}{2}$

9. $\text{Sec}^{-1}x + \text{Csc}^{-1}x = \dfrac{\pi}{2}$

For Integrals **A:** $\sin mx\sin nx = \dfrac{1}{2}[\cos(m - n)x - \cos(m + n)x]$

                 B: $\cos mx\cos nx = \dfrac{1}{2}[\cos(m + n)x + \cos(m - n)x]$

                 C: $\sin mx\cos nx = \dfrac{1}{2}[\sin(m + n)x + \sin(m - n)x]$

Change of base formulas

For Exponents	For logarithms
1. $b^x = a^{x \log_a b}$ 2. $b^x = e^{x \log_e b} = e^{x \ln b}$	$\log_b M = \dfrac{\log_c M}{\log_c b}$

Inverse Cofunction Identities

1. $\mathrm{Sin}^{-1}x + \mathrm{Cos}^{-1}x = \frac{\pi}{2}$, $[-1, 1]$; **2.** $\mathrm{Tan}^{-1}x + \mathrm{Cot}^{-1}(x) = \frac{\pi}{2}$, $(-\infty, \infty)$;

3 $\mathrm{Sec}^{-1}x + \mathrm{Csc}^{-1}x = \frac{\pi}{2}$, $|x| \geq 1$

Negative angle relationships for inverse trig functions

1. $\mathrm{Sin}^{-1}(-x) = -\mathrm{Sin}^{-1}x$; **3.** $\mathrm{Tan}^{-1}(-x) = -\mathrm{Tan}^{-1}x$; **5.** $\mathrm{Sec}^{-1}(-x) = \pi - \mathrm{Sec}^{-1}x$;

2. $\mathrm{Cos}^{-1}(-x) = \pi - \mathrm{Cos}^{-1}x$; **4.** $\mathrm{Cot}^{-1}(-x) = \pi - \mathrm{Cot}^{-1}x$; **6.** $\mathrm{Csc}^{-1}(-x) = -\mathrm{Csc}^{-1}x$

Range of Inverse Trig Functions when $x \geq 0$

$0 \leq \mathrm{Sin}^{-1}x \leq \frac{\pi}{2}$ --Ist quadrant

$0 \leq \mathrm{Cos}^{-1}x \leq \frac{\pi}{2}$ --Ist quadrant

$0 \leq \mathrm{Tan}^{-1}x < \frac{\pi}{2}$ --Ist quadrant

$0 < \mathrm{Cot}^{-1}x \leq \frac{\pi}{2}$ --Ist quadrant

$0 \leq \mathrm{Sec}^{-1}x < \frac{\pi}{2}$ --Ist quadrant

$0 < \mathrm{Csc}^{-1}x \leq \frac{\pi}{2}$ --Ist quadrant

Range of Inverse Trig Functions when $x < 0$

$-\frac{\pi}{2} \leq \mathrm{Sin}^{-1}x < 0$ <--4th quadrant

$\frac{\pi}{2} < \mathrm{Cos}^{-1}x \leq \pi$ <-- 2nd quadrant

$-\frac{\pi}{2} < \mathrm{Tan}^{-1}x < 0$ <--4th quadrant

$\frac{\pi}{2} < \mathrm{Cot}^{-1}x < \pi$ <--2nd quadrant

$\frac{\pi}{2} < \mathrm{Sec}^{-1}x \leq \pi$ <--2nd quadrant

$-\frac{\pi}{2} \leq \mathrm{Csc}^{-1}x < 0$ <--4th quadrant

www.ingramcontent.com/pod-product-compliance
Lightning Source LLC
Chambersburg PA
CBHW081758200326

41597CB00023B/4070